Aṣṭāṅgahṛdaya

Worldwide interest in Ayurveda is on the rise, ever since the World Health Organization adopted the Alma Ata Declaration in 1978. Ayurveda is increasingly being adopted and many phytotherapy schools in Europe and the Americas teach Ayurveda as a wellness system. Considering the prominent position that *Aṣṭāṅgahṛdaya* occupies in Ayurveda, a scientific synopsis of this masterpiece is now presented before the world of Ayurveda.

In eight comprehensive chapters, *Aṣṭāṅgahṛdaya: A Scientific Synopsis of the Classic Ayurveda Text* presents a lucid summary of the teachings of Vāgbhaṭa. Ayurvedic view of the human body, basic principles of Ayurveda, surgical armamentarium, diagnosis and treatment of diseases, herbs and other medicinal substances used in the preparation of various dosage forms, ayurvedic pharmaceuticals, ayurvedic approach to food fortification, salient features of *Aṣṭāṅgahṛdaya* and a road-map for the future are among the topics discussed.

Key Features:

- Presents the ayurvedic view of conception and the human body.
- Discusses aspects of ayurvedic pharmaceuticals.
- Examines diagnosis of diseases, lines of treatment, prognosis of diseases, signs of imminent death, management of mishaps and treatment of new diseases.

The voluminous information pertaining to the subjects of the various chapters is presented in the form of many tables for ready reference and quick survey. This book provides a helping hand to those interested in rediscovering the teachings of Vāgbhaṭa and is a great resource for researchers of medicine, traditional and alternative medicine, pharmacology and drug discovery.

Aṣṭāṅgahṛdaya

A Scientific Synopsis of the Classic Ayurveda Text

G.R. Arun Raj, N.K.M. Ikbal, and D. Suresh Kumar

CRC Press
Taylor & Francis Group
Boca Raton London New York

CRC Press is an imprint of the
Taylor & Francis Group, an informa business

First edition published 2023
by CRC Press
6000 Broken Sound Parkway NW, Suite 300, Boca Raton, FL 33487-2742

and by CRC Press
4 Park Square, Milton Park, Abingdon, Oxon, OX14 4RN

CRC Press is an imprint of Taylor & Francis Group, LLC

Library of Congress Cataloging-in-Publication Data
Names: Raj, G. R. Arun, 1985- author. | Ikbal, N. K. Muhammed, 1964-
author. | Suresh Kumar, D., 1949- author.
Title: Aṣṭāṅgahṛdaya : a scientific synopsis of the classic Ayurveda
text / G.R. Arun Raj, N.K.M. Ikbal, D. Suresh Kumar.
Description: First edition. | Boca Raton : CRC Press, 2023. |
Includes bibliographical references and index.
Identifiers: LCCN 2022035532 (print) | LCCN 2022035533 (ebook) |
ISBN 9780367708689 (hardback) | ISBN 9780367685720 (paperback) |
ISBN 9781003148296 (ebook)
Subjects: LCSH: Medicine, Ayurvedic–Early works to 1800.
Classification: LCC R127.2 .R25 2023 (print) | LCC R127.2 (ebook) |
DDC 615.5/38–dc23/eng/20220913
LC record available at https://lccn.loc.gov/2022035532
LC ebook record available at https://lccn.loc.gov/2022035533

ISBN: 978-0-367-70868-9 (hbk)
ISBN: 978-0-367-68572-0 (pbk)
ISBN: 978-1-003-14829-6 (ebk)

DOI: 10.1201/9781003148296

Typeset in Times
by codeMantra

Lambaśmaśru kapālam aṃbujanibha cchāyādyutīm vaidyakān

Antēvāsina Indu-Jajjaṭa- mukhān adhyāpayantaṃ sadā

Āgulphāmala kañcukāñcitadarā lakṣyōpavītōjvalaṃ

Kaṇṭhastāgarusāraṃ añjitadṛśaṃ dhyāyē dṛḍhaṃ Vāgbhaṭaṃ

(We pay homage to our guru Vāgbhaṭa, who is long-bearded, having bright eyes and the complexion of clouds, who teaches his physician-disciples Indu and Jajjaṭa, wearing a white robe that flows up to the feet, and wearing around his neck, a garland made of sweet-smelling aguru beads!)

Contents

Preface

It is traditionally believed and understood that Ayurveda was "recollected" by Brahmā. Brahmā passed on the knowledge to Dakṣaprajāpati, who in turn transmitted it to the celestial twins Aśvinikumāra. Indra received the knowledge from the Aśvinīkumāra and handed it over to Sage Bharadvāja from whom it went in a chain to Atri and then to Ātrēya. Considering the extensive use of symbolism in Hindu religious literature, the legend about the divine origin of Ayurveda can be given a logical interpretation. Brahmā is the symbol of knowledge and Dakṣaprajāpati is considered to be his mind-created son, who is the progenitor of all animate and inanimate beings. The Aśvinīkumāra are symbols of *vijñānamayakōśa* (seat of awareness, insight and consciousness) and *ānandamayakōśa* (the pure and radiant bliss body), which are two of the five aspects of the subtle body of all living beings. Indra is the symbolic entity that controls functions of mind. As Ayurveda is eternal like any other form of knowledge, having no beginning or end, it is logical to assume that like modern-day scientists, Sage Bharadvāja might have discovered the principles of Ayurveda through keen observation of nature and yogic introspection. The link between deities and humans can be better explained thus.

Ātrēya Punarvasu, who was a descendant of Atri and whose greatness is extolled in Hindu epics, discussed Ayurveda with scholar-sages at a series of symposia convened in the Himalayan region in northern India. He is believed to have formulated the basic concepts of Ayurveda. His illustrious disciple Agnivēśa was the one who documented his guru's teachings in *Agnivēśa Tantra*, which was the celebrated and most popular text in ancient days, representing the school of Ātrēya. Agnivēśa's treatise was later revised and enlarged by Caraka, whose compendium is known as *Caraka Samhita*, which is a milestone in *kāyacikitsa*, one of the eight branches of Ayurveda, namely *kāyacikitsa* (internal medicine), *śalya* (surgery), *śālākya* (diseases of supra-clavicular region), *kaumārabhṛtya* (pediatrics including obstetrics and gynecology), *agadatantra* (toxicology), *bhūtavidya* (treatment of spiritual afflictions), *rasāyana* (rejuvenation therapy) and *vājīkaraṇa* (virilification).

The school of surgery is believed to have begun with King Divōdāsa of Kāśi (present-day Benares), who is regarded as the reincarnation of Lord Dhanvantari, the patron-deity of Ayurveda. He taught surgery to disciples such as Suśruta, Aupadhēnava, Aurabhra, Bhōja, Galava, Gārgya, Gōpuraraksita, Kāṅkāyana, Nimi, Pauṣkalavata, Karavīra and Vaitaraṇa. All of them composed treatises on the seven limbs of Ayurveda. Nevertheless, the one by Suśruta is the only surviving and most famous one. It is an outstanding treatise of Ayurveda, with special emphasis on *śalya* and *śālākya*.

For many centuries the schools of Ātrēya and Suśruta remained as two separate and distinct streams, until Vāgbhaṭa arrived on the scene. He synthesized the teachings of these two schools and composed *Aṣṭāṅgasaṃgraha*. As Vāgbhaṭa himself claimed, the *Aṣṭāṅgasaṃgraha* was born from the eight-fold medical lore, just as ambrosia was obtained from churning of the great ocean of milk. However, in order to help aspirants incapable of great effort, *Aṣṭāṅgahṛdaya* was separately composed on the basis of *Aṣṭāṅgasaṃgraha*. The poetic nature of *Aṣṭāṅgahṛdaya* was one reason for its increased popularity even today. *Aṣṭāṅgahṛdaya* owes its fame over earlier treatises to the charm of its verses, style of condensation, arrangement of subjects, clarity of explanation and several other merits. Because of its erudition, it stays at par with the teachings of Caraka and Suśruta, forming the Great Triad.

Worldwide interest in Ayurveda has been on the rise, ever since the World Health Organization adopted the Alma Ata Declaration in 1978. Ayurveda is increasingly being adopted in the Western world and is said to be the fastest growing alternative medical system in Austria, Germany, Switzerland and several other countries. Many phytotherapy schools in Europe and the Americas teach Ayurveda as a wellness system. These favorable developments call for scholarly works explaining the tenets of Ayurveda to the English-speaking world. Considering the position that

Aṣṭāṅgahṛdaya occupies in Ayurveda, we have attempted to prepare a scientific synopsis of this Ayurveda masterpiece.

Chapter 1 presents an introduction to the subject, followed by a chapter that deals with the ayurvedic view of conception and the human body. Chapter 3 explains the basic principles of Ayurveda, including desirable code of conduct, seasonal regimen, dietetics, ethical considerations and principles of therapeutics. The surgical armamentarium of Ayurveda is presented in Chapter 4. Surgical instruments, bloodletting, extraction of sharp foreign bodies, various aspects of surgery and cauterization are described in this chapter. Diseases are diagnosed and treated in Ayurveda based on the wholistic doctrine of *tridōṣa*. Aspects such as diagnosis of diseases, lines of treatment, prognosis of diseases, signs of imminent death, management of mishaps and treatment of new diseases are included in Chapter 5. Numerous herbs, animal products and minerals are used in the preparation of dosage forms such as paste, juice, powder, decoction, fermented liquids, medicated oils, medicinal clarified butter, electuary, collyria and pill. Aspects of ayurvedic pharmaceutics are dealt with in Chapter 6. Fortification of food and beverages is now being adopted globally as a measure to stay healthy and to prevent the appearance of diseases. It is noteworthy that the concept of food fortification was already inbuilt in Ayurveda. The ayurvedic approach to food fortification is the theme of Chapter 7. The book discusses some of the salient features of *Aṣṭāṅgahṛdaya* and concludes with a roadmap for the future. The voluminous information pertaining to the subjects of the various chapters is presented in the form of many tables, so as to assist the reader with more information. All references to the verses of *Aṣṭāṅgahṛdaya* are based on the machine-readable transcription of Vāgbhaṭa's *Aṣṭāṅgahṛdayasaṃhitā*, compiled and edited by Rahul Peter Das and Ronald Eric Emmerick (2009). It served as a good reference standard of *Aṣṭāṅgahṛdaya*.

The authors have taken great care to present the teachings of *Aṣṭāṅgahṛdaya* faithfully. Nevertheless, all the chapters were read critically by experts of Ayurveda. We are thankful to Prof. Rahul Peter Das, Germany; Dr. Nishanth Gopinath, Nagarjuna Herbal Concentrates Ltd, Thodupuzha; Dr. M. Sujalakshmi, Ayursoukhya Ayurveda, Kottayam; Dr. Nirmal Narayanan, Vishnu Ayurveda College, Shoranur; Prof. Sanjeev Rastogi, State Ayurveda College and Hospital, Lucknow; Prof. Wandee Gritsanapan, Phyto Product Research, Bangkok, Thailand; Dr. V. Remya, Vaidyaratnam Oushadhasala Pvt Ltd, Trichur; and Dr. Diego Tresinari dos Santos, State University of Campinas, Brazil for reading critically Chapters 1–8, respectively, and giving valuable comments. Photographs of sepulcher of Vāgbhaṭa were kindly provided by Muhammad Murshid of Pulamanthole, Kerala. Photographs of various equipment used in ayurvedic treatment and photographs of a few ayurvedic measures were kindly provided by Dr. A. P. Raman Nambudiri, Divakara Concepts, Coimbatore; Nagarjuna Ayurvedic Centre Ltd, Kalady; and Dr. M. Sujalakshmi, Ayursoukhya Ayurveda, Kottayam. Figures of blunt and sharp instruments used in surgery were drawn by Antony Selvaraj, Max Computers, Bangalore, who also prepared all the graphics used in this book. We express our gratitude to all of them for their cooperation and thoughtfulness. Stephen Zollo and Laura Piedrahita of CRC Press, Boca Raton, encouraged us constantly to bring this book project to fruition, especially during the turbulent COVID-19 pandemic period. We record our indebtedness to them. *Aṣṭāṅgahṛdaya* is an Ayurveda classic that has withstood the test of time. The authors hope that *Aṣṭāṅgahṛdaya: A Scientific Synopsis of the Classic Ayurveda Text* will lend a helping hand to aspirants desirous of rediscovering the teachings of Vāgbhaṭa.

G.R. Arun Raj
N.K.M. Ikbal
D. Suresh Kumar

Authors

Dr. G.R. Arun Raj was born on 19 November 1985 at Kollam, Kerala State, India. He graduated in Bachelor of Ayurvedic Medicine and Surgery from Sree Narayana Institute of Ayurvedic Studies and Research, Kollam, in 2009. He secured M.D. (Ayu.) degree from Rajiv Gandhi University of Health Sciences, Bangalore in 2014. He has published about 87 papers in various national and international journals and is the author of chapters in *Nutraceuticals and Functional Foods* (Studium Press LLC, Houston, 2016) and *Ayurveda in the New Millennium* (CRC Press, Florida, 2021). He was identified as expert by Ministry of AYUSH, Government of India for developing Technical Guidelines for Integration in March 2021. He is the principal investigator of a major research project on spastic cerebral palsy sponsored by Rajiv Gandhi University of Health Sciences. Currently, he is working as Assistant Professor and pediatric consultant, Department of Kaumarabhritya (Ayurveda Pediatrics), Parul Institute of Ayurveda and Research, Parul University, Vadodara district, Gujarat, India. He is also member of Institutional research committee, Parul Institute of Ayurveda and Research. His research interests include application of traditional medicine in public health, developmental pediatrics and infant and child nutrition. E-mail: drdarunraj26@gmail.com

Dr. N.K.M. Ikbal was born on 14 September 1964 at Palakkad in Kerala. He graduated with a Bachelor of Ayurvedic Medicine and Surgery degree from Vaidyaratnam P.S. Variar Ayurveda College, Kottakkal in 1988. He successfully completed a Masters in Ayurveda (M.D.) from Government Ayurveda Medical College, Bangalore University in 1992 and completed doctoral studies leading to Ph.D. from Sree Sankaracharya University, Kalady, Kerala in 2000. Dr. Ikbal served a short stint (1992) as Lecturer at Ayurveda College, Coimbatore, later joining Vaidyaratnam Ayurveda College, Thrissur, in 1993. He is presently Professor and Head of the Department of Śalyatantra with 30 years of active presence in the academy, research and profession. He has worked as Visiting Professor and External Examiner for the Ayurveda Degree Program at Thames Valley University, London (1999–2006) and Visiting Faculty at Chung Bo Institute, Seoul, South Korea (2006). He has published original research papers and serves as the Executive Editor of *Vaidyaratnam Journal of Ayurvedic Medicine* (V-JAM), which is published quarterly. Dr. Ikbal is the founder member of National Suśruta Association (Association of Ayurvedic Surgeons of India). He has attended national and international seminars as key speaker on topics of interest ranging from immunology, rheumatology and surgical specialties in Ayurveda. He contributed the opening chapter "What We Learn from the History of Ayurveda" in *Ayurveda in the New Millennium: Emerging Roles and Future Challenges*, published by CRC Press, Boca Raton, Florida (2021). E-mail: ikbalnk@hotmail.com

D. Suresh Kumar was born on 21 September 1949 in the southern Indian province of Kerala, where he received his early education. He obtained a B.Sc. degree in Zoology from the University of Kerala (1969) and earned M.Sc. (1972) and Ph.D. degrees (1977) from Banaras Hindu University, Varanasi. His doctoral thesis was on the hormonal control of oxidative metabolism in reptiles. Thereafter, he spent two years as a postdoctoral fellow in the Department of Biological Sciences, University of Aston in Birmingham, England, investigating the pancreatic physiology of the rainbow trout. He returned to India in 1980 and joined the Department of Zoology, University of Calicut, Kerala, as pool officer in the scientist pool of the Council of Scientific and Industrial Research, New Delhi. During his stay there, a chance encounter with some religious persons introduced him to the study of Ayurveda, the traditional medical system of India. He undertook a survey of the state of Ayurveda in the province and published his findings in provincial and national weeklies. In 1986, he joined the International Institute of Ayurveda, Coimbatore, as a research officer in the Department of Physiology. From 1986 to 2003, he conducted research on various aspects of Ayurveda. In collaboration with Dr. Y.S. Prabhakar, formerly of C.D.R.I. Lucknow, he proposed the first mathematical model for the ayurvedic concept of *tridōṣa* in the disease state. He also offered a novel definition for the ayurvedic class of medicine *arka*, based on his study of the Sanskrit text *Arkaprakāśa*. In 2003, he joined Sami Labs Ltd, Bangalore, as senior scientist in the R&D laboratory. He spent several years in the company working on various aspects of new product development. From 2012 to 2015, he worked as head of the R&D laboratory of the Ayurveda consortium, Confederation for Ayurveda Renaissance-Keralam Ltd, Koratty, Kerala. In January 2016, he joined Cymbio Pharma Pvt Ltd, Bangalore, as head of new product development, and since July 2019 he has been Ayurveda consultant. He is the author of *Herbal Bioactives and Food Fortification: Extraction and Formulation* (2016) and editor of *Ayurveda in the New Millennium* (2020) published by CRC Press. E-mail: dvenu21@yahoo.com

Abbreviations

AH	*Aṣṭāṅgahṛdaya*
Ci.	*Cikitsāsthānam*
Ka	*Kalpasthānam*
Ni.	*Nidānasthānam*
Śa.	*Śārīrasthānam*
Sū.	*Sūtrasthānam*
Ut.	*Uttarasthānam*

TRANSLITERATION OF DEVANĀGARI ALPHABET

Dēvanāgari	Transliteration Key
अ	A
आ	Ā ā
इ	I
ई	Ī ī
उ	U
ऊ	Ū ū
ऋ	Ṛ ṛ
ए	e
–	Ē ē
ऐ	ai
ओ	o
–	Ō ō
औ	au
अं	*ṃ*
अः	ḥ
क	ka
ख	kha
ग	ga
घ	gha
ङ	ṅa
च	ca
छ	cha
ज	ja
झ	jha
ञ	Ñ ña
ट	Ṭ ṭa
ठ	ṭha
ड	ḍa
ढ	ḍha
ण	ṇa
त	ta
थ	tha
द	da
ध	dha
न	na
प	pa
फ	pha
ब	ba
भ	bha
म	ma
य	ya
र	ra
ल	la
व	va
श	Ś śa
ष	Ṣ ṣa
स	sa
ह	ha
ळ	ḷa
क्	kṣa

1 Introduction

1.1 ORIGIN OF AYURVEDA

Ayurveda or "the sacred knowledge of longevity" dates back at least 2,000 years in its codified form. However, it has roots that are much deeper. In modern times, it stretches well beyond the boundaries of its homeland and is fast becoming a transnational and multicultural phenomenon. Ayurveda is regarded to be a secondary (*upavēda*) of *Atharvavēda* (Thakar, 2010). Ayurveda is inextricably connected to Hindu culture, and it is believed that this knowledge was conceived by Brahmā, the creator of the universe and all forms of knowledge (Sharma, 1981). As the medical knowledge advanced gradually, Ayurveda was grouped into eight specialties. They are *kāyacikitsa* (general medicine), *śalya* (surgery), *śālākya* (diseases of supra-clavicular region), *kaumārabhṛtya* (gynecology, obstetrics and pediatrics), *agadatantra* (toxicology), *bhūtavidya* (treatment of spiritual afflictions), *rasāyana* (rejuvenation therapy) and *vājīkaraṇa* (virilification). Many treatises were composed on each of these branches, making all of them full-fledged specialties. Nevertheless, two among them emerged as distinct schools – the Ātrēya school (general medicine) and the Dhanvantari school (surgery). *Caraka Saṃhita* and *Suśruta Saṃhita* respectively became their authentic texts (Sharma, 1981).

1.2 EMERGENCE OF *AṢṬĀṄGAHṚDAYA*

During a certain period in the history of north-western India, a need was felt to compose a text integrating the teachings of *Caraka Saṃhita* and *Suśruta Saṃhita*. It is believed that *Aṣṭāṅgasamgraha* emerged to fulfill that need. However, as *Aṣṭāṅgasamgraha* was too elaborate, composed in prose and poetry and difficult to memorize, a shorter version of *Aṣṭāṅgasamgraha* was produced and that is the celebrated *Aṣṭāṅgahṛdaya*.

1.3 AUTHORSHIP

It is traditionally believed that Vāgbhaṭa is the author of *Aṣṭāṅgasamgraha* and *Aṣṭāṅgahṛdaya*. However, Hoernle opined that the author of *Aṣṭāṅgasamgraha* was Vṛddha Vāgbhaṭa who lived in the 7th century (A.D. 625), while the author of *Aṣṭāṅgahṛdaya* was Laghu Vāgbhaṭa who perhaps lived a century later (Subhaktha et al., 2009). Nevertheless, by a line-by-line comparison of the *Cikitsāsthānam* and *Kalpasthānam* of both texts with each other and with the relevant passages of *Caraka Saṃhita*, Hilgenberg and Kirfel (1941) came to the conclusion that *Aṣṭāṅgasamgraha* and *Aṣṭāṅgahṛdaya* are two different recensions of the same text, one of which appears enlarged or abridged as against the other (Vogel, 1965). This suggests that both the texts were composed by one author. That *Aṣṭāṅgahṛdaya* is the abridgement of *Aṣṭāṅgasamgraha* is stated by the author himself in *Uttarasthānam* (*Ut.* 40: 79):

> Aṣṭāṅga vaidyaka mahōdadhi manthanēna
> Yō f ṣṭāṅgasaṃgraha mahāmṛta rāśir āptaḥ
> Tasmād analpaphalam alpa samudyamānāṃ
> Prītyartham ētad uditaṃ pṛthag ēva tantram

The *Aṣṭāṅgasamgraha* was born from the eight-fold medical lore, just as *amṛt* (ambrosia) was obtained from churning of the great ocean. In order to help aspirants incapable of great effort, this short treatise was separately composed on the basis of the *Aṣṭāṅgasamgraha*.

(*Rao, 1985*)

DOI: 10.1201/9781003148296-1

A similar statement is found in the *Sūtrasthānam* (*Sū*. 1: 4–6):

> Tēbhyō f tiviprakīrṇēbhyaḥ prāyaḥ sāratarōccayaḥ
> Kriyatē f ṣṭāṅgahṛdayaṃ nātisaṃkṣēpavistaram

"The works of Agnivēśa and others being too widely scattered, there is now made from them, as a collection for the most part of very essential matter, the *Aṣṭāṅgahṛdaya*, without too much brevity or prolixity" (Vogel, 1965).

It should be noted that Aruṇadatta, Indu and Candranandana, famous commentators of *Aṣṭāṅgahṛdaya*, leave no doubt that they consider authors of *Aṣṭāṅgasamgraha* and *Aṣṭāṅgahṛdaya* identical (Vogel, 1965). The ultimate aim of all ancient bodies of Hindu knowledge is salvation from the cycle of birth and death (*mōksa*). Persons who excel in these spheres ultimately transform themselves into humble individuals without ego or the awareness of "I". That is why very little is known about authors of ancient Sanskrit texts. The author of *Aṣṭāṅgahṛdaya* was one such noble soul. Therefore, he did not care to shed much light on himself.

1.4 DATE OF AUTHOR

Once the theory of a senior and junior Vāgbhaṭa is deprived of its basis, the question remains to be answered how else the term Vṛddha Vāgbhaṭa can be understood. Hilgenberg and Kirfel (1941) remark that analogous cases will be of help. In the *Catalogus Catalogorum*, Aufrecht lists a Vṛddhāryabhaṭa besides an Āryabhaṭa, the *Vṛddhayavanajātaka* of a Vṛddhayavanācārya besides a *Yavanajātaka* of a Yavanācārya, a *Vṛddhagārgyasaṃhita* besides a *Gārgyasaṃhita* and a *Vṛddhayōgaśataka* besides a *Yōgaśataka*. It is noteworthy that Bhāvamiśra mentions among his sources not only a Vāgbhaṭa and a Vṛddhavāgbhaṭa but also a Suśruta and a Vṛddhasuśruta. Ṭōḍaramalla mentions an Ātrēya and a Vṛddhātrēya, a Hārīta and Vṛddhahārīta and a Vṛddhabhōja. It is reasonable to infer that the attribute *Vṛddha* signifies not a senior writer or an older book, as against a junior writer or younger work, but the author of an enlarged recension as against that of a shorter original. That this is the only interpretation possible for the aphorisms, going by the name of Vṛddha Cānakya has been demonstrated irrefutably by Kressler (Vogel, 1965).

Vogel (1965) states that Vāgbhaṭa's date may be fixed with considerable accuracy, once the two-author theory is discarded. The well-known Chinese pilgrim I-ching who stayed in India from 672 to 688 A.D. mentioned in his book of travels "the eight sections of medical science". He further states in his book: "*ssu chih pa shu hsien wei pa pu. chin-jih yu jén lŭeh wei i chia. wu-tien chih ti hsien hsi ch'iu hsiu. tan ling chieh chê wu pu shih-lu*". Following is the translation of these Chinese lines:

> These eight arts formerly existed in eight books, but lately a man epitomized them and made them into one bundle. All physicians in the five parts of India practice according *to this* book, and any physician who is well versed in it never fails to live by the official pay.

Although neither name nor title is mentioned, there can be no doubt that this passage refers to Vāgbhaṭa, the author of *Aṣṭāṅgahṛdaya* who must consequently have flourished near the middle of the 7th century (650 A.D.) (Vogel, 1965).

I-ching's narration agrees with the fact that the Persian physician Alī ibn Sahl Rabban aṭ-Ṭabarī included a "survey of the Indian system of medicine" in his *Firdaus al-Hikma* (Paradise of Wisdom) dated 849/850 A.D. Aṭ-Ṭabarī names Ğarak, Susrud, *Aštānqahradī* and *Nidān* as his sources. If *Aṣṭāṅgahṛdaya* were so popular in Persia during that period as to be put on a par with Caraka and Suśruta, Vogel (1965) is of the opinion that it must have been composed during the 7th century at the latest.

1.5 ABOUT VĀGBHAṬA

Nothing else is known about Vāgbhaṭa's life other than the short autobiographical note mentioned in *Aṣṭāṅgasamgraha*:

> Bhiṣagvarō Vāgbhaṭa ity abhūn me
> Pitāmahō nāmadharō f smi yasya
> Sutō f bhavat tasya ca Siṃhaguptas
> Tasyāpy ahaṃ Sindhuṣu labdhajanmā
> Samadhigamya gurōr Avalōkitāt
> Gurutarāc ca pituḥ pratibhāṃ mayā
> Subahubheṣajaśāstravilōcanāt
> Suvihitō f ṅgavibhāgavinirṇayaḥ

My paternal grandfather, whose namesake I am, was the eminent physician Vāgbhaṭa. His son was Siṃhagupta and his son again am I. I was born among the people of Sind. Having obtained my knowledge from the venerable Avalōkita and my even more venerable father, whose eye represents medical science in a very high degree, there was well made by me this complete exposition, arranged according to the eight parts of medicine.

(*Vogel, 1965*)

Vogel (1965) notes that in the colophon of Jajjaṭa's *Suśrutaṭīkā*, Vāgbhaṭa is given the appellation Mahājahnupati (Lord of Mahājahnu), which is identified with the present-day Manjhand, a small township and subdivision (*tehsil*) in Jamshoro District, Sindh, Pakistan. It is situated about 50 miles north of Hyderabad, on the west bank of the River Indus. Vogel (1965) considers it not altogether impossible that this was Vāgbhaṭa's abode. The old village was washed away many centuries ago and the present Manjhand stands on its remains. A crumbling Siva temple is the only ancient structure that speaks of its glorious past. However, remnants of an ancient civilization were unearthed from an archaeological exploration in December 2017 (Anonymous, 2022a, b) (Figure 1.1).

1.6 RELIGIOUS BELIEF OF VĀGBHAṬA

Vāgbhaṭa's writings demonstrate Buddhist leanings, as evidenced by a prayer to Buddha in *Aṣṭāṅgasamgraha*:

Ōm namō bhagavatē bhaiṣajyaguravē vaiḍūryaprabhārājāya tathāgatāyārhatē samyak saṃbuddhāya ("Om! Reverence to the victorious one, the medicine master, the cat's- eye-splendored king, the thus-gone one, the saint, the fully enlightened one!") (Vogel, 1965). The same *mantra* is mentioned in *Sūtrasthānam* of *Aṣṭāṅgahṛdaya*, while explaining the fortification of emetic medicine (*Sū*. 18: 17):

> Ōm namō bhagavatē bhaiṣajyaguravē vaiḍūryaprabhārājāya
> Tathāgatāyārhatē samyak saṃbuddhāya tadyathā
> Ōm bhaiṣajyē bhaiṣajyē mahā bhaiṣajyē samudgatē svāhā.

However, in the anterior part of the same *mantra* can be found references to Hindu deities such as Brahma, Dakṣa, Aśvinīkumāra, Rudra and Indra (*Sū*. 18: 16–17). One also finds in *Aṣṭāṅgahṛdaya* features typical of Hindu *dharma*, like relating the story of fire emanating from the third eye of Rudra with origin of fever (*Ni*. 2: 1), reference to the churning of the milky ocean and legend of garlic sprouting from the drops of blood falling down from the wounded neck of Rāhu who had stealthily drunk *amṛt* (ambrosia) (*Ut*. 39: 111–112). While describing seasonal regimen (*ṛtucarya*),

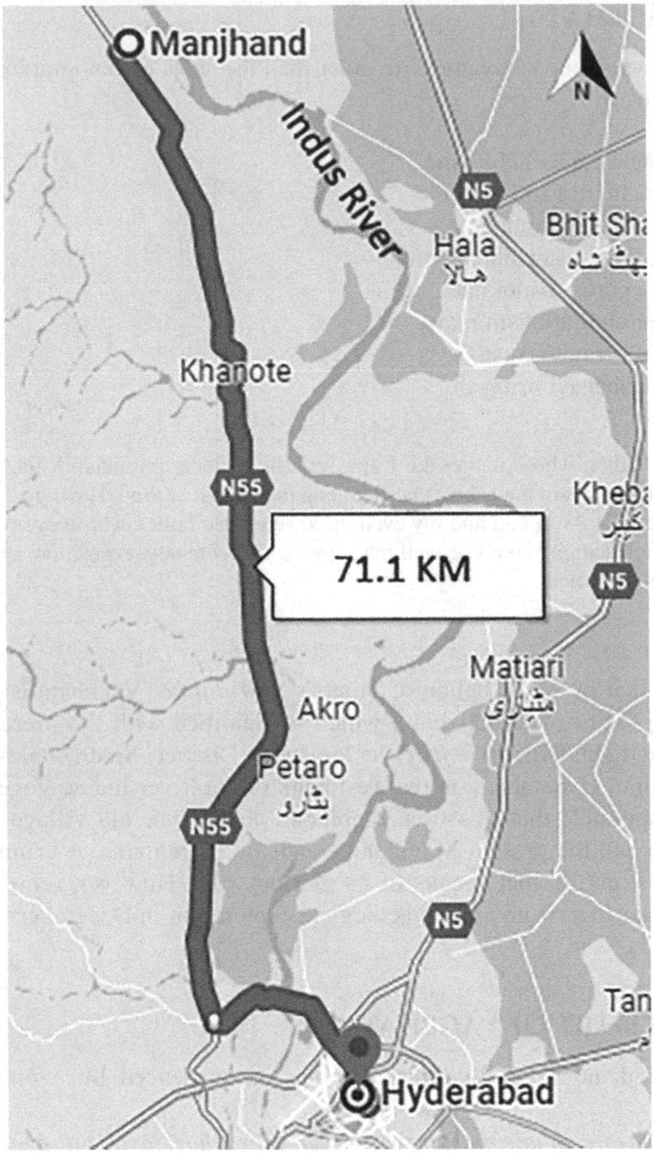

FIGURE 1.1 Location of Manjhand on Google map. The distance from Hyderabad to Manjhand by road is 71.1 km. (Photo courtesy: Google Maps.)

Aṣṭāṅgahṛdaya recommends variously cooked meat preparations to be consumed in *vasanta ṛtu* (*Sū*. 3: 20), *varṣa ṛtu* (*Sū*. 3: 45) and *śarad ṛtu* (*Sū*. 3: 51). Moreover, the medicinal qualities of many kinds of fermented liquids (*Sū*. 5: 62–78), meat (*Sū* 6: 43–71) and animal parts are described in *Sūtrasthānam*. These pieces of evidence indicate that Vāgbhaṭa practiced a blend of Hindu and Buddhist traditions.

1.7 VĀGBHAṬA IN KERALA CULTURE

It is believed that after a long journey through northern India, Vāgbhaṭa finally arrived in Kerala, which was part of the Tamil country at that time. He spent the remaining years there, popularizing the knowledge enshrined in his treatise. The *Aṣṭavaidya* (eight physician) tradition of Kerala is believed to have originated from Vāgbhaṭa. Indu and Jajjaṭa, famous commentators of *Aṣṭāṅgahṛdaya*, are said to have been his pupils (Variar, 1985, 1987; Menon, 2019). The *Aṣṭavaidya*s of Kerala chant every evening the following prayer (*dhyāna śḷōka*) during their worship of Vāgbhaṭa:

Lamba śmaśru kapālam ambudanibha chāyādyutīm vaidyakān
Antēvāsina indu jajjaṭa mukhān adhyāpayantam sadā
Āgulphāmala kañcukāñcita tanum lakṣyōpavītōjvalam
Kaṇṭhasthāgarusāram añcita dṛśam vandē gurum vāgbhaṭām!

(I pay obeisance to my guru Vāgbhaṭa, who is long-bearded, having bright eyes and the complexion of clouds, who teaches his physician-disciples Indu and Jajjaṭa, wearing a white robe that flows up to the feet, and wearing around his neck, a garland made of sweet-smelling *Aguru* wood!) (Ikbal et al., 2021).

Members of the Pulamanthole Mooss family in northern Kerala believe that the sepulcher in their estate which is still worshipped contains the entombed body of Vāgbhaṭa (Figure 1.2). The stone lamps over the sepulcher are lit every evening, and special *pūja* is held on all full moon days (Anonymous, 2022c).

Some linguistic evidence is also available to adduce Vāgbhaṭa's impact on south Indian medical tradition. *Vākaṭam* is a word commonly employed in Tamil medical books. *Piḷḷaippiṇi Vākaṭam* and *Kuḻantai Vākaṭam* (Book of Diseases of Children) are examples (see Kasinathan, 1955). According to the Tamil-English medical dictionary of T.V. Sambasivam Pillai, the word *Vākaṭam* means a book that describes perfectly the measures to cure ailments (Pillai, 1931). Dr. Asko Parpola, the Finnish Tamil scholar, opines that the word *Vākaṭam* is derived from the Sanskrit word Vāgbhaṭa (Dr. Asko Parpola, personal communication). *Aṣṭavaidya* physicians of Kerala refer to Vāgbhaṭa as *Vāhaṭa* or *Vāhaṭan* (Mooss, 1982). Interestingly, at the end of the fourth chapter of *Niratarapada vyākhya* composed by Jajjaṭa, he clearly states that he was the disciple of "Mahājahnupati Śrī Vāhaṭa" (Sastry and Pandey, 1975). It is possible that the Tamil word *Vākaṭan* might have evolved into Vāhaṭa or Bāhaṭa.

FIGURE 1.2 The sepulcher that exists in the estate of the Pulamanthole Mooss family in northern Kerala is believed to contain the mortal remains of Vāgbhaṭa. (Photo courtesy: Muhammed Murshid, Pulamanthole.)

1.8 ORGANIZATION OF TEXT

Aṣṭāṅgahṛdaya follows essentially the arrangement of information adopted in *Caraka Samhita*. It is narrated in six *sthāna* (sections), each *sthāna* consisting of varying number of chapters. However, unlike *Caraka Samhita*, there are no *Vimānasthānam* and *Indriyasthānam* (Table 1.1).

The information contained in them is presented in *Śārīrasthānam* and *Sūtrasthānam*. In total, the text has 120 chapters running into 7,462 *śḷōka* (quatrains) (Tables 1.2 and 1.3).

Sūtrasthānam explains the basic principles of Ayurveda, daily regimen, seasonal regimen and the fundamentals of health and disease. This section deals with qualities and sites of predominance of *tridōṣa*, circadian, circannual rhythms of *tridōṣa*, high and low stages of activities of *tridōṣa*, *dhātu* (tissue elements), *mala* (waste products); effects of six basic taste modalities on *tridōṣa*, qualities of matter, types of land and their relation to health and disease, *pañcakarma* therapy, surgical measures, dosage forms and essential groups of herbs to be used in various diseases. It is traditionally believed that a very intelligent person can learn all aspects of Ayurveda and become an accomplished physician just by studying diligently the *Sūtrasthānam* alone.

TABLE 1.1
Organization of Sections in *Aṣṭāṅgahṛdaya* Compared to *Caraka Samhita* and *Aṣṭāṅgasamgraha*

		Number of Chapters		
Sl. No.	**Title of Section**	***Caraka Samhita***	***Aṣṭāṅgasamgraha***	***Aṣṭāṅgahṛdaya***
1	*Sūtrasthānam*	30	40	30
2	*Nidānasthānam*	8	16	16
3	*Vimānasthānam*	8	–	–
4	*Śārīrasthānam*	8	12	6
5	*Indriyasthānam*	12	–	–
6	*Cikitsāsthānam*	30	24	22
7	*Kalpasthānam*	12	8	6
8	*Siddhisthānam*	12	–	–
9	*Uttarasthānam*	–	50	40
Total		120	150	120

TABLE 1.2
Plan of the Text[a]

Sl. No.	**Title of *sthānam***	**No. of Chapters**	**Total No. of *Śḷōka***
1	*Sūtrasthānam*	30	1,599
2	*Śārīrasthānam*	6	557
3	*Nidānasthānam*	16	785
4	*Cikitsāsthānam*	22	1,978
5	*Kalpasthānam*	6	304
6	*Uttarasthānam*	40	2,239
Total		120	7,462

[a] Das and Emmerick (2009).

TABLE 1.3
Chapters and *Śḷōka* in *Aṣṭāṅgahṛdaya*[a]

Sl. No.	Title of Chapter	Total No. of *Śḷōka*
	1 *Sūtrasthānam*	
1	*Āyuṣkāmīyādhyāyaḥ prathamaḥ*	48
2	*Dinacaryādhyāyo dvitīyaḥ*	48
3	*Ṛtucaryādhyāyas tṛtīyaḥ*	58
4	*Rōgānulpādanīyādhyāyaś caturthaḥ*	36
5	*Dravadrayavijñānīyādhyāyaḥ pañcamaḥ*	84
6	*Annasvarūpavijñānīyādhyāyaḥ ṣaṣṭaḥ*	172
7	*Annarakṣādhyāyaḥ saptamaḥ*	77
8	*Mātrāśitīyādhyāyo 'ṣṭamaḥ*	55
9	*Dravyādivijñānīyādhyāyo navamaḥ*	28
10	*Rasabhēdīyādhyāyaḥ*	44
11	*Dōṣādivijñānīyādhyāyaḥ*	45
12	*Dōṣabhēdīyādhyāyaḥ*	78
13	*Dōṣōpakramaṇīyādhyāyaḥ*	41
14	*Dvidhōpakramaṇīyādhyāyaḥ*	37
15	*Śōdhanādigaṇa saṅgrahādhyāyaḥ*	47
16	*Snēhādhyāyaḥ*	46
17	*Svēdavidhyadhyāyaḥ*	29
18	*Vamanavirēcanavidhir adhyāyaḥ*	60
19	*Vastividhir adhyāyaḥ*	87
20	*Nasyavidhir adhyāyaḥ*	39
21	*Dhūmapānavidhir adhyāyaḥ*	22
22	*Gaṇḍūṣādividhir adhyāyaḥ*	34
23	*Āścōtanāñjanavidhir adhyāyaḥ*	30
24	*Tarpaṇapuṭapākavidhir adhyāyaḥ*	22
25	*Yantravidhir adhyāyaḥ*	42
26	*Śastravidhir adhyāyaḥ*	55
27	*Sirāvyadhavidhir adhyāyaḥ*	53
28	*Śalyāharaṇavidhir adhyāyaḥ*	47
29	*Śastrakarmavidhir adhyāyaḥ*	80
30	*Kṣārāgnikarmavidhir adhyāyaḥ*	53
Total		1,599
	2 *Śārīrasthānam*	
1	*Garbhāvakrāntir adhyāyaḥ*	100
2	*Garbhavyāpadvidhir adhyāyaḥ*	62
3	*Aṅgavibhāgaśārīrādhyāyaḥ*	120
4	*Marmavibhāgaśārīrādhyāyaḥ*	70
5	*Vikṛtivijñānīyādhyāyaḥ*	132
6	*Dūtādvijñānīyādhyāyaḥ*	73
Total		557
	3 *Nidānasthānam*	
1	*Sarvarōganidānādhyāyaḥ*	23
2	*Jvaranidānādhyāyaḥ*	79
3	*Raktapittakāsanidānādhyāyaḥ*	38

(*Continued*)

TABLE 1.3 (*Continued*)
Chapters and *Śḷōka* in *Aṣṭāṅgahṛdaya*[a]

Sl. No.	Title of Chapter	Total No. of *Śḷōka*
4	*Śvāsahidhmānidānādhyāyaḥ*	31
5	*Rājayakṣmādinidānādhyāyaḥ*	57
6	*Madātyayanidānādhyāyaḥ*	41
7	*Arśōnidānādhyāyaḥ*	59
8	*Atisāragrahaṇīrōgayōr nidānādhyāyaḥ*	30
9	*Mūtrāghātanidānādhyāyaḥ*	40
10	*Pramēhanidānādhyāyaḥ*	41
11	*Vidradhinidānādhyāyaḥ*	63
12	*Udaranidānādhyāyaḥ*	46
13	*Pāṇḍuśōphavisarpanidānādhyāyaḥ*	67
14	*Kuṣṭhanidānādhyāyaḥ*	56
15	*Vātavyādhinidānādhyāyaḥ*	56
16	*Vātaśōṇitanidānādhyāyaḥ*	58
Total		785
	4 *Cikitsāsthānam*	
1	*Jvaracikitsādhyāyaḥ*	177
2	*Raktapittacikitsādhyāyaḥ*	50
3	*Kāsacikitsādhyāyaḥ*	180
4	*Śvāsahidhmācikitsādhyāyaḥ*	59
5	*Rājayakṣmacikitsādhyāyaḥ*	83
6	*Chardyādicikitsādhyāyaḥ*	84
7	*Madātyayacikitsādhyāyaḥ*	115
8	*Arśaścikitsitādhyāyaḥ*	164
9	*Atisāracikitsādhyāyaḥ*	124
10	*Grahaṇīdōṣacikitsādhyāyaḥ*	93
11	*Mūtrāghātacikitsādhyāyaḥ*	63
12	*Pramēhacikitsādhyāyaḥ*	43
13	*Vidradhivṛddhicikitsādhyāyaḥ*	55
14	*Gulmacikitsādhyāyaḥ*	129
15	*Udaracikitsādhyāyaḥ*	131
16	*Pāṇḍurōgackitsādhyāyaḥ*	58
17	*Śvayathucikitsādhyāyaḥ*	42
18	*Visarpacikitsādhyāyaḥ*	38
19	*Kuṣṭhacikitsādhyāyaḥ*	98
20	*Śvitrakṛmicikitsādhyāyaḥ*	35
21	*Vātavyādhicikitsādhyāyaḥ*	83
22	*Vātaśōṇitacikitsādhyāyaḥ*	74
Total		1,978
	5 *Kalpasthānam*	
1	*Vamanakalpādhyāyaḥ*	47
2	*Virēcanakalpādhyāyaḥ*	62
3	*Vamanavirēcanavyāpatsiddhir adhyāyaḥ*	39
4	*Dōṣaharaṇasākalyavastikalpādhyāyaḥ*	73
5	*Vastivyāpatsiddhir adhyāyaḥ*	54
6	*Bhēṣajakalpādhyāyaḥ*	29
Total		304

(*Continued*)

TABLE 1.3 (*Continued*)
Chapters and *Śḷōka* in *Aṣṭāṅgahṛdaya*[a]

Sl. No.	Title of Chapter	Total No. of *Śḷōka*
	6 *Uttarasthānam*	
1	*Bālōpacaraṇīyādhyāyaḥ*	49
2	*Bālāmayapratiṣēdhādhyāyaḥ*	77
3	*Bālagrahapratiṣēdhādhyāyaḥ*	60
4	*Bhūtavidyājñānādhyāyaḥ*	44
5	*Bhūtapratiṣēdhādhyāyaḥ*	53
6	*Unmādapratiṣēdhādhyāyaḥ*	60
7	*Apasmārapratiṣēdhādhyāyaḥ*	37
8	*Vartmarōgavijñānīyādhyāyaḥ*	27
9	*Vartmarōgapratiṣēdhādhyāyaḥ*	41
10	*Sandhisitāsitarōgavijñānīyādhyāyaḥ*	31
11	*Sandhisitāsitarōgapratiṣēdhādhyāyaḥ*	58
12	*Dṛṣtirōgavijñānīyādhyāyaḥ*	33
13	*Timirapratiṣēdhādhyāyaḥ*	100
14	*Liṅganāśapratiṣēdhādhyāyaḥ*	32
15	*Sarvākṣirōgavijñānīyādhyāyaḥ*	24
16	*Sarvākṣirōgapratiṣēdhādhyāyaḥ*	67
17	*Karṇarōgavijñānīyādhyāyaḥ*	26
18	*Karṇarōgapratiṣēdhādhyāyaḥ*	66
19	*Nāsārōgavijñānādhyāyaḥ*	27
20	*Nāsārōgapratiṣēdhādhyāyaḥ*	25
21	*Mukharōgavijñānādhyāyaḥ*	69
22	*Mukharōgapratiṣēdhādhyāyaḥ*	111
23	*Śirōrōgavijñānādhyāyaḥ*	32
24	*Śirōrōgapratiṣēdhādhyāyaḥ*	60
25	*Vraṇavijñānīyapratiṣēdhādhyāyaḥ*	68
26	*Sadyōraṇapratiṣēdhādhyāyaḥ*	57
27	*Bhaṅgapratiṣēdhādhyāyaḥ*	41
28	*Bhagandarapratiṣēdhādhyāyaḥ*	44
29	*Granthyarbudaślīpadāpacīnāḍī vijñānādhyāyaḥ*	31
30	*Granthyarbudaślīpadāpacīnāḍī pratiṣēdhādhyāyaḥ*	40
31	*Kṣudrarōgavijñānādhyāyaḥ*	34
32	*Kṣudrarōgapratiṣēdhādhyāyaḥ*	33
33	*Guhyarōgavijñānādhyāyaḥ*	52
34	*Guhyarōgapratiṣēdhhyāyaḥ*	67
35	*Viṣapratiṣēdhādhyāyaḥ*	73
36	*Sarpaviṣapratiṣēdhādhyāyaḥ*	94
37	*Kīṭalūtādiviṣapratiṣēdhādhyāyaḥ*	86
38	*Mūṣikāḷarkaviṣapratiṣēdhādhyāyaḥ*	40
39	*Rasāyanādhyāyaḥ*	181
40	*Vājīkaraṇādhyāyaḥ*	89
Total		2,239
Grand total		7,462

[a] Das and Emmerick (2009).

Śārīrasthānam explains fundamental aspects of human fertility, treatment of infertility and conception. Interestingly, this section describes some measures to beget a son or daughter according to one's choice. Management of pregnancy, complications arising in pregnancy, parts of body, vital points in the human body (*marma*) and characteristics of constitutional types are described. The chapter entitled *aṅgavibhāgaṃ* describes anatomy and physiology. Chapters entitled *vikṛtivijñānīyaṃ* and *dūtādvijñānīyaṃ* describe, respectively, signs of imminent death and application of astrology in diagnosis, treatment and prognosis of diseases.

Causative factors, symptoms and signs and classification of 24 major diseases are presented in *Nidānasthānam*. *Cikitsāsthānam* describes the treatment of these diseases. Maximum number of verses are devoted to describing the treatment of cough (*kāsa*, 180), followed by fever (*jvara*, 177) and hemorrhoids (*arśas*, 164).

Kalpasthānam describes the various herbs to be used in emesis (*vamana*), purgation (*virēcana*), medicated enema (*vasti*), complications that can arise from these measures, formulae of medicated enema that pacify all diseases, collection of herbs, methods of preparation of major dosage forms, weights and measures.

Uttarasthānam deals with the diagnosis and treatment of pediatric diseases, mental disorders, epilepsy, diseases of eye, ear, nose, mouth and head, wounds, fractures, anal fistula; diseases of genitals, bites of snakes, rats, rabid dogs, stings of insects; rejuvenation therapy (*rasāyanavidhi*) and virilification (*vājīkaraṇavidhi*). Comparing the *Cikitsāsthānam* with *Uttarasthānam*, it is observed that *rasāyanavidhi* is the chapter for which maximum number of verses (181) are devoted in *Aṣṭāṅgahṛdaya*.

On closer observation of *Cikitsāsthānam* and *Uttarasthānam*, it can be said that *kāyacikitsa*, one of the eight branches of Ayurveda, is described in *Cikitsāsthānam*, while the other seven limbs of *Aṣṭāṅga Ayurveda* (*śalya*, *śālākya*, *kaumārabhṛtya*, *agadatantra*, *bhūtavidya*, *rasāyana*, *vājīkaraṇa*) are described in *Uttarasthānam*. In modern parlance, *Cikitsāsthānam* deals with general medicine and *Uttarasthānam* with various specialties.

1.9 THE NOVELTY OF *AṢṬĀṄGAHṚDAYA*

Vāgbhaṭa occupies a prominent position in Ayurveda literature as a great physician, master teacher and a poet *par excellence*. Every verse in *Aṣṭāṅgahṛdaya* is pleasing to the ear. Perhaps this must have been a reason for its wide acceptance in the country. Ancient texts such as *Caraka Samhita* and *Suśruta Samhita* are lengthy and difficult to understand. Vāgbhaṭa offered a solution to this problem by restructuring complex themes lucidly. The literary craftsmanship of Vāgbhaṭa offered cohesion and compactness to *Aṣṭāṅgahṛdaya*. He extracted the gist of voluminous texts that had come down over the centuries. He separated the non-essential from the essential and the central from the peripheral. He excluded concepts based on *Sāmkhya* and *Vaiśēṣika* philosophies on the dynamics of the material world, logical parameters of debate, rituals associated with initiation of the disciple, mythological background of diseases and obsolete formulations. Specialized surgical procedures were briefly outlined. However, no disease described in *Caraka Samhita* was left out and the voluminous information on signs of imminent death was included. Vāgbhaṭa declared in the last chapter of *Uttarasthānam* (*Vājīkaraṇādhyāyaḥ*) that *Aṣṭāṅgahṛdaya* is a lucid version of *Aṣṭāṅgasamgraha* that emerged from the churning of the ocean of *Aṣṭāṅgavaidyam*. It is intended to enlighten less knowledgeable physicians (*Ut*. 40: 78–80). The value of this great work can be understood from the following statement: "The teachings of *Aṣṭāṅgahṛdaya* born from the *vēda* bring forth discernible results. Therefore, this work can be used just like a *mantra*. One should not doubt that it will ever fail!" (*Ut*. 40: 81).

1.10 SALIENT ELEMENTS OF POETRY

1.10.1 Prosody in Aṣṭāṅgahṛdaya

Vāgbhaṭa stands out among authors of ancient Ayurveda texts, not only as a guru, but also as a poet blessed with exceptional literary and poetical skills. Vāgbhaṭa's prowess as a poet is evident from the many similes he used in the work. For example, before enumerating the signs of imminent death, he states:

puṣpaṃ phalasya dhūmō'gnēr varṣasya jaladōdayaḥ
yathā bhaviṣyatō liṅgam riṣṭam mṛtyōs tathā dhruvam.

(Just as flower, smoke and gathering clouds predict the appearance of fruit, fire and rain, respectively, *riṣṭam* foretells imminent death (*Śa.* 5: 1).) While discussing ethics he advises "*Ātmavat satataṃ paśyēd api kīṭa pipīlikam*" (even bugs and ants are to be viewed as replicas of one's own self) (*Sū.* 2: 23). The effectiveness of an herb is illustrated using a simile:

Kōkilākṣaka niryūhaḥ pītas tacchāka bhōjīnā
Kṛpābhyāsa iva krōdhaṃ vātaraktaṃ niyacchati

(Consumption of *kvātha* of *Kōkilākṣa* and its leaves as food cures *vātaśōṇita*, just as compassion pacifies anger) (*Ci.* 22: 18–19).

Vāgbhaṭa used rhymes prolifically, improving the sweetness of his poetry and making memorizing easier. Description of *śanairmēha* is an example. "*Śanaiḥ śanaiḥ śanair mēhi mandaṃ mandaṃ pramēhati*" (One who is affected with *śanairmēha* voids urine very slowly and sluggishly) (*Ni.* 10: 13). While listing the names of diseases that respond to *Tiktaka ghṛta*, it is stated "*Bhagandaram apasmāram udaraṃ pradaraṃ garam*" (anal fistula, epilepsy, enlargement of abdomen, dysfunctional bleeding and homicidal poisoning) (*Ci.* 19: 7). Rhyming is applied while describing the rejuvenating effect of *Bhallātaka*. "*Vapuṣkaram aruṣkaraṃ parama mēdhyam āyuṣkaram*" (Consumption of *Bhallātaka* strengthens body, confers intelligence and longevity) (*Ut.* 39: 80).

Various meters (*chandas*) are used in composition. *Chandas* refers to the study of poetic meters and verse in Sanskrit language. It is one of the six pillars of Vedic studies such as *śikṣa*, *chandas*, *vyākaraṇa*, *nirukta*, *kalpa* and *jyōtiṣa* (Lochtefeld, 2002). Knowledge of meters helps in reciting and memorizing poems of Ayurveda literature. 95.11% of quatrains of *Aṣṭāṅgahṛdaya* are composed in *Anuṣṭup* meter. All the verses of *Nidānasthānam* and *Kalpasthānam* are in *Anuṣṭup* meter, while *Cikitsāsthānam* employs 32 different meters. Most varied types of meters are utilized in Chapter 19 of *Cikitsāsthānam* and Chapter 15 of *Sūtrasthānam* (Yerawar et al., 2019). The rejuvenating property of *Bījaka* is presented in *Svāgat*a meter thus:

Bījakasya rasam aṅguli hāryaṃ
Śarkarāṃ madhu ghṛtam triphalāṃ ca
Śīlayatsu puruṣēṣu jarattā
Svāgatāpi vinivartata ēva

(*Kvātha* of *Bījaka* thick enough to be lifted by finger should be mixed with sugar, honey, clarified butter and *Triphala*. Daily consumption of it wards off aging) (*Ut.* 39: 153). The *vāta*-lowering property of *Sahacarādi kvātha* is described in *Drutavilambita* meter:

Sahacaram suradāru sanāgaraṃ
Kvathitam ambhasi taila vimiśritam
Pavana pīḍita dēha gatiḥ piban
Drutavilambitagō bhavaticchayā

(*Sahacara*, *Suradāru* and *Nāgara* made into a *kvātha* mixed with sesame oil and consumed by one suffering from bodily infirmity due to *vāta* offers him quick and slow pace as he wishes) (*Ci*. 21: 56). *Puṣpitāgra* meter is applied in the description of virilific agents:

Madhu mukham iva sōtpalam priyāyāḥ
Kalaraṇanā parivādinī priyēva
Kusuma caya manōramā ca śayyā
Kisalayinī latikēva puṣpitāgrā

(Desire of Kāmadēva or Cupid will be fulfilled by grape wine with fragrance of *Utpala* flowers, lute sounding like sweetheart's voice, a soft bed, a quiet place, disease-free body and aphrodisiacs) (*Ut*. 40: 46). Vāgbhaṭa skillfully embedded the names of these meters (*Svāgata*, *Drutavilambita* and *Puṣpitāgra*) in each of these quatrains (Yerawar et al., 2019).

1.10.2 Maxims

The charm and usefulness of Sanskrit works on Ayurveda are enhanced by the ingenious application of several techniques. *Nyāyā*s or maxims are foremost among them. Maxims are worldly sayings, based on stories or legends popular in the society. They are familiar instances quoted to explain similar cases. The literal meaning of the word *nyāya* is "justice". However, it also means "logic". Authors of ancient Ayurveda treatises made use of different maxims to drive home the significance of concepts. The various maxims incorporated in *Aṣṭāṅgahṛdaya* were not mentioned in the text but identified and elaborated by commentators such as Aruṇadatta and Hēmādri. Rajkumar et al. (2018) provide a list of the maxims adopted in *Aṣṭāṅgahṛdaya*. They include 19 maxims such as *Siṃhāvalōkana nyāya*, *Kākāṣigōḷa nyāya*, *Aśmaśastra nyāya* and so on (Shreedevi and Shreevathsa, 2017).

Pāṭhādi cūrṇa is recommended in the treatment of *śḷēṣmātisāra*. But there is no mention of *anupāna* (post-prandial drink). Aruṇadatta solved this problem by following the *Siṃhāvalōkana nyāya*. A lion always looks forward and backward after killing a prey, to make sure that there is no rival to snatch the catch. Based on *Siṃhāvalōkana nyāya*, Aruṇadatta recommends *kōṣṇavāri* (luke-warm water) as *anupāna*. Body channels carrying the tissue element *rasa* are obstructed in *rājayakṣma* patient by excessive *kapha*, resulting in just maintenance (*dhātu sthiti*) and no nourishment of *dhātu* (*dhātu puṣṭi*). Aruṇadatta interprets this phenomenon on the strength of *Abhinava mṛtkumbha jalasyanda nyāya*. A newly made mud pot, when filled with water for the first time, has water filled in all of its minute pores, with no water leaking out through the pores. Similarly, the *dhātu*s in *rājayakṣma* patient receive no nutrition because of the blocking of the *rasa*-containing body channels with *kapha* (Shridevi and Shreevathsa, 2017).

1.11 VĀGBHAṬA'S INNOVATIVE APPROACH

1.11.1 Introduction of Novel Theories

Ayurveda explains the stages of pathogenesis based on a concept known as *rōganidāna*. The stages in evolution of diseases are collectively called *samprāpti*. There are general concepts (*sarvarōganidāna*) and specified subgroupings of individual diseases. The method of study of evolution of diseases adopted by Caraka, Suśruta and Vāgbhaṭa are similar in most instances. Nevertheless, while endorsing the views of earlier authorities, Vāgbhaṭa omitted certain vital postulates and offered explanations for hitherto unknown new groups included among disease entities in *Aṣṭāṅgahṛdaya*.

1.11.2 Omission of Tenets

Rōganidāna (etiopathogenesis) is illustrated with five attributes such as *nidāna* (etiology), *pūrvarūpa* (premonitory symptoms), *rūpa* (well-defined symptoms), *upaśaya* (confirmation of diagnosis through food and medicine by trial and error) and *samprāpti* (pathogenesis) (*Ni.* 1: 1–11). Vāgbhaṭa did not follow the earlier concept of *ṣaṭkriyākāla* (six specific and distinct stages of development of a disease related to aggravation of *dōṣas*) advanced by Suśruta. Another select set of terms connoting vitiation and mitigation, such as *caya*, *prakōpa* and *prasara*, are mentioned in terms of climatic impacts on *dōṣas* (*Sū.* 12: 23–29). The concept of progressive stages such as *sañcaya*, *prakōpa*, *prasara*, *sthānasamśraya*, *vyakti* and *bhēda* put forth by Suśruta is not upheld by Vāgbhaṭa, while he endorses most of the anatomical and surgical tenets described in *Suśruta Samhita*.

1.11.3 Newer Explanations

The last chapter of *Nidānasthānam* is devoted to a new disease category named *vātaśōṇita*, placed after the highly structured chapter entitled *Vātavyādhinidānādhyāyaḥ*, describing diseases with dominance of *vātadōṣa* (*Ni.* 16: 1–58). *Vātaśōṇita* is described in *Caraka Samhita* with its basic characteristics. The prefix *vāta* attached to the word *śōṇita* (*rakta*) indicates the crucial role of *vāta* in the causation of this disease. Vāgbhaṭa did not include this disease entity in the previous chapter on *vāta* diseases. The precedence of naming the disease, as in the case of the third chapter *Raktapittakāsa nidānādhyāyaḥ*, was intentional in recognizing the dominant role of *pitta* over *rakta* in that context. On the same scale, *vātaśōṇita* qualifies to be included in the chapter on *vātavyādhi*. However, Vāgbhaṭa opted to introduce a new disease spectrum by adding a new chapter entitled *Vātaśōṇita nidānādhyāyaḥ* (*Ni.* 16: 1–58). Looking at the nature of the disease, it becomes apparent that the role of *śōṇita* is fundamental and decisive over *vāta*. It is worth recalling that the previous generation of physicians in Kerala preferred to call the disease *raktavāta*.

1.11.4 Introduction of New Concept

After describing the etiopathogenesis of *vātaśōṇita* (*Ni.* 16: 1–30), Vāgbhaṭa introduces a group of diseases laced with a new concept called *āvaraṇa* meaning "envelopment" (*Ni.* 16: 31–58). It is argued that when a disease is manifested with overlapping symptoms, a new approach can be conceived with the idea of considering one *dōṣa* to be covering or concealing the other, by applying the equation of *dōṣa* involvement. There can be an instance of *vāta* vitiation, but the encirclement by *kapha* makes the symptoms of *vāta* less evident. This concept was later expanded, extrapolating *dōṣa* subtypes and the *dhatu* variants, totaling 15. The scheme of *āvaraṇa* was imagined through multiple propositions, and Vāgbhaṭa finally concluded the discussion by authorizing the physician to derive an appropriate assessment of *dōṣa* involvement in each clinical situation, considering the merit of the case. Vāgbhaṭa opined that *āvaraṇa* can be endless in number, as each context can be specific to be concluded on a logical basis (*Ni.* 16: 53–56). It is evident that the new principle of *āvaraṇa* in the domain of pathogenesis is open-ended, neither precise nor finite.

It makes sense that the discussion on many diseases not described in *Aṣṭāngahṛdaya* can be complex, when viewed in the framework of specific disease entities categorized in the *Nidānasthānam*. The concept of *āvaraṇa* is a window opened to encourage thinking out of bounds. Vāgbhaṭa's attempt was unprecedented. It was the first step of identifying *vātaśōṇita* disease types as separate from *vātavyādhi*, further exploring the concept of *āvaraṇa* to explain diseases that cannot be included in foregoing regular groups of diseases and to widen the scope of including new diseases that would appear in future!

TABLE 1.4
Commentaries and Translations of *Aṣṭāṅgahṛdaya*

Sl. No.	Commentaries and Translations
	English
1	Murthy K.R.S. 1995. *Aṣṭāṅgahṛdayasaṃhitā* – Vāgbhaṭa's *Aṣṭāṅga Hṛdayam* (Text, English translation, notes, appendix and indices). Varanasi: Krishnadas Academy.
2	Vidyanath R. 2013. Illustrated *Aṣṭāṅgahṛdaya* of Vāgbhaṭa (*Sūtrasthānam*). Varanasi: Chaukhamba Surbharati Prakashan.
	German
3	Luise Hilgenberg, Willibald Kirfel. 1941. Vāgbhaṭa's *Aṣṭāṅgahṛdayasaṃhitā - ein altindisches Lehrbuch der Heilkunde* (aus dem Sanskrit ins Deutsche übertragen mit Einleitung, Anmerkungen und Indices). Leiden: E.J. Brill.
	Tibetan
4	Vogel C. 1965. Vāgbhaṭa's *Aṣṭāṅgahṛdayasaṃhita*: The first five chapters of its Tibetan version, edited and rendered into English along with the original Sanskrit. Wiesbaden: Kommissionsverlag Franz Steiner GmbH.
	Sanskrit
5	Anonymous. 1933. *Aṣṭāṅgahṛdayasaṃhitā* – Vidvadvaraśrīmad Aruṇadattakṛtā Sarvāṅgasundarāṭīkā Śrīmadācāryamaudgalyakṛtā maudgalyaṭippaṇī ca. Lavapurī: Motīlāl Banārsīdās.
6	Das R.P. and R.E Emmerick. 2009. *Aṣṭāṅgahṛdaya* (Romanized version of the text with a complete word index of the text). Groningen: Egbert Forsten.
7	Harisasty. 1982. *Aṣṭāṅgahṛdayam* with Sarvāngasundarā commentary of Aruṇadatta and *Āyurvēdarasāyana* commentary of Hemādri. Varanasi: Chowkhamba Sanskrit Series Office.
8	Pandit Bhiṣagāchārya Hariśāstrī Parādkar Vaidya 1939. *Aṣṭāṅgahṛdayasaṃhitā* with the commentaries (Sarvāngasundarā) of Aruṇadatta and of Hemādri, collated by the late Dr. Aṇṇā Moreśwar Kunte and Kṛiṣṇa Rāmchandra Śātrī Navre. Bombay: Nirṇaya-sāgar Press.
9	Rājyavaidya pam Rāmaprasādjī Śarmā 1928. *Aṣṭāṅgahṛdayasaṃhitā* – (Sūtrasthāna), Śrī Aruṇadatta kṛta Sarvāṅgasundarā Śrī Candranandana kṛta padārthacandrokā ca Śrī Hemādri kṛta āyurveda rasāyana ṭīkā. kṛta ṭippaṇī sahita. Bambaī: Śrīveṅkaṭeśvar Sṭīm Press.
10	Raman Menon P. P. 2012. *Sārārthabōdhini* Sanskrit commentary of *Aṣṭāṅgahṛdaya*. Chalakudy: VKRT Foundation for Ayurvedic Studies.
11	Śrījyotiṣacandra Devaśarma. 1942. *Aṣṭāṅgahṛdayasaṃhitā* – Aṣṭāṅgahṛdaya Saṃhitāyā Uttaraṃ Tantram, śrīmat śivadāsasena viracitayā tattvabodha vyākhyāyopetam, śrīmaj jyotiṣacandra sarasvatī mahodayair nibaddhenopodghātena sambalitam. Calcutta: Paṇḍita Śyāmasundaraśarma.
12	Śrīkṛṣṇalālbharatiyā and Vaidya rāja paṃ. Bhikkhīlāljī Śarmā 1956. Aṣṭāṅgahṛdayasaṃhitā – śrīmadvāgbhaṭācārya viracita Aṣṭāṅgahṛdaya arthāt Vāgbhaṭa. anuvādaka mathurā nivāsī svargīya śrīkṛṣṇalālbharatiyā, sampādaka tathā bhāṣyakāra govardhana nivāsī vaidya rāja paṃ. bhikkhīlāljī śarmā tasya ātmaja kavirāja paṃ. gopālprasādjī "kauśik" vaidya. Mathurā : Govardhan Pustakālaya.
13	Taradutt Pant.2016. *Aṣṭāṅgahṛdayam*. Varanasi: Chowkhamba Sanskrit Series Office.
14	Vaidyaratna paṃ. Rāmprasādjī and Śivśarmā-āyurvedācārya 1929. Aṣṭāṅgahṛdayasaṃhitā – Śrīvāgbhaṭācārya-viracita Aṣṭāṅgahṛdaya Saṃhitā, Śivadīpikā' bhāṣāṭīkāsahit. Bambaī : Śrīveṅkaṭeśvar Sṭīm Pres.
	Hindi
15	Upadhyay Y. 1962. Aṣṭāṅgahṛdayasaṃhitā – Śrīmadvāgbhaṭaviracitam Aṣṭāṅgahṛdayam, 'vidyotinī' bhāṣāṭīkā-vaktavya-pariśiṣṭasahitam, ṭīkākāra kavirāja śrī atrideva gupta, saṃpādaka vaidya śrī yadunandana upādhyāya. Vārāṇasī: Chowkhamba Sanskrit Series Office.
16	Śāstrī K. 1980. *Aṣṭāṅgahṛdayasaṃhitā* with critical *vidvan-manorañjinī* Hindi commentary, (*sūtrasthānam*). Vārāṇasī: Chowkhamba Sarasvatībhavan.

(*Continued*)

TABLE 1.4 (*Continued*)
Commentaries and Translations of *Aṣṭāṅgahṛdaya*

Sl. No.	Commentaries and Translations
	Marathi
17	Garde G. K. 1954. *Sārth Vāgbhaṭa*. Ashtangahridya with Marathi translation. Pune: Aryabhushana Mudranalaya.
	Gujarathi
18	Vijaysankar Dhanshankar Munshi. 2017. *Aṣṭāṅga Hṛdaya*. Ahmedabad: Sastu Sahitya Mudranalaya Trust.
	Tamil
19	Pandit M. Durai Swami Aiyangar, A.V.D. 1935. *Aṣṭāṅga Hridaya*. Translated from Sanskrit original into Tamil with notes, introduction, index etc. Madras: The Vaidyakalanidhi Office.
	Telugu
20	Anonymous. 1949. *Aṣṭāṅgahṛdaya*. Madras: Vavila Ramaswamy Sastrulu & Sons.
	Kannada
21	Vaidyagaru Anantacharya Adya. 1955. *Ashtanga Hridaya of Vagbhata,* Part 1 *(Sutrasthan)* A translation with text, copious and explanatory notes in Kannada, Arya Vaidyak Granthamala No. 4. Vijayapura: Karnataka Kudos Press.
	Malayalam
22	Cheppat Achyutha Variar 1962. *Aṣṭāṅgahṛdaya*. Quilon: Sree Ramavilasam Book Depot.
23	Dhanwantharidas C.N. Narayanan Moos 1940. *Aṣṭāṅgahṛdayasaṃhitā* by *Vāhaṭācārya* with the commentary of Sreedasapandita, Part I. Trivandrum: V.V. Press.
24	Govindan Vaidyan P.M. 1990. *Aruṇōdayam* Commentary of *Aṣṭāṅgahṛdaya*. Kodungallur: Devi Book Stall.
25	Kaikulangara Ramavariar. 2017. *Bhāvaprakāśa* Commentary of *Aṣṭāṅgahṛdaya*. Trichur: Sahithi Books.
26	Mooss N.S. 1942. *Aṣṭāṅgahṛdayasaṃhitā* – Vahata's *Ashtangahridaya, Uttarasthāna*, with the *Kairali* commentary. Kottayam: Vaidyasarathy Press.
27	Mooss N.S. 1968. *Aṣṭāṅgahṛdayasaṃhita* with the *Vākyapradīka* commentary of Parameswara. Kottayam: Vaidyasarathy Press.
28	Mooss N.S. 1978. *Aṣṭāṅgahṛdayasaṃhitā* with the *Śaśilekhā* commentary of Indu, Parts 1, 3 and 4. Kottayam: Vaidyasarathy Press.
29	Narayana Pillai 1950. *Aṣṭāṅgahṛdayasaṃhitā* with the commentary *Hṛdayabodhikā* of Śrīdāsapaṇḍita, Part II (*Sūtrasthāna* Ch. XVI to XXX). Trivandrum: University of Travancore.
30	Rāghavan Pilllai K. 1962. Aṣṭāṅgahṛdayasaṃhitā of Vāhaṭa with hṛdaya-bodhikākhyayā of śrīdāsapaṇḍitaviracitayā, Part III (*śārīrasthānaṃ* and *Nidānasthānaṃ*). Quilon: Śrīrāmavilāsa Press.

1.12 COMMENTARIES ON *AṢṬĀṄGAHṚDAYA*

Aṣṭāṅgahṛdaya is unique in having the highest number of commentaries than any other Sanskrit Ayurveda text. Though about 30 commentaries are known, most of them are either lost or remaining in manuscript form in manuscript libraries of India and other countries. Only six are available in print, one completely and the rest in parts. The only commentary available in full and in print is *Sarvāṅgasundarā* by Aruṇadatta. Aruṇadatta, the son of Mṛgāṅkadatta, lived probably in North India in the 12th century (Vogel, 1965). *Sarvāṅgasundarā* is held in high esteem because of its easy diction and lucid expansion. Wherever necessary he commented with the help of his own verses for the benefit of readers. Other commentaries are *Āyurvēdarasāyana* by Hēmādri, *Padārthacandrika* by Candranandana, *Hṛdayabōdhika* of Śrīdāsa Paṇḍita, *Nidānacintāmaṇi* of Ṭōḍaramalla Kanhaprabhu, *Tatvabōdha* of Śivadāsasēna and *Śaśilēkha* of Indu. Published versions of commentaries and translations of *Aṣṭāṅgahṛdaya* are listed in Table 1.4.

1.13 CONCLUSION

Aṣṭāṅgahṛdaya composed by Vāgbhaṭa is a landmark contribution to Ayurveda. It is considered to be a shorter version of its predecessor, *Aṣṭāṅgasamgraha*. Its style of presentation, literary qualities, logical arrangement of the content and clarity of explanation make it endearing to students of Ayurveda, accounting for its popularity all over India and especially in the south. Narration of medical concepts, therapeutic measures and medicinal formulae in *Aṣṭāṅgahṛdaya* are traceable to *Caraka Samhita* and *Suśruta Samhita*. However, *Aṣṭāṅgahṛdaya* is unique in its synthesis of teachings of Caraka and Suśruta. Ever since the World Health Organization adopted the Alma Ata Declaration in 1978, interest in Ayurveda is on the rise globally (Hall and Taylor, 2003). On account of the strength of the *tridōṣa* doctrine, Ayurveda is able to diagnose any disease and treat it successfully. This is substantiated by the statement of Vāgbhaṭa that a physician can manage a disease successfully after studying the nature of the underlying cause, even if he cannot assign a name to it (*Sū*. 12: 64–66). Thus, Ayurveda offers hope to the science of medicine in its fight against diseases that continue to cause misery to the world. Being the best textbook of Ayurveda, *Aṣṭāṅgahṛdaya* can enlighten the world on the intricacies of the "sacred knowledge of longevity".

REFERENCES

Anonymous. 2022a. Memories of Manjhand. https://www.dawn.com/news/1043022 (accessed April 18, 2022).

Anonymous. 2022b. Huge archaeological site discovered near Manjhand. https://www.dawn.com/news/1376903 (accessed April 18, 2022).

Anonymous. 2022c. Vāgbhaṭa Samādhi. https://www.facebook.com/srdayurveda/posts/2303453639879509 (accessed February 15, 2022).

Das, R.P. and R.E. Emmerick. 2009. *Aṣṭāṅgahṛdaya*. (Romanized version of the text with a complete word index of the text). Groningen: Egbert Forsten.

Hall, J.J. and R. Taylor. 2003. Health for all beyond 2000: The demise of the Alma-Ata Declaration and primary health care in developing countries. *Global Health* 178(1):17–19.

Hilgenberg, L. and W. Kirfel. 1941. *Vāgbhaṭa's Aṣṭāṅgahṛdayasaṃhitā - ein altindisches Lehrbuch der Heilkunde (aus dem Sanskrit ins Deutsche übertragen mit Einleitung, Anmerkungen und Indices).* Leiden: E.J. Brill.

Ikbal, N.K., D. Induchoodan and D.S. Kumar. 2021. What we learn from the history of Ayurveda. In *Ayurveda in the New Millennium*, 1–20. Boca Raton: CRC Press.

Kasinathan, R. 1955. *Piḷḷaippiṇi Vākaṭam*, 1–96. Chennai: Government Oriental Manuscripts Library and Research Centre.

Lochtefeld, J.G. 2002. *Chhandas*. In *The Illustrated Encyclopedia of Hinduism*, Vol 1, A-M, 140. New York: Rosen Publishing Group.

Menon, I. 2019. The ashtavaidyans of Malabar. In *Hereditary Physicians of Kerala*, 151–216. Oxon: Routledge.

Mooss, N.S. 1982. Identification and cultivation of medicinal plants mentioned in Ayurvedic classics. *Ancient Science of Life* 1:224–228.

Pillai, T.V.S. 1931. *Tamil-English Dictionary of Medicine, Chemistry, Botany and Allied Sciences*, Vol 5, 1036. Madras: The Research Institute of Siddhar's Science.

Rajkumar, C., S. Kamble, A.S. Baghel, H. Vyas and N.N.L. Bhagavathi. 2018. Significance of *nyaya*s (maxims) in understanding philosophical aspects of Ayurveda: A critical review. *The Journal of Research and Education in Indian Medicine* 24:81–92.

Rao, S.K.R. 1985. Ashtānga Hrdaya. In *Encyclopaedia of Indian Medicine*, Vol 1, 21–23. Bombay: Popular Prakashan.

Sastry, V.V.S. and A.N. Pandey. 1975. Jajjata. *Bulletin of The Indian Institute of History of Medicine* 5(3):115–122.

Sharma, P.V. 1981. Introduction. In *Caraka Saṃhita*, Vol 1, v–xxxii. Varanasi: Chaukhamba Orientalia.

Shridevi, T.N. and Shreevathsa. 2017. Nyayas found in Ashtanga Hridaya Samhitha. *International Ayurvedic Medical Journal* 5(10).

Subhaktha, P.K.J.P., M.S. Gundeti and A. Narayana. 2009. Vagbhata – His contribution. *Journal Indian Medical Heritage* 39:111–136.

Thakar, V.J. 2010. Historical development of basic concepts of Ayurveda from *Veda* up to *Samhita*. *Ayu* 31:400–402.

Variar, P.R. 1985. The Ayurvedic heritage of Kerala. *Ancient Science of Life* 5:54–64.

Variar, P.R. 1987. Ayurveda in Kerala. In *Encyclopaedia of Indian Literature*, ed. A. Datta, Vol 1, 311–313. New Delhi: Sahitya Akademi.

Vogel, C. 1965. Introduction. In *Vāgbhaṭa's Aṣṭāṅgahṛdayasaṃhita: The First Five Chapters of Its Tibetan Version, Edited and Rendered into English along with the original Sanskrit*, 1–43. Wiesbaden: Kommissionsverlag Franz Steiner GmbH.

Yerawar, P., A. Tak and S. Thakur. 2019. Critical review of Ashtanga Hridayam as per Chanda shastra (Sanskrit prosody). *World Journal of Pharmaceutical Research* 8:1705–1710.

2 The Human Body

2.1 INTRODUCTION

The first chapter of *Śārīrasthānam* (*Garbhāvakrāntir adhyāyaḥ*) deals with conception of the human body. The reproductive elements *śukra* and *ārtava* unaffected by vitiated *tridōṣa* unite together to form an embryo. It feeds on the essence of the five *bhūta*s derived from the food of the mother and grows (*Śā*. 1: 1–3). A boy child results if the element of *śukra* is dominant during conception. The dominance of *ārtava* facilitates the development of a girl child. An offspring of neuter gender is formed if *śukra* and *ārtava* are in equal strength. Because of the changes caused by vitiated *tridōṣa*, improperly formed fetuses also appear (*Śā*. 1: 5–6). In women, blood named *rajas* starts to exude for three days. This bleeding starts at the 12th year, reduces by the 50th year and gradually ceases (*Śā*. 1: 7). A son is born out of the union of reproductive elements (*śukra* and *ārtava*) of a sixteen-year-old young lady with clean uterus, heart and a twenty-year-old youth. Union of couples below these age groups can give birth to offspring with diseases and other defects (*Śā*. 1: 8–9).

Conception may not occur sometimes, even if sexual union takes place. The reason for this is the unhealthy nature of the *śukra* and *ārtava* due to the effects of vitiated *tridōṣa*. Conditions like *vātaśukḷa*, *pittaśukḷa*, *kaphaśukḷa*, *kuṇapaśukḷa*, *granthiśukḷa*, *pūyaśukḷa*, *ksīṇaśukḷa*, *mūtraśukḷa*, *purīṣaśukḷa*, *vātārtava*, *pittārtava*, *kaphārtava*, *kuṇapārtava*, *granthyārtava*, *pūyārtava*, *kṣīṇārtava*, *mūtrārtava* and *purīṣārtava* develop with their characteristic signs and symptoms (*Śā*. 1: 10–11).

Medicines for the respective *dōṣa*s are to be administered in *śukḷa* and *ārtava* vitiated by *vāta*, *pitta* and *kapha*. *Kuṇapaśukḷa* is to be cured by administering *ghṛta* prepared with *Woodfordia fruticosa*, *Acacia catechu*, *Punica granatum* and *Terminalia arjuna*. *Ghṛta* prepared with herbs of *Asanādi gaṇa* also cures *kuṇapaśukḷa* (*Śa*. 1: 12–13). *Ghṛta* prepared with *Rotula aquatica* and ash of *Butea monosperma* pacifies *granthiśukḷa*. *Pūyaśukḷa* is cured by the administration of *ghṛta* prepared with *Parūṣakādi gaṇa* or *Nyāgrōdhādi gaṇa*. Virilification measures are useful in the cure of *ksīṇaśukḷa* (*Śa*. 1: 14). Patient suffering from *purīṣaśukḷa* is to be treated at first with appropriate elimination measures. It is to be followed by the consumption of *ghṛta* prepared with *Ferula fetida*, *Vetiveria zizanioides* and *Plumbago zeylanica* (*Śa*. 1: 15).

Decoction of *Cyclea peltata*, *Trikaṭu* (*Piper longum*, *Piper nigrum*, *Zingiber officinale*) and *Holarrhena antidysenterica* cures *granthyārtava*. Decoction of *Santalum album* is to be administered in *kuṇapārtava* and *pūyārtava*. All treatments and measures recommended in chapter on the disease of genitals (*Guhyarōgapratiṣēdhādhyāyaḥ*) are to be adopted along with *uttaravasti* in all diseases of *śukḷa* and *ārtava* (*Śa*. 1: 15–16).

The husband and wife should consume *snēha* such as *Jīvanīya ghṛta*, *Mahākalyāṇaka ghṛta* and *Phala ghṛta* that cause unctuousness in body. Following this, *dōṣa*s in the body are to be eliminated with *vamana*, *virēcana* and *vasti*. After performing elimination measures, the husband should consume *Kākōḷyādi ghṛta* or *Jīvantyādi ghṛta* and milk. Wife should consume sesame oil, food prepared with black gram and *pitta*-increasing food such as fish and sesame seeds. These measures are intended to improve the volume of blood. The couple should enter into sexual union on the fourth day of menstruation. During sexual union, they should visualize the features and qualities of the son who is to be born to them. Signs such as feeling of satisfaction, heaviness in abdomen, feeling of pulsation in the precordial area, drowsiness, thirst, tiredness and horripilation indicate successful sexual union (*Śā*. 1: 18–36).

DOI: 10.1201/9781003148296-2

2.2 SEX TRANSFORMATION

Seven days after sexual union, *śukra* and *ārtava* unite to form a fetus, the sex of which remains undifferentiated for a month. During that period measures can be adopted to beget a child of one's choice. Chapter 1 describes several *puṃsavana prayōga* or medicines and measures that can offer a child with a gender of one's choice (*Śā*. 1: 37–42). A few of such techniques are given below.

1. On the day of *Puṣya* star, a figurine of a man made of gold, silver or iron is placed in a furnace and on turning red hot is plunged into a bowl containing milk. Drinking of 1 *añjali* (4 *pala* or 4×48 ml) of this treated milk by the pregnant mother will cause the birth of a boy child (*Śā*.1: 38–39).
2. The pregnant mother should herself grind the root of s *Śvēta bṛhati* (*Solanum indicum*) in milk and instill the medicated milk into her right nostril to beget a boy child and the left nostril to beget a girl child (*Śā*.1: 40–41).
3. Root of *Lakṣmaṇa* (*Ipomoea sepiaria*) is to be ground and suspended in milk. Drinking of this milk or instillation into nose facilitates the birth of a boy child. Eight buds of *Ficus benghalensis* (*vaṭaśṛṅgāṣṭakam*) can also be used in this way (*Śā*.1: 41–42).

2.3 PREGNANCY CARE

After performing the sex transformation measure, the pregnant mother should consume nourishing medicines. *Abhyaṅga* should be carried out with *taila* having roborant qualities. She should consume during pregnancy pleasing food of her choice. She should be assisted in all physical activities by husband and servants. Consumption of butter, clarified butter and milk during pregnancy will be beneficial. The pregnant mother should avoid indulging in activities that cause exertion, thick blankets, keeping awake at night, melancholy, anger, fear, getting startled, restraining defecation and micturition reflexes, fasting, consumption of wine, meat and lying in supine position (*Śā*. 1: 37–47).

Consumption of *vāta*-enraging food may cause the birth of a hunchback, blind, dull-witted or dwarf child. *Pitta*-increasing food can be the cause of birth of a bald child or one with reddish brown hair. A child afflicted with vitiligo or pallor may be born due to indulgence in *kapha*-increasing food. Discomforts experienced by pregnant mothers are to be pacified with soft and comfortable medicines. Milk porridge mixed with clarified butter is to be consumed in the eighth month of pregnancy. *Snēhavasti* and *nirūhavasti* are also recommended. Unctuous food mixed with meat soups and gruel containing copious amount of clarified butter are the choice food during the ninth month of pregnancy. *Snēhavasti* also may be performed. From ninth month onward, a piece of cloth soaked in medicated oil (*picu*) is to be placed every day in the vagina. Water boiled with *vāta*-lowering herbs is to be used for bathing. *Snēha* should be administered internally or applied topically every day (*Śā*. 1: 48–69).

Appearance of breast milk in right breast, tendency to perform actions with right hand, seeing men in dreams and protrusion of right side of abdomen indicate that the child *in-utero* is a boy. Opposite signs such as appearance of breast milk in left breast, tendency to perform actions with left hand, seeing women in dreams, protrusion of left side of abdomen, interest to wear fragrant floral garlands and passion for music and dance suggest that a girl child will be born to that mother (*Śā*. 1: 69–71).

A well-furnished room is to be set aside in the ninth month itself for facilitating child birth. Tiredness, looseness of abdomen and eyes, heaviness in lower abdomen, distaste for food, excessive salivation, voiding of large volume of urine, pain in thighs, abdomen, hip, back, precordial area, bladder region and groins, splitting pain and feeling of pulsations in vagina accompanied by exudation of fluid indicate that child birth is imminent. Physician and experienced midwives are to assist in child birth. Post-partum mother is to be looked after with great care, as disorders affecting her are difficult to be cured (*Śā*. 1: 73–100).

2.3.1 Management of Mishaps

Due to dietary and behavioral indiscretions, pregnant ladies may encounter many disorders such as *upaviṣṭakam*, *nāgōdaram*, *līna* and *mūḍhagarbham*. Bleeding may sometimes happen after the fetus has started growing. In that case the fetus stays inside the womb pulsating, but without growing. The girth of abdomen does not increase on this account. This condition is known as *upaviṣṭakam*. Grief, fasting and consumption of oil-free food cause bleeding, followed by enragement of *vāta*. This results in leaning of the fetus, diminishing the girth of abdomen. There will be delayed pulsation as well. This condition is known as *nāgōdaram*. *Ghṛta*, milk and meat soups processed with sweet, roborant and *vāta*-pacifying herbs are to be administered in both these conditions. Afterward, the body of the pregnant mother is to be shaken by traveling in chariots and on horseback (*Śā*. 2: 14–18).

The fetus does not pulsate in the pregnancy mishap known as *līna*. Soup prepared from the meat of kite, cow, fish, sea eagle, peacock, black gram and radish is to be administered with copious amount of clarified butter. Paste of unripe fruit of *Aegle marmelos* and powders of sesame seeds, black gram and parched rice are to be suspended in milk and consumed. Blend of meat soup and wine brewed from honey is also beneficial. Waist of the pregnant mother is to be massaged regularly, after applying medicinal oil. *Udāvarta* is a disorder caused by retention of feces. *Udāvarta* experienced by pregnant mothers is to be pacified by administration of oleaginous substances and appropriate medicinal enemas. Constipation is to be prevented. Stillbirths (*mūḍhagarbham*) may happen sometimes due to dietary or behavioral indiscretions. The physician should at first try to expel the stillbirth by applying herbal pastes inside vagina, failing which he should remove it surgically, after obtaining permission from king (*Śā*. 2: 18–26).

2.4 THE HUMAN BODY

Ayurveda considers the human body to be grossly made up of the head, thorax, two hands and two legs. Organs like eyes and heart are subdivisions of the body. Sound, tactile sensation, form, taste modalities and odor are, respectively, the qualities of *ākāśa*, *vāyu*, *agni*, *ap* and *pṛdhvi bhūta*. All the channels of the body (*srōtāmsi*) faculty of auditory system, sound and emptiness are derived from *ākāśa*. *Vāyu bhūta* gives rise to faculty of tactile sensation and driving force. *Agni bhūta* is the basis of vision, form and transformations. Faculty of gustatory sensation, taste modalities and all secretions are based on *ap bhūta*. Finally, the faculty of olfaction, odor and bones arise from *pṛdhvi bhūta*. All soft organs of the body like *rakta* (blood), *māṃsa* (muscle), *majja* (bone marrow) and *guda* (rectum and anus) are inherited from the mother. *Śukra*, *dhamani* (artery) and *kacādika* (hair) which are rich in *sthira guṇa* (firmness) are derived from the father. Mind (*citta*), faculties of sense organs (*akṣa*) and various incarnations originate from *ātmā* (causal body). From homologation (*sātmya*) arises longevity, freedom from diseases, enthusiasm without laziness, luster and stamina of body. The body, its existence, growth and generosity (*alōlata*) are derived from the essence of digestion. The qualities of a human being are derived from the three fundamental qualities (*guṇa*): *sattva*, *rajas* and *tamas*. Thus cleanliness, religious nature, interest in ethical values and intelligence stem from *sattva guṇa*. Pride, anger, vanity and ego arise from *rajōguṇa*. *Tamōguṇa* is the causative agent of fear, ignorance, sleep, laziness and melancholy. Therefore, it can be concluded that the human body is a product of the *pañcabhūta* (*Śā*. 3: 1–8).

2.4.1 Anatomical Details

The embryo formed from the union of *śukra* and *ārtava* slowly grows into a full-fledged organism with several anatomical elements. This growth continues after parturition. The third chapter of *Śārīrasthānam* (*Aṅgavibhāgaśārīrādhyāyaḥ*) deals with the anatomical aspects of the human body from the perspective of Ayurveda. Seven layers of skin are formed from the metabolism of blood. Similarly, the moisture present in *dhātus* and organs is processed by their own heat and get transformed into structures called *kala*, which get covered with *kapha* and tendinous, waxy substance (*snāyu*) (*Śā*. 3: 8–9).

The human body contains seven *āśaya* (containers or receptacles) for holding *rakta* (*raktāśaya*), *kapha* (*kaphāśaya*), *āma* (*āmāśaya*), *pitta* (*pittāśaya*), digested food (*pakvāśaya*), *vāyu* (*vātāśaya*) and *mūtra* (*mūtrāśaya*). In women, the eighth *āśaya* or *garbhāśaya* is situated between *pittāśaya* and *pakvāśaya* (*Śā*. 3: 10–11). *Kōṣṭhāṅga* (organs in the chest and abdomen) are the following: *hṛdaya* (heart), *kḷōma*, *phusphusa* (lungs), *yakṛt* (liver), *plīha* (spleen), *uṇḍuka* (cecum), *vṛkka* (kidneys), *nābhi* (umbilicus), *diṃbha* (uterus), *āntra* (intestine) and *vasti* (urinary bladder) (*Śā*. 3: 10–12). *Kḷōma* is variously described as trachea, right lung, gall bladder and pancreas.

Jīvitadhāma (seat of life) are *śiras* (head), *rasana bandhana* (attachment of the tongue), *kaṇḍha* (throat), *asra* (blood), *hṛdaya* (heart), *nābhi* (umbilicus), *vasti* (urinary bladder), *śukra* (reproductive element), *ōjas* (essence of *dhātu* cycle) and *guda* (rectum and anus). There exist in the body 16 *jāla* (net-like structures) and 16 *kaṇḍara* (large tendons) separately, 6 *kūrca* (brush-like structures) and 7 *sīvani* (sutures, raphae) located near the penis, tongue and skull. These are not to be touched with sharp instruments. In addition to these, there are 4 *māṃsarajju* (muscular ropes), 14 *asthisaṅghāta* (confluence of bones) and 18 *sīmanta* (demarcations). There are 360 *asthi* (bones), including teeth and nails. There are 210 *sandhi* (bony joints) and 900 *snāyu* (tendons). There are 500 *pēśi* (muscles) in men and 520 in women, including those located in the vagina and breasts (*Śā*. 3: 13–18).

2.4.1.1 Vessels

Ten *mahāsira* (root veins) are located in the *hṛdaya*. They supply *rasa* (tissue fluid) and *ōjas* to entire body. All bodily activities are dependent on them. They are big at roots and small at tips. They appear like veins on a leaf. Thus divided, they become 700 in number. In each limb, there are 100 *sira*. Among them, one named *jālandhara* and three located deep inside should not be cut during venesection. In the pelvis (*śrōṇi*), there are 32 *sira*. Out of them, two each in both *vaṅkṣaṇa* (groin) and two each in both *kaṭīkataruṇa* (crest of pelvic bones) are not to be touched by sharp instruments. In the *pārśva* (flanks), there are 16 *sira*. Of them, one on either side going upward should be left untouched. In the *pṛṣṭha* (back), there are 24 on either side of the vertebral column. Among them two each on either side going upward should not be cut. In the *jaṭhara* (abdomen), their number is the same as in the back. Two of them on either side, located above the penis, on either side of line of hair should be left untouched by sharp instruments (*Śā*. 3: 18–24).

In the *uras* (chest), there are 40 *sira*, and out of them, 2 are in each *stanarōhita* (upper border of chest), 2 are in each *stanamūla* (lower border of chest), 2 in *hṛdaya*, 1 each in *apastambha* (left and right borders of chest) and *apalāpa* (left and right borders of the back). These should not be cut. In the *grīva* (neck), their number is same as in the back. Out of them, 2 *nīla*, 2 *manya*, 2 *kṛkaṭika*, 2 *vidhura* and 8 *mātṛka* (total 16) are to be avoided. In *hanu* (lower jaw), there are 16 *sira*. Out of them, the two which bind the lower jaw with skull should not be cut (*Śā*. 3: 25–27).

In the *jihva* (tongue), there are 16 *sira*. Out of them, two lower ones involved in taste perception and two related to speech should not be cut. In the nose, there are 24 *sira*. The two responsible for olfaction and the one located in the palate should not be cut. In the eyes, there are 56 *sira*. Among them, two each involved in opening and closing of the eyes and two in each *apāṅga* (outer canthus of eye), totaling six, are to be protected from any damage. In the forehead, there are 60 *sira*. Among them, the one located at the *sthāpani*, two at the *āvarta* and four at the borderline of hair should be protected from cuts. In the ear, there are 16, and the 2 responsible for auditory function should not be cut. In the temple (*śaṅkha*) also, there are 16 *sira*. The two located at the joint of the temple should be protected. In the head, there are 12 and 8 *sira* including one each located at the *utkṣēpa* (border line of hair), *sīmanta* (sutures in the skull) and *adhipati* (bregma in the center of skull) should not be cut. These 98 *sira* are not to be cut for venesection. Apart from these, *sira* that are fused together, shaped like lumps, very small, curved and located at joints are also not to be cut (*Śā*. 3: 28–34). An overview of *sira* is presented in Table 2.1.

TABLE 2.1
***Sira* in the Human Body**

Sl. No.	Part of Body	No. of *sira*
1	Four limbs (*śākha*)	400
2	Hip (*śrōṇi*)	32
3	Flanks (*pārśva*)	16
4	Back (*pṛṣṭha*)	24
5	Abdomen (*jaṭhara*)	24
6	Thorax (*uras*)	40
7	Neck (*grīva*)	56
8	Tongue (*jihva*)	36
9	Nose (*nāsā*)	24
10	Eyes (*nayanam*)	32
11	Ears (*karṇam*)	16
Total		700

One-fourth of the *sira* carry separately blood vitiated by *vāta*, *pitta*, *kapha* and pure blood. Vessels which pulsate are small and bluish-red in color carry blood mixed with *vāta*. Those which are warm, containing bluish-yellow blood carry *pitta*-dominant blood. Those which are white, smooth, static and cold carry *kapha*-dominant blood. *Sira* which are deep seated, smooth, evenly placed and of red color carry pure blood. There are 24 *dhamanis* (arteries) attached to the umbilicus (*nābhi*). They are arranged like the spokes of a wheel, with the *nābhi* at the center. The vessels radiating from the *nābhi* nourish the body (*Śā*. 3: 35–40).

2.4.1.2 Orifices

There are nine *srōtas* (orifices) in human body. They are two in nostrils, two ears, two eyes, anus, mouth and urinary orifice. They are large and open to the exterior. They are also known as *sthūlasrōtas* (large orifices), *bāhyasrōtas* (external orifices), *navadvāra* (nine openings) and *navacchidra* (nine slits). In women, there are two more – vagina and two in breasts. There are 13 *ābhyantarasrōtas* (internal orifices). They respectively serve as conduits for *prāṇa*, seven *dhātu*, three *mala*, water and food. Diseases appear when these *ābhyantarasrōtas* get vitiated due to dietary and behavioral indiscretions. Vitiation of *ābhyantarasrōtas* is induced by food and behavior having qualities similar to those of the *dōṣās* and causative agents having qualities opposite to those of the *dhātus*. Vitiation of *srōtāmsi* (plural of *srōtas*) can be inferred from increased, decreased or obstructed movement of substances through these *srōtāmsi*, exemplified by conditions like dysentery, bleeding, constipation and so on. Fainting, shivering, enlargement of abdomen, vomiting, fever, delirium, pricking pain, obstruction of feces, urine and even death may occur when *srōtāmsi* are injured. Therefore, an intelligent physician should carefully remove the object that caused the wound and treat for its quick healing (*Śā*.3: 40–48).

2.4.1.3 Vital Points

Marma (vital points) are defined as junctions in the body, where two or more muscles, vessels, bones, ligaments and joints meet. There are 107 *marma* in the human body. Their distribution is as follows: four limbs – 44, abdomen – 3, thorax – 9, back – 14 and neck – 37 (*Śā*. 4: 1–2). These *marma* are classified into six groups: *māṃsamarma*, *asthimarma*, *snāyumarma*, *dhamanimarma*, *sirāmarma* and *sandhimarma*. Life force is said to reside in them, and when injured, they cause death (*Śā*. 4: 38). The effects of damage to these *marma* are described in Chapter 4 (*Marmavibhāgasārīrādhyāyaḥ*) (*Śā*. 4: 47–51) (Table 2.2).

TABLE 2.2
Effects of Injury to *marma*

Marma	Effects of Injury	Reference
Māṃsamarma	Continuous exudation of pale red blood, pallor, improper functioning of organs, death	*Śā*. 4: 47
Asthimarma	Frequent exudation of clear blood laced with bone marrow, excruciating pain	*Śā*. 4: 48–49
Snāyumarma	Abnormal movement of muscles, jerking, arrest of movement of joints, intense pain, inability to stand, walk or sit down; crippling or death	*Śā*. 4: 48–49
Dhamanimarma	Fainting, exudation of warm blood with froth	*Śā*. 4: 48–49
Sirāmarma	Exudation of thick blood, excessive thirst, giddiness, respiratory distress, mental confusion, hiccup, death	*Śā*. 4: 50
Sandhimarma	Pricking pain in wound, paralysis of hand or leg on healing of the wound, loss of stamina, loss of movements, leaning, edema in joints	Śā. 4: 51

Injuries to body parts other than *marma* are not fatal. But injuries to *marma* are fatal. If the patient survives by the expertise of the physician, he will live for the rest of his life as a cripple. Therefore, alkalis, poisons and fire are not to be applied on *marma*. Even a small injury to *marma* causes great pain. Thus, injuries to *marma* are to be treated with great care (*Śā*. 4: 66–70).

2.5 DIGESTION OF FOOD

Ingested food on reaching the *kōṣṭha* (alimentary canal) is acted upon by *pācakapitta*. It is held in the stomach by a structure called *grahaṇi*, which acts as a window to intestine, so that longevity, health and luster are bestowed on the individual. A healthy *grahaṇi*, which is the seat of *agni*, holds the food in *āmāśaya* (stomach), allows it to undergo transformation and then releases into the *pakvāśaya* (intestine). But when the *grahaṇi* is unhealthy, it releases the untransformed food into *pakvāśaya*, causing many diseases (*Śā*. 3: 49–53).

Food ingested at the appropriate time is propelled into the stomach by *prāṇa*. There the food is liquefied and transformed by *jaṭharāgni*, which is stimulated by *samāna vāyu*. The food which is made up of six tastes first acquires sweet (*madhura*) taste and then turns *amḷa* (sour) in taste. As it reaches the *pakvāśaya*, water gets removed and the digested food becomes a solid mass and *kaṭu* (pungent) in taste. Thereupon, five *agni*, each designated for one *bhūta*, act upon the digested food and assimilate it into corresponding *bhūta* present in the body. The clear portion of the digested food becomes urine, and the solid part becomes *purīṣa* (feces). The essential products of digestion are thereafter acted upon by *dhātvagni* and transformed sequentially into *rasa*, *rakta*, *māmsa*, *mēdas*, *asthi*, *majja* and *śukra* (*Śā*. 3: 55–63). The end product of *dhātu* cycle is known as *ōjas*. Though it is concentrated in *hṛdaya* (heart), it circulates in the body, imparting vitality. It is cold (*śīta*), unctuous (*snigdha*) and slightly reddish-yellow in color. Diminution of *ōjas* (*ōjakṣaya*) results in death (*Sū*. 11: 37–39). *Ōjakṣaya* is caused by anger, appetite, worry, grief and bodily exertion. *Ōjakṣaya* causes fear, loss of stamina, loss of luster of body, increased dryness and negative mental attitude (*Sū*. 11: 39–41). *Kapha* (mucus), *pitta* (bile), excretions of the nine orifices, sweat, nails, hairs, oily matter from eyes, skin, *purīṣa* and *ōjas* are respectively the *mala* of the *dhātus*. The seven *dhātus* circulate in the body under the influence of *vyāna vāyu*. Diseases arise when this circulation is disrupted (*Śā*. 3: 63–69).

The activities of *dhātvagni* and *bhūtāgni* are dependent on *jaṭharāgni*, which is the pivotal factor. Changes in its activity will reflect in the activities of *dhātvagni* and *bhūtāgni*, causing diseases. When *samāna vāyu* stays at its own seat without increase or decrease in activity, *jaṭharāgni* will also be in steady state (*samāgni*). When *samāna vāyu* is deflected from its normal direction, *jaṭharāgni* becomes irregular (*viṣamāgni*). When *pitta* is associated with *samāna vāyu*, the *jaṭharāgni* becomes sharp (*tīkṣṇāgni*). When *samāna vāyu* is influenced by *kapha*, the *jaṭharāgni* becomes dull (*mandāgni*). Thus, *jaṭharāgni* exists as *samāgni*, *viṣamāgni*, *tīkṣṇāgni* and *mandāgni*. *Samāgni* transforms food in the desired way. Delayed transformation takes place due to *viṣamāgni*. *Tīkṣṇāgni* transforms fast even heavy food consumed in excess. *Mandāgni* causes delayed transformation of even food consumed in small measures causing symptoms like dryness of mouth, borborygmi, enlargement and heaviness of abdomen (*Śā*. 3: 71–76).

2.5.1 Stamina

Stamina (*balam*) is of three types: *sahaja* (innate), *kālaja* (related to time) and *yuktikṛta* (related to logic). *Sahaja balam* originates in the body due to its innate composition. *Kālaja balam* is related to age and season. *Yuktikṛta balam*, on the other hand, arises from food and exercise. Stamina bestowed by roborant medicines of the *rasāyana* and *vājīkaraṇa* type can also be considered to be of the *yuktikṛta* variety (*Śā*. 3: 77–78).

2.5.2 Quanta of Body Elements

Some information is also provided on the quantity of the body elements. For example, the amount of *majja dhātu* present in the human body is one *prasṛta*. The quantities of the others are as follows: *mēdas* 2 *prasṛta*, *mūtra* 4 *prasṛta*, *pitta* 5 *prasṛta*, *kapha* 6 *prasṛta*, *purīṣa* 7 *prasṛta*, *rakta* 8 *prasṛta*, *rasadhātu* 9 *prasṛta* and *jalam* (water) 10 *prasṛta*. The volume of fluid that can be held in two palms held together like a bowl, with fingers slightly curved inward, is considered to be the unit known as *prasṛta* and two *prasṛta* make one *añjali*. If the physician wants to assess the quanta of *dhātus* in an individual, that individual's palms kept like a bowl are to be considered as the unit *prasṛta*. *Ōjas*, brain (*mastiṣka*) and *śukra* are present in a body at the rate of one *prasṛta* each. Breast milk is usually present in a volume of two *añjali* and menstrual blood has a volume of four *añjali*. These values are indicated for a person with *tridōṣa* in steady state and vary according to derangement of *tridōṣa* (*Śā*. 3: 80–82).

2.6 THE PHYSICAL ENVIRONMENT

The physical environment also exerts considerable influence on human biology. Land endowed with less water, vegetation and hills is known as *jāṅgala dēśam*. Contrarily, terrain blessed with plenty of water, trees and hills is called *ānūpa dēśam*. The intermediate type of land is known as *sādhāraṇa dēśam*. *Jāṅgala dēśam* causes minor diseases and never becomes a threat to population. However, *ānūpa dēśam* generates many kinds of diseases. *Sādhāraṇa dēśam* causes diseases moderately and is beneficial to health (*Śā*. 3: 79).

2.7 CONSTITUTIONAL TYPES

Due to predominance of *tridōṣa* in reproductive elements, diet, behavior of pregnant mother, uterus and the season in which conception occurs, offspring of three basic constitutions (*prakṛti*) are born. Their combinations give rise to *vāta-pittaja*, *pitta-kaphaja*, *vāta-kaphaja* and *vāta-pitta-kaphaja* (*sannipātaja*) *prakṛti*. In total, there are seven *prakṛti* (*Śā*. 3: 83–104). The characteristics of *vātaja prakṛti*, *pittaja prakṛti* and *kaphaja prakṛti* are presented in Tables 2.3–2.5). *Prakṛti* is important in the diagnosis and treatment of diseases (*Sū*.12: 67).

For describing the characteristics of the three basic *prakṛti*, Vāgbhaṭa assigns to them 22 symbols like tiger, mouse, cat, lion, vulture and so on. This is an ingenious method, the rationale of which is evident from Table 2.6. Perhaps Vāgbhaṭa attempts here to unravel the truth that all living beings including humans are made up of the same *pañcabhūta* and *tridōṣa* and that they share

TABLE 2.3
Characteristics of *vātaja prakṛti* (*Śā*. 3: 85–89)

Sl. No.	Characteristics
1	Low or mean nature
2	Split and broken hairs and smoky complexion
3	Aversion toward coolness
4	Unstable and fluctuating courage, memory, activities, affection, eye movements and walking
5	Garrulous nature
6	Reduced wealth, stamina, life span and sleep
7	Words spoken will be tired-sounding, broken or interrupted
8	Agnosticism and hyperphagia
9	Interest in singing, laughter, hunting and disputes
10	Feeling comfort in consuming warm food and beverages of sweet, acidic tastes; interest in them
11	Lean and tall body
12	Poor self-control
13	Uncultured behavior
14	Disliked by spouse
15	Blessed with many children
16	Dry, round, dust-colored eyes like those of a dead man; half-open while sleeping
17	Dreams of flying in sky over hills and trees
18	Less fortunate, unethical and suffering from jealousy
19	Symbolized by dog, jackal, camel, vulture, mouse and crow

TABLE 2.4
Characteristics of *pittaja prakṛti* (*Śā*. 3: 90–95)

Sl. No.	Characteristics
1	Intense thirst and hunger
2	Fair complexion and warm body
3	Redness in palm, sole and face
4	Valiant, proud, virtuous and clean nature
5	Copper-colored hair, premature graying, aging and tendency to become bald
6	Interest to wear fine clothes, ornaments and fragrances
7	Kindness toward dependents and followers
8	Blessed with wealth, adventurous nature and intelligence
9	Willingness to give asylum even to enemies
10	Loose joints
11	Unpopular among women
12	Low virility and sexual urge
13	Interest to consume in large quantities astringent, sweet, bitter and cold food
14	Disinterest in warmth
15	Hyperidrosis and body odor
16	Short temper and jealousy
17	Dreaming of trees with bright, red flowers (*Erythrina variegata*, *Butea frondosa*), lightning, fire and sun
18	Small, copper-colored, vibrant eyes with thin eyelashes and having tendency to turn red quickly due to anger, drinking wine and on exposure to hot sun
19	Medium longevity and stamina
20	Scholarly bend of mind and not having ability to face difficulties
21	Symbolized by tiger, bear, monkey, cat and *yakṣa*

TABLE 2.5
Characteristics of *kaphaja prakṛti* (*Śā*. 3: 96–103)

Sl. No.	Characteristics
1	Soft-spoken nature with soft and well-built body and golden complexion
2	Ability to tolerate hunger, thirst and grief
3	Virtuous and ethical nature, possessing good intelligence, honest, not having greed
4	Long hands stretching up to knees, high forehead and black luxuriant hair
5	Blessed with great sexual vigor, many children, servants, wealth, good memory long life and farsighted vision of life
6	Not speaking harsh words but keeping in mind confidentially for a long time, the insult received from someone
7	Walking with elephant-like gait
8	Reverberating voice like that of thunder, roaring of sea and lion
9	Ready to do anything, but humble in nature and not crying too much even in childhood
10	Eating small quantities of bitter, astringent, pungent, dry, warm food, but having good stamina
11	Bright eyes with thick eyelashes
12	Free from jealousy and anger, but speaking very few words
13	Religious, handsome, benevolent, honest, cultured, having great patience, far-sighted and generous
14	Sleeping for long hours and slow in actions
13	Grateful, straightforward, deeply affectionate and valuing long-standing friendship
16	Dreams of clouds and lakes filled with lotus flowers and birds
17	Devotion to preceptors
18	Symbolized by Brahmā, Rudra, Indra, Varuṇa, Garuḍa, swan, elephant, lion, horse, cow and bull

TABLE 2.6
Prakṛti and Symbolism

Sl. No.	Symbol	Associated Qualities
		Vātaja prakṛti
1	Dog	Quarrelsome, wanderer, jealousy, interrupted sleep (*śvānanidra*)
2	Jackal	Cunningness, opportunism
3	Camel	Dry, rough and dust-colored body
4	Vulture	Unethical, bent upon exploiting others at any cost
5	Mouse	Slender body, restless behavior, short life span, fearful and seclusive nature, high breeding rate
6	Crow	Black color, unclean lifestyle, erratic movements of body and eyes
		Pittaja prakṛti
1	Tiger	Valor, readiness to take on challenges
2	Bear	Irritable temper, anger
3	Monkey	Obstinacy (*markaṭamuṣṭi*), moving in group
4	Cat	Egotic, appreciates attention
5	Yakṣa	Anger, vanity, benevolence
		Kaphaja prakṛti
1	Brahmā	Creativity, knowledge
2	Rudra	Terrible nature if provoked
3	Indra	Presence of mind
4	Varuṇa	Liked by all, coolness
5	Garuḍa	Royal stature
6	Swan	Heavy body, pleasing movements (*haṃsagamanam*), white color, grace, loyalty and honor
7	Elephant	Large body, slow walking, great memory, long life span, powerful roaring sound
8	Lion	Straightforwardness, golden color, dignified behavior, kingly nature
9	Horse	Well-shaped muscular body, stamina, virility, graceful movements
10	Cow	Generosity, gentleness, calmness, slow walking
11	Bull	Readiness to undertake any hard work, ability to tolerate hunger and thirst

common behavioral characteristics. He implies that their *prakṛti* also can be determined when their language and natural history are understood well. This view is very similar to the concept of *collective unconscious* proposed by Carl Gustav Jung, according to whom all living creatures exhibit similar behavior, as collective unconscious pervades in them (Jung, 1981).

2.8 CONCLUSION

As a child grows his *dhātus*, faculties of sense organs and *ōjas* grow steadily. This process continues up to seventy years. However, after seventy years of age, *dhātus* start diminishing gradually (*Śā*. 3: 105). A person dominated by *sattva guṇa* enjoys happiness without pride and overcomes difficult times without being desperate. One with personality controlled by *rajōguṇa* goes through these two phases overpowered by them. A person dominated by *tamōguṇa*, on account of being ignorant, does not discriminate between happiness and sorrow. Qualities such as generosity, good conduct, kindness, truthfulness, gratitude and virtuousness endow one with providential blessing and longevity (*Śā*. 3: 119–120).

REFERENCE

Jung, C.G. 1981. The concept of the collective unconscious. In *The Archetypes and the Collective Unconscious*, 42–53. London: Routledge.

3 Fundamental Principles of Longevity

3.1 INTRODUCTION

Long life is essential to realize the four goals of life viz., leading a virtuous life (*dharma*), acquisition of wealth (*artha*), fulfillment of desires (*kāma*) and attainment of salvation (*mōkṣa*). Longevity is dependent on health and health can be maintained by the practice of Ayurveda. People from all walks of life should strive to learn and practice the sacred knowledge of longevity, so as to safeguard their health. Therefore, after paying homage to the great physician, the Almighty, who has conquered the six pollutants of mind or *rāgadidōṣān* such as *kāma* (lust), *krōdha* (anger), *lōbha* (greed for wealth), *mōha* (delusion), *mada* (arrogance) and *mātsarya* (envy), Vāgbhaṭa reveals the key to maintain long life (*Sū* 1: 2).

3.2 THE *TRIDŌṢA*

All life processes are regulated by *vāta*, *pitta* and *kapha*, collectively known as *tridōṣa*. Diseases are caused when they are destabilized and health results when they exist in steady state (*Sū*. 1: 6–7, 20). Though the *tridōṣa* are present all over the body, they are predominant in certain areas of the body. Thus, *vāta* is dominant in area below the navel and *kapha* in area above heart. The middle area is largely the seat of *pitta* (*Sū*. 1: 7). If life span, day, night and period between consumption of food and completion of digestion are divided into three equal parts: the first part will be dominated by *kapha*, the second by *pitta* and the third by *vāta*. Because of the influence of *vāta*, *pitta* and *kapha*, *jaṭharāgni* (abdominal fire) becomes respectively irregular, sharp or dull (*Sū*. 1: 8). In turn the *kōṣṭha* (alimentary tract) becomes hard (*krūra*), soft (*mṛdu*) or medium (*madhya*), respectively (*Sū*. 1: 9). Because of the dominance of *tridōṣa* in reproductive elements of parents (*śukra*–masculine aspect, *ārtava*=feminine aspect), offspring are born in body constitution (*prakṛti*) dominated by *vāta* (*vātaja prakṛti*), *pitta* (*pittaja prakṛti*) or *kapha* (*kaphaja prakṛti*). Among them *vātaja prakṛti* is inferior, *pittaja prakṛti* is medium and *kaphaja prakṛti* is superior (*Sū*. 1: 9–10).

3.2.1 VĀTA

The seats of *vāta* are intestine, hip, legs, auditory organ, bones and skin. *Pitta* resides in navel, stomach, sweat, saliva, blood, *rasa dhātu*, eyes and skin. Thorax, throat, head, joints, stomach, *rasa dhātu*, adipose tissue, olfactory organs and gustatory system are seats of *kapha* (*Sū*. 12: 1–3). On functional grounds, *vāta* is divided into five subgroups viz., *prāna*, *udāna*, *vyāna*, *samāna* and *apāna*. Among them *prāna* resides in the head and controls spitting, sneezing, belching, inspiration, expiration and deglutition. *Udāna* resides in the thorax and controls speaking, bodily activities, serenity of mind, stamina, complexion and memory. The seat of *vyāna* is heart and controls walking, turning of body, closing and opening of eyes and all physical activities. *Samāna* is intimately connected to abdominal fire (*jaṭharāgni*) and is active in the entire alimentary tract. It receives food brought in by *prana*, transforms it and turns it into products of digestion and waste products. Seated in the rectal area, *apāna* causes the movement of seminal fluid, menstrual blood, stool, urine and feces (*Sū*. 12: 4–9).

Vāta causes effects in the body because of its inherent qualities such as dryness (*rūkṣa*), lightness (*laghu*), coolness (*śīta*), roughness (*khara*), subtlety (*sūkṣma*) and mobility (*cala*) (*Sū*. 1: 11). In

DOI: 10.1201/9781003148296-3

the undisturbed steady state, *vāta* blesses the body with enthusiasm, inspiration, expiration, body movements, voiding of feces and urine, optimum transformation of body elements and satisfactory performance of sense organs (*Sū.* 11: 1–2). Increased *vāta* causes leaning of body, black coloration, interest in warmth, shivering of body, bloating of abdomen with gas, constipation, loss of stamina, insomnia, weakening of organs, garrulousness, giddiness and melancholy (*Sū.* 11: 6). Decreased *vāta* results in tiredness, inability to talk or indulge in activities, loss of consciousness, dull appetite, exudation of mucus, laziness, heaviness of body, pallor, feeling of cold, looseness of limbs, respiratory distress, cough and hypersomnia (*Sū.* 11: 8, 15).

3.2.2 Pitta

Pitta is slightly unctuous (*sasnēha*), sharp (*tīkṣṇa*), hot (*uṣṇa*), light (*laghu*), foul smelling (*visra*), flowing (*sara*) and liquid (*drava*) (*Sū.* 1: 11). In the undisturbed state, *pitta* facilitates digestion of food, warmth of body, vision, hunger, thirst, gustation, luster of body, comprehension, intelligence, valor and softness of body (*Sū.* 11: 2–3). Increased *pitta* causes yellowing of feces, urine, eyes and skin; hunger, thirst, burning sensation and reduced sleep (*Sū.* 11: 7). Lowering of *pitta* results in dull appetite, lowering of body temperature and loss of luster of body (*Sū.* 11: 16). *Pitta* is grouped into five types such as *pācaka pitta*, *rañjaka pitta*, *sādhaka pitta*, *ālōcaka pitta* and *bhrājaka pitta*, which are involved, respectively, in digestion, hemopoiesis, mental activities, vision and luster of body (*Sū.* 12: 10–14).

3.2.3 Kapha

Kapha is unctuous (*snigdha*), cold (*śīta*), heavy (*guru*), slow (*manda*), soft (*śḷakṣṇa*), sticky (*mṛtsna*) and firm (*sthira*) (*Sū.* 1: 12). On account of these innate qualities, *kapha* imparts to the body stability, unctuousness, efficient joints and stamina (*Sū.* 11: 3). When enraged, *kapha* causes lowering of abdominal fire, exudation of mucus, laziness, heaviness of body, pallor, lowering of body temperature, looseness of limbs, respiratory distress, cough and hypersomnia (*Sū.* 11: 7–8). Giddiness, feeling of emptiness of thorax, nausea and loosening of joints are caused by lowering of *kapha* (*Sū.* 11: 16).

Kapha also falls into five groups: *avalaṃbaka kapha*, *kḷēdaka kapha*, *bōdhaka kapha*, *tarpaka kapha* and *śḷēṣaka kapha*. *Avalaṃbaka kapha* is located in chest and the meeting place of shoulder, neck and back. By its innate strength and the power of products of digestion, it lubricates and nourishes the organs. It also influences the activity of the other four groups of *kapha*. *Kḷēdaka kapha* resides in stomach. It moistens hard food and aids in digestion. *Bōdhaka kapha* stays in tongue and is involved in gustation. *Tarpaka kapha* nourishes the sense organs by staying in head. *Śḷēṣaka kapha* is located in joints of bones (*Sū.* 12: 15–18).

Due to dietary and behavioral indiscretions and effects of season, the *tridōṣa* undergo *caya* (increasing in own sites), *kōpa* (migration to various sites) and *śamana* (returning to pacified state). *Caya* and *kōpa* states generate many diseases, and they are to be cured with appropriate food, beverages, medicines and measures (*Sū.* 12: 18–32). The *tridōṣa* can increase or decrease in various magnitudes, and these changes are broadly grouped into 63 combinations (*Sū.* 12: 74–78) (Table 3.1).

3.3 *DHĀTU*

Rasa (tissue fluid), *rakta* (blood), *māṃsa* (muscle), *mēdas* (adipose tissue), *asthi* (bone), *majja* (bone marrow) and *śukra* (reproductive element) are the seven *dhātu* (tissue elements) (*Sū.* 1: 13). The functions of seven *dhātus* (*rasa*, *rakta*, *māmsa*, *mēdas*, *asthi*, *majja* and *śukra*) are, respectively, offering nourishment, maintenance of health, envelopment of body, imparting oiliness, holding of body, filling of space inside bones and reproduction (*Sū.* 11: 4). Like *tridōṣa*, the *dhatus* also have high and low states of activity. Increased *rasa dhātu* produces symptoms such as lowering of abdominal fire, exudation of mucus, laziness, heaviness of body, pallor, lowering of body temperature,

TABLE 3.1
Sixty-Three Combinations of *tridōṣa* States

Type of Combination	*Tridōṣa* States
***Sannipāta* (Simultaneous Aggravation of All Three *dōṣas*) – 13 Types**	
Two *dōṣa*s aggravated in excess and remaining one aggravated mildly	1. Vāta ativṛddha-Pitta ativṛddha-Kapha vṛddha 2. Pitta ativṛddha-Kapha ativṛddha-Vāta vṛddha 3. Kapha ativṛddha-Vāta ativṛddha-Pitta vṛddha
One *dōṣa* aggravated in excess and remaining two aggravated mildly	1. Vāta ativṛddha-Pitta vṛddha-Kapha vṛddha 2. Pitta ativṛddha-Vāta vṛddha-Kapha vṛddha 3. Kapha ativṛddha-Vāta vṛddha-Pitta vṛddha
One *dōṣa* aggravated mildly and remaining aggravated moderately and in excess	1. Vāta vṛddha-Pitta vṛddhataraḥ-Kapha vṛddhatamaḥ 2. Vāta vṛddha-Kapha vṛddhataraḥ-Pitta vṛddhatamaḥ 3. Pitta vṛddha-Kapha vṛddhataraḥ-Vāta vṛddhatamaḥ 4. Pitta vṛddha-Vāta vṛddhataraḥ-Kapha vṛddhatamaḥ 5. Kapha vṛddha-Vāta vṛddhataraḥ-Pitta vṛddhatamaḥ 6. Kapha vṛddha-Pitta vṛddhataraḥ-Vāta vṛddhatamaḥ
Aggravation of all the three *dōṣa*s in same degree	1. Vāta-Pitta-Kapha
***Saṃsarga* (Simultaneous Aggravation of Any Two *dōṣas*) – Nine Types**	
One *dōṣa* aggravated in excess	1. Pitta vṛddha-Vāta ativṛddha 2. Pitta vṛddha-Kapha ativṛddha 3. Vāta vṛddha-Pitta ativṛddha 4. Vāta vṛddha-Kapha ativṛddha 5. Kapha vṛddha-Vāta ativṛddha 6. Kapha vṛddha-Pitta ativṛddha
Two *dōṣa*s aggravated in same degree	1. Vāta vṛddha-Pitta vṛddha 2. Pitta vṛddha-Kapha vṛddha 3. Kapha vṛddha-Vāta vṛddha
***Ēkadōṣa-vṛddha* (Aggravation of Only One *dōṣa*) – Three Types**	
Aggravation of only one *dōṣa*	1. Vāta vṛddha 2. Pitta vṛddha 3. Kapha vṛddha
***Sannipāta* (Simultaneous Diminution of All Three *dōṣas*) – 13 Types**	
Two *dōṣa*s diminished in excess and remaining one diminished mildly	1. Vāta atikṣīṇa-Pitta atikṣīṇa-Kapha kṣīṇa 2. Pitta atikṣīṇa-Kapha atikṣīṇa-Vāta kṣīṇa 3. Kapha atikṣīṇa-Vāta atikṣīṇa-Pitta kṣīṇa
One *dōṣa* diminished in excess and remaining two diminished mildly	1. Vāta atikṣīṇa-Pitta kṣīṇa-Kapha kṣīṇa 2. Pitta atikṣīṇa-Vāta kṣīṇa-Kapha kṣīṇa 3. Kapha atikṣīṇa-Vāta kṣīṇa-Pitta kṣīṇa
One *dōṣa* diminished mildly and remaining diminished moderately and in excess	1. Vāta kṣīṇa-Pitta kṣīṇataraḥ-Kapha kṣīṇatamaḥ 2. Vāta kṣīṇa-Kapha kṣīṇataraḥ-Pitta kṣīṇatamaḥ 3. Pitta kṣīṇa-Kapha kṣīṇataraḥ-Vāta kṣīṇatamaḥ 4. Pitta kṣīṇa-Vāta kṣīṇataraḥ-Kapha kṣīṇatamaḥ 5. Kapha kṣīṇa-Vāta kṣīṇataraḥ-Pitta kṣīṇatamaḥ 6. Kapha kṣīṇa-Pitta kṣīṇataraḥ-Vāta kṣīṇatamaḥ
Diminution of all the three *dōṣa*s in same degree	1. Vāta-Pitta-Kapha

(*Continued*)

TABLE 3.1 (*Continued*)
Sixty-Three Combinations of *tridōṣa* States

Type of Combination	*Tridōṣa* States
***Saṃsarga* (Simultaneous Diminution of Any Two *dōṣas*) – Nine Types**	
One *dōṣa* diminished in excess	1. Pitta kṣīṇa-Vāta atikṣīṇa 2. Pitta kṣīṇa-Kapha atikṣīṇa 3. Vāta kṣīṇa-Pitta atikṣīṇa 4. Vāta kṣīṇa-Kapha atikṣīṇa 5. Kapha kṣīṇa-Vāta atikṣīṇa 6. Kapha kṣīṇa-Pitta atikṣīṇa
Two *dōṣas* diminished in same degree	1. Vāta kṣīṇa-Pitta kṣīṇa 2. Pitta kṣīṇa-Kapha kṣīṇa 3. Kapha kṣīṇa-Vāta kṣīṇa
***Ēkadōṣa-kṣaya* (Diminution of Only One *dōṣa*) – Three Types**	
Diminution of only one *dōṣa*	1. Vāta kṣaya 2. Pitta kṣaya 3. Kapha kṣaya
12 Permutations and Combinations of *dōṣas* Where *vṛddhi-kṣaya* Take Place Simultaneously	
One *dōṣa* is aggravated (*vṛddha*), second is in normal state (*samata*), and third is in the state of diminution (*kṣaya*)	1. Vāta vṛddha-Pitta samata-Kapha kṣaya 2. Pitta vṛddha-Kapha samata-Vāta kṣaya 3. Kapha vṛddha-Pitta samata-Vāta kṣaya 4. Vāta vṛddha-Kapha samata-Pitta kṣaya 5. Pitta vṛddha-Vāta samata-Kapha kṣaya 6. Kapha vṛddha-Vāta samata-Pitta kṣaya
Two *dōṣas* aggravated mildly and one *dōṣa* is in the state of diminution	1. Vāta vṛddha-Pitta vṛddha-Kapha kṣaya 2. Pitta vṛddha-Kapha vṛddha-Vāta kṣaya 3. Kapha vṛddha-Vāta vṛddha-Pitta kṣaya
One *dōṣa* aggravated mildly and other two are in the state of diminution	1. Vāta vṛddha-Pitta kṣaya-Kapha kṣaya 2. Pitta vṛddha-Vāta kṣaya-Kapha kṣaya 3. Kapha vṛddha-Vāta kṣaya-Pitta kṣaya
Tridōṣa in steady state (Health)	1. Vāta-Pitta-Kapha

looseness of limbs, respiratory distress, cough and hypersomnia. Increased *rakta* causes cellulitis, splenomegaly, abscess, skin diseases, *vāta*-induced blood disorders, hemorrhages of obscure origin, *gulma*, chronic gingivitis, jaundice, black pigmentation on face, loss of appetite, fainting and reddish tinge in skin, eyes and urine. Heightened activity of *māṃsa* causes swelling; tumor; glandular swelling; enlargement of cheeks, thighs, and abdomen; and thickening of muscles in neck and armpits. Similar symptoms are observed in increased *mēdas* as well. Increased *mēdas* also causes panting, respiratory distress, sagging of buttocks, breasts and abdomen. Abnormal growth of bones and teeth are caused by increased *asthi dhātu*. Heightened activity of *majja* causes heaviness in eyes and limbs and reddish, obstinate boils with thick base in joints. Increase in *śukra dhātu* induces heightened desire in sexual intercourse and the disorder known as *śukrāsmari* (*Sū.* 11: 8–12).

Decrease in *rasa dhātu* causes dryness of body, giddiness, leaning, tiredness and intolerance to sound. *Rakta kṣāya* causes interest in sour and cold substances, looseness of veins and dryness of body. *Māṃsa dhātu* in low stage of activity causes tiredness, sunken cheeks, wasting of buttocks and pain in joints. Diminished sensation around waist, splenomegaly and leaning of body ensue from low stage of activity of *mēdas*. Diminution of *asthi dhātu* causes pricking pain in bones;

breaking of teeth, hair and nails; increased porosity of bones; giddiness; and blackouts. Reduction in *śukra dhātu* results in emission of blood instead of semen during copulation, pricking pain in scrota and burning sensation in penis (*Sū.* 11: 17–20).

3.3.1 Mala

The seven *dhātu* undergo sequential transformation. The essence of digestion becomes *rasa*, which sequentially transforms into *rakta, māṃsa, mēdas, asthi, majja* and *śukra* (*Sū.* 11: 34–35, *Śā.* 3: 55–63). *Mūtra* (urine), *purīṣa* (feces) and *svēda* are the waste products of digestion, and they are collectively called *mala* (*Sū.* 1: 13). The three *mala* act as support to the body, eliminate water and hold hairs (*Sū.* 11: 5).

Mala also have high and low states of activity. Increase in *purīṣa* causes bloating of abdomen, borborygmi, heaviness and pain in lower abdomen. Pricking pain in urinary bladder and feeling of incomplete voiding of urine ensue from increase in *mūtra*. Increase in *svēda* results in excessive sweating, body odor and itching (*Sū.* 11: 13–14). Diminution of *purīṣa* causes air to bind the intestine and move around in stomach causing rumbling sound. After tormenting heart and the flanks, it goes out as eructation. A person with reduced *mūtra* voids with difficulty, small quantities of discolored or blood-stained urine. Reduced *svēda* causes hair loss, stiffness of hair and cracks on skin (*Sū.* 11: 21–22).

Among *dhātu* and *mala*, *vāta* resides in bone (*asthi*), *pitta* in sweat (*svēda*) and blood (*rakta*). All the other *dhātu* and *mala* are rich in *kapha*. When there is an increase in *pitta* and *kapha* in body, there is also increase of *dhātu* and *mala* associated with them. The same rule applies when *pitta* and *kapha* decrease in body. However, an inverse relationship exists between *vāta* and *asthi*. When *vāta* increases in body, there is a reduction in bone and vice versa. Diseases arising from increase of *rasa, rakta, māmsa, mēdas, majja* and *śukra* should be treated with *laṅghana* therapy, which thins body and reduces quantities of these *dhātu*. Diseases arising from decrease of these six *dhātus* are to be treated with *bṛṃhaṇa*, which increases bulk of body. However, in the case of *vāta*, the reverse approach is adopted. Increase in *vāta* is to be treated with *bṛṃhaṇa* therapy, as *asthi* is reduced. Nevertheless, *laṅghana* therapy is ideal, when *vāta* is decreased (*Sū.* 11: 26–29).

3.3.2 Dhātu Cycle

The essence of digestion (*rasa dhātu*) enters the *dhātu* cycle. Under the influence of the principles of transformation (*agni*) existing in each *dhatu*, it sequentially transforms into *rakta, māmsa, mēdas, asthi, majja* and *śukra*. Sequential transformation of the seven *dhātus* results in the production of an end product known as *ōjas* (shiny essence), which circulates in the body, imparting vigor and vitality. It is unctuous, cold and of reddish-yellow color. Loss of *ōjas* results in death. Reduction of *ōjas* arises from anger, hunger, mental worries, grief and exertion. One with reduced *ōjas* becomes weak, fearful, thoughtful, evil-minded and with dry body. Diminution of *ōjas* can be countered by consumption of revitalizing medicines, milk and meat soups. Increase in *ōjas* bestows happiness and strength to body. Increase of *tridōṣa* in body generates interest in food, beverages and environment having qualities opposite to those of the *dōṣa* in question. Contrarily, their decrease encourages the adoption of food, beverages and environment having similar qualities. For example, *vāta* has qualities such as dryness, lightness, coolness, roughness, subtlety and mobility. Increase of *vāta* in body encourages adoption of food, beverages and environment with opposing qualities like unctuousness, heaviness, warmth, smoothness, bulkiness and immobility. *Kapha* has qualities such as unctuousness, coolness, heaviness, slowness, softness, stickiness and firmness. Similarly, lowering of *kapha* in body induces the individual to choose food, beverages and environment having qualities similar to those of *kapha*, so that *kapha* is brought back to normal state. A wise physician should be aware of this interaction among *tridōṣa*, food, beverages, environment and human behavior (*Sū.* 1: 11–12, *Sū.* 11: 37–43, *Śā.* 3: 57–63).

TABLE 3.2
Qualities of *dravya*

Sanskrit Term	English Equivalent	Sanskrit Term	English Equivalent
Guru	Heavy	Laghu	Light
Manda	Slow	Tīkṣṇa	Penetrating
Śīta	Cold	Uṣṇa	Hot
Snigdha	Unctuous	Rūkṣa	Dry
Ślakṣṇa	Smooth	Khara	Rough
Sāndra	Thick	Drava	Liquid
Mṛdu	Soft	Kaṭhina	Hard
Sthira	Firm	Cala	Moving
Sūkṣma	Subtle	Sthūla	Gross
Viśada	Non-slimy	Picchila	Slimy

3.4 MATTER

Dravya (all forms of matter) are made up of varying combinations of the five primordial elements (*bhūta*) such as *pṛdhvi* (earth), *ap* (water), *tējas* (fire), *vāyu* (air) and *ākāśa* (sky) collectively called *pañcamahābhūta* (*Sū*. 9: 1–2). Qualities (*guṇa*) of a *dravya* are dependent on its predominant *bhūta*. For example, if a *dravya* is rich in the element of earth, it will be heavy, gross and firm. It imparts to the body, heaviness, stability, compactness and growth. Water-dominant matter will be liquid, cold, heavy, unctuous, slow and dense. It produces unctuousness, exudation of *kapha*, happiness and possesses holding property. Matter rich in the element of fire is dry, penetrating, hot, shiny and subtle. It confers heat, beauty, color, brightness and ability to digest or transform. Air-dominant matter turns out to be dry, shiny and light. It produces effects such as dryness, lightness, clarity, morbidity and lassitude. Matter with predominance of the element of sky is subtle, shiny and light. It produces porosity and lightness of organs of body. Thus, all forms of matter are formed by innumerable combinations of *bhūta* and have a multitude of actions in a living body, on account of their ability to modulate *tridōṣa*. Therefore, there is no *dravya* in this world that is not medicinal. *Dravya* rich in the elements of fire, air and sky have tendency to work upward and are light in nature. On the other hand, matter having predominance of earth and water work downward and are heavy in nature (*Sū*. 9: 5–11). *Dravya* possess numerous qualities. Nevertheless, for the sake of convenience, only 20 opposing qualities in ten pairs are considered (*Sū*. 1: 18). They are listed in Table 3.2.

All forms of matter have the property of potency (*vīrya*), which may be either *uṣṇa vīrya* (hot potency) or *śīta vīrya* (cold potency). Among them, matter with hot potency causes giddiness, thirst, lassitude, sweating and burning sensation. It causes transformation quickly. On the basis of its qualities, it pacifies *vāta* and *kapha*. Matter with cold potency produces happiness, increase in *ōjas* and coolness. Such matter pacifies vitiation of *rakta* and *pitta* (*Sū*. 9: 12–19).

3.4.1 THE SIX BASIC TASTES

Dravya have six basic tastes (*rasa*) such as *madhura* (sweet), *amḷa* (sour), *lavaṇa* (salty), *kaṭu* (pungent), *tikta* (bitter) and *kaṣāya* (astringent). On account of its formation from five elements, a form of matter will have several tastes. It is assigned one taste based on its predominant taste (*Sū*. 9: 3–4). The ability of tastes to impart strength to body is in the descending order. Consequently, sweetness is most powerful in improving strength whereas astringent has the least ability in this regard (*Sū*. 1: 14–15). These six tastes result from the combinations of two *bhūta*s such as earth-water (sweet), fire-earth (sour), water-fire (salty), sky-air (bitter), fire-air (pungent) and earth-air (astringent). Among them, sweetness produces happiness. Sour taste removes pollutants from mouth, causes

TABLE 3.3
Effects of Taste Modalities on *tridōṣa*

Taste Modality	Effects on *tridōṣa* Vāta	Pitta	Kapha
Madhura (Sweet)	Pacifies	Pacifies	Aggravates
Amḷa (Sour)	Pacifies	Aggravates	Aggravates
Lavaṇa (Salty)	Pacifies	Aggravates	Aggravates
Kaṭu (Pungent)	Aggravates	Aggravates	Pacifies
Tikta (Bitter)	Aggravates	Pacifies	Pacifies
Kaṣāya (Astringent)	Aggravates	Pacifies	Pacifies

horripilation, gnashing of teeth and contraction of eyebrows. Salty taste induces fluid to exude from mouth and causes burning sensation in cheeks and throat. Bitter taste dampens gustatory ability of tongue. Pungent taste causes the tip of tongue to shiver as if with fear. It creates the feeling of small insects creeping inside mouth. Pungent taste induces the flow of fluid from eyes, nostrils and mouth. Cheeks feel like being burnt. Astringent taste makes the tongue immobile and obstructs throat (*Sū.* 10: 1–6). Out of these six tastes, *madhura*, *amḷa* and *lavaṇa* pacify *vāta*. The others enrage *vāta*. *Kaṭu*, *tikta* and *kaṣāya* pacify *kapha*. But *madhura*, *amḷa* and *lavaṇa* increase it. *Kaṣāya*, *tikta* and *madhura* pacify *pitta*. However, *amḷa*, *lavaṇa* and *kaṭu* increase *pitta* (*Sū.* 1: 15–16). The effects of the six taste modalities on *tridōṣa* are presented in Table 3.3.

Sweet taste strengthens *dhātu*s. It improves complexion, *ōjas*, hair growth and sensory abilities in children, the aged, injured and exhausted persons. It improves quality of voice, strengthens body, heals wounds, induces production of breast milk and improves longevity. *Pitta*, *vāta* and effects of poisoning are pacified by its use. However, excessive consumption of food rich in sweetness causes obesity, lowering of abdominal fire, coma, polyuric diseases, swelling in neck and tumor (*Sū.* 10: 6–9).

Sour taste kindles abdominal fire. It is unctuous, pleasing, tasty and digestive. It is of hot potency and increases *kapha* and *pitta*. It also enables obstructed *vāta* to move in its natural direction (*vātānulōmata*). Over indulgence in sour food and beverages loosens the body and generates disorders such as cataract, giddiness, itching, pallor, cellulitis, edema, eruptions on body, thirst and fever (*Sū.* 10: 10–12).

Salty taste relieves rigidity, clears obstructions of body channels, improves digestive efficiency, produces unctuousness, causes sweating, penetrates deep into *dhātu*s and improves taste. Excessive consumption of salty food causes *vāta*-induced blood disorders, baldness, graying of hair, appearance of wrinkles, thirst, skin diseases and cellulitis. It reduces stamina as well (*Sū.* 10: 12–14).

Bitter taste is generally not appreciated. Nevertheless, it pacifies distaste for food, worm infestation, thirst, effects of poisoning, skin diseases, fainting, fever, burning sensation and enragement of *pitta* and *kapha*. It reduces fat, stool and urine. It is light, cold and dry. Bitter taste clears breast milk and throat. Excessive indulgence in bitter-tasting food and beverages disrupts *dhātu*s and causes the appearance of *vāta* diseases (*Sū.* 10: 14–16).

Pungent taste pacifies throat diseases, urticaria, skin diseases, wounds and edema. It reduces unctuousness, fat and exudation of fluid. It is digestive, improves taste, reduces *kapha* and opens up obstructed body channels. When used in excess, pungent taste causes thirst, sexual debility, loss of stamina, fainting, muscle cramps, shivering and pain in hip and back of body (*Sū.* 10: 17–19). Astringent taste pacifies *pitta* and *kapha*. It is heavy, cold and dry in nature, heals wounds and reduces fat. Excessive use of astringent taste causes bloating of abdomen, constipation, precordial pain, thirst, leaning, sterility and obstruction of body channels (*Sū.* 10: 20–22).

Sweet, sour, salty, pungent, bitter and astringent tastes undergo transformation during digestion. This process of acquiring an after-taste is called *vipāka*, Sweet and salty tastes turn sweet in *vipāka*. Sour taste remains as sour. Pungent, bitter and astringent tastes acquire pungent post-digestive taste (*Sū*. 1: 17, *Sū*. 9: 21). *Dravya*s also possess another characteristic known as *prabhāva* (specific action). It can be explained in the following way. Two *dravya*s have same *rasa, guṇa, vīrya* and *vipāka*. However, one of these *dravya*s produces a special effect which cannot be explained on the basis of their *rasa, guṇa, vīrya* or *vipāka*. This is known as the *prabhāva* of that *dravya*. For example, *Nāgadanti* (*Baliospermum montanum*) is similar to *Citraka* (*Plumbago zeylanica*) in *rasa, guṇa, vīrya* and *vipāka*. However, unlike *Citraka, Nāgadanti* causes laxation. Similarly, on the basis of *rasa, guṇa, vīrya* and *vipāka, Madhuka* (*Glycyrrhiza glabra*) and *Mṛdvīka* (*Vitis vinifera*) are identical. Nevertheless, *Mṛdvīka* has the unique property of causing laxation. Clarified butter (*ghṛta*) and milk have same four-fold characteristics. Notwithstanding, clarified butter is carminative, while milk is not. *Dravya*s bring forth their beneficial or harmful effects on account of their *rasa, guṇa, vīrya, vipāka* or *prabhāva*. *Vipāka* is more powerful than *rasa*, and these two are surpassed by *vīrya*. The strength of *rasa, guṇa, vīrya* and *vipāka* are excelled by *prabhāva*. Therefore, the therapeutic effects of *dravya*s cannot be explained on the basis of their *rasa* alone (*Sū*. 9: 22–28).

Based on taste, important medicinal substances are classified into six groups such as *madhura gaṇa, amḷa gaṇa, lavaṇa gaṇa, kaṭu gaṇa, tikta gaṇa* and *kaṣāya gaṇa* (*Sū*. 10: 22–32) (Table 3.4). The 6 tastes form 63 combinations (*Sū*. 10: 39–43) (Table 3.5). These 63 combinations are important in therapeutics. They are to be considered carefully for pacifying the 63 states of *tridōṣa* (*Sū*. 10: 44) (see Table 3.1). *Dravya* fall into three classes: *śamana, kōpana* and *svasthahita*. *Kōpana* and *śamana dravya* respectively enrage or pacify *dōṣa*s. The *svasthahita* type neither enrages nor pacifies but maintains steady state of *dōṣa*s (*Sū*. 1: 16). This three-fold grouping of *dravya* can be understood from their tastes.

3.5 PREVENTION OF DISEASES

The human body is engaged in many voluntary and involuntary activities which are essential for the maintenance of health. These activities give rise to 14 natural urges such as liberation of flatus, belching, defecation, urination, sneezing, thirst, hunger, sleep, cough, panting, yawning, weeping, vomiting and ejaculation. Provocation or retainment of these natural urges causes many discomforts and diseases. Most of these ill effects are caused by the enragement of *vāta*. Therefore, *vāta*-lowering food, beverages and medicines are to be used for pacifying them (*Sū*. 4: 1–23). The effects of restrainment of these natural urges and their cure are listed in Table 3.6.

Those who long for happiness in this life and the after-life should not leave their sense organs unbridled. They should constantly strive to restrain the urges of lust, anger, greed for wealth, delusion, arrogance and envy. *Dōṣa*s that increase seasonally are to be eliminated from the body in that season itself, adopting appropriate elimination measures. *Dōṣa*s that are pacified by fasting and digestive (*pācana*) medicines may get enraged later. However, *dōṣa*s that are eliminated from body never arise later. *Dhātu*s may sometimes be negatively affected by elimination measures and later be a cause of pathology. Therefore, after eliminating the *dōṣa*s, an intelligent physician should administer time-tested rejuvenators and aphrodisiacs. Topical application of *taila* all over body (*abhyaṅga*), massaging with dry herbal powders (*udvartana*), bathing and medicinal enemas (*snēhavasti, nirūhavasti*) are to be adopted in the event of adverse effects on body due to medication or therapeutic measures. Pleasing and carminative medicines are to be administered alongside. Body should be strengthened with *Śāli, Ṣaṣṭika* rice, wheat, green gram, meat and clarified butter. By doing so, there will be enhancement of various *agni* (principles of transformation) such as *jaṭharāgni, dhātvagni* and *bhūtāgni*; intelligence, complexion, functioning of sense organs, virility and longevity (*Sū*. 4: 24–30).

Diseases caused by spirits or invisible beings (*bhūta*), poison, air, fire, wounds, fractures, greed, anger and fear are of exogenous origin. Giving up of evil conduct, control of sense organs, increased

TABLE 3.4
Medicinal Substances Grouped on the Basis of Their Tastes (*Sū.* 10: 22–32)

	Medicinal Substances	
Sl. No.	**Sanskrit Name**	**Latin/English Equivalent**
I	**Sweet Group (*Madhura gaṇa*)**	
1	Abhīru	*Asparagus racemosus* Willd.
2	Akṣōta	*Juglans regia* L.
3	Atibala	*Abutilon indicum* G. Don.
4	Bala	*Sida rhombifolia* ssp. *retusa* (L.) Boiss.
5	Bimbi	*Coccinia grandis* (L.) Voigt.
6	Drākṣa	*Vitis vinifera* L.
7	Ghṛta	Clarified butter
8	Gōkṣura	*Tribulus terrestris* L.
9	Guḍa	Jaggery
10	Hēma	Gold
11	Ikṣu	*Saccharum officinarum* L.
12	Jīvakaḥ	*Malaxis acuminata* D. Don.
13	Jīvanti	*Holostemma annulare* (Roxb.) K. Schum.
14	Kākōḷi	*Fritillaria roylei* Hook.
15	Kāśmari	*Gmelina arborea* Roxb.
16	Kṣīrakākōḷi	*Lilium polyphyllum* D. Don.
17	Kṣīraśukḷa	*Pueraria tuberosa* (Willd.) DC
18	Kṣīram	Milk
19	Kṣudrasaha	*Barleria strigosa* Willd.
20	Kṣoudra	Honey
21	Madhuka	*Glycyrrhiza glabra* L.
22	Madhūkaḥ	*Madhuca longifolia* (Koenig.) Macb.
23	Mahāmēda	*Polygonatum verticillatum* (L.) Allioni
24	Mahāsaha	*Barleria cristata* L.
25	Mahāśrāvaṇi	*Sphaeranthus africanus* Jacq.
26	Māṣaparṇi	*Phaseolus sublobatus* Roxb.
27	Mēda	*Polygonatum cirrhifolium* Royle
28	Mōca	*Musa paradisiaca* L.
29	Mudgaparṇi	*Phaseolus adenanthus* G.F. Meyer
30	Nāgabala	*Sida vernonicaefolia* Lam.
31	Panasa	*Artocarpus heterophyllus* L.
32	Parūṣaka	*Grewia asiatica* L.
33	Pṛśniparṇi	*Pseudarthria viscida* W. & A.
34	Rājādana	*Manilkara hexandra* (Roxb.) Dubard
35	Ṛṣabhakaḥ	*Malaxis muscifera* (Lindley) O. Kuntze
36	Śālaparṇi	*Desmodium gangeticum* DC
37	Śrāvaṇi	*Sphaeranthus indicus* L.
38	Tāla	*Borassus flabellifer* L.
39	Tugākṣīri	*Maranta arundinacea* L.
40	Vidāri	*Ipomoea digitata* L.
41	Vīra	*Lasia spinosa* (L.) Thw.
II	**Sour Group (*Amḷa gaṇa*)**	
1	Amḷavētasa	*Solena amplexicaulis* (Lam.) Gandhi
2	Āmra	*Mangifera indica* L.

(*Continued*)

TABLE 3.4 (*Continued*)
Medicinal Substances Grouped on the Basis of Their Tastes (*Sū*. 10: 22–32)

Sl. No.	Medicinal Substances	
	Sanskrit Name	**Latin/English Equivalent**
3	Āmrātaka	*Spondias pinnata* (L.f.) Kurz
4	Bhavya	*Averrhoa carambola* L.
5	Dadhi	Curd
6	Dāḍima	*Punica granatum* L.
7	Dhātri	*Emblica officinalis* Gaertn.
8	Kapittham	*Feronia elephantum* Correa
9	Karamardaka	*Carissa carandas* L.
10	Mātuḷuṅga	*Citrus medica* L.
11	Pālēvata	*Diospyros malabarica* (Desr.) Kostel
12	Phalāmḷika	*Tamarindus indica* L.
13	Rajata	Silver
14	Śukta	Vinegar
15	Takra	Buttermilk
III	**Salty Group (*Lavaṇa gaṇa*)**	
1	Audbhidam	Salt prepared from vegetable alkali
2	Kṛṣṇam	Salt prepared by specific method
3	Kṣāra	Caustic alkalis
4	Pāmsujam	Earth salt
5	Rōmakam	Sambhar salt
6	Sāmudram	Sea salt
7	Sauvarcalam	Sonchal salt
8	Sīsam	Tin
9	Varam	Rock salt
10	Viḍam	Salt prepared by specific method
IV	**Pungent Group (*Kaṭu gaṇa*)**	
1	Aruṣkaram	*Semecarpus anacardium* L.
2	Cavika	*Piper chaba* Hunter
3	Citraka	*Plumbago zeylanica* L.
4	Hiṅgu	*Ferula foetida* Regel
5	Kṛmijit	*Embelia ribes* Burm. f.
6	Kuṭhēra	*Orthosiphon aristatus* (Blume) Miq.
7	Marica	*Piper nigrum* L.
8	Mūtra	Urine
9	Nāgara	*Zingiber officinale* Rosc.
10	Pippali	*Piper longum* L.
11	Pippalīmūlam	*Piper longum* L. root
12	Pitta	Bile
V	**Bitter Group (*Tikta gaṇa*)**	
1	Aguru	*Aquilaria agallocha* Roxb.
2	Apamārga	*Achyranthes aspera* L.
3	Aṭarūṣakam	*Adhatoda vasica* Nees.
4	Ativiṣa	*Aconitum heterophyllum* Wall.
5	Ayaḥ	Iron
6	Bhūnimba	*Andrographis paniculata* (Burm.f.) Nees

(*Continued*)

TABLE 3.4 (*Continued*)
Medicinal Substances Grouped on the Basis of Their Tastes (*Sū.* 10: 22–32)

Sl. No.	Medicinal Substances	
	Sanskrit Name	**Latin/English Equivalent**
7	Bṛhati	*Solanum indicum* L.
8	Candana	*Santalum album* L.
9	Dhanvayāṣakam	*Tragia involucrata* Linn. var. *angustifolia* Hook. f.
10	Duṇḍukaḥ	*Oroxylum indicum* Vent.
11	Dvirajani	*Coscinium fenestratum* (Gaertn.) Colebr.
12	Dvirajani	*Curcuma longa* L.
13	Guḍūci	*Tinospora cordifolia* (Willd.) Miers. ex Hook. f. & Th.
14	Kāmsya	Bronze
15	Kāśmaryaḥ	*Gmelina arborea* Roxb.
16	Kaṭuka	*Picrorhiza kurroa* Royle.
17	Mūrvā	*Marsdenia volubilis* T. Cooke
18	Musta	*Cyperus rotundus* L.
19	Naktamāla	*Pongamia glabra* Vent.
20	Nimba	*Azadirachta indica* A. Juss.
21	Pāṭala	*Stereospermum tetragonum* DC
22	Pātha	*Cyclea peltata* Diels.
23	Paṭōlaḥ	*Trichosanthes cucumerina* L.
24	Śvēta bṛhati	*Solanum xanthocarpum* Sch. &Wendl.
25	Tagara	*Valeriana wallichii* DC
26	Takkāri	*Premna serratifolia* L.
27	Trāyanti	*Bacopa monnieri* (L.) Pennel
28	Uśira	*Vetiveria zizanioides* Nash.
29	Vaca	*Acorus calamus* L.
30	Vāḷakam	*Coleus vettiveroides* K.C. Jacob
31	Vatsaka	*Holarrhena antidysenterica* (Heyne ex Roth) A. DC
32	Vilva	*Aegle marmelos* Corr.
33	Viśāla	*Citrullus colocynthis* Schrad.
VI	**Astringent Group (*Kaṣāya gaṇa*)**	
1	Akṣa	*Terminalia belerica* Roxb.
2	Añjana	Antimony sulphide
3	Bālam kapittham	Unripe fruit of *Feronia elephantum* Correa
4	Gairika	Red ochre
5	Kadamba	*Anthocephalus indicus* A. Rich.
6	Khadira	*Acacia catechu* Willd.
7	Kharjūra	*Phoenix dactylifera* L.
8	Madhu	Honey
9	Mukta	Pearl
10	Padmam	*Nelumbo nucifera* Gaertn.
11	Pathya	*Terminalia chebula* Retz.
12	Pravāḷa	Coral
13	Śirīṣa	*Albizia lebbeck* (L.) Benth,
14	Udumbara	*Ficus glomerata* Roxb.
15	Utpala	*Monochoria vaginalis* (Burm. f.) C. Presl. ex Kunth.
16	Visa	Leaf stalk of *Nelumbo nucifera* Gaertn.

TABLE 3.5
Sixty-Three Combinations of Taste Modalities

Combination	Sl. No.	Combinations
Individual tastes	1	*Madhura* (sweet)
	2	*Amḷa* (sour)
	3	*Lavaṇa* (salty)
	4	*Kaṭu* (pungent)
	5	*Tikta* (bitter)
	6	*Kasāya* (astringent)
Combination of two tastes	1	Madhura+Amḷa
	2	Madhura+Tikta
	3	Madhura+Kasāya
	4	Madhura+Lavaṇa
	5	Madhura+Kaṭu
	6	Amḷa+Lavaṇa
	7	Amḷa+Tikta
	8	Amḷa+Kaṭu
	9	Amḷa+Kasāya
	10	Lavaṇa+Tikta
	11	Lavaṇa+Kaṭu
	12	Lavaṇa+Kasāya
	13	Tikta+Kaṭu
	14	Tikta+Kasāya
	15	Kaṭu+Kasāya
Combination of three tastes	1	Madhura+Amḷa+Lavaṇa
	2	Madhura+Amḷa+Tikta
	3	Madhura+Amḷa+Kaṭu
	4	Madhura+Amḷa+Kasāya
	5	Madhura+Lavaṇa+Tikta
	6	Madhura+Lavaṇa+Kaṭu
	7	Madhura+Lavaṇa+Kasāya
	8	Madhura+Tikta+Kaṭu
	9	Madhura+Tikta+Kasāya
	10	Madhura+Kaṭu+Kasāya
	11	Amḷa+Lavaṇa+Tikta
	12	Amḷa+Lavaṇa+Kaṭu
	13	Amḷa+Lavaṇa+Kasāya
	14	Amḷa+Tikta+Kaṭu
	15	Amḷa+Tikta+Kasāya
	16	Amḷa+Kaṭu+Kasāya
	17	Lavaṇa+Tikta+Kaṭu
	18	Lavaṇa+Tikta+Kasāya
	19	Lavaṇa+Kaṭu+Kasāya
	20	Tikta+Kaṭu+Kasāya

(*Continued*)

TABLE 3.5 (*Continued*)
Sixty-Three Combinations of Taste Modalities

Combination	Sl. No.	Combinations
Combination of four tastes	1	Madhura+Amḷa+Lavaṇa+Tikta
	2	Madhura+Amḷa+Lavaṇa+Kaṭu
	3	Madhura+Amḷa+Lavaṇa+Kasāya
	4	Madhura+Amḷa+Tikta+Kaṭu
	5	Madhura+Amḷa+Tikta+Kasāya
	6	Madhura+Amḷa+Kaṭu+Kasāya
	7	Madhura+Lavaṇa+Tikta+Kaṭu
	8	Madhura+Lavaṇa+Tikta+Kasāya
	9	Madhura+Lavaṇa+Kaṭu+Kasāya
	10	Madhura+Tikta+Kaṭu+Kasāya
	11	Amḷa+Lavaṇa+Tikta+Kaṭu
	12	Amḷa+Lavaṇa+Tikta+Kasāya
	13	Amḷa+Lavaṇa+Katu+Kasāya
	14	Amḷa+Tikta+Kaṭu+Kasāya
	15	Lavaṇa+Tikta+Kaṭu+Kasāya
Combination of five tastes	1	Amḷa+Lavaṇa+Tikta+Kaṭu+Kasāya
	2	Madhura+Lavaṇa+Tikta+Kaṭu+Kasāya
	3	Madhura+Amḷa+Tikta+Kaṭu+Kasāya
	4	Madhura+Amḷa+Lavaṇa+Kaṭu+Kasāya
	5	Madhura+Amḷa+Lavaṇa+Tikta+Kasāya
	6	Madhura+Amḷa+Lavaṇa+Tikta+Kaṭu
Combination of six tastes	1	Madhura+Amḷa+Lavaṇa+Kaṭu+Tikta+Kasāya

awareness, knowledge about the environment, period of time and the causal body (*ātma*) are essential for the prevention and cure of all diseases of endogenous (*nija rōga*) and exogenous (*āgantu rōga*) origin. *Dōṣas* that undergo *caya* in pre-winter and winter seasons are to be eliminated in spring. Those which undergo *caya* in summer require elimination in monsoon and autumn is the appropriate time to eliminate *dōṣas* that undergo *caya* in monsoon. He who does so will never be affected by diseases that arise due to change of seasons. One who desires a disease-free life should get used to food, beverages and behavior that do not provoke *dōṣas*. He should be very careful while performing all actions and should not be carried away by inputs of sense organs. He should be generous, possess equanimity, forbearance and honesty. Above all, he should follow the footsteps of virtuous predecessors (*Sū*. 4: 31–36).

3.6 OBJECTIVE OF AYURVEDA

Ayurveda is intended to guide mankind on management of health. Health is maintained by preventing the appearance of diseases and curing the diseases that appear. Therefore, the sacred knowledge of longevity deals with *svasthavṛtta* (management of health) and *āturavṛtta* (management of patients). *Svasthavṛtta* is the protocol to prevent the occurrence of diseases. *Āturavṛtta* deals with the human body, medicinal substances, diseases and treatment. AH is all about elucidation of *svasthavṛtta* and the four aspects of *āturavṛtta*. As health is dependent on food and behavior, which are connected to everyday life, *svasthavṛtta* begins with daily regimen (*dinacarya*).

TABLE 3.6
Effects of Restrainment of Natural Urges (*Sū*. 4: 1–21)

S. No.	Natural Urge That Is Restrained	Illnesses Caused	Remedial Measure
1	Voiding of flatus	*Gulma*, *udāvarta*, pain, tiredness	Introduction of suppository in rectum, *abhyaṅga*, *avagāha*, *svēdana* and *vasti*
2	Voiding of stool	Cramps in calf muscle, rhinitis, headache, belching, cutting pain in stomach, discomfort in precordial area, halitosis, *gulma*, *udāvarta*, pain, tiredness	Introduction of suppository in rectum, *abhyaṅga*, *avagāha*, *svēdana* and *vasti*; food and beverages
3	Voiding of urine	Racking of limbs, urinary stone; pain in phallus, urinary bladder and groin; illnesses caused by restrainment of urge to void flatus and stool	Introduction of suppository in rectum, *abhyaṅga*, *avagāha*, *svēdana* and *vasti*; consumption of *ghṛta* before food and after digestion of food
4	Belching	Distaste for food, shivering of body, heaviness in precordial area and thorax, bloating of abdomen, cough, hiccup	Treatment recommended for hiccup (*hidhma*)
5	Sneezing	Headache, weakness of limbs, difficulty in turning neck, facial paralysis	*Dhūmapāna*, *añjana* and *nasya* of *tīkṣṇa* nature; smelling of *tīkṣṇa* herbs, staring at bright sun, *snēhana* and *svēdana*
6	Thirst	Drying of mouth, flaccidity of limbs, impaired hearing, stupor, giddiness, heart disease	Application of cold measures
7	Hunger	Racking of limbs, anorexia, lassitude, emaciation, pricking pain in stomach, giddiness	Consumption of small quantities of light, oily and warm food
8	Sleep	Stupor, heaviness of head and eyes, laziness, yawning, pressing pain in body	Sleeping and massage
9	Cough	Increased frequency of coughing, respiratory distress, anorexia, heart disease, drying of mouth, hiccup	Treatment recommended for *kāsa*
10	Panting	*Gulma*, *hṛdrōga*, stupor	Taking rest, *pitta*-pacifying measures
11	Yawning	Headache, weakness of limbs, difficulty in turning neck, facial paralysis	*Vāta*-pacifying measures
12	Weeping	Catarrh, eye diseases, diseases of head, heart disease, difficulty in turning head, anorexia, giddiness, *gulma*	Sleeping, drinking wine and listening to interesting stories
13	Vomiting	Cellulitis, rashes, skin diseases, itching in eyes, morbid pallor, fever, cough, respiratory distress, nausea, black pigmentation, edema	*Gaṇḍūṣa*, *dhūmapāna*, fasting, eating dry food and vomiting it out, exercise, venesection, *virēcana*, *abhyaṅga* with *taila* containing alkali and salt
14	Ejaculation	Oozing of seminal fluid, pain and edema in pelvic area, fever, precordial pain, dysuria, racking of limbs, hydrocele, *śukḷāśmari* and impotency	Consumption of cooked chicken meat, *Sura* wine, *Śāli* rice, *vasti*, *abhyaṅga*, *avagāha*, milk boiled with herbs that cleanse the bladder, sexual intercourse

3.7 DAILY REGIMEN

A healthy person is expected to wake up during the auspicious hour of Brāhma *(Brāhma muhūrta)*. It begins one hour and thirty-six minutes before sunrise and ends forty-eight minutes later. After voiding bladder and bowel, one should brush teeth using thin and soft twigs of trees or bush such as *Calotropis gigantea*, *Ficus benghalensis*, *Acacia catechu*, *Pongamia glabra* and *Terminalia arjuna*, having astringent, pungent or bitter tastes. Thereafter, collyrium (*añjana*) is to be applied in eyes. Daily application of *añjana* bestows brightness to eyes and improves acuity of vision. Nasal instillation of medicines (*nasya*), holding of medicines in mouth (*gaṇḍūṣa*), inhalation of medicinal smoke (*dhūmapāna*) and chewing of betel leaves (*tāmbūla carvaṇa*) can also be performed. Application of medicated oil over body (*abhyaṅga*) is to be carried out regularly, as it pacifies *vāta*, tiredness and retards aging. *Abhyaṅga* is to be performed especially on head, ears and soles of feet. *Abhyaṅga* may be followed by exercise, as it strengthens body, reduces unwanted fat and stimulates digestion. However, children, the aged and those suffering from indigestion or *vāta-pitta* diseases should not exercise. Persons with stamina can exercise in pre-winter (*hēmanta*), winter (*śiśira*) and spring (*vasanta*) seasons expending half of their stamina. In other seasons, one should exercise mildly. Body is to be pressed lightly after exercise, till it cools down. Over-exercising can cause polydipsia, emaciation, respiratory distress, hemorrhages of obscure origin, cough, fever, vomiting and tiredness. Excessive indulgence in exercise, keeping awake at night, walking long distances, sexual intercourse, laughter and speech is to be avoided, as they harm the body (*Sū*: 2: 1–14).

Massaging (*udvartana*) reduces *kapha*, liquefies excess fat, brightens skin and strengthens body. Bathing stimulates digestive efficiency, sexual vigor, calms the mind and improves stamina. It also allays itching, generation of body wastes, tiredness, sweating, lassitude, thirst and body heat. Stamina is improved by bathing in warm water. Nevertheless, pouring warm water over head weakens hair and eyes. Bath is forbidden immediately after food and in facial palsy (*ardita*); diseases of eyes, mouth and ear; dysentery; bloating of abdomen; rhinitis; and indigestion. Following bath, pleasing food is to be consumed in moderation. Bowel and bladder should not be voided forcefully. These urges are also not to be restrained, as they can cause many disorders (see Table 3.6) (*Sū*: 2: 15–19).

3.7.1 Dos and Don'ts

All actions of living beings are directed toward achieving happiness. Happiness does not arise in the absence of *dharma* or proper conduct conforming to one's duty and nature. Therefore, one should be ethical, interacting only with good people. He should stay away from morally bad people. The ten sinful acts such as hurting others, stealing, fornication, gossip-mongering with intent to create discord between individuals, scolding with hatred and anger, speaking untruth, ridiculous speech, harboring malicious intentions, uncontrolled desire for other people's wealth and hostility to religious beliefs are not to committed by thought, speech or action. Destitutes, sick persons and those who suffer from other causes of grief should be helped. Even bugs and ants should be viewed as replicas of one's own self. Deities, cows, scholars, physicians, elderly persons, kings and guests should be respected or revered. Seekers of help should not be turned away or humiliated. One should possess equanimity of mind, unaffected by prosperity or disaster. A person should not disclose to anyone that he is an enemy of someone. He should equally refrain from expressing his animosity toward others. Similarly, disgraceful experiences are also not to be revealed to anyone (*Sū*: 2: 20–28).

Moderation is to be maintained while speaking, without giving room for disagreement. Happiness is not to be enjoyed alone. Therefore, one needs to be compassionate to others. Organs of body, including sense organs, are not to be put to too much or too little use. One shall follow a middle path in all religious and profane matters. Hair, nails and beards shall be short, one's feet and secretory

TABLE 3.7
Actions One Should Refrain from Committing (*Sū*. 2: 33–44)

Sl. No.	The Forbidden Actions
1	Treading over the shadows of sacred trees, revered persons, flagpole, mean persons, sacred ash, husk of grain, unclean objects, gravel, clay bricks and spots where sacred rites were performed
2	Trying to cross a river by swimming with arms alone
3	Facing a conflagration or firestorm
4	Getting into a canoe that may capsize, climbing a decayed tree, trying to ride an unfamiliar horse
5	Sneezing, laughing and yawning without covering face; squatting for long time
6	Continuing bodily and mental activities even after feeling tired
7	Sitting or lying down for long time, keeping knees upward
8	Sitting underneath a tree at night
9	Spending time at night beneath a peepul tree, staying in the premises of a temple or standing at intersections of three or four pathways
10	Sitting in an abandoned house, in a burial ground or at a place where people were murdered
11	Looking at sun at any time
12	Carrying heavy loads on head
13	Always looking at very small, unclean and unpleasant objects
14	Brewing, drinking, offering and selling wine
15	Getting exposed to easterly wind, hot sun, dust, dew and storm
16	Sneezing, belching, coughing, sleeping, eating and indulging in sexual intercourse, while keeping body in awkward position
17	Staying in the proximity of trees on river banks, shadow of a slope, enemy of king, beasts of prey, fanged animals, horned cattle; wicked, dishonorable and over-shrewd people; quarreling with good persons
18	Eating, sleeping, studying and indulging in sexual intercourse at dusk; eating food provided by enemy, innkeeper, strangers, prostitutes and traders; partaking of admixture of compatible and incompatible food
19	Imitating sounds of musical instruments using body, face and finger nails
20	Walking through the midst of water bodies, fire and group of venerable persons
21	Getting exposed to smoke emanating from a funeral pyre
22	Having too much interest in drinking wine
23	Putting faith in women

paths clean. One should bathe regularly and wear good clothes. Body should be made fragrant and adorned with precious stones. Regular chanting of *mantra* (words of power) related to a deity of choice is always beneficial. Umbrella and footwear are to be used while walking during daytime. However, walking along with assistants is essential at night. Headgear is to be worn and a walking stick shall be handy. AH also lists certain actions one should refrain from committing (Table 3.7).

World is the teacher to an intelligent person. Therefore, in matters regarding good conduct, one should learn from the world. For example, the fox teaches us that indulging in stealing and causing misery to others is indeed a detestable activity. Similarly, the cow educates us how to be of service to the society, by providing sustenance. Taking cue from the lives of birds which wake up early in the morning and enliven the world with their whistling, warbling and chirping, one should get up early in the morning, contemplate upon the Supreme Being for a moment and live in the company of virtuous persons, serving the world at large. Compassion, charity, equanimity and helping attitude are the traits of a virtuous person. One who leads a righteous life earns good health, prosperity, fame, longevity and freedom from the cycle of birth and death (*Sū*: 2: 45–48).

A house rests on pillars, and it will collapse if the pillars crumble. In the same way, the human body is maintained on three pillars such as food, sleep and libido. Optimization of these three components is pivotal to good health (*Sū*: 7: 52).

3.7.2 Food

At the stipulated time, one should be served pleasing, light, warm, oily food having all the six taste modalities, but predominating in sweetness. Ancestors, deities, guests, masters and children should be pleased with appropriate rites and food. Pets and domestic animals such as horse and cow are also to be fed. After washing palms, feet and face, one should sit in the company of friends and consume food neither fast nor very slowly. Food that was cooked long time ago and re-heated should not be served. So is very hot and salty food. Leafy vegetables are to be consumed in moderation. Cheese, curd, condensed milk, vinegar, dried meat; meat of pig, sheep, cow and buffalo; fish, black gram, dried leaves and thickened sugarcane juice (*phāṇita*) are not to be consumed regularly. On the contrary, red rice (*Śāli*), wheat, barley, meat of animals from dry land (*jāṅgala dēśa*), *Terminalia chebula*, *Emblica officinalis* berries, *Vitis vinifera* fruits, green gram, sugar, clarified butter, milk, honey, *Punica granatum* fruits and rock salt can be part of everyday food. *Triphala cūrṇa* mixed with honey and clarified butter can be consumed every day at night for conserving eye health. While taking food, heavy, oily and sweet food is to be served first and light food such as broth (*yūṣa*) is to be consumed in the end. Salty or sour food needs to be consumed in the middle (*Sū*: 8: 35–46).

Half volume of stomach is to be filled with food, one quarter with beverages and the other quarter left empty. Water is post-prandial drink (*anupāna*) following consumption of barley, wheat, curd, wine (*madya*) and honey. Warm water is ideal *anupāna* in the case of food prepared with rice. Curd, water, buttermilk and fermented rice water (*kāñjika*) are *anupāna* following consumption of leafy vegetables and green gram. *Sura* wine is beneficial for lean persons and honey water pacifies obesity. Meat soup is appropriate *anupāna* for patients of kingly consumption (*rājayakṣma*) and *madya* as *anupāna* stimulates abdominal fire of persons suffering from dull appetite (*agnimāndya*). Milk is comparable to ambrosia in the case of children, aged persons and those suffering from tiredness due to disease, medication, exertion, fasting, excessive sexual intercourse and exposure to hot sun. *Anupāna* loosens food and helps in digestion. *Anupāna* is not advisable after singing or speaking too much or when suffering from lassitude of voice (*svarasāda*). It is also contra-indicated in diseases above the collar bone, respiratory distress (*śvāsa*), cough (*kāsa*), injury to thorax (*uraḥkṣata*), catarrh (*pīnasa*), polyuric diseases (*pramēha*), eye diseases, sores, ulcers and diseases of throat. One should not indulge in talking, walking, sleeping, getting exposed to hot sun, traveling, jumping and running, immediately after consuming food or beverages. The ideal time for consuming food is when the bowels are cleared and good appetite is felt (*Sū*: 8: 46–55).

Consumption of admixture of compatible and incompatible food is termed as *samaśanaṃ*. Eating food again before the previous meal is digested is called *adhyaśanaṃ*. Consumption of food in large or small measures at inappropriate time is called *viṣamāśanam*. *Samaśanaṃ*, *adhyaśanaṃ* and *viṣamāśanam* cause dreadful diseases and even death (*Sū*: 8: 33–35).

3.7.3 Sleep

By sleeping well, an individual gets comfort and nourishment. Unhappiness and leaning of body result in its absence. Happiness and health are harmed by sleeping too much and too little or by sleeping at odd hours. Keeping awake at night makes the body dry (*rūkṣa*), and sleeping during daytime is unctuous (*snigdha*) in nature. Sleeping during daytime is beneficial in summer. However, it enrages *kapha* and *pitta* in all other seasons. The damaging effects of talking excessively, walking long distances, riding too much on horseback, carrying heavy loads, drinking spirituous liquors, immoderate sexual intercourse, anger, grief and fear are counteracted by sleeping during daytime. Daytime sleep is also beneficial to children, the aged, weak and tired persons and those suffering from respiratory distress (*śvāsa*), hiccup (*hidhma*), dysentery (*atisāra*), indigestion (*ajīrṇa*), physical trauma (*abhighāta*), polydipsia (*tṛṣṇa*), insanity (*unmāda*) and pricking pain (*śūla*). Persons suffering from enragement of *kapha* and adipose tissue (*mēdas*) and those who are administered oily substances (*snēha*) internally or topically are not to sleep during daytime. Persons affected by poisoning due to animal bites or ingestion of poisons and those suffering from diseases of throat should keep awake even during night (*Sū*: 7: 53–60).

Sleep at odd times causes the appearance of mental confusion, fever, rhinitis, headache, cough, nausea, obstruction of body channels and lowering of abdominal fire. Treatments for such disorders are fasting, emesis, sudation and nasal instillation of medicines. Effects of excessive sleeping (*atinidra*) can be pacified by emesis with *tīkṣṇa* herbs, application of collyrium, nasal instillation of medicines, fasting, mental stress, indulging in excessive sexual intercourse, grief, fear and anger. *Kapha* is reduced significantly by these measures. Insomnia (*nidrānāśa*) causes heaviness of body, pressing pain all over body, lassitude, yawning, giddiness, slow digestion and drowsiness. Therefore, one should get accustomed to sleeping at regular time. Persons suffering from lack of sleep should consume milk, *madya*, meat soup and curd. *Abhyaṅga*, *udvartana*, bathing, *śirōvasti*, *karṇapūraṇam* and *akṣitarpaṇam* are also beneficial. A contented and happy person, who has control over his mind and is disinterested in women, sleeps at appropriate time (*Sū*: 7: 61–68).

3.7.4 Sexual Intercourse

One should not indulge in sexual intercourse with a woman who is obese, lean and unable to lie in supine position. Pregnant or menstruating women, women other than one's own wife, one suffering from vaginal diseases or a woman performing religious vows are to be avoided. Bestiality is forbidden. A couple should copulate with phallus and vagina; involving other parts of body is also prohibited. Coitus should not take place in the house of guru, temple premises, royal palace, underneath sacred trees, on burial ground, places where murders took place and intersections of three or four paths. Coitus should be avoided during daytime, on full moon day (*pourṇami*) and the first day of the solar month (*saṅkrānti*). Head and bosom are not to be pressed during sexual union. A young boy, old man, fearful person, one who has eaten in overabundance, one who is hungry, thirsty or sick and a cripple should refrain from indulging in sexual intercourse. One who has consumed sufficient quantity of aphrodisiacs can indulge *ad libitum* in sexual activity during pre-winter and winter seasons. Once in three days is ideal in spring and autumn seasons. Once in fifteen days is recommended in rainy season and summer (*Sū*: 7: 69–73).

Violation of these mandatory guidelines can cause giddiness, tiredness, weakness of thighs, loss of stamina, sterility and even death. One who indulges in sexual intercourse properly will have good memory, comprehension skills, health, stamina, virility, delayed aging and long life. After sexual union one should take bath and apply fragrant pastes over body, consume food and beverages such as *Khaṇḍakhādya*, cold water, milk, meat soup, broth, *Sura* and *Prasanna*. Shortly thereafter, he should go to bed (*Sū*: 7: 74–77).

Sexual intercourse with one's own wife is the only ethically permitted option. Nevertheless, immoral and perverted sexual practices were rampant in ancient days, just as they are prevalent in modern times. Bestiality, anal sex, oral sex and copulation in open air were practiced in those days also, as is evident from the injunctions given in Chapter 7 of *Sūtrasthānam* (*Sū*: 7: 69–74). In the chapter on diseases of the genitals (*Guhyarōgavijñānādhyāyaḥ*), AH states that *upadamśa* (venereal ulcers) arise from several causes including bestiality, incest and oral sex (*Ut*. 33: 1–5). *Vāgbhaṭa* knew very well that such abominations are difficult to be curbed. The mandatory guidelines are intended to warn the society of the inherent dangers, as well as to prevent mishaps and misery.

3.8 SEASONAL REGIMEN

Śiśira (winter), *vasanta* (spring), *grīṣma* (summer), *varṣa* (monsoon), *śarat* (autumn) and *hēmanta* (pre-winter) are the six seasons. *Śiśira*, *vasanta* and *grīṣma* are collectively called *uttarāyana* (northward movement of sun). As *uttarāyana* reduces stamina, it is called *ādānaṃ* (debilitation). Due to northward movement of sun, sun and air turn to be highly *tīkṣṇa*, *uṣṇa* and *rūkṣa*, weakening the pleasing qualities permeating in the world. Consequently, *tikta*, *kaṣāya* and *kaṭu* tastes dominate, due to the *uṣṇa* nature of *uttarāyana*. *Varṣa*, *śarat* and *hēmanta* constitute *dakṣiṇāyana* (southward movement of sun). *Dakṣiṇāyana* is also known as *visarga* (liberation), as it liberates strength in man. Because of the increase in coolness, the moon turns stronger. As a result, *madhura*, *amḷa* and *lavaṇa* tastes predominate. Stamina of humans is high in *hēmanta* and *śiśira*, low in *grīṣma* and *varṣa* and medium in *vasanta* and *śarat* seasons (*Sū*: 3: 1–6).

3.8.1 Pre-winter and Winter Regimen

As body defends the cold environment, internal fire gets stronger in pre-winter season. In the event of lowered food consumption, the internal fire starts degrading *dhātus*. Therefore, food and beverages predominantly *madhura*, *amḷa* and *lavaṇa* are to be consumed in this season. After completing morning rituals such as voiding of bowel and bladder, one should get subjected to *abhyaṅga* and massage with *vāta*-lowering *taila*. He can also take part in wrestling. Bath is to be taken after exercise, followed by application of fragrant pastes over body and fumigation with heartwood of *Aguru* (*Aquilaria agallocha*). Food should consist of meat soups, clarified butter, wines and delicacies prepared with wheat, rice, black gram, sugarcane juice and milk. Animal fat, sesame oil and freshly harvested rice are also to be included in food. Warm water is to be used for bath and washing body parts. Bed should be covered with various kinds of bed sheets such as *Prāvāra* (woolen blanket), *Ājina* (pleasant-to-touch hairy skin of tiger), *Kauśēya* (silk), *Pravēṇi* (colored woolen cloth) and *Kaucava* (goat's hair sheet). Sunbath and sudation are also recommended. Feet are to be protected with footwear. Embrace of friendly, warm-bodied, voluptuous, scented and young buxom women wards off cold. One shall also stay in a house heated with a fire place. The same regimen is to be adopted in winter season as well (*Sū*: 3: 7–17).

3.8.2 Spring Regimen

Kapha accumulated in winter season gets liquefied by heat of spring, lowering abdominal fire and generating diseases. Therefore, *kapha* is to be eliminated expeditiously, with *vamana* and *nasya* of *tīkṣṇa* nature. Food needs to be *laghu* and *rūkṣa*. Exercise, *udvartana* and foot massage also reduce *kapha*. After performing these *kapha*-lowering measures, fragrant pastes are to be applied on body. Thereafter, one can partake of food prepared with aged barley and wheat. Grilled meat of *jāṅgala* animals can be consumed, mixed with honey. Various wines such as *Āsava*, *Ariṣṭa*, *Mṛdvīka* and *Mādhava* can also be consumed in the company of friends. Water boiled with dry ginger, heartwood of *Pterocarpus marsupium*, tubers of *Cyperus rotundus* or honey water can be drunk instead of plain water. Afternoons are to be spent in cool and romantic places, in the company of women. Siesta is to be avoided and so is consumption of *guru*, *śīta*, *snigdha*, *amḷa* and *madhura* food, which invariably increase *kapha* (*Sū*: 3: 18–26).

3.8.3 Summer Regimen

Vāta increases in summer season, with concomitant decrease in *kapha*. Therefore, *lavaṇa*, *kaṭu*, *amḷa* tastes, exercise and hot sun are to be avoided. Light, oily, cold, liquid food having sweet taste is recommended in summer. Cold water is to be used for bath. *Madya* is to be consumed, diluted with water, failing which, edema, flaccidity, burning sensation and stupor may arise. *Śāli* rice and *jāṅgala* meat are to be included in food. Low fat meat soup and beverages such as *Rasāḷa*, *Rāga* and *Khāṇḍava* are to be consumed. Drinking of cold and sweetened buffalo milk is beneficial at night. Midday and night are to be spent in cool places (*Sū*: 3: 26–41).

3.8.4 Monsoon Regimen

All measures that pacify *tridōṣa* and stimulate abdominal fire are to be adopted in monsoon period. Body is to be conditioned with *vamana, virēcana* and *nirūha vasti*. Aged grains and cooked *jāṅgala* meat are to be included in food, which should be dry, oily, light and mixed with honey. Boiled water, *yūṣa*, aged *madya*, whey containing sonchal salt and water boiled with *Pañcakōla* herbs are to be used as beverages. Fumigated clothes are to be worn always. River water, parched rice powder suspended in water, siesta, exertion and hot sun are to be avoided (*Sū*: 3: 42–48).

3.8.5 Autumn Regimen

Pitta increases in autumn and needs to be eliminated through *virēcana* and *raktamōkṣa*. Internal consumption of *Tiktaka ghṛta* is recommended for pacifying *pitta*. On achieving good appetite, one should consume light food having *madhura, tikta* and *kaṣāya* tastes. Consumption of *Śāli* rice, green gram, sugar, gooseberry, snake gourd, honey and *jāṅgala* meat is advised. Paste of *Santalum album, Vettiveria zizanioides* and *Cinnamomum camphora* is to be applied on body. Eating in overabundance, getting exposed to dew, alkalis, curd, sesame oil, animal fat, hot sun, strong *madya*, siesta and easterly wind are to be avoided. One should get accustomed to *madhura, amḷa, lavaṇa* in pre-winter, winter, monsoon seasons; *kaṭu, tikta, kaṣāya* in spring; *madhura* in summer and *madhura, tikta, kaṣāya tastes* in autumn seasons. However, food should always have the six tastes (*Sū*: 3: 49–57).

3.9 ORIGIN OF DISEASES

Tridōṣa increase in the body or undergo *vṛddhi* with food, beverages, behavior and environment having similar qualities. However, opposite qualities cause their decrease (*kṣāya*). In addition to these two states, the *tridōṣa* can also undergo *caya* (increasing in own sites), *kōpa* (migration to various sites) and *śamana* (returning to pacified state). *Vāta* undergoes *caya* when its innate qualities combine with *uṣṇa* (warmth). *Kōpa* of *vāta* results when these qualities interact along with *śīta* (cold). Qualities opposite to *vāta* bring about its *śamana*. *Pitta* undergoes *caya* when it is exposed to factors having *śīta* quality. *Uṣṇa*, on the other hand, causes its *kōpa*. When opposite qualities combine with *śīta*, *pitta* attains the state of *śamana*. The inherent qualities of *kapha* work with *śīta* to cause its *caya*. When the same qualities come in contact with *uṣṇa*, *kapha* undergoes *kōpa*. Opposite qualities bring about *śamana* of *kapha* (*Sū*. 12: 19–23).

Vāta undergoes *caya* in summer, *kōpa* in monsoon and *śamana* in autumn. *Pitta* undergoes *caya, kōpa* and *śamana* in monsoon, autumn and pre-winter seasons, respectively. *Kapha* undergoes *caya* in winter, *kōpa* in spring and *śamana* in summer. However, food, beverages, behavior and environment can cause *caya* and *kōpa* of *tridōṣa* any time. *Dōṣa*s which undergo *kōpa* spread all over the body from head to toe, causing various diseases with many symptoms and signs (*Sū*. 12: 24–25).

On the basis of quality and magnitude, any voluntary activity can be of four types: *hīnayōga* (inadequate), *mithyāyōga* (improper or perverse), *atiyōga* (excessive) and *samyakyōga* (proper, optimum). For example, eating very small quantity of food is the *hīnayōga* of the action of eating. Eating too much food is *atiyōga*. Eating at odd times and eating incompatible food can be considered to be *mithyāyōga*. Eating permissible food at appropriate time in moderate quantity can be considered to be *samyakyōga*. *Samyakyōga* promotes health, whereas *hīnayōga, mithyāyōga* and *atiyōga* enrage *tridōṣa* (*Sū*. 1: 19, *Sū*. 12: 40–44). The fundamental causes of *kōpa* of dōṣas are *hīnayōga, atiyōga* and *mithyāyōga* of sense organs, season and actions of the individual mediated by thought, speech or action (*Sū*. 12: 24–42).

*Dōṣa*s which have undergone *kōpa* spread to *śākha* (branches), *kōṣṭha* (alimentary canal) and *asthisandhi* (joints of bones). *Rakta, māmsa, mēdas, asthi, majja, śukra* and skin constitute branches (*śākha*). That is the external route of diseases. Elevated mole, black pigmentation on face, swelling, hard or closed pustule, tumor and diseases like hemorrhoids, *gulma* and edema which appear on the surface of body are dependent on *śākha*. Alimentary canal is the seat of many diseases such as vomiting, dysentery, cough, respiratory distress, enlargement of abdomen, fever, edema, hemorrhoids, *gulma*, cellulitis and abscess. It is the internal route of diseases. Head, heart and urinary bladder are the components of *asthisandhi*. Veins, sinews and tendons are dependent on it. It is the central route of diseases. Diseases such as kingly consumption; hemiplegia; facial palsy; diseases of head, ears and eyes; pricking pain in joints, bones and sacral area are dependent on *asthisandhi* (*Sū*. 12: 44–49).

Vāta enragement causes various kinds of pain such as stretching, pricking, crushing, splitting types; tingling, numbness, obstruction, constriction, thirst, tremors, roughness, porosity, dryness, leaning, pulsations, rigidity, astringent taste in mouth and bluish red color. Enragement of *pitta* is reflected in symptoms and signs such as burning sensation, redness, heat, suppuration, sweating, secretions, fainting, stupor, sensations of pungent and sour tastes in mouth and discoloration excluding pallor and bluish redness. When enraged, *kapha* produces symptoms and signs such as oiliness, hardness, itching, coldness, heaviness, obstruction, pasting of layers inside body channels, immobility, edema, lowering of abdominal fire, hypersomnia, pallor, sensations of sweetness and saltiness in mouth and slow progression of the enragement. These symptoms and signs enable the physician to study the perturbation of *dōṣa*s in patients (*Sū*. 12: 49–55). *Vāta*, *pitta* and *kapha* undergo increase and decrease in 63 combinations. A physician should consider these factors also before arriving at a diagnosis (see Table 3.1).

When viewed from the angle of importance, diseases are of two types such as primary disease (*svatantra rōga*) and secondary disease (*paratantra rōga*). Secondary disease can either be the prodromes (*pūrvarūpa*) of the primary disease or the problems that arise after the appearance of the full-fledged disease entity. Primary disease makes its appearance after following a definite course of development. It will have distinct symptoms and signs. Nevertheless, secondary disease does not have these characteristics and may appear in the form of several discomforts. Secondary disease is usually pacified when the primary disease is cured. Any secondary disease that remains even after the primary one is cured should be treated separately (*Sū*. 12: 60–63).

Ten factors are to be considered in detail in diagnosis and treatment. They include the *dhātu*s that are affected (*dūṣyaṃ*), the overall condition of body and habitat of the patient (*dēśaṃ*), severity of the disease and stamina of patient (*balaṃ*), stage of disease and season of year (*kālam*), digestive efficiency (*anala*), constitution (*prakṛti*), age of patient (*vayaḥ*), mental make-up (*satvaṃ*), likes, dislikes, habits and addictions (*sātmyaṃ*) and food habits (*āhāraṃ*). Because of the mental make-up, satisfactory or poor stamina of patient, there may occur differences in the intensity of symptoms and signs. A physician who mistakes a serious disease for a less intense one or a minor one for a major ailment is certainly at fault. Therefore, he shall examine the patient carefully and arrive at a proper diagnosis and treatment protocol (*Sū*. 12: 67–73).

3.10 THE FOUR PILLARS

AH lays great emphasis on ethical principles. The physician, medicines, attendant and the patient are the four pillars of treatment of diseases. Physician should be competent to do the right thing at the right time. He should have learned the healing art from a guru and must have good experience in performing various therapeutic measures. The medicines which fall into categories like decoctions, powders, medicinal oils and so on should be effective in various disease conditions. They should be potent and efficacious on account of quality of ingredients and processing. Attendant should be clean, tidy, good at heart, intelligent, affectionate toward the patient and experienced in patient care. The patient should be rich and willing to follow diligently the instructions of the physician. He should inform the physician periodically about the progress of treatment (*Sū*. 1: 27–29).

Ethical conduct of physician is also advocated in AH. For example, if urinary stone is not curable by therapeutic measures, the physician should obtain permission from the king to resort to surgery, as surgery may sometimes result in the death of the patient. On receiving permission, he should perform surgery and remove the urinary calculus (*Ci.* 11: 43–63). Royal permission is also essential in the case of surgical removal of stillbirth (*Śā.* 2: 26). *Udara* is a disease affecting liver and or spleen, resulting in enlargement of abdomen. If untreated, the disease finally reaches the advanced stage (*sannipātōdara*). If stamina and digestive efficiency have not reduced in such a patient, the physician may at first decline to treat him. Nevertheless, on request of the patient's relatives, he should administer orally, oil extracted from the seeds of *Baliospermum montanum* and *Croton tiglium*. If the patient fails to respond to such conventional therapy, the physician should seek consent from the patient's relatives to try more powerful, but dangerous treatment. On receiving their permission, he should administer pastes of *Abrus precatorius* seeds and root of *Nerium indicum*, suspended in wine. Poisons of herbal or mineral origin may also be administered, mixed with food and beverages. In such case, the patient may or may not survive (*Ci.* 15: 76–81). Sometimes enlargement of abdomen caused by *baddhōdara, chidrōdara* and *jalōdara* may not respond to medicines. In such cases, the physician should resort to surgery, after obtaining permission from the patient's relatives and the king (*Ci.* 15: 107).

Medicinal enema (*vasti*) is an elimination measure that needs to be performed with great care. Therefore, the physician should confer with his peers and arrive at a consensus on the composition of the pastes, decoctions, other ingredients and their volumes. *Vasti* is to be performed with their cooperation as well (*Sū.* 19: 36–38). The physician should strive diligently to pacify the effects of excessive application of elimination measures. If patient does not vomit after administration of emetic medicine, the physician should find out the reason for the failure, make the patient unctuous again and administer the emetic medicine. Similarly, if the purgative induces emesis instead of purgation, he should administer palatable and safer purgatives (*Ka.* 3: 1–5). If the patient loses blood, feels thirsty or faints due to excessive purgation, the physician should apply treatments that pacify hemorrhages of obscure origin and dysentery. Freshly collected blood from elephant, cow, buffalo or goat is to be administered. He should continue his efforts to the last breath of the patient (*Ka.* 3: 35–37). When the patient exhibits weakness, loss of stamina and digestive efficiency following *vasti*, the physician should resort to all measures to save his life (*Ka.* 5: 51–54).

Diseases caused by a single *dōṣa* and acute in nature can be cured easily. Diseases that need to be treated with surgery, alkalis and cauterization (*agnikarma*) are difficult to cure. Diseases characterized by anxiety, loss of consciousness, disinclination to do anything (*arati*) and having characteristic signs of imminent death are impossible to be cured. A physician should not treat a patient despised by the king. He should also avoid one who speaks ill of the king or the physician. Disobedient, fearful, ungrateful, one with violent temperament and who pretends to be knowledgeable about healing arts should also not be treated (*Sū.* 1: 30–35).

3.11 DIAGNOSIS

A physician should examine the patient by *darśana* (observation), *sparśana* (palpation) and *praśna* (interrogation). He shall evaluate the disease based on *nidāna* (etiology), *pūrvarūpa* (prodromes), *lakṣaṇa* (signs and symptoms), *upaśaya* (diagnosis by observing the effect of certain, food, medicine or measures) and *samprāpti* (pathogenesis) (*Sū.* 1: 22).

3.12 TREATMENT OF DISEASES

3.12.1 Pacification of *Dōṣas*

Enragement of *dōṣa*s is the cause of all diseases. Therefore, pacification of *dōṣa*s is the way to cure diseases. Consumption of oleaginous substances; sudation; laxation (*mṛdu saṃśōdhanaṃ*); adoption of food with sweet, sour and salty tastes; pressing or tightly binding the body after applying *taila* all over the body (*abhyaṅga*); startling; pouring *taila* or decoctions in a thin stream on body;

drinking wines brewed with rice powder or jaggery; medicinal enemas (*vasti*) with unctuous and pungent medicines; regular application of alternating medicinal enemas such as *nirūhavasti* and *snēhavasti*; staying without physical exertion; and internal or topical application of medicinal oils and clarified butter prepared with carminative and digestive properties are the remedies for pacifying enraged *vāta*. The most important treatment of *vāta* involves consumption of various kinds of meat, meat soups, sesame oil and *snēhavasti* (*Sū.* 13: 1–3).

Pitta is pacified by consumption of clarified butter; purgation with sweet and cool medicinal substances; adoption of food and medicines possessing sweet, bitter and astringent tastes; smelling fragrant; pleasing and cool odoriferous substances; applying frequently over body pastes of camphor, sandalwood and *khus* roots (*Vetiveria zizanioides*); and spending evenings in cool homes or woods in the company of loveable women. Milk, clarified butter and purgation are the important remedies for pacifying enragement of *pitta* (*Sū.* 13: 4–9).

Treatment of *kapha* enragement involves properly performed emesis and purgation; food that is dry, slightly piercing and hot in nature and having pungent, bitter and astringent tastes; aged wine; excessive sexual intercourse; keeping awake at night; various kinds of physical exercises; mental worries; and massaging without applying oil. The major remedies that pacify *kapha* are soups of green gram and horse gram, honey, medicines that reduce obesity, inhalation of medicinal smoke (*dhūmapāna*), gargles (*gaṇḍūṣa*) and physical exertion. These measures may be combined judiciously for curing *saṃsargaṃ* or combinations of enraged *vāta*, *pitta* and *kapha*, Summer, spring and autumn regimens are to be adopted in *saṃsargaṃ of vāta-pitta*, *kapha-vāta* and *kapha-pitta*, respectively (*Sū.* 13: 10–14).

Physical exertion, increased body temperature, consumption of improper food and stimulus provided by *vāta* induce *tridōṣa* to move from alimentary canal to *śākha* and *asthisandhi*. On account of opening of body channels, flowing nature of the *dōṣas*, absence of transformation and lack of *vāta*-stimulation, these *dōṣas* move from *śākha*, *asthisandhi* and return to alimentary canal, waiting for favorable conditions. At the opportune moment, they acquire strength, move to sites of other *dōṣas* (*sthāni*) and get enraged. When this newly arrived *dōṣa* (*āgantu*) is weak, treatment for the *sthāni* is to be carried out. However, if the *āgantu* acquires strength and impairs the activity of the *sthāni*, the *āgantu* is to be treated first and the *sthāni* later (*Sū.* 13: 17–20).

Blockage of channels, loss of stamina, heaviness of body, obstruction of movement of *vāta* (*anilamūḍhata*), laziness, indigestion, frequent spitting, constipation, distaste for food and tiredness are the signs of *dōṣa* associated with *āma*. The term *āma* means "undigested", and it denotes the *dhātus* and *mala* remaining untransformed due to lack of principles of transformation (*agni*), food remaining in the stomach and vitiated essence of digestion (*rasa dhātu*). *Vāta*, *pitta* and *kapha* associated with *āma*, *dhātus* vitiated by them and the ensuing diseases are called *sāma* or "with *āma*". *Tridōṣa* polluted with *āma* spread all over the body and tend to damage the body. These *āma*-tainted elements are to be liquefied at first with oleation procedures that are carminative and digestive. Thereafter, they are to be eliminated from body at the appropriate time, using suitable elimination measures. Emetic medicine administered through mouth and nostrils eliminate *dōṣas* from stomach. *Dōṣas* staying in throat and head are eliminated with *nasya* medicines administered through nostrils. *Dōṣas* lodged in the intestines are eliminated with the help of medicinal enemas administered through rectum and purgatives introduced through mouth. On complete elimination of *dōṣa*, the patient is fed with food that is easily digestible and suitable for the medical condition (*Sū.* 13: 23–36).

3.12.2 Administration of Curative Medicines

Śamana (curative) medicines are to be administered after elimination of *dōṣas*. They are administered in ten different modes such as without food, in the beginning of eating, in the middle of eating, soon after completion of eating, in between each morsel of food, concealed in each morsel of food, mixed frequently with food, before and after food and at night. Medicine is to be administered without food in *kapha* diseases, full-blown diseases and in patients having stamina. Medicine is administered in the beginning of eating, when *apāna vāyu* is enraged and in the middle, when *samāna vāyu* is enraged. For pacifying *vyāna vāyu* and *udāna vāyu*, medicine is to be administered soon after completing breakfast and dinner, respectively. In vitiation of *prāṇa vāyu*, medicine is administered in two modes – concealed in each morsel and in between morsels. This is recommended when the disease is stronger than the patient or when disease and patient have equal strength. Frequent administration is recommended in poisoning, vomiting, hiccup, thirst, respiratory distress and cough. Distaste for food is treated by administering medicine mixed with various kinds of food. Medicine is to be administered before and after food, in treatment of shivering, convulsions and hiccup. Light food is to be consumed in such cases. Administering medicines before retiring to bed is ideal in the treatment of diseases above the collar bone. *Dōṣōpakramaṇīyādhyāyaḥ* (chapter on pacification of *dōṣas*) contains 41 quatrains, and on that basis, some commentators remark that curative medicines are to be administered consecutively for forty-one days (*Sū.* 13: 37–41).

3.13 THE TWO THERAPEUTIC MEASURES

Treatment is generally aimed at the disease or vitiation of *dōṣas* (*dōṣavaiṣamyam*). As *dōṣavaiṣamyam* is either increase or decrease of *dōṣas*, treatment of diseases is also broadly classified into two types such as *santarpaṇa* (*bṛṃhaṇa*) and *apatarpaṇa* (*laṅghana*). That which causes nourishment (strengthening) of body is called *bṛṃhaṇa* therapy and that which causes leaning of body is called *laṅghana* therapy. Matter that causes *bṛṃhaṇa* is predominant in the elements of earth and water. *Laṅghana* is caused by matter having predominance of fire, air and sky. *Laṅghana* is further subdivided into *śōdhana* (eliminative) and *śamana* (curative) measures. That which causes elimination of *dōṣas* from body is *śōdhana*. It is of five types such as medicinal enema with decoctions (*nirūhavasti*), emesis (*vamana*), purgation (*virēcana*), elimination of *dōṣas* from head (*nasya*) and bloodletting (*raktamōkṣa*). These five measures are collectively known as the five actions (*pañcakarma*). Curative measures do not eliminate *dōṣas* or enrage *dōṣas* staying in steady state. They bring enraged *dōṣas* to steady state. Curative measures comprise seven components such as medicines that stimulate digestion (*dīpana*) and enhance transformation (*pācana*), fasting (*upavāsa*), restrainment of thirst (*tṛṣṇānigraha*), physical exercises (*vyāyāma*) and exposure to hot sun and breeze (*ātapasēvana, mārutasēvana*). *Bṛṃhaṇa* is also a curative measure for enragement of *vāta* or *vāta-pitta* (*Sū.* 14: 1–7) (Figure 3.1).

Subjects who qualify for *bṛṃhaṇa* include those who are exhausted due to diseases or treatment measures, alcoholism, excessive indulgence in sexual intercourse, grief, carrying heavy loads, walking long distances and trauma of thorax. Along with them children, the aged, pregnant and post-partum mothers, dry-bodied, lean, weak persons and those with enragement of *vāta* are to be subjected to *bṛṃhaṇa* is summer season. *Bṛṃhaṇa* is to be carried out by feeding of meat, milk, sugar and clarified butter; medicinal enema with sweet and unctuous ingredients; *abhyaṅga*; bathing; inducing sound sleep; and sexual pleasure (*Sū.* 14: 8–10).

Patients of polyuric diseases; *āmadōṣa*; fever; stiffness of thighs; skin diseases; cellulitis; abscess; splenomegaly; diseases of head, throat and eyes; and excessively unctuous and obese persons are to be regularly subjected to *laṅghana*. Among them, those suffering from *āmadōṣa*, fever, vomiting, dysentery, heart disease, constipation, heaviness of body, belching and nausea are to be subjected to *pañcakarma*, if they have predominance of obesity, stamina, *pitta* and *kapha*. Those who have medium obesity, stamina, *pitta* and *kapha* should be subjected to *laṅghana* with *pācana*, *dīpana* medicines. *Laṅghana* with fasting and restraint of thirst is to be carried out in those who have low degrees of obesity, stamina, *pitta* and *kapha*. Patients with medium predominance of *dōṣas* are to be subjected to *laṅghana* with exposure to hot sun, breeze and physical exercises, if they have good stamina. Patients with mild enragement of *dōṣas* are to be subjected to *laṅghana* with diet control and mild exposure to hot sun, breeze and exercise. *Bṛṃhaṇa* measures are not to be performed on subjects eligible for *laṅghana*. But those who are eligible for *bṛṃhaṇa* may be subjected to mild *laṅghana* (*Sū.* 14: 10–16).

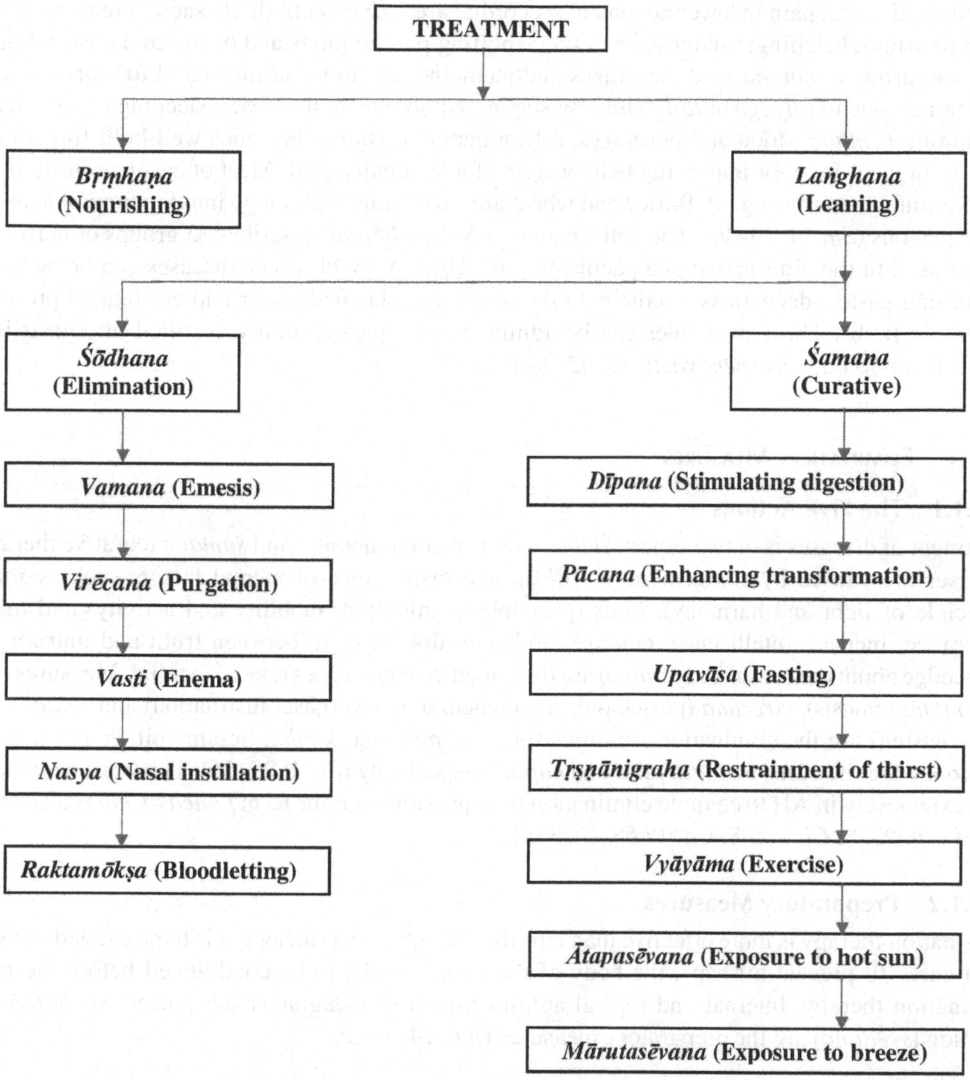

FIGURE 3.1 Scheme showing various treatment modes such as *bṛṃhaṇa*, *laṅghana*, *śōdhana* and *śamana*.

Performance of optimum *bṛṃhaṇa* (*samyakyōga*) strengthens body and cures the disease. Optimum performance of *laṅghana* results in proper functioning of sense organs, good bowel movements, lightness of body, interest to consume food, occurrence of hunger and thirst at same time, clearing of throat, cure of the disease in question, enthusiasm and lack of laziness. Performance of *bṛṃhaṇa* and *laṅghana* without considering dose and other factors may cause obesity and excessive leaning. Excessive performance of *bṛṃhaṇa* (*atiyōga*) causes obesity, swelling in neck, polyuric diseases, fever, enlargement of abdomen, anal fistula, cough, coma, dysuria, *āmadōṣa*, skin diseases and many other dreadful ones. Horse gram, corn, little millet, barley, green gram, honey water, whey, buttermilk, fermented medicines, mental worries, keeping awake at night and such herbs, medicinal substances and measures that reduce fat, *vāta* and *kapha* are useful in the treatment of obesity. Pastes of *Triphala*, *Tinospora cordifolia*, *Terminalia chebula* or *Cyperus rotundus* mixed with honey pacify obesity (*Sū.* 14: 16–22).

Atiyōga of *laṅghana* causes excessive leaning; giddiness; cough; excessive thirst; distaste for food; loss of unctuousness of body; digestive efficiency; sleep; vision; hearing; virility; *ōjas*;

appetite and voice; pain in lower abdomen, precordial area, head, calf, thigh, sacral area and flanks; fever; delirium; belching; tiredness; vomiting; splitting pain in joints and bones; and obstruction of stool and urine. *Bṛṃhaṇa* food, beverages and medicines are to be administered for curing excessive leaning due to *atiyōga* of *laṅghana*. By staying without mental worries, sleeping too much and consuming *bṛṃhaṇa* food and beverages, a lean person certainly becomes well-built like a boar. Meat is an ideal food for improving body weight of a lean individual. Meat of carnivorous animals is more suitable in this regard. Barley and wheat are two grains that can go into the food of lean and obese persons (*Sū.* 14: 29–36). The 15th chapter of *Sūtrasthānam* describes 33 groups of herbs that can be used in the elimination and pacification of *dōṣas*. Very obstinate diseases can be cured by consuming pastes, decoctions, medicinal oils, medicated clarified butter and electuaries prepared with these herbs. These medicines can be administered topically, orally, instilled into nostrils or introduced into body as *snēhavasti* (*Sū.* 15: 1–47).

3.13.1 Elimination Measures

3.13.1.1 The Five Actions

Treatment of diseases is of two types: *śōdhana* (elimination therapy) and *śamana* (curative therapy). Diseases are caused by enragement of *tridōṣa* and disturbance of mental factors such as *sattva* (principle of light and harmony), *rajas* (principle of initiation, mobility and activity) and *tamas* (ignorance, inertia). Intelligence, courage, ability to discriminate between truth and untruth and knowledge about the causal body (*ātma*) are the remedies for diseases related to mind. Measures such as *vamana* (emesis), *virēcana* (purgation), *vasti* (enema), *nasya* (nasal instillation) and *raktamōkṣa* (bloodletting) are the elimination measures for *vāta*, *pitta* and *kapha*. Sesame oil, clarified butter and honey are the cures for *vāta*, *pitta* and *kapha*, respectively (*Sū.* 1: 25–26). The word *śōdhana* is used extensively in AH to connote elimination therapy. However, the term *pañcakarma* is also made use of (*Sū.* 27: 8, *Ci.* 7: 108, *Ci.* 10: 65, *Ut.* 34: 62).

3.13.1.2 Preparatory Measures

Elimination therapy is more effective than curative therapy. Nevertheless, it is to be carried out with great care. To prevent mishaps, the body of the patient needs to be conditioned before the main elimination therapy. Internal and topical administration of oleaginous substances (*snēhana*) and sudation (*svēdana*) are the preparatory measures (*pūrvakarma*).

3.13.1.3 Oleation

Clarified butter from milk (*ghṛta*), bone marrow (*majja*), visceral fat (*vasa*) and oil from sesame seeds (*taila*) are the excellent oleaginous substances (*snēha*). Among them, *ghṛta* is the ideal lipid. *Ghṛta* pacifies *vāta* and *pitta*, *majja* and *vasa* calm *vāta*, and *taila* appeases *vāta-kapha*. Mixture of two lipids is called *yamaka*, three is known as *tṛvṛta*, and four bears the name *mahāsnēha* (*Sū.* 16: 1–4). Oleation (*snēhana*) is recommended for those who are eligible for sudation and elimination therapy; persons habituated to drinking alcoholic beverages, fornication and exercise; those continuously exposed to mental stress; aged persons; individuals with weak, dry and lean bodies; persons suffering from *vāta* diseases, conjunctivitis and cataract; and persons with lowered sexual vigor. However, *snēhapāna* (internal consumption of *snēha*) is contra-indicated in stiffness of thigh muscles, dysentery, indigestion, diseases of throat, poisoning, enlargement of abdomen, vomiting, anorexia, *kapha* diseases, polydipsia, alcoholism, depressed or exaggerated appetite, obesity and debility (*Sū.* 16: 5–8).

Ghṛta is suitable for improving intelligence, memory, comprehension and digestive efficiency. *Taila* is recommended for those suffering from tumors, sinuses and ulcers, *vāta* diseases and those who desire to have light and sturdy body. Individuals who experience constipation can also benefit from its use. Physically exhausted individuals and those who suffer from excessive appetite can consume *majja* and *vasa*. *Vasa* also pacifies pain in joints, bones, vital points and abdomen. *Taila* is to be administered in rainy season, *ghṛta* in autumn and *vasa* and *majja* in spring. Oleation is to be carried out at night during summer. In *pitta* diseases and *pitta*-dominant, *pittavāta* diseases also oleation is advised at night. *Snēha* is not to be administered at night in *kapha* diseases, *kapha*-dominant conditions and in winter, to prevent the appearance of *vāta-kapha* diseases. *Pitta* diseases will appear if *snēhana* is carried out during day time, in *pitta*-vitiated conditions and in summer. The four types of fats are to be ingested as such. Nevertheless, patients may find it difficult to consume the fat alone. In such cases, the physician can administer *snēha* through food, *vasti*, *nasya*, *abhyaṅga* (application all over body), *gaṇḍūṣa* (holding medicinal fluid in mouth), *mūrdhatarpaṇa* (*śirōvasti*), *karṇatarpaṇa* (pouring of oil into auditory canal) and *akṣitarpaṇa* (keeping *taila* or *ghṛta* over eyes). Consuming *snēha* alone (*acchapēya*) is considered to be the best mode (*Sū.* 16: 8–17).

Three doses (*mātra*) are specified for *snēhana*. They are known as *hrasva mātra* (low dose), *madhyama mātra* (medium dose) and *uttama mātra* (ideal dose). The volume of *snēha* that can be digested by a person in six hours is the low dose for him. *Snēha* in ideal dose is administered as *acchapēya* in the morning for cleansing the body. *Snēha* in medium dose is administered in the morning, when the patient feels hungry. This mode is for pacifying *dōṣas*. *Snēha* in low dose is administered along with meat soup, *madya* and food. This mode is invigorating and is ideal for children, the aged, those who have excessive thirst, who hate fatty substances, consume *madya* regularly, indulge in sexual intercourse daily and use oily substances every day. Pleasure-seeking persons and those suffering from dull appetite are also benefited by this approach. This mode is suitable for summer season as well. *Snēha* can be consumed in three ways. The first one is to consume *snēha* before food. This is called *prāgbhakta*. The second way is to have some food, consume *snēha* and then continue eating (*madhyabhakta*). The third one is known as *uttarabhakta* or consuming *snēha* after food. Consuming *snēha* in low dose in *prāgbhakta*, *madhyabhakta* and *uttarabhakta* modes cures, respectively, diseases of lower, middle and upper parts of body. These three body segments are strengthened also by this measure (*Sū.* 16: 17–22).

After consuming *snēha* as *acchapēya*, the patient needs to drink warm water. Warm water can be drunk again to confirm whether the *snēha* has been digested, in which case eructation will be odorless. After that he may consume warm, semi-solid, oil-free food in small measure. Warm water is to be used for bath. He should sleep well at night. Exercise, anger, grief, walking, traveling, sitting at one place for long time, siesta, getting exposed to smoke and dust and using too big or very small-sized pillows are to be avoided during all the days of oleation and those many days after the end of the therapeutic measure. Under the influence of *pitta*, the alimentary canal of individuals turns into *mṛdukōṣṭha* (soft alimentary tract). *Krūrakōṣṭha* (hard alimentary tract) results from *vāta* vitiation. Patients with *mṛdukōṣṭha* are to consume *snēha* as *acchapēya* for three days. Those with *krūrakōṣṭha* should consume it for seven days. If optimum unctuousness is not attained, *snēhapāna* is to be continued till that condition appears. Normal movement of *vāta* (*vātānulōmata*), satisfactory digestive efficiency, soft stool, aversion toward *snēha* and tiredness are the signs of attaining optimum unctuousness. Excessive *snēhana* causes pallor of skin and exudation of fluid from nose, mouth and anus (*Sū.* 16: 23–31).

Consumption of *snēha* unsuitable for the disease state, incorrect dosage, wrong time of administration and incompatible food or behavior causes edema, hemorrhoids, drowsiness, cessation of body movements, erratic sensory perception, itching, skin disease, fever, exudation of fluid, pricking pain, bloating of abdomen and giddiness. These undesirable effects are known as *snēha vyāpat* (complications arising from improper *snēhana*). AH *recommends* several measures to overcome these situations. Tolerating hunger and thirst; vomiting; sudation; consumption of oil-free food, beverages and medicines; *Takrāriṣṭa* (buttermilk fermented with herbs, barley, little millet – *Panicum miliare*); Kodo millet (*Paspalum scrobiculatum*); *Pippali* (*Piper longum*); *Triphala*; honey; *Pathya* (*Terminalia chebula*); cow urine; guggul gum; and medicines prescribed for these disorders are the countermeasures (*Sū.* 16: 32–35).

On achieving optimum unctuousness, the patient should spend three days undergoing sudation and consuming unctuous, liquid and hot food, along with meat soup. He can be subjected to *virēcana* on the fourth day. Patient who is to undergo *vamana* should spend one day undergoing sudation and above-mentioned food, after achieving optimum unctuousness. On the next day, *kapha* in the body should be liquefied and brought to the alimentary canal using appropriate medicines and food. He should be subjected to *vamana* on the third day (*Sū*. 16: 36–37).

Children and aged persons unable to observe dietary and behavioral restrictions, on account of inability to comply with, can be administered seven food that do not cause aversion and produce immediate optimum unctuousness:

i. Soup of meat of rooster or boar that is rich in body fat
ii. Warm gruel mixed with *snēha*
iii. Powder of sesame seeds mixed with *snēha* and thickened sugarcane syrup (*phāṇita*)
iv. Rice cooked with sesame seeds and blended with *snēha* and *phāṇita*
v. Warm gruel cooked with rice and milk, containing copious amount of *snēha*
vi. Cream with jaggery
vii. Gruel named *Pañcaprasṛta* prepared with one *prasṛta* each of rice, clarified butter, *majja*, *vasa* and *taila* (one *prasṛta* = 96 g).

Jaggery, *ānūpa* meat (meat of animals living in marshy land), milk, sesame seeds, black gram, *Sura* wine and curd are not to be included in the preparation of *snēha* to be administered to patients of skin diseases, edema and polyuric diseases. *Triphala*, *Piper longum*, *Terminalia chebula* and guggul gum can be suitably formulated with *snēha* that do not evoke reactions for use in the treatment of these diseases. He who uses *snēha* judiciously, lives for hundred years with sound appetite, clean bowel, healthy tissue elements (*dhātu*), sound sense organs, stamina and complexion (*Sū*. 16: 39–46).

3.13.1.4 Sudation

After oleation, the patient is to be subjected to sudation (*svēdana*), before undergoing elimination therapy. There are four kinds of sudation: *tāpasvēda*, *upanāhasvēda*, *ūṣmasvēda* and *dravasvēda*. Sudation with the help of fire, heated clothes, iron plates and palm of hand is known as *tāpasvēda*. *Upanāhasvēda* is carried out by covering the affected part with a paste of medicinal substances and bandaging with *snigdha*, *uṣṇa vīrya*, soft and odorless leather straps, and in their non-availability, with *vāta*-pacifying leaves or silk. The bandage is tied at night and removed in the morning. Alternately, the bandage can be tied during day and removed at night. In *vāta* disease without the association of *pitta* or *kapha*, the paste is prepared with Salt, *snēha*, vinegar, buttermilk, milk, *Acorus calamus*, sediment from the bottom of the tank in which *Sura* wine was brewed, *Anethum graveolens*, *Cedrus deodara*, all kinds of grains and odoriferous substances, *Alpinia galanga*, *Ricinus communis* and meat. In *kapha-vāta* diseases, the paste is prepared with herbs of *Surasādi gaṇa*. However, in *pitta-vāta* vitiation, *upanāha* is to be carried out frequently with herbs of *Padmakādi gaṇa* and a formulation known as *Sālvaṇa*, which is a mixture of all *vāta*-pacifying herbs, *snēha*, acidic substances, *ānūpa* fish, meat and salt applied with tolerable warmth (*Sū*. 17: 1–5).

Ūṣmasvēda is performed with *Utkārika* (balls of rice paste cooked in boiling water), clay bricks, broken pieces of clay pots, stones, powders, chopped leaves of herbs, grains, dried cow dung powder, sand and rice husk. Boluses of these substances are warmed and pressed on the body. *Dravasvēda* is of two types: *pariṣēka* and *avagāha*. In *pariṣēkasvēda*, chopped leaves of *Moringa oleifera*, *Crataeva religiosa*, *Ocimum sanctum* and so on, ingredients of *upanāhasvēda*, *ānūpa* fish and meat, *Daśamūla* herbs, *Sura* wine, vinegar, water, milk and *snēha* compatible to the deranged *dōṣa* in question are boiled. The fluid is poured in a thin stream, with tolerable warmth, on the part of body, after covering it with a cloth. The fluid obtained from boiling the above mixture is poured into a large, cylindrical metal tank. The patient subjected to *abhyaṅga* is advised to sit in the tank immersing his body in the fluid. This type of sudation is called *avagāhasvēda*. It is beneficial in *vāta* vitiation all over body (*sarvāṅga vāta*), hemorrhoids and dysuria. *Avagāhasvēda* pacifies pain (*Sū*. 17: 6–11).

Based on the principles of Ayurveda, certain treatment measures were developed in the province of Kerala in medieval times. One of them is known as *elakizhi* or sudation with boluses of herbs. Grated coconut, chopped leaves of fresh herbs such as *Ocimum sanctum*, *Moringa oleifera*, *Calotropis procera* and so on are sautéed in medicinal oil and placed on pieces of cloth having an area of 50 square inches. The four ends of the cloth are bundled together and tied firmly in the form of a bolus. This bolus is dipped in warm medicinal oil and placed over parts of body. Slight massaging is also done with these boluses. The warmth of the oil causes sudation. This is a form of *piṇḍasvēda* (Figure 3.2).

A variant of *dravasvēda* or *pariṣēka* was also developed in Kerala. This is the pouring of warm medicinal oil on the whole or parts of body. After performing certain sacred rites, the patient lies on a wooden basin or (*tailadrōṇi*). Two or four masseurs sit, on either side of the *tailadrōṇi*. They dip pieces of clean cloth in the warm oil and squeeze the cloth over the body with their hands. The warmed medicinal oil may be poured, using small pots known as *kiṇḍi* or the cut halves of coconut shells. The oil is poured at a medium speed and from a moderate height. The oil that flows down is collected, warmed and again poured. This measure is known as *piẓiccil* (Mooss, 1983a) (Figures 3.3 and 3.4).

Patient is subjected to sudation after making him unctuous by feeding *snēha* and applying *taila* all over body. On complete digestion of the ingested food, the patient is seated in a room without airflow and subjected to sudation according to any one of the modes, such as *madhya*, *vara* and *avara*. Making the patient sweat profusely is known as *vara*. Sweating slightly is *avara* and intermediate mode is called *madhya*. The right mode is selected on the basis of the patient's stamina, season and type of land where the treatment is carried out. In *kapha* diseases, the patient is subjected to *rūkṣasvēda* (*tāpasvēda*) without administration of *snēha*. *Rūkṣa-snigdha svēda* (*ūṣmasvēda*) is recommended in *kapha-vāta* vitiation. When *vāta* is enraged in stomach, *rūkṣasvēda* is carried out first, followed by *snigdha svēda*. Similarly, when *kapha* is vitiated in small intestine, *snigdha svēda* is performed first, followed by *rūkṣasvēda*. Sudation is made *snigdha* or *rūkṣa* by including such substances in the medicinal mix. Eyes, scrota and precordial area are not to be subjected to sudation. Signs of optimum sudation are reduction in feeling cold and pain and softness of limbs. Thereafter, the body is wiped dry and the patient is allowed to have bath. After sudation, patient needs to observe dietary and behavioral restrictions (*Sū*. 17: 12–15). Enragement of *pitta* and

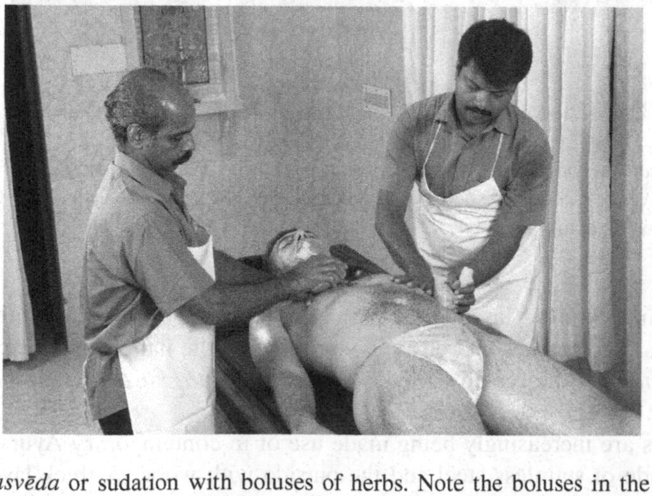

FIGURE 3.2 *Piṇḍasvēda* or sudation with boluses of herbs. Note the boluses in the hands of masseurs. (Photo courtesy: Nagarjuna Ayurvedic Centre Ltd, Kerala) (www.nagarjunaayurveda.com).

FIGURE 3.3 Wooden basin (*tailadrōṇi*) on which patient lies during the performance of various therapeutic measures. (Photo courtesy: Divakara Concepts, Coimbatore.)

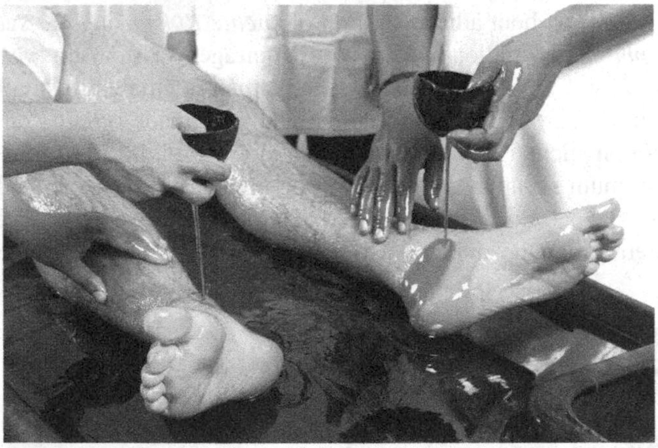

FIGURE 3.4 *Piẓiccil*, a variant of *dravasvēda* or *pariṣēka*. Warm medicinal oil being poured over body, using cut halves of coconut shells. (Photo courtesy: Nagarjuna Ayurvedic Centre Ltd, Kerala) (www.nagarjunaayurveda.com).

rakta, excessive thirst, fainting, lassitude of voice, tiredness, giddiness, discomforts in joints, fever, appearance of bluish-red patches on body and vomiting are the indicators of excessive sudation. In such cases, *stambhana* is the remedy. Medicines having *tikta*, *kaṣāya* and *madhura* tastes cause *stambhana* (*Sū.* 17: 16–17).

Steam chambers are increasingly being made use of in contemporary Ayurveda. The inside of the chamber is made of stainless steel and the outside with wood. A thick layer of glass wool is packed in between to prevent loss of heat. Steam from a boiler is passed into the chamber through

FIGURE 3.5 Steam chamber for carrying out whole body sudation. (Photo courtesy: Divakara Concepts, Coimbatore.)

pipes. After conditioning the body with oleation, the patient sits on a stool placed inside the chamber with his head projecting out. Sudation is started and after a specific period of time, the chamber is opened and the patient moves out (Figure 3.5).

Obese, weak persons, those suffering from diseases such as cataract, enlargement of abdomen, cellulitis, skin diseases, edema, rheumatoid arthritis, jaundice, morbid pallor and polyuric diseases are not to be subjected to sudation. So are individuals with enragement of *pitta*; pregnant, post-partum and menstruating women. Appropriate categories of sudation using suitable ingredients are to be carried out in respiratory distress, cough, catarrh, hiccup, bloating of abdomen, constipation, hoarseness of voice, facial palsy, *vāta* diseases and so on. Sudation devoid of heat is essential when *vāta* is enveloped (*āvaraṇa*) by *mēdas* and *kapha*. Ten means such as fully closed room, exercise, thick blankets, fear, *upanāhasvēda*, wrestling, anger, excessive consumption of *madya*, hunger and exposure to hot sun induce sweating without the involvement of fire. *Dōṣas* residing in alimentary canal, tissue elements, body channels, limbs and bone gets loosened, liquefied and reach alimentary canal under the influence of sudation. Properly performed *śōdhana* measures eliminate them effectively (*Sū.* 17: 21–29).

3.13.1.5 Emesis

Patients suffering from *kapha* diseases, *kapha-vāta* vitiation and *kapha-pitta* derangement are to be subjected to *vamana* (emesis). Especially, patients suffering from acute fever, dysentery, hemorrhages of obscure origin appearing in lower part of body, kingly consumption, leprosy and other skin diseases, polyuric diseases, cervical glandular swellings, glandular swellings, elephantiasis, lunacy, cough, respiratory distress, nausea, cellulitis, vitiation of breast milk and diseases appearing in the thoracic area and above are ideal subjects for performing emesis (*Sū.* 18: 1–3). Table 3.8 lists the diseases and conditions in which *vamana* is indicated.

Emesis is contra-indicated in many conditions and diseases. They include pregnant women, children, the aged, lean or obese individuals, those who vomit frequently, patients suffering from dysuria,

TABLE 3.8
Diseases and Conditions in Which *vamana* Is Indicated

Sl. No.	Diseases and Conditions: Sanskrit Term	Diseases and Conditions: English Equivalent	Reference
1	Agnimāndya	Dull abdominal fire	*Ka*. 1: 14
2	Akālanidra	Discomforts arising from sleeping at odd times	*Sū*. 7: 62
3	Ākhuviṣa	Rat bite	*Ut*. 38: 22–23
4	Āmaśayagata pakvaśayagata saviṣānna	Poisoned food in stomach and intestine	*Sū*. 7: 25
5	Antarvidradhi	Internal abscess	*Ka*. 1: 9
6	Apaci	Cervical glandular swelling	*Ka*. 1: 11, *Ut*. 30: 13
7	Apasmāra	Epilepsy	Ut. 7: 17
8	Arbuda	Tumor	*Ka*. 1: 11
9	Arōcaka	Distaste for food	*Ci*. 5: 50–53, *Ka*. 1: 11, 22, 32
10	Arūmṣika	Pityriasis	*Ut*. 24: 25
11	Bādhiryam	Deafness	*Ut*. 18: 22
12	Chardi	Vomiting	*Ka*. 1: 12–13, 27
13	Dūṣīviṣa	Poison from inanimate, animate sources or poison retained in body after partial digestion	*Ut*. 35: 38
14	Gaḷagaṇḍa	Swelling in neck	*Ka*. 1: 31, *Ut*. 22: 71
15	Gaḷarōga	Diseases of throat	*Ka*. 1: 32
16	Gara/Viṣa	Poisoning with intent to influence an individual subliminally or to cause death	*Ka*. 1: 27, 30, 31, 35–36, 39, 41
17	Granthi	Glandular swelling	*Ka*. 1: 11, *Ut*. 30: 2, *Ut*. 30: 2
18	Gulma	Abdominal mass characterized by tumor-like hard, round mass, unstable in size and consistency, moving or immobile, situated in the bowel	*Ka*. 1: 9, 31, 33, 35, 41, *Ci*. 14: 27–28, *Ci*. 14: 76
19	Hikka	Hiccup	*Ci*. 4: 3–4, *Ka*. 1: 19–20
20	Hṛddāha	Burning sensation in precordial area	*Ka*. 1: 12–13, 38
21	Hṛdrōga	Heart disease	*Ci*. 6: 49–50, *Ka*. 1: 44
22	Jālakagardabha	Lymphangitis	*Ut*. 32: 6
23	Jvara	Fever	*Ci*. 1: 6–8, *Ka*. 1: 9–10, 11, 19, 25, 26, 27, 28, 32, 33, 44
24	Kapharōga	All diseases caused by enragement of *kapha*	*Ka*. 1: 35–36
25	Kaphōtkḷiṣṭa	Disease of eyelid characterized by stickiness of lid margin and heaviness of lid	*Ut*. 9: 23
26	Karṇanādam	Tinnitus	*Ut*. 18: 22
27	Kāsa	Cough	*Ci*. 3: 25, *Ci*. 3: 42, *Ka*. 1: 19–20, 22, 27, 29, 35, 38
28	Kṣīrālasaka	Post-partum mother and infant suffering from vitiation of breast milk	*Ut*. 2: 23
29	Kukūṇakam	Neonatal conjunctivitis	*Ut*. 9: 24–29
30	Kuṣṭha	Leprosy and other skin diseases	*Ci*. 19: 21, *Ka*. 1: 30, 41, 43, 44
31	Madātyaya	Alcoholism	*Ci*. 7: 22–23, 33, 107–108
32	Mahōdara/udara	Enlargement of abdomen	*Ka*. 1: 11
33	Mukhadūṣika	Pimples	*Ut*. 32: 4
34	Nāsārōga	Diseases of nose	*Ut*. 20: 1

(*Continued*)

TABLE 3.8 (*Continued*)
Diseases and Conditions in Which *vamana* Is Indicated

Sl. No.	Diseases and Conditions: Sanskrit Term	Diseases and Conditions: English Equivalent	Reference
35	Pakṣmarōdha	Trichiasis	*Ut.* 9: 34
36	Pakṣmaśatana	Madarosis	*Ut.* 9: 19
37	Pāṇḍu	Morbid pallor	*Ci.* 16: 5, *Ka.* 1: 22, 30, 41
38	Plīha	Enlargement of spleen	*Ka.* 1: 41
39	Pratiśyāya	Common cold	*Ka.* 1: 9
40	Prasēka	Exudation of mucus from mouth	*Ka.* 1: 12–13
41	Pūtivadana	Bad breath	*Ut.* 22: 79
42	Rājayakṣma	Kingly consumption	*Ci.* 5: 1–2, *Ka.* 1: 20–22
43	Śvāsa	Respiratory distress	*Ci.* 4: 3–4, *Ka.* 1: 12–13, 19–20, 27, 29
44	Ślīpada	Elephantiasis	*Ka.* 1: 31
45	Śōpha	Edema	*Ka.* 1: 41
46	Śōṣa	Leaning	*Ka.* 1: 14
47	Sthaulya	Obesity	*Ka.* 1: 16–17
48	Tṛṣṇa	Polydipsia	*Ci.* 6: 72, 80–81
49	Unmāda	Insanity	*Ka.* 1: 39–40, *Ut.* 6: 19
50	Valmīka	Actinomycosis	*Ut.* 32: 9
51	Vasantacarya	Part of regimen for spring season	*Sū.* 3: 19
52	Vātapittarōga	*Vāta-pitta* afflictions	*Ka.* 1: 23–25
53	Vātasōṇita	*Vāta*-induced blood disorders	*Ci.* 22: 15, *Ka.* 1: 44
54	Vātavyādhi	*Vāta* diseases	*Ci.* 21: 14
55	Vidagdhājirṇa	*Pitta*-dominant indigestion	*Sū.* 8: 27
56	Viruddhādhyaśana	Consumption of incompatible food and overeating	*Sū.* 8: 15
57	Visarpa	Cellulitis	*Ci.* 18: 1, *Ka.* 1: 44
58	Vraṇa	Wound, sore, ulcer	*Ut.* 25: 23–24
59	Yōnirōga	Gynecological diseases	*Ut.* 34: 27

enlargement of abdomen, *gulma* and so on. Elimination measures such as *vamana*, *virēcana*, *vasti*, *nasya* and *dhūmapāna* are strictly forbidden to pregnant women, hungry and melancholic individuals, children, the aged, lean, weak or obese persons, and those suffering from heart disease, injury to thorax and *āmajvara* (fever arising from impaired digestion) (*Sū.* 18: 3–7).

Days in *Śrāvaṇa* (July to third week of August), *Kārtika* (overlapping October and November) and *Caitra* (March or April) months are usually selected for performing emesis. Patient is subjected to oleation and sudation. On the next day, he consumes at dinner time food prepared with fish, black gram and sesame seeds. These foods loosen, liquefy the *kapha* in body and bring it to the alimentary canal. The patient sleeps soundly during night. The next day morning the physician confirms whether the dinner of previous night has been digested fully. Thereafter, the patient is administered gruel (*pēya*) mixed with clarified butter, so as to induce unctuousness. This step is optional and depends on the physical condition of the patient. Thereafter, the emetic medicine, potentiated by chanting a *mantra* (word of power), is mixed with honey, rock salt and administered to the patient, who sits on a knee-high stool, facing the east. Following this, he drinks *madya*, sugarcane juice or meat soup in overabundance and rests for nearly one *muhūrta* (forty-eight minutes). As nausea is felt and salivation starts, vomiting is induced by putting into his throat, his two fingers or a soft plant stem, without causing pain. As the vomiting reflex sets in, the patient vomits a large volume of vomitus. The physician confirms that all the *dōṣa* have been eliminated, by stimulating fauces and inducing vomiting reflex again. As the patient vomits, his flanks and forehead are to be supported by two assistants. Similarly, the navel area and back of body are to be massaged from top to downward, against the direction of vomiting. In *kapha* disease, the emetic medicine needs to be of *tīkṣṇa*, *uṣṇa* and *kaṭu* nature. Emesis is to be induced in *pitta*-dominant conditions, with *madhura* and *śīta vīrya* herbs. In *vāta-kapha*-dominant situation, *snigdha*, *amḷa* and *lavaṇa* medicines are to be administered. After all the *kapha dōṣa* has been eliminated, bile will make its appearance and elimination of *kapha* can be inferred to be completed (*Sū*. 18: 12–22).

Sometimes the vomiting reflex may be slow. In that situation, the patient can be persuaded to vomit several times by drinking powder of *Piper longum*, *Emblica officinalis*, *Brassica juncea* and rock salt suspended in warm water. Absence of vomiting reflex, obstructed vomiting and vomiting only the emetic medicine are signs of improper *vamana* (*ayōga*). The patient keeps on spitting and feels itching. Red patches may appear on body along with fever. In optimum emesis (*samyagyōga* of *vamana*), *kapha*, *pitta* and *vāta* move in the normal direction, evidenced by belching. *Atiyōga* of *vamana* (excessive emesis) causes vomiting of froth and blood, and the vomitus may be multi-colored like peacock feather. Burning sensation, drying of throat, blackout, giddiness and signs of intense *vāta* enragement such as various kinds of pain, dyspnoea, hiccup and muscle twitches may ensue. Death of the patient may also happen. On completion of satisfactory *vamana*, the patient inhales medicinal smoke (*dhūmapāna*) in the appropriate mode. Following that dietary and behavioral restrictions recommended in *snēhana* are to be observed. On the next day or the morning after, he takes bath in warm water and consumes cooked *Śāli* rice. To stimulate digestive efficiency, the patient is advised to follow *pēyādi krama* or consuming first *maṇḍa* (water decanted from cooked rice), followed by *pēya* (gruel with more *maṇḍa* and less rice grains) and *vilēpi* (gruel having good amount of cooked rice) (*Sū*. 18: 23–29).

The topic of *vamana* is discussed in *Kalpasthānam* as well. The chapter entitled *Vamanakalpādhyāyaḥ* describes the procedures for preparing emetic medicines and medicinal formulations to be used in carrying out this measure. Several medicated soups, *mantha* (a liquid obtained by churning any grain or millet after adding 14 times of water), *ghṛta*, medicinal milk and curd are recommended to initiate *vamana* comfortably. There is also mention of some interesting formulae that are handy in *vamana* therapy. For example, emesis can be induced by smelling a floral garland dipped in juices of flower and fruit of *Lagenaria siceraria* (*Ka*. 1: 34). The diseases that can be cured with *vamana* include *kapha jvara*, *kapha-vāta jvara*, *kaphaja chardi*, *pitta-kapha jvara*, *pitta jvara*, *kaphaja kāsa*, *kaphaja śvāsa*, common cold, *gulma*, internal abscess, distaste for food, glandular swelling, cervical glandular swelling, tumor, enlargement of abdomen, burning sensation in precordial area, heart disease, *tamaka śvāsa*, lowering of abdominal fire, morbid pallor, leprosy and other skin diseases, swelling in neck, elephantiasis, lunacy, cellulitis and *vāta*-induced blood disorders (*Ka*. 1: 1–47). Management of mishaps encountered in *vamana* is described in chapter entitled *Vamanavirēcanavyāpatsiddhir* adhyāyaḥ (*Ka*. 3: 1–39).

3.13.1.6 Purgation

Virēcana (purgation) is to be carried out in diseases having predominance of *pitta*, *pitta-vāta* and *pitta-kapha* (*Sū*. 18: 1–3). *Virēcana* cures *gulma*, hemorrhoids, chronic fever, abscesses, *vāta*-induced blood disorders and many others. However, patients suffering from acute fever, indigestion, dysentery and injury to anus are to be excluded from performing *virēcana* (*Sū*. 18: 8–11). Patient subjected to *vamana* is made unctuous again by internal and topical administration of *taila* and or *ghṛta*. Thereafter purgation is induced, after monitoring the condition of the alimentary canal. Patient with *mṛdukōṣṭha* and *pitta* enragement is administered milk to cause purgation. Powder of *tīkṣṇa* herb such as *Operculina turpethum* is administered in *krūrakōṣṭha*, due to *vāta* vitiation. Purgation is induced in *pitta* enragement using medicines having *kaṣāya* and *madhura* tastes. In predominance of *kapha*, medicine with *kaṭu* taste is employed. Medicines with *snigdha*, *uṣṇa* qualities and *lavaṇa* taste are handy to cause purgation in *vāta*-predominant diseases. Warm water is to be drunk if purgation is slow. If purgation does not take place properly, *virēcana* medicine can be administered next day also. If it is confirmed that it is because of the non-unctuous nature of the alimentary canal, oleation and sudation are to be carried out as before. After ten days, the patient is to consume a well-formulated medicine to cause purgation (*Sū.* 18: 33–38).

Feeling of heaviness in abdomen and precordial area, itching, papules, running nose and constipation are the signs of improper *virēcana*. Voiding of white, black, blood-stained and foul-smelling fluid may happen after the passing of normal stool. This is a sign of excessive purgation. Rectal prolapse, thirst, giddiness, sunken eyes and signs of excessive *vamana* may also be evident. On completion of *virēcana* in normal manner, the patient is to adopt *pēyādi krama*. He can consume normal food on regaining digestive efficiency (*Sū.* 18: 38–43).

Several decoctions, powders, electuaries, pills, fermented liquids and *ghṛta* that are useful in *virēcana* are described in *Kalpasthānam* (*Ka.* 2: 1–62). Management of emergencies met with in purgation therapy is detailed in the chapter entitled *Vamanavirēcanavyāpatsiddhir* adhyāyaḥ (*Ka.* 3: 1–39). Table 3.9 lists the diseases and conditions in which *virēcana* is indicated.

Oleation and sudation are to be performed between emesis and purgation. After performing elimination therapy, oleation is to be applied again. Elimination of *dōṣas* after applying oleation and sudation measures, cleanses the body excellently. Carrying out elimination measures without oleation and sudation damages health. Properly performed elimination therapy improves intelligence, memory and comprehension. *Dhātu*s achieve stability, digestive efficiency is enhanced, and the subject leads a heathy, long life (*Sū.* 18: 57–60).

3.13.1.7 Medicated Enema

Medicated enema (*vasti*) is suitable in the treatment of *vāta*-dominant diseases. It is also the best among all treatment modes. *Vasti* is of three kinds: *nirūhavasti* (*āsthāpanavasti*, *kaṣāyavasti*), *anuvāsanavasti* (*snēhavasti*) and *uttaravasti*. It is used in the treatment of *gulma*, distension of abdomen, *vāta*-induced blood disorders, enlargement of spleen, dysentery un-associated with other diseases, pricking pain in stomach, chronic fever, common cold, urinary stone and many *vāta* diseases. Patients with injury to thorax, *āmātisāra* (dysentery caused by vitiated mucus in abdomen and characterized by hard and fetid stool), vomiting, respiratory distress, cough, pregnant women and so on are not eligible for *nirūhavasti*. Patients suffering from *gulma*, distension of abdomen, excessive appetite and *vāta* diseases are to be treated with *anuvāsanavasti*. *Anuvāsanavasti* is contra-indicated in morbid pallor, jaundice, polyuric diseases, catarrh and several others. A unique *vasti* is described in the treatment of *atisāra* (dysentery). Certain herbs are processed according to the protocol specified for *puṭapāka* (see 3.13.3.4). The juice obtained from the *puṭapāka* is administered as *vasti*, and this measure is called *picchāvasti* (mucilaginous enema) (*Ci.* 9: 72–75; *Ci.* 9: 95–96; *Ci.* 9: 118). It is applied in the treatment of dysentery, prolapse of rectum, bloody diarrhea and fever (*Ci.* 8: 125–128). The equipment for administering the enema (*vastiyantra*) is fabricated with urinary bladder of goat, sheep or buffalo (*Sū.* 19: 1–17).

Patient eligible for *nirūhavasti* is first of all subjected to oleation and sudation. After improving stamina, he is at first subjected to *snēhavasti*, which is performed during day time in pre-winter, winter and spring seasons. Night time is ideal for *snēhavasti* in summer, monsoon and autumn seasons. After applying *taila* all over body, the patient is given a bath. Thereafter, he consumes a small volume of thin rice gruel, followed by post-prandial drink. Bowels are cleared if necessary, and the patient lies on a cot without using pillows. He lies turned leftward, folding the right leg. His rectum is smeared with *taila* and tube of the *snēha*-filled *vastiyantra* is slowly introduced inside. Thereafter, the bladder of the *vastiyantra* is pressed gradually so as to administer the *snēha* into the rectum. Care is taken to ensure that the air left in the bladder does not get trapped inside (Figures 3.6 and 3.7). After introduction of *snēha*, the patient lies in supine position. He is advised to hit the buttocks with clenched fist and heels. This may also be performed by assistants. The bottom side of the cot is then raised and lowered three times. After that the patient lies with his hands

TABLE 3.9
Diseases and Conditions in Which *virēcana* Is Indicated

	Diseases and Conditions		
Sl. No.	Sanskrit Term	English Equivalent	Reference
1	Agnimāndya	Lowering of abdominal fire	*Ka*. 2: 22–23
2	Ākhuviṣa	Rat bite	*Ut*. 38: 23
3	Alaji	Hard or closed pustule	*Ka*. 2: 55–56
4	Alarkaviṣa	Bite of rabid dog	*Ut*. 38: 36
5	Āmaśayagata pakvaśayagata saviṣānna	Poisoned food in stomach and intestine	*Sū*. 7: 25
6	Antarvidradhi	Internal abscess	*Ci*. 13: 9
7	Apaci	Cervical glandular swelling	*Ut*. 30: 13
8	Apasmāra	Epilepsy	*Ut*. 7: 16–17
9	Arōcaka	Distaste for food	*Ka*. 2: 15–16
10	Arśas	Hemorrhoid	*Ci*. 8: 139, *Ka*. 2: 20, 55–56, 58–61
11	Aśmari	Urinary stone	*Ci*. 11: 41
12	Bhagandara	Anal fistula	*Ka*. 2: 20, 53–55
13	Bhrama	Giddiness	*Ka*. 2: 22
14	Chardi	Vomiting	*Ci*. 6: 2–3, *Ka*. 2: 22
15	Chardi vēgavidhāraṇāt	Disorders arising from restrainment of urge to vomit	*Sū*. 4: 18
16	Dhūmara	Smoky vision	*Ut*. 13: 91
17	Dūṣīviṣa	Poison from inanimate, animate sources or poison retained in body after partial digestion	*Ut*. 35: 38, *Ka*. 2: 43–44
18	Gara	Poisoning with intent to influence an individual subliminally or to cause death	*Ka*. 2: 22–23, 43, 49–50
19	Grahaṇi	Bowel disease characterized by voiding of loose and hard stool	*Ka*. 2: 20, 58–61
20	Granthi	Glandular swelling	*Ut*. 30: 2, 3
21	Gulma	Abdominal mass characterized by tumor-like hard, round mass, unstable in size and consistency, moving or immobile, situated in the bowel	*Ci*. 14: 43–44, *Ci*. 14: 88, 119–120, *Ka*. 2: 15–16, 20, 43, 50–51, 53–54, 55–57
22	Halīmaka	Subtype of *pāṇḍu* characterized by green, dark brown or yellow color; giddiness, thirst, fever, loss of stamina and lowering of abdominal fire	*Ka*. 2: 15–16
23	Hikka	Hiccup	*Ci*. 4: 6–7
24	Hṛdrōga	Heart disease	*Ci*. 6: 44, *Ka*. 2: 30, 50–51
25	Jālakagardabha	Lymphangitis	*Ut*. 32: 6
26	Jvara	Fever	*Ci*. 1: 98–100, *Ka*. 2: 11, 22, 30
27	Kakṣadāha	Burning sensation in arm pits	*Ka*. 2: 55–56
28	Kāmala	Jaundice	*Ka*. 2: 20
29	Kapharōga	*Kapha* diseases	*Ka*. 2: 7–9
30	Kāsa	Cough	*Ci*. 3: 3, 27, 42, 151–152, *Ka*. 2: 15–16, 22
31	Kōṭha	Erythema	*Ka*. 2: 58–61
32	Kṛmi	Worm infestation	*Ci*. 20: 21, *Ka*. 2: 53–55
33	Kuṣṭha	Leprosy and other skin diseases	*Ci*. 19: 20, *Ka*. 2: 20, 58–61

(*Continued*)

TABLE 3.9 (*Continued*)
Diseases and Conditions in Which *virēcana* Is Indicated

Sl. No.	Diseases and Conditions: Sanskrit Term	Diseases and Conditions: English Equivalent	Reference
34	Mada	Stupor	*Ci.* 7: 107–108
35	Mahōdara/udara	Enlargement of abdomen	*Ci.* 15: 1, *Ka.* 2: 20, 43, 49–50, 53–54
36	Madhumēha	Diabetes mellitus	*Ka.* 2: 43, *Ka.* 2: 52–54, 55–56
37	Mūtrakṛchra	Dysuria	*Ka.* 2: 22
38	Pakṣmarōdha	Trichiasis	*Ut.* 9: 34
39	Pāṇḍu	Morbid pallor	*Ci.* 16: 5, *Ka.* 2: 20, 23, 43–44, 53–55, 58–61
40	Pittarōga	*Pitta* diseases	Ka. 2: 7–9
41	Pḷīhōdara	Enlargement of spleen	Ka. 2: 15–16
42	Rājayakṣma	Kingly consumption	*Ci.* 5: 1–3, *Ka.* 2: 22
43	Śaradcarya	Part of regimen for autumn season	*Sū.* 3: 50
44	Śōpha	Edema	*Ka.* 2: 43, 49–50
45	Śōṣa	Leaning	*Ka.* 2: 22
46	Śukḷarōga	Diseases of sclera	*Ut.* 11: 29, 42
47	Śukḷasaṅga	Obstruction of seminal fluid	*Ka.* 2: 55–57
48	Śvāsa	Respiratory distress	*Ci.* 4: 6–7
49	Śvitra	Vitiligo	*Ci.* 20: 2–3
50	Timira	Cataract	*Ut.* 13: 47
51	Tṛṣṇa	Polydipsia	*Ci.* 10: 89, *Ka.* 2: 11
52	Udāvarta	Abdominal disease due to retention of feces	*Ka.*2: 30, 34
53	Unmāda	Insanity	*Ut.* 6: 19, *Ka.* 2: 43–44
54	Ūrdhvaga raktapitta	Hemorrhages of obscure origin directed upward	*Ci.* 2: 5
55	Valmīka	Actinomycosis	*Ut.* 32: 9
56	Vatasoṇita	*Vata*-induced blood disorders	*Ci.* 22: 5, *Ka.* 2: 30
57	Vātātmaka stanya	Post-partum mother suffering from *vāta*-vitiation of breast milk	*Ut.* 2: 11
58	Vātavyādhi	*Vāta* diseases	*Ci.* 21: 10–11, *Ka.* 2: 7, 55–56
59	Vibandha	Constipation	*Ka.* 2: 55–56
60	Vidradhi	Abscess	*Ka.* 2: 55–56
61	Visarpa	Cellulitis	*Ci.* 18: 1, *Ka.* 2: 43, 55–56
62	Viṣūci	Cholera	*Sū.* 8: 17
63	Vraṇa	Wound, sore, ulcer	*Ut.* 25:23
64	Yōnirōga	Gynecological diseases	*Ut.* 34: 26–27

stretched away from body and resting is head on pillow. The physician hits the patient's heels with his fist. Pain felt can be assuaged by massaging with *taila*. These steps are intended to keep the *snēha* inside, without leaking out. If *snēha* flows out, fresh *snēha* in introduced again. The *vasti* is concluded when the *snēha* stays inside for stipulated period and is drained out. He is served light food at dinner time, if appetite is satisfactory. The *snēha* administered through *vasti* is to flow out within three *yāma* (nine hours). If it stays inside beyond this duration, it is to be left as such and *phalavarti* (suppository) or *vasti* with *tīkṣṇa* medicines is to be applied for its complete evacuation. Patient needs to fast for the whole day, and next day morning, he can drink lukewarm decoction of ginger and coriander (*Sū.* 19: 20–33).

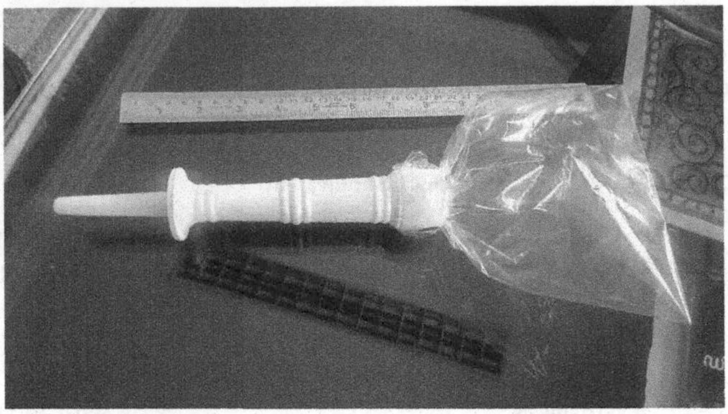

FIGURE 3.6 Equipment for performing *nirūhavasti* or medicated enema with decoctions of herbs. (Photo courtesy: Divakara Concepts, Coimbatore.)

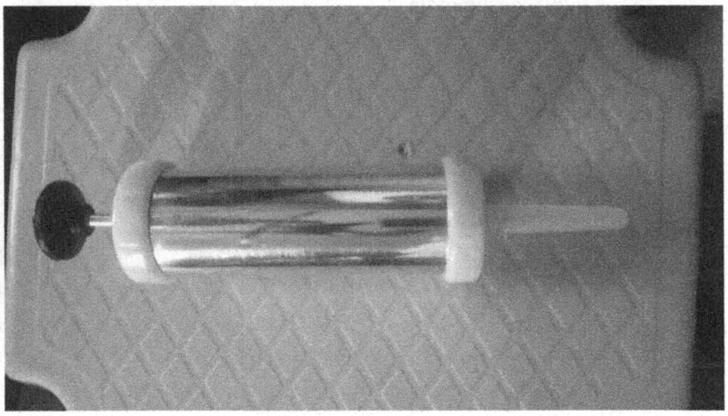

FIGURE 3.7 Equipment for performing *snēhavasti* or medicated enema with oleaginous substances. (Photo courtesy: Divakara Concepts, Coimbatore.)

Snēhavasti is performed again on the third or fifth day. *Vasti* is to be performed again if oleation is insufficient. Because of this reason, *snēhavasti* is to be performed every day in the case of patients with high vitiation of *vāta*, those who exercise regularly, those with good appetite and persons with *rūkṣa* body. By performing three or four *vasti* in this order, body attains good unctuousness. Thereafter, one *nirūhavasti* is to be performed for cleansing body channels. The decoction is prepared with herbs suitable for the disease and eight fruits of *Randia dumetorum*. Specified quantity of *snēha* is added to the decoction. The quantity of *snēha* varies in *vāta*-, *pitta*- and *kapha*-dominant conditions. Paste of herbs, jaggery, honey and rock salt are added to the mixture. The warmed blend of medicinal fluids is introduced into the rectum. This blend should have optimum degree of qualities such as warmth, unctuousness, sharpness and softness. On complete introduction of the medicinal fluid into rectum, patient lies in supine position, his head resting on pillow. He should void his bowels on feeling defecation reflex (*Sū.* 19: 33–47).

The optimum, improper and excessive modes of *nirūhavasti* are similar to those described for *virēcana*. In *kapha* diseases, one to three *snēhavasti* are to be performed. In *pitta* diseases, five to seven, and in *vāta* diseases, nine to eleven *snēhavasti* are to be performed. Food is served after performing *nirūhavasti*. Patients suffering from *kapha*-, *pitta*- and *vāta*-dominant diseases are to be served food along with *yūṣa*, milk and meat soup, respectively. In *vāta* diseases, the *vasti* fluid should consist of decoction of *vāta*-pacifying herbs, paste of *Operculina turpethum*, rock salt, meat soup and more of *snēha*. The *vasti* fluid in this case needs to have qualities such as *madhura*, *amḷa*, *snigdha* and *uṣṇa*. One *vasti* is to be performed in *vāta* diseases. In *pitta* diseases, the *vasti* fluid is to contain decoction of *Nyāgrōdhādi gaṇa*, paste of *Padmakādi gaṇa*, sugar, clarified butter, milk, sugarcane juice and honey. Two such *vasti* with *vasti* fluid having *madhura* taste and cold potency are to be administered. Three *vasti* with decoction of *Āragvadhādi gaṇa*, paste of *Vatsakādi gaṇa*, honey and cow urine are required in *kapha* diseases. The fluid should have *kaṭu* taste and qualities such as *rūkṣa*, *tīkṣṇa* and *uṣṇa*. Three *vasti* are required in *sannipātaja* diseases (*Sū*. 19: 50–60).

Snēhavasti and *nirūhavasti* are also performed in alternating sequence. First of all, a *snēhavasti* is carried out. Thereafter, 12 *snēhavasti* and 12 *nirūhavasti* are performed in alternating sequence such as *snēha-nirūha*, *snēha-nirūha* and so on. In the end, five *snēhavasti* are also applied. This combination of 30 *vasti* is known as *karmavasti*. *Kālavasti* is a combination of 15 *snēha-nirūha vasti*. One *snēhavasti* and three *snēhavasti* are performed in the beginning and end. Six *snēhavasti* and five *nirūhavasti* alternate in between. In *yōgavasti*, three *nirūhavasti* and three *snēhavasti* are in the middle, with one *snēhavasti* each in the beginning and end. *Snēhavasti* performed with a volume of *snēha* equivalent to the low dose of *snēhapāna* is known as *mātrāvasti*. This can be performed always and is suitable for children, the aged, persons habituated to fornication and exercise, those continuously exposed to mental stress, kings, princes and pleasure-seekers. *Mātrāvasti* can be performed without dietary and behavioral restrictions. It improves stamina and prevents constipation (*Sū*. 19: 63–69). Table 3.10 lists the diseases and conditions in which *vasti* is indicated.

Chapter entitled *Dōṣaharaṇasākalyavastikalpādhyāyaḥ* of *Kalpasthānam* describes 28 formulae of *vasti* fluids that successfully eliminate *dōṣas* (Table 3.11). A wise physician selects appropriate formula, modifies it slightly, if necessary, and uses it in *vasti*, after assessing various aspects of the disease, involvement of *dōṣas*, season of the year, constitution of the patient and his stamina. *Vasti* performed with such a medicinal formulation cures the disease successfully (*Ka*. 4: 1–73). Measures to pacify problems that arise due to improper performance of *vasti* are described in *Vastivyāpatsiddhir adhyāyaḥ* (*Ka*. 5: 1–54).

3.13.1.8 Urethral and Vaginal Enema

Urethral and vagina enema (*uttaravasti*) is performed to cure urological and gynecological diseases. It is carried out in men through the urethral route and in women through the vagina. The *vasti* is performed after cleansing the alimentary canal with two or three *nirūhavasti*. The tube of the *vastiyantra* is 12 *aṅgula* (finger-breadth) long, with reference to patient's finger and having limiting ridges at base and midportion. The tip has an aperture through which a mustard seed can pass. The *vastiyantra* is made up of metals such as gold or silver. The bladder is to be light and soft. The volume of the *vasti* fluid is one *śukti*, equivalent to 24 ml. After bath and having food, the patient is seated on a knee-high stool. A thin probe is introduced into the erect penis, and on confirming the absence of obstructions, the tube of the *vastiyantra* is introduced nearly six finger-breadth inside. *Vasti* fluid is slowly introduced by pressing the bladder. Rest of the procedure is similar as in the case of *snēhavasti*. Three or four *uttaravasti* are performed in this way (*Sū*. 19: 70–77).

Uttaravasti is carried out in women generally during the menstrual period. However, it can be performed any time in conditions such as prolapse of vagina, pricking pain in vagina, dysfunctional bleeding and other gynecological disorders. The patient lies in supine position, with both legs folded at the knees and spread wide apart. Three or four *vasti* are to be performed in a day and night. In this way, *vasti* is to be carried out for three days, slowly increasing the dose. After taking rest for three days, *vasti* is performed again for three days (*Sū*. 19: 77–82).

Patient subjected to *vamana* in the proper way can undergo *virēcana* after fifteen days. Fifteen days later, he can be subjected to *nirūhavasti*. He who has undergone *nirūhavasti* can be subjected to *snēhavasti* straight away. Anyone who has undergone *virēcana* can be subjected to *snēhavasti* after seven days. *Dōṣas* liquefied by oleation and sudation are eliminated from body by performing *vasti*. All diseases of limbs, body elements (*dhātu*), alimentary canal, vital points and the entire body arise due to vitiation of *vāta*. *Vāta* is the factor responsible for the abnormal and normal movements of *kapha*, *pitta* and *mala*. For pacifying enraged *vāta*, there is no better remedial measure

TABLE 3.10
Diseases and Conditions in Which *vasti* Is Indicated

Sl. No.	Sanskrit Term	English Equivalent	Reference
	Diseases and Conditions		
1	Abhiṣyanda	Conjunctivitis	*Ka*. 4: 23–24
2	Āḍhyavāta	Rheumatic palsy on the loins	*Ka*. 4: 62–66
3	Agnimāndya	Lowering of abdominal fire	*Ka*. 4: 17–19
4	Ānāha	Distension of abdomen	*Ka*. 4: 24–25, 37–43, 62–66
5	Annavidvēṣa	Aversion to food	*Ka*. 4: 17–19
6	Antarvidradhi	Internal abscess	*Ci*. 13: 9
7	Āntravṛddhi	Hernia	*Ka*. 4: 27–28
8	Aparāpātanam	Expulsion of placenta	*Śā*. 1: 88–90
9	Arśas	Hemorrhoid	*Ci*. 8: 88–89, 93–94, 100, 125–129, *Ka*. 4: 7–10, 27–28, 37–43, 62–66
10	Aśmari	Urinary stone	*Ci*. 11: 41, *Ka*. 4: 7–10, 37–43, 62–66
11	Aṣṭamamāsōpacāra	Pregnancy care in eighth month	*Śā*. 1: 64–65
12	Atisāra	Dysentery	*Ci*. 9: 50–52, 72–76, *Ka*. 4: 12–16, 37–43
13	Bhaṅga	Fracture	*Ka*.4: 49–52, *Ut*. 27: 33
14	Bhēṣajakṣaya	Mishaps in treatment	*Sū*. 4: 29
15	Dāha	Burning sensation	*Ka*. 4: 12–16
16	Ēkāṅgarōga	Hemiplegia	*Ci*. 3: 6–9
17	Garbhiṇyāḥ udāvarta	*Udāvarta* experienced by pregnant mother	*Śā*. 2: 21–22
18	Grahaṇi	Bowel disease characterized by voiding of loose and hard stool	*Ci*. 10: 23–25, *Ka*. 4: 7–10
19	Guhyaśūla	Pricking pain in genital area	*Ka*. 4: 7–10
20	Gulma	Abdominal mass characterized by tumor-like hard, round mass, unstable in size and consistency, moving or immobile, situated in the bowel	*Ci*. 14: 2–8, *Ka*. 4: 7–10, 12–16, 27–28, 37–43, 62–66
21	Hṛchūla	Pricking pain in precordial area	*Ka*. 4: 7–10
22	Hṛdrōga	Heart disease	*Ci*. 6: 27–28, 49, *Ka*. 4: 12–16
23	Jaṅghāśūla	Pricking pain in calf muscle	*Ka*. 4: 7–10, 29–30
24	Jvara	Fever	*Ci*. 1: 106, 116–117, 119–121, 121–122, 123–125, *Ci*. 8: 125–129, *Ka*. 4: 12–16, 37–43
25	Kāmala	Jaundice	*Ka*. 4: 12–16
26	Kaphavātaja rōga	*Kapha-vāta* diseases	*Ka*. 4: 7–10
27	Kaphavyādhi	*Kapha* diseases	*Ka*. 4:17–19, 34–36, 62–66, 66–67
28	Kāsa	Cough	*Ci*. 3: 2, *Ci*. 3: 6–9, *Ka*. 4: 37–43
29	Kaṭīruja	Pain in hip	*Ka*. 4: 37–43
30	Kōṣṭhaśūla	Pricking pain in stomach	*Ka*. 4: 7–10, 37–43
31	Kṛmi	Worm infestation	*Ci*. 20: 19–21, *Ka*. 4: 23–24, 27–28
32	Kṣīṇabala	Loss of stamina	*Ka*. 4: 49–52, 59–62
33	Kuṣṭha	Leprosy and other skin diseases	*Ka*. 4: 23–24
34	Mada	Stupor	*Ci*. 7: 107–108
35	Madātyaya	Alcoholism	*Ci*. 7: 52

(*Continued*)

TABLE 3.10 (*Continued*)
Diseases and Conditions in Which *vasti* Is Indicated

Sl. No.	Diseases and Conditions: Sanskrit Term	Diseases and Conditions: English Equivalent	Reference
36	Manyāruja	Pain in neck	*Ka.* 4: 37–43
37	Mēhanaśūla	Pricking pain in phallus	*Ka.* 4: 29–30
38	Mōha	Mental confusion	*Ka.* 4: 37–43
39	Mūtragraha	Dysuria	*Ka.* 4: 12–16, 37–43, 49–52
40	Mūtrarōdha	Discomforts arising from restrainment of urge to void urine	*Sū.* 4: 5
41	Navamamāsōpacāra	Pregnancy care in ninth month	*Śa.* 1: 67
42	Pādaśūla	Pricking pain in leg	*Ka.* 4: 7–10
43	Pāṇḍu	Morbid pallor	*Ci.* 16: 55, *Ka.* 4: 12–16, 34–36
44	Pāyuśūla	Pricking pain in anus	*Ka.* 4: 29–30
45	Pittarōga	*Pitta* diseases	*Ka.* 4: 11, 12–16
46	Plīhōdara	Enlargement of spleen	*Ci.* 3: 6–9, *Ka.* 4: 62–66
47	Pradara	Dysfunctional bleeding	*Ka.* 4: 12–16, 37–43
48	Pramēha	Polyuric diseases	*Ka.* 4: 23–24, 27–28, 37–43, 62–66
49	Pṛṣṭhaśūla	Pricking pain in back of body	*Ka.* 4: 7–10
50	Rājayakṣma	Kingly consumption	*Ci.* 5: 69, 83
51	Raktapitta	Hemorrhages of obscure origin	*Ci.* 2: 47, *Ka.* 4: 12–16
52	Sadyōvraṇa	Traumatic wound	*Ut.* 26: 12
53	Sarvajatrūrdhva rōga	All diseases above collar bone	*Ut.* 24:47–49
54	Sirāgranthi	Varicose vein	*Ut.* 30: 7
55	Śiraḥkampa	Shivering of head	*Ci.* 3: 6–9
56	Śirōruja	Headache	*Ka.* 4: 37–43, *Ut.* 24: 12
57	Śōpha	Edema	*Ka.* 4: 37–43
58	Śrōtraruja	Pain in ear	*Ka.* 4: 37–43
59	Śukḷakṣaya	Sexual debility	*Ka.* 4: 25–26, 34–36, 43–44, 45–46, 49–52, 59–61
60	Śukḷārtavadōṣa	Male and female infertility	*Śa.* 1: 16
61	Timira	Cataract	*Ut.* 13: 47
62	Tṛkaśūla	Pricking pain in sacral area	*Ka.* 4: 7–10
63	Udara	Enlargement of abdomen	*Ci.* 15: 51
64	Udāvarta	Abdominal disease due to retention of feces	*Ka.* 4: 62–66
65	Ūrdhvānila	Belching	*Ci.* 3: 6–9
66	Unmāda	Insanity	*Ka.* 4: 37–43, *Ut.* 6: 19
67	Ūruśūla	Pricking pain in thigh	*Ka.* 4: 7–10, 29–30, 37–43
68	Vaṅkṣaṇa vēdana	Pain in groins	*Ci.* 3: 6–9, *Ka.* 4: 37–43
69	Varṣacarya	Part of regimen for monsoon season	*Sū.* 3: 45
70	Vastiśūla	Pricking pain in urinary bladder	*Ka.* 4: 29–30
71	Vastivyāpat	Mishaps encountered while performing *vasti*	*Ka.* 5: 9–12, 18–21, 28–36, 38–42, 47–49
72	Vastyāṭōpa	Swelling in urinary bladder	*Ka.* 4: 34–36
73	Vātapittakōpa	Enragement of *vāta* and pitta	*Ka.* 4: 59–62
74	Vātaśōṇita	*Vāta*-induced blood disorders	*Ci.* 22: 13–14, *Ka.* 4: 37–43, 49–52
75	Vātātmaka stanya	Post-partum mother suffering from *vāta*-vitiated breast milk	*Ut.* 2: 12

(*Continued*)

TABLE 3.10 (*Continued*)
Diseases and Conditions in Which *vasti* Is Indicated

Sl. No.	Diseases and Conditions: Sanskrit Term	Diseases and Conditions: English Equivalent	Reference
76	Vātavyādhi	*Vāta* diseases	*Ci.* 21: 1–3, *Ka.* 4: 4, 21–22, 49–52, 54–58
77	Vibandha	Constipation	*Ka.* 4: 7–10, 24–25, 37–43, 49–52
78	Virēcanavyāpat	Mishaps encountered while performing *virēcana*	*Ka.*3: 9–10, 13, 37–38
79	Visarpa	Cellulitis	*Ka.* 4: 37–43
80	Viṣūci	Cholera	*Ka.* 4: 34–36
81	Vṛddhi	Inguinoscrotal swelling	*Ci.* 13: 30, *Ka.* 4: 7–10, 37–43, 62–66
82	Vṛṣaṇaśūla	Pricking pain in scrota	*Ka.* 4: 29–30
83	Yōnirōga	Diseases of vagina	*Ut.* 34: 27, 51–54
84	Yōnivēdana	Pain in vagina	*Ci.* 3: 6–9

than *vasti*. Because of this reason, *vasti* can be considered to represent half of the treatment of any disease. Some authorities remark that *vasti* itself constitutes treatment of any disease in its entirety (*Sū.* 19: 83–87).

3.13.1.9 Nasal Instillation

Nasal instillation of medicines (*nasya*) is desirable in diseases of throat and head. Medicines administered through nostrils spread to all parts of throat and head, promoting the elimination of *dōṣa*s lodged therein. *Nasya* is of three types: *virēcana nasya*, *bṛṃhaṇa nasya* and *śamana nasya*. *Virēcana nasya* is essential in headache, heaviness or weakness of head, diseases of throat, edema, swelling in neck, worm infestation, glandular swelling, skin diseases and epilepsy. *Bṛṃhaṇa nasya* cures *vāta*-dominant headache; migraine; lassitude of voice; diseases of nose; drying of mouth; speech defects; eye disease characterized by stiffness, painful eyelids and difficulty in opening eyes; and paralysis of arms. *Śamana nasya* is beneficial in naevi, black pigmentation on face and appearance of lines in sclera. *Virēcana nasya* is carried out with specific formulae of herbs processed as *snēha*, pastes and decoctions; honey, salt or fermented liquids. Meat juice and blood of *jāṅgala* animals are the substances used in *bṛṃhaṇa nasya*. Medicinal substances used in *śamana nasya* are *snēha*, pastes, decoctions, *jāṅgala* meat soup, blood, milk and water. *Nasya* with *snēha* is subdivided into *marśa nasya* and *pratimarśa nasya*, based on dose. *Nasya* in high dose is called *marśa nasya*, and the one with low dose is called *pratimarśa nasya*. *Nasya* performed with pastes and decoctions of *tīkṣṇa* nature is intended for elimination of *dōṣa*s from ear, nose and throat. It is called *avapīḍa nasya*. *Nasya* with medicinal powders, which also works like *avapīḍa nasya*, is known as *dhmāna nasya*. Medicinal powder taken in a pipe of reed open at both ends is blown into the nostrils. A large volume of *dōṣa* is eliminated because of the powdery nature of the medicine. A drop falling from an index finger dipped in a liquid is called *bindu* (drop). Ten, eight and six *bindu* are, respectively, the excellent, medium and low doses of *marśa nasya*. These doses are applicable to *nasya* with pastes, decoctions, *snēha*, decoctions and so on (*Sū.* 20: 1–11).

Nasya is forbidden to persons who have consumed water, *madya*, poison, alkalis or clarified butter; post-partum women; and those who have been subjected to *vamana*, *virēcana* and *vasti*. *Nasya* is not to be performed on cloudy days. This measure is contra-indicated also in respiratory distress and cough. For the cure of *kapha*-, *pitta*- and *vāta* diseases, *nasya* is to be carried out in morning, midday and at night, respectively. *Nasya* is recommended to healthy persons as well. This measure

TABLE 3.11
Formulae of *vasti* Which Successfully Eliminate *dōṣas*

Sl. No.	Name of Formula	Indications	Reference
1	Balāguḍūcyādi vasti	Suitable for healthy persons, enhances *ōjas*, roborant	*Ka*. 4: 1–3
2	Daśamūlādi vasti	All *vāta* diseases	*Ka*. 4: 4
3	Balāpaṭōlyādi vasti	Improves abdominal fire, stamina and visual acuity, roborant	*Ka*. 4: 5–6
4	Ēraṇḍamūlādi vasti	Pain in calf, thigh, legs, backbone, alimentary canal, heart and pelvic area; heaviness of body, constipation, *gulma*, urinary stone, inguinoscrotal swelling, *grahaṇi*, hemorrhoids and *kapha-vātaja* diseases	*Ka*. 4: 7–10
5	Yaṣṭyāhvādi vasti	*Pitta* diseases	*Ka*. 4: 11
6	Rāsnāvṛṣādi vasti	Burning sensation, dysentery, dysfunctional bleeding, hemorrhages of obscure origin, heart disease, morbid pallor, irregular fever, *gulma*, dysuria, jaundice, all *pitta* diseases	*Ka*. 4: 12–16
7	Kōśātakadi vasti	*Kapha* disease, dull abdominal fire, distaste for food	*Ka*. 4: 17–19
8	Vātahara vasti 1	Enragement of *vāta*, improves stamina and complexion	*Ka*. 4: 21
9	Vātahara vasti 2	Enragement of *vāta*	*Ka*. 4: 22
10	Paṭōlanimbādi vasti	Conjunctivitis, worm infestation, leprosy and other skin diseases, polyuric diseases	*Ka*. 4: 23–24
11	Tailagōmūtrādi vasti	Constipation, bloating of abdomen	*Ka*. 4: 24–25
12	Payasyādi vasti	Impotency	*Ka*. 4: 25–26
13	Mādhutailika vasti	Polyuric diseases, hemorrhoids, worm infestation, *gulma*, hernia	*Ka*. 4: 27–28
14	Yāpana vasti	Pain in anus, calf, thigh, scrota, lower abdomen and penis	*Ka*. 4: 29–30
15	Yuktaratha vasti	Enragement of *vāta, pitta* and *kapha*	*Ka*. 4: 31–32
16	Dōṣahara vasti	Enragement of *vāta, pitta* and *kapha*	*Ka*. 4: 32–33
17	Siddha vasti	Improves stamina and vigor	*Ka*. 4: 33–34
18	Dvipañcamūlādi vasti	*Kapha* diseases, morbid pallor, cholera, obstruction of seminal fluid and *vāta*	*Ka*. 4: 34–36
19	Mustāpāṭhādi yāpanavasti	*Vāta*-induced blood disorders, mental confusion, polyuric diseases, hemorrhoids, *gulma*, constipation, dysuria, irregular fever, cellulitis, hydrocele, bloating of abdomen, diarrhea; pain in groins, thigh, hip, stomach, neck, ear and head; dysfunctional bleeding, lunacy, edema, cough and urinary stone	*Ka*. 4: 37–43
20	Mahāsnēhavasti	Enragement of *vāta*, impotency	*Ka*. 4: 43–44
21	Mayūramāmsa vasti	Improves stamina and vigor	*Ka*. 4: 45–47
22	Gōdhādi vasti	Pain in thorax, diminution of *ōjas*, *vāta*-induced blood disorders, *vāta* diseases, sexual debility	*Ka*. 4: 49–52
23	Daśamūlādi snēhavasti	All *vāta* diseases	*Ka*. 4: 54–57
24	Vātahara snēhavasti	All *vāta* diseases	*Ka*. 4: 58
25	Saindhavakṛta snēhavasti	Enragement of *vāta*	*Ka*. 4: 59
26	Jīvantyādiyamaka snēhavasti	Leaning, enragement of *vāta-pitta*, improves stamina and vigor	*Ka*. 4: 59–62
27	Saindhavādi snēhavasti	*Kapha* diseases, hydrocele, *udāvarta*, *gulma*, hemorrhoid, enlargement of spleen, polyuric diseases, *āḍhyavāta*, bloating of abdomen, urinary stone	*Ka*. 4: 62–66
28	Pañcamūlādi snēhavasti	*Kapha* diseases	*Ka*. 4: 66–67

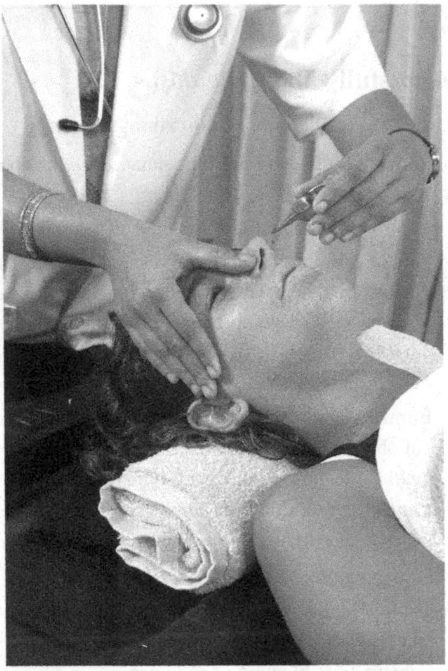

FIGURE 3.8 Nasal instillation of medicines (*nasya*). (Photo courtesy: Nagarjuna Ayurvedic Centre Ltd, Kerala) (www.nagarjunaayurveda.com).

is ideal in mornings during autumn and spring seasons. Midday is suitable in pre-winter and winter seasons. *Nasya* in evening is suitable in summer and during sunny days in monsoon season. Patients suffering from *vāta* enragement in head, hiccup, Stokes-Adams syndrome, inability to turn neck and altered voice quality are to be subjected to *nasya* every morning and evening. In other diseases, *nasya* on alternate days will suffice. *Nasya* can be performed continuously for seven days (*Sū.* 20: 11–17).

Head of the patient is subjected to *snēhana* and sudation before performing *nasya*. After clearing bowels and brushing teeth, the patient is taken to a wind-free room. The area above collar bone is to be subjected again to sudation. Thereafter, patient lies on bed with hands and legs stretched. Slightly warmed medicine is introduced as drops into the right nostril, keeping the left one closed. Following this, medicine is instilled into the left nostril keeping the right one closed (Figure 3.8). After instillation of the medicine, sole of feet, shoulder, palms and ear lobe are to be massaged. Thereafter, the patient makes growl-like sounds with his mouth, so as to agitate phlegm, turns head to both sides and spits out the phlegm and instilled medicine. *Nasya* is to be performed in this way two or three times. After conclusion of the measure, the patient lies in supine position for 100 *mātra* (one *mātra* is the time taken to close and open eyes). Following this, he inhales medicinal smoke to clear throat and holds mouthful of warm water for a few minutes (*Sū.* 20: 18–23).

Unctuousness caused by *nasya* with *snēha* induces effortless expiration, sound sleep, waking up early in morning and improved efficiency of sense organs. Absence of unctuousness, on the other hand, causes lowered efficiency of sense organs, drying of nostrils and throat. Itching, heaviness of head, exudation of phlegm, distaste for food and rhinitis are caused by excessive unctuousness. Optimum performance of *virēcana nasya* causes lightness of organs and improves quality of voice. Its excessive application results in discomforts to head. *Pratimarśa nasya* is suitable for children, the aged, pleasure-seekers and individuals with injured thorax. On account of its low dose, *pratimarśa nasya* is not worthy of application in pernicious rhinitis, alcoholism and in those with weak auditory system. *Nasya* is forbidden to children below seven years and those beyond eighty years of age (*Sū.* 20: 23–31).

Pratimarśa nasya, like *mātrāvasti*, can be performed any time between birth and death. It can be carried every day without any dietary or behavioral stipulations. *Nasya* in a healthy person can be performed every day with sesame oil. Other *snēha* are not recommended. *Aṇutaila* is a reputed *taila* recommended for *nasya*. One who performs *nasya* regularly is endowed with well-built and raised neck, shoulders, chest, pleasant countenance and bright skin. His sense organs will be very well organized, and he will never suffer from graying of hair (*Sū*. 20: 30–39). Table 3.12 lists the diseases and conditions in which *nasya* is indicated. Details of *raktamōkṣa*, the fifth component of *pañcakarma*, are described in Chapter 4.

3.13.2 Ancillary Procedures

Emesis, purgation, medicinal enemas, nasal instillation of medicines and bloodletting effectively eliminate from body the *dōṣas* responsible for the various pathological conditions. There are a few more measures that are recommended to be performed along with the main *pañcakarma*. At times, they can be performed independently as well.

3.13.2.1 Inhalation of Medicinal Smoke

Inhalation of medicinal smoke (*dhūmapāna*) is recommended for the elimination of *kapha-vāta* vitiation in throat and head and to prevent their resurgence. Nevertheless, it is to be carried out carefully. *Dhūmapāna* is classified into *snigdha*, *madhyama* and *tīkṣṇa* types. They are to be applied in predominance of *vāta*, *vāta-kapha* and *kapha*, respectively. This measure is forbidden in several conditions including cataract, distension of abdomen, enlargement of abdomen, polyuric diseases and morbid pallor. It is also forbidden to patients with enragement of *pitta* and *rakta* and those who have consumed fish, *madya*, curd, milk, honey, *snēha* or poison. *Dhūmapāna* carried out in excess or at inappropriate times causes the appearance of enragement of *rakta* and *pitta*, blindness, deafness, excessive thirst and mental confusion (*Sū*. 21: 1–7). Specifications of the tube used for inhaling the smoke are also available (*Sū*. 21: 7–8).

Sitting straight, the patient fixes his attention on *dhūmapāna*. He inhales the medicinal smoke through left nostril, keeping the right one closed. Thereafter, inhalation is carried out through the right nostril, keeping the left one closed and keeping mouth open. In this way, three cycles of *dhūmapāna* are to be repeated. Smoke is to be inhaled through mouth and then through nostrils to cause excitation and elimination of *dōṣas*. When *dōṣas* are already excited, inhalation can be carried out through nose and then through mouth. Inhalation of smoke through mouth is recommended to expel *dōṣa* residing in throat. Whether smoke is inhaled through mouth or nostrils, exhalation is to be carried out through mouth. Exhalation through nostrils can result in damage to eyes (*Sū*. 21: 9–11).

Ingredients of *snigdha dhūma* are *Aquilaria agallocha*, guggul gum, *Cyperus rotundus*, *Vettiveria zizanioides*, black gram, barley, sesame seeds, animal fat, bone marrow, clarified butter, various vegetable oils and several other herbs. *Madhyama dhūma* is carried out using ingredients including lac, *Elettaria cardamomum*, *Glycyrrhiza glabra*, lotus flower and sugar. Components of *tīkṣṇa dhūma* are *Celastrus paniculatus*, *Curcuma longa*, *Saussurea lappa*, *Valeriana wallichii*, herbs of *Vēllādi gaṇa* and a few others. A 12 finger-breadth long blade of *Darbha* grass (*Desmostachya bipinnata*) is left in water for a day and night. Thereafter, it is taken out, paste of the ingredients smeared over it and dried five times, so that it acquires the thickness of thumb finger. It is then dried under shade and the blade removed, to make it resemble a hollow wick. It is then smeared with a *snēha* appropriate for the disease condition, fixed to the base of smoking pipe, lit and the medicinal smoke inhaled (*Sū*. 21: 19–20).

Diseases and conditions such as cough; respiratory distress; catarrh; altered voice quality; inflammations in nasal cavity and mouth; pallor; hair problems; exudation of fluid from ear, mouth and eyes; itching; various kinds of pain; drowsiness; and hiccup will not affect a person who practices *dhūmapāna* (*Sū*. 21: 13–22). *Dhūmapāna* is not to be performed in persons below eighteen years of age (*Sū*. 20: 23–31). Table 3.13 lists the diseases and conditions in which *dhūmapāna* is indicated.

TABLE 3.12
Diseases and Conditions in Which *nasya* Is Indicated

	Diseases and Conditions		
Sl. No.	Sanskrit Term	English Equivalent	Reference
1	Ajaka	Coppery red eruptions on cornea, with slimy surface and bloody discharge resembling droppings of goat in color and appearance	*Ut.* 11: 57–58
2	Akālaśayana	Discomforts arising from sleeping at odd times	*Sū.* 7: 62
3	Ākhuviṣa	Rat bite	*Ut.* 38: 24
4	Āmāśayagata pakvāśayagata saviṣānna	Poisoned food in stomach and intestine	*Sū.* 7: 26
5	Apaci	Cervical glandular swelling	*Ut.* 30: 14
6	Apasmāra	Epilepsy	*Ut.* 5: 6–8, *Ut.* 7: 29–31
7	Atinidra	Discomforts arising from sleeping excessively	*Sū.* 7: 63
8	Bhaṅga	Fracture	*Ut.* 27: 33
9	Dantacāla	Looseness of teeth	*Ut.* 22: 15
10	Dantanāḷi	Dental abscess	*Ut.* 22: 42
11	Dāruṇaka	Dandruff	*Ut.* 24–26
12	Dhūmara	Smoky vision	*Ut.* 13: 93
13	Dōṣāndham	Ability to see objects only during day	*Ut.* 13: 90
14	Ēkāṅgarōga	Hemiplegia	*Ci.* 3: 6–9
15	Gaḷagaṇḍa	Swelling in neck	*Ut.* 22: 71
16	Giḷābuka	Globus pharyngeus	*Ut.* 22: 63
17	Grahabādha	Spiritual afflictions	*Ut.* 5: 6
18	Hidhma	Hiccup	*Ci.* 4: 47–48
19	Hṛdrōga	Heart disease	*Ci.* 6: 27–28
20	Indralupta	Alopecia	*Ut.* 24: 33–38
21	Jatrūrdhvarōga	Diseases above collar bone	*Ut.* 24: 46–49
22	Jvara	Fever	*Ci.* 1: 125–126, 150, 161–162
23	Kaphōṣṭhakōpa	Swollen lips with red pustules	*Ut.* 22: 8
24	Kaphōtkḷiṣṭa	Disease of eyelid with stickiness of lid margin and heaviness of lid	*Ut.* 9: 23
25	Karṇabādhirya	Deafness	*Ut.* 18: 24
26	Karṇakaṇḍu	Itching in ear	*Ut.* 18: 34
27	Karṇaśūla	Pricking pain in ear	*Ut.* 18: 9
28	Kāsa	Cough	*Ci.* 3: 6–9, 41–42
29	Khaṇḍōṣṭha	Cleft lip	*Ut.* 22: 2
30	Kṛchrōnmīla	Eye disease characterized by stiff and painful eyelids, sensation of eyes filled with sand, difficulty in opening eyes and moving the lid upward	*Ut.* 9: 1
31	Kṛmi	Worm infestation	*Ci.* 20: 28
32	Kṛmidanta	Dental caries	*Ut.* 22: 19
33	Kṣut vidhāritāt	Disorders arising from restrainment of urge to sneeze	*Sū.* 4: 9

(*Continued*)

TABLE 3.12 (*Continued*)
Diseases and Conditions in Which *nasya* Is Indicated

Sl. No.	Sanskrit Term	English Equivalent (Diseases and Conditions)	Reference
34	Mada	Stupor	*Ci.* 7: 105
35	Mukhadūṣika	Pimples	*Ut.* 32: 4, 30
36	Mukhapāka	Inflammation of mouth	*Ut.* 22: 75
37	Navārbuda	Recently formed tumor in mouth	*Ut.* 22: 79
38	Nīlika	Naevi	*Ut.* 32: 30
39	Pakṣmaśatana	Madarosis	*Ut.* 9: 19
40	Pāḷīśōṣam	Thinning of ear lobule	*Ut.* 18: 38
41	Palita	Graying of hair	*Ut.* 24: 45, *Ut.* 32: 30
42	Pīnasa	Catarrh	*Ut.* 20: 10
43	Pittakaṇṭaka	Red eruptions on tongue	*Ut.* 22: 43
44	Plīhōdara	Enlargement of spleen	*Ci.* 3: 6–9
45	Pumsavana	Prenatal rite through which sex of the fetus is programmed according to one's choice	*Śa.* 1: 40–42
46	Pūtyāsyata	Bad breath	*Ut.* 22: 79
47	Rājayakṣma	Kingly consumption	*Ci.* 5: 69
48	Raktapitta	Hemorrhages of obscure origin	*Ci.* 2: 47–49
49	Śālūka	Adenoids	*Ut.* 22: 63
50	Sannyāsa	Coma	Ci. 7: 110
51	Sarpaviṣa	Snake bite	*Ut.* 36–72
52	Śiraḥkampa	Shivering of head	*Ci.* 3: 6–9, *Ut.* 24: 19
53	Śiraḥstāpa	Headache	*Ut.* 24: 4, 10–15
54	Śītāda	Gingivitis	*Ut.* 22: 28
55	Śītadantam	Teeth sensitivity to cold	*Ut.* 22: 13
56	Śuddhaśukra	Eye disease characterized by white cornea and slight pain	*Ut.* 11: 42
57	Suṣira	Dental cavity	*Ut.* 22: 37
58	Śuṣkākṣipāka	Eye disease with pricking pain, difficulty to open and close eyes, swollen lids, dryness and suppuration	*Ut.* 16: 28
59	Svarasāda	Lassitude of voice	*Ci.* 5: 35, 37–38, 41
60	Tālupāka	Inflammation of tongue	*Ut.* 22: 54
61	Tilakāḷakam	Non-elevated mole	*Ut.* 32: 30
62	Timira	Cataract	*Ut.* 13: 46–47
63	Tṛṣṇa	Polydipsia	*Ci.* 6: 66
64	Tuṇḍikēri	Tonsilitis	*Ut.* 22: 63
65	Unmāda	Insanity	*Ut.* 5: 7–8, *Ut.* 6: 19
66	Unmantha	Edema with itching on ear lobule	*Ut.* 18: 46
67	Ūrdhvānila	Belching	*Ci.* 3: 6–9
68	Vali	Wrinkles	*Ut.* 32: 30
69	Vaṅkṣaṇavēdana	Pain in groins	*Ci.* 3: 6–9
70	Vasanta carya	Part of regimen for spring season	*Sū.* 3: 19
71	Vātavyādhi	*Vāta* diseases	*Ci.* 21: 17
72	Vṛnda	Edema on either side of neck, associated with fever	*Ut.* 22: 63
73	Vyaṅga	Black pigmentation on face	*Ut.* 32: 30
74	Yōnivēdana	Pain in vagina	*Ci.* 3: 6–9

TABLE 3.13
Diseases and Conditions in Which *dhūmapāna* Is Indicated

	Diseases and Conditions		
Sl. No.	**Sanskrit Term**	**English Equivalent**	**Reference**
1	Dantanāḷi	Dental abscess	*Ut*. 22: 41–42
2	Hidhma	Hiccup	*Ci*. 4: 10–14
3	Jvara	Fever	*Ci*. 1: 127–130, 143
4	Kaphaja gaḷagaṇḍa	Swelling in neck, with predominance of *kapha*	*Ut*. 22: 71
5	Kaphōṣṭhakōpa	Swollen lips with red pustules	*Ut*. 22: 8
6	Kāsa	Cough	*Ci*. 3: 18, 67–69, 147–151
7	Kṛchrōnmīla	Eye disease characterized by stiff and painful eyelids, sensation of eyes filled with sand, difficulty in opening eyes and moving the lid upward	*Ut*. 9: 1
8	Kṛmija śirahstāpa	Headache due to worm infestation	*Ut*. 24: 16, 18
9	Kumbhikā vartma	Meibomian cyst	*Ut*. 9: 9
10	Nasārōga	Diseases of nose	*Ut*. 20: 1
11	Pīnasa	Catarrh	*Ut*. 20: 7–8, 15–17
12	Pūtivadana	Bad breath	*Ut*. 22: 79
13	Raktaja karṇaśūla	Pricking pain in ear due to vitiation of *rakta*	*Ut*. 18: 17
14	Rājayakṣma	Kingly consumption	*Ci*. 5: 35, 48, 69
15	Sannyāsa	Coma	*Ci*. 7: 110, 114
16	Śvāsa	Respiratory distress	*Ci*. 4: 10–14
17	Tataḥ param dhūma ācarēt	*Dhūmapāna* is to be performed regularly after application of *añjana*	*Sū*. 2: 6
18	Unmāda	Insanity	*Ut*. 6: 43
19	Viśēṣāt ślēṣmarōgāṇām nivartayē tataḥ param dhūma ācarēt	*Dhūmapāna* is remedy for discomforts arising from vitiation of *kapha*	*Sū*.13: 12

3.13.2.2 Gargles

Holding fluids such as fats, oils and decoctions in mouth (gargle) is known as *gaṇḍūṣa*. It is of four types: *snigdha gaṇḍūṣa*, *śamana gaṇḍūṣa*, *śōdhana gaṇḍūṣa* and *rōpaṇa gaṇḍūṣa*. They are applied in *vāta-*, *pitta-*, *kapha* diseases and sores, respectively. *Snigdha gaṇḍūṣa* is performed using *snēha* prepared with herbs having *madhura*, *amḷa* and *lavaṇa* tastes. *Tikta*, *kaṣāya* and *madhura* ingredients are used in *śamana gaṇḍūṣa*. Ingredients of *śōdhana gaṇḍūṣa* have *tikta*, *kaṭu*, *amḷa*, *lavaṇa* tastes and hot potency. Medicine for *rōpaṇa gaṇḍūṣa* is prepared with *kaṣāya* and *tikta* herbs. Liquids such as *snēha*, milk, honey water, vinegar, *madya*, meat soup and cow urine are also added to the medicine mix intended for these four types of *gaṇḍūṣa*. *Gaṇḍūṣa* is carried out with medicine mix having tolerable warmth (*Sū*. 22: 1–4).

Paste of sesame seeds suspended in cold or warm water is ideal in looseness and gnashing of teeth. Sesame oil or meat soup can be used for performing *gaṇḍūṣa* every day. Clarified butter or milk is to be used in inflammation of mouth with burning sensation, wounds in mouth due to exogenous cause and burn injury caused by poison, alkali or fire. *Gaṇḍūṣa* with honey cleanses buccal cavity, heals wounds and pacifies burning sensation and excessive thirst. *Gaṇḍūṣa* with *Dhānyāmḷa* (sour gruel made by fermenting rice water) cures bad taste in mouth and bad breath. *Dhānyāmḷa* mixed with a little bit of rock salt applied as *gaṇḍūṣa* cures drying of mouth. On a sunny day, after subjecting the patient to sudation of shoulder and neck, *gaṇḍūṣa* is to be performed in a wind-free room. After turning the head slightly upward and holding the medicinal fluid in mouth, the patient

TABLE 3.14
Diseases and Conditions in Which *gaṇḍūṣa* Is Indicated

	Diseases and Conditions		
Sl. No.	Sanskrit Term	English Equivalent	Reference
1	Dantabhēda	Toothache	*Ut*. 22: 13–14
2	Dantacāla	Looseness of teeth	*Ut*. 22: 14
3	Dantaharṣa	Gnashing of teeth	*Ut*. 22: 13–14
4	Dantanāḷi	Sinus of gums	*Ut*. 22: 42
5	Dantasuṣira	Dental cavity	*Ut*. 22: 35–36
6	Dantavidradhi	Dental abscess	*Ut*. 22: 33
7	Dantōddhāraṇam	Extraction of tooth	*Ut*. 22: 24
8	Dinacarya	*Gaṇḍūṣa* is to be performed, after application of *añjana*	*Sū*. 2: 6
9	Gaḷaśuṇṭhika	Pharyngitis	*Ut*. 22: 46
10	Gaḷavidradhi	Swelling in neck with pain and exudation of pus	*Ut*. 22: 64
11	Giḷābuka	Globus pharyngeus	*Ut*. 22: 63
12	Jihvākaṇṭaka	Glossitis	*Ut*. 22: 43
13	Jvara	Fever	*Ci*. 1: 127
14	Kaṇṭharōga	Diseases of throat	*Ut*. 22: 54, 59–62
15	Kaphōṣṭhakōpa	Swollen lips with red pustules	*Ut*. 22: 8
16	Kṛmidanta	Dental caries	*Ut*. 22: 19
17	Kṛmiśūla	Acute dental pain	*Ut*. 22: 22
18	Kumbhikā vartma	Meibomian cyst	*Ut*. 9: 9
19	Mukhapāka	Inflammation of mouth	*Ut*. 22: 73–74
20	Nāsārōga	Diseases of nose	*Ut*. 20: 1
21	Navārbuda	Recently formed tumor in mouth	*Ut*. 22: 78
22	Pitta-kapha-raktaja abhiṣyanda	Conjunctivitis caused by *pitta*, *kapha* and *rakta*	*Ut*. 16: 1
23	Pūtivadana	Bad breath	*Ut*. 22: 79–81
24	Raktaja karṇaśula	Pricking pain in ear due to vitiation of *rakta*	*Ut*. 18: 17
25	Śalūka	Adenoids	*Ut*. 22: 63
26	Śītāda	Gingivitis	*Ut*. 22: 28
27	Tālupāka	Inflammation of tongue	*Ut*. 22: 51–52
28	Tāluśōṣa	Xerostomia	*Ut*. 22: 53
29	Tatra sēvyādyaiḥ gaṇḍūṣaḥ	*Gaṇḍūṣa* of *Sēvya*, *Candana* and *Padmaka* is to be kept in mouth, in the event of mouth coming in contact with poisoned food	*Sū*. 7: 22
30	Tuṇḍikēri	Tonsilitis	*Ut*. 22: 63
31	Upakuśa	Chronic gingivitis	*Ut*. 22: 31
32	Vidarbha	Traumatic injury to gums	*Ut*. 22: 39
33	Viśēṣāt ślēṣmarōgāṇām nivartayē tataḥ param gaṇḍūṣam ācarēt	*Gaṇḍūṣa* is remedy for discomforts arising from vitiation of *kapha*	*Sū*. 13: 12
34	Vṛnda	Edema on either side of neck, associated with fever	*Ut*. 22: 63

should wait till mucus secretion starts or fluid appears in nostrils and eyes. Care should be taken to avoid drinking the medicinal fluid. Holding fluid fully in the buccal cavity is called *gaṇḍūṣa*. On the other hand, keeping a small quantity of medicinal fluid that can be agitated in the mouth is known as *kabaḷa*. Disorders affecting neck; diseases of head, ear, mouth, eyes, throat; exudation of mucus; drying of mouth; nausea; drowsiness; and rhinitis can be cured by *kabaḷa* (*Sū*. 22: 5–12). Table 3.14 lists the diseases and conditions in which *gaṇḍūṣa* is indicated.

3.13.2.3 Facial Creams

Kapha diseases manifesting in mouth and eyes are cured by applying medicines and draining the fluid. This procedure is called *pratisāraṇa*. Herbs recommended for *gaṇḍūṣa* are processed appropriately as paste, *rasakriya* (paste mixed with honey) or powder and applied in mouth and over the eyes. Cream applied on face (*mukhalēpa*) is of three types. It can be applied with a thickness of ¼, 1/3 and ½ finger-breadth. *Mukhalēpa* pacifies *dōṣa*s, removes toxins and improves complexion. The cream is applied with tolerable warmth in *vāta-*, *kapha-* and *vāta-kapha* conditions. Cold creams are recommended in vitiation of *pitta*, *rakta* and *rakta-pitta*. The cream should remain moist. Dried up cream needs to be removed after moistening with milk. A *taila* should be massaged over face, after removal of the cream. Siesta, long talk, exposure to fire and hot sun, melancholy and anger are to be avoided. *Mukhalēpa* is contra-indicated in catarrh, indigestion, lockjaw and distaste for food. It is also to be avoided while keeping awake at night. Proper application of *mukhalēpa* cures premature graying, black pigmentation, naevi, wrinkles and cataract. Regular use of this measure improves visual acuity and complexion (*Sū.* 22: 13–23).

3.13.2.4 Application of Oil on Head

Applying medicated oil on head is known as *mūrdhataila*. It has four subdivisions such as *abhyaṅga*, *sēka*, *picu* and *vasti*. *Abhyaṅga* is anointing with *taila*, the entire body including head, auditory canals and soles of feet. This measure improves vision, stamina, longevity, quality of sleep and texture of skin. However, *abhyaṅga* is forbidden to those suffering from indigestion, enragement of *kapha* and those subjected to emesis and purgation (*Sū.* 2: 8–9). *Abhyaṅga* is essential when body experiences dryness, itching and exudation of waste products from ear, eye and nose (*Sū.* 22: 24–25). This is an important therapeutic measure and is applied in many diseases.

Pouring of *taila* in a thin stream over the body parts is known as *sēka*. *Sēka* on head is indicated in pityriasis, headache, burning sensation, suppuration and wounds (*Sū.* 22: 24–26). The term is also applicable to pouring of various medicinal fluids on head or other parts of body. *Śiraḥsēka* or *śirodhāra* is also a measure developed in medieval Kerala. The patient lies in the *tailadrōṇi*, after anointing his body and head with appropriate medicinal oil (*taila*). Head should be slightly raised by placing on a pillow. *Śirodhāra* begins after performance of the sacred rites. Two attendants are required to perform this measure. One supports the wide-mouthed vessel from which the *taila* drips (*dhāra* pot) and the other for collecting the *taila* that falls down and pouring it back into the pot (Figure 3.9). The *dhāra* pot is hung over the head of the patient by means of a tough rope. A wick hanging from the bottom of the *dhāra* pot should be 3 inches away from the forehead of the patient. The *taila* chosen for the treatment is poured into the *dhāra* pot and allowed to flow through the wick in a stream on to the upper part of the forehead (Figure 3.10). The *dhāra* pot is kept refilled with the drippings collected from the sink in the *tailadrōṇi*, on which the patient lies. *Śirodhāra* is continued for one and half hours, with the patient lying in supine position throughout the period. This treatment is carried out between 7 and 10 A.M., for seven to fourteen days (Mooss, 1983b).

Placing on head a strip of cloth soaked in *taila* is called *picu*. Application of *ghṛta* or *taila* in orifices of body in a similar way is also known as *picu*. *Picu* is generally indicated in breaking or losing of hair, feeling of smoke emanating from head, difficulty in moving eye balls, burning sensation and discomforts in eyes. *Picu* is recommended to be performed on infants, as soon as they are born (*Ut.* 1: 7–8). It is one of the measures to be applied in *ardita* (facial palsy) (*Ci.* 21: 42–43). The inflammation of lips and gynecological diseases are also treated with *picu* (*Ut.* 22: 3, *Ut.* 34: 27, *Ut.* 34: 51–54).

Śirōvasti is indicated in numbness of scalp, *ardita*, insomnia, dryness of nose, xerostomia, cataract and severe diseases of head (*Sū.* 22: 26–27). The patient is subjected to oleation, sudation, emesis, purgation and seated on a knee-high stool. Thereafter, a strip of moderately tough, clean and dry cloth, about three finger-breadth wide and sufficiently long to go six or seven times around the head is smeared on both sides with paste of black gram prepared with hot water. The cloth is then

FIGURE 3.9 *Dhāra* pot for performing *śirodhāra*. (Photo courtesy: Divakara Concepts, Coimbatore.)

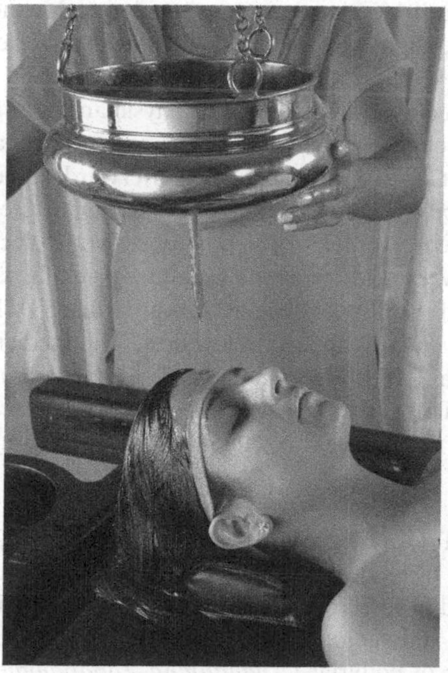

FIGURE 3.10 *Śirodhāra* or pouring of medicated oil on forehead in a thin stream. Note the wick hanging from the bottom of the vessel. (Photo courtesy: Nagarjuna Ayurvedic Centre Ltd, Kerala) (www.nagarjunaayurveda.com).

wound round the head, passing over the ears and forehead, leaving a finger space between the lower edge of the bandage and eyebrows. Over this is placed a 12 finger-breadth high cap-like structure, without a top, made of hide of ox or buffalo. Another band of cloth smeared with black gram paste is wound again on the lower end of the cap, passing over the first band. This is to fix the cap in position and to prevent the leakage of the medicinal oil (*Sū*. 22: 27–29, Mooss, 1983c).

The prescribed *taila* is warmed to body temperature and poured into the cap. The *taila* is poured till it rises and remains at a level of one finger-breadth above the scalp. The *taila* is retained in the cap till mucus oozes out of nose and mouth. In *vāta-*, *pitta-* and *kapha* diseases the *taila* is to remain for 10,000, 8,000 and 6,000 *mātra*, respectively. On the appearance of signs of optimum oleation, the *taila* is drained out. Following this the patient's shoulders, neck, back and forehead are massaged gently. *Śirōvasti* may be performed continuously for a maximum period of seven days (*Sū*. 22: 29–32).

Taila is to be poured into ear and retained for 100 *mātra*. Back of the ear lobe, behind the antitragus is to be massaged gently during this period. *Mūrdhataila* cures hair loss, graying, browning and breaking of hair. It cures all *vāta* diseases of head, enhances the efficiency of sense organs and strengthens voice, temples and head (*Sū*. 22: 32–34).

3.13.3 Special Ophthalmic Procedures

AH describes several measures that are to be made use of exclusively in the prevention and treatment of eye diseases. They are unique procedures having a wide range of application in the management of eye diseases. They include *sēka*, *āścyōtana*, *añjana*, *tarpaṇa*, *puṭapāka* and *biḍālaka*.

3.13.3.1 Eye Drops

Āścyōtana is the process of irrigating the eye with drops of decoction or applying colloidal preparations in eyes. Paste of the ingredient herbs can be applied over the eyelids as well. In such cases, it is called *biḍālaka*. *Āścyōtana* is to be applied in the early stages of eye diseases. This measure pacifies pricking pain, pain, itching, lachrymation, burning sensation and redness. In *vāta* and *kapha* vitiation, the medicine should have slight warmth. Cold medicine is preferred in enragement of *pitta* and *rakta* (*Sū*. 23: 1–2).

The patient lies on a cot in a wind-free room. The physician raises his eyelid with left hand and with the right hand pours through a wick, 10–12 drops of medicine, above the inner angle of eye (inner canthus) from a height of two finger-breadth. Afterward, the eye is wiped dry with clean cloth. Mild fomentation is applied on eyelids in *vāta* and *kapha* diseases, with a cloth dipped in warm water and then squeezed. *Āścyōtana* which is excessively *uṣṇa* and *tīkṣṇa* in nature causes pain, redness and blindness. Very cold *āścyōtana* generates pricking pain, difficulty in moving eyeballs and pain. Heaviness and inability to open eyelids are caused by high dose of *āścyōtana*. The disease progresses when *āścyōtana* is applied in low dose. The medicine applied in eyes enters the channels situated in the canthi (palpebral fissure), head, nostrils and mouth. The secretions inside these channels are removed. This is why *āścyōtana* is performed in the initial stages of eye diseases (*Sū*. 23: 2–7). Table 3.15 lists the diseases and conditions in which *āścyōtana* is indicated.

3.13.3.2 Collyrium

Āścyōtana causes ripening of *dōṣa*s, indicated by reduction in edema, excessive itching and sliminess. There will be lowering of redness and lachrymation, followed by the appearance of rheum. Collyrium (*añjana*) is applied in such conditions. This measure is indicated in eye diseases caused by vitiation of *pitta*, *kapha* and *rakta*. Special collyrium is applied in *vāta*-dominant eye diseases. There are three types of collyria: *lēkhanāñjana*, *rōpaṇāñjana* and *dṛṣṭiprasādanāñjana*. *Lēkhanāñjana* is prepared with herbs having astringent, acidic, salty and pungent tastes. It is applied in pterygium. The ingredient herbs of *rōpaṇāñjana* have bitter taste. It is applied in conjunctivitis. *Prasādanāñjana* is intended to improve vision and is prepared with herbs having sweet taste and

TABLE 3.15
Diseases and Conditions in Which *āścyōtana* Is Indicated

Sl. No.	Diseases and Conditions: Sanskrit Term	Diseases and Conditions: English Equivalent	Reference
1	Abhiṣyanda	Conjunctivitis	*Ut.* 16: 8, 11, 13, 16
2	Adhimandha	Glaucoma	*Ut.* 16: 21
3	Alpaśōphākṣipāka	Similar to *saśōphākṣipāka*, but having slight edema	*Ut.* 16: 33
4	Anyatōvāta	Neuralgia of the fifth cranial nerve	*Ut.* 16: 21
5	Arjuna	Phlyctenular conjunctivitis	*Ut.* 11: 11
6	Kumbhikā vartma	Meibomian cyst	*Ut.* 9: 14
7	Liṅganāśam	Blurred vision (near blindness)	*Ut.* 14: 26–27
8	Parvaṇi	Eye disorder affecting the sclero-corneal junction	*Ut.* 11: 4
9	Saśōphākṣipāka	Suppurated eye with pain, edema; warm, cold or slimy tears and reddish sclera	*Ut.* 16: 33

cold potency. Ten finger-breadth long metal rod, thin at the center and with flower bud-like tip, is used for applying collyrium. *Lēkhanāñjana* is applied using a rod made of copper. Index finger and rod made of iron are used for applying *rōpaṇāñjana*. *Prasādanāñjana* is applied with a rod made of gold or silver. The physician holds the rod with thumb and index finger of right hand. Raising the eyelid with left hand, he applies collyrium from inner or nasal canthus to outer canthus. Based on consistency, collyria are of three varieties such as solid (*piṇḍa*), paste (*rasakriya*) and powder (*cūrṇa*). *Piṇḍāñjana* is applied when *dōṣa*s are highly aggravated. *Cūrṇāñjana* is applied when there is less *dōṣa* involvement. *Rasakriya* is handy in medium aggravation (*Sū.* 23: 8–14).

Collyrium is to be applied in morning and evening. Nevertheless, it is not recommended when the weather is too hot or too cold. Collyrium is not to be applied in frightened, melancholic, angry or tired persons. It is also contra-indicated in fever and indigestion. After application of collyrium, the patient should close eyes and roll the eye balls leftward and rightward. The eyelids are to be slightly moved with fingers so as to spread the collyrium well. Eye should not be pressed or washed. When all discomforts due to the application subside, eyes are to be washed with decoctions of appropriate herbs (*Sū.* 23: 14–30). Table 3.16 lists the diseases and conditions in which *añjana* is indicated

3.13.3.3 Refreshing the Eyes

Tarpaṇa is a therapeutic measure that refreshes the eyes. It is applied on eyes that are tired, dry, suffering from madarosis, pain, redness, lachrymation and dim vision. It is also indicated in eye diseases such as *kṛchrōnmīla*, *sirāharṣam*, redness and inflammation of eyes, cataract, phlyctenular conjunctivitis, conjunctivitis, glaucoma, *anyatōvāta*, *vātaparyaya* and diseases of sclera. This measure is carried out in the morning and evening in a wind-free room. A dough prepared by grinding barley and black gram is placed around both eyes in the shape of a two finger-breadth high ring. *Ghṛta* compatible to the condition is melted over warm water and slowly poured into the ring, till the eyelashes are submerged (visceral fat of animals is used in the case of night blindness, *vāta*-dominant cataract and *kṛchrōnmīla*). As the patient opens his eyes, the physician counts the *mātra*. In diseases of eyelids, inner canthus of the eye, sclera and bulbar conjunctiva, cornea and macula, the medicine is to be retained for 100, 300, 500, 700 and 800 *mātra*, respectively. The retention time in glaucoma is 1,000 *mātra*, *vāta*-dominant diseases 1,000 *mātra* and *pitta* vitiation 600 *mātra*. Duration of 500 *mātra* is sufficient for healthy persons and those with *kapha*-dominant eye diseases. After the stipulated period, the medicine is drained by making a slit on the ring at the outer canthus. *Dhūmapāna* is to be performed on completion of *tarpaṇa*. The patient should avoid looking at sky and bright objects (*Sū.* 24: 1–9).

TABLE 3.16
Diseases and Conditions in Which *añjana* Is Indicated

	Diseases and Conditions		
Sl. No.	Sanskrit Term	English Equivalent	Reference
1	Ākhuviṣa	Rat bite	*Ut.* 38: 24
2	Alpaśōphākṣipāka	Similar to *saśōphākṣipāka* but having slight edema	*Ut.* 16: 36–39
3	Āmāśayagata pakvaśayagata saviṣānna	When poisoned food is retained in stomach and intestine, *añjana* of a specific combination of herbs is to be applied	*Sū.* 7: 25–26
4	Arjuna	Phlyctenular conjunctivitis	*Ut.* 11: 12
5	Armaśēṣa	Residual pterygium remaining after surgery	*Ut.* 11: 25
6	Bhūtabādha	Spiritual afflictions	*Ut.* 5: 7, 13–14, 36–37, 41–47
7	Dhūmara	Smoky vision	*Ut.* 13: 92
8	Dinacarya	*Sauvīrāñjana* is to be applied in eyes after brushing teeth	*Sū.* 2: 5
9	Dōṣāndham	Ability to see objects only during day	*Ut.* 13: 84
10	Jvara	Fever	*Ci.* 1: 130, 161, 164
11	Kāca	Eye disease with progressive loss of vision	*Ut.* 13: 82–83
12	Kāmala	Jaundice	*Ci.* 16: 44
13	Kaphōtkḷiṣṭa	Disease of eyelid with stickiness of lid margin and heaviness of lid	*Ut.* 9: 23
14	Kṛchrōnmīla	Eye disease characterized by stiff and painful eyelids, sensation of eyes filled with sand, difficulty in opening eyes and moving the lid upward	*Ut.* 9: 1
15	Kukūṇakam	Neonatal conjunctivitis	*Ut.* 9: 33
16	Liṅganāśam	Blurred vision (near blindness)	*Ut.* 14: 29
17	Mada/Mūrcha	Stupor, fainting	*Ci.* 7: 109
18	Pakṣmarōdha	Trichiasis	*Ut.* 9: 39
19	Pūyālasa	Inflammation of inner canthus of eye	*Ut.* 11: 5–6
20	Sannyāsa	Coma	*Ci.* 7: 110
21	Sapta rātrē vyatītē srāvaṇārthē rasāñjanam yōjayēt	*Rasāñjana* is to be applied after seven days for elimination of *kapha*	*Sū.* 2: 6
22	Sarpaviṣa	Bite of serpents	*Ut.* 36: 60, 71, 72, 75, 77, 81
23	Saśōphākṣipāka	Suppurated eye with pain, edema; warm, cold or slimy tears and reddish sclera	*Ut.* 16: 34–35
24	Sirāharṣa	Redness and inflammation of eye leading to impaired vision	*Ut.* 11: 11
25	Sirōtpāta	Redness and inflammation of eye	*Ut.* 11: 10
26	Sthāvara viṣa	Inanimate poison	*Ut.* 35: 18
27	Śukḷarōga	Diseases of sclera	*Ut.* 11: 25, 29, 32, 44–48, 53–54
28	Śuṣkākṣipāka	Eye disease with pricking pain, difficulty to open and close eyes, swollen lids, dryness and suppuration	*Ut.* 16: 29–31
29	Timira	Cataract	*Ut.* 13: 47
30	Unmāda	Insanity	*Ut.* 6: 20, 38–40

Tarpaṇa is performed daily in *vāta* diseases and on alternate days in the case of *pitta* vitiation. In *kapha* diseases and in healthy persons, *tarpaṇa* is to be carried out with a gap of two days. Thus, this measure is to be continued till signs of optimum performance appear. After optimum performance of *tarpaṇa*, the patient does not show any hesitation to look at bright light and feels comfortable. Eyes become bright and feel light. Opposite signs are manifested in improper performance. Excessive performance of *tarpaṇa* generates *kapha*-related discomforts (*Sū.* 24: 10–11).

3.13.3.4 Heating in Closed Container

As body experiences fatigue after oleation, eyes also get exhausted following *tarpaṇa. Puṭapāka* (heating in closed container) is the countermeasure that is to be carried out after *tarpaṇa.* It is of three varieties such as *snēhana puṭapāka, lēkhana puṭapāka* and *prasādaṇa puṭapāka. Snēhana puṭapāka* is ideal when eyes are dry under the influence of *vāta. Lēkhana puṭapāka* is carried out when unctuousness due to *kapha* is to be curbed. *Prasādaṇa puṭapāka* is recommended when eyes are weak and when *pitta* and *rakta* are aggravated. It is suitable for healthy persons as well. The medicinal mix for *snēhana puṭapāka* consists of fat, bone marrow and meat of animal varieties such as *vilēśaya* (animals that live in burrows), *prasaha* (creatures that grab and tear off their food), *mahāmṛga* (large animals), *apcara* (birds that live around or on the surface of water bodies), *matsya* (animals that live under water) and herbs of *Jīvanīya gaṇa* ground with milk. Liver and meat of *jāṅgala* animals and birds, pearl, iron, copper, rock salt, stibynite, conch shell, cuttlefish bone and orpiment are ground to paste and processed to perform *lēkhana puṭapāka.* The ingredients of medicinal mix for *prasādaṇa puṭapāka* are liver, bone marrow, visceral fat, heart and meat of *jāṅgala* creatures and herbs having sweet taste. They are ground to a fine paste with clarified butter, breast milk and cow milk (*Sū.* 24: 12–17).

The combined pastes of animal products and herbs are rolled into large balls weighing nearly 100 g. Balls of pastes intended for *snēhana-, lēkhana-* and *prasādaṇa puṭapāka* are rolled in leaves of *Ricinus communis, Ficus benghalensis* and *Nelumbo nucifera*, respectively. They are then covered with fine paste of clay, dried and embedded in red-hot cinders. They are assumed to be cooked well when the covering of clay turns red hot. The balls are taken out of fire and allowed to cool. Thereafter, the cooked contents of the balls are taken out, squeezed and the juice collected. This medicinal fluid is applied on eyes, as was done in the case of *tarpaṇa.* The fluid is to be retained in *lēkhana-, snēhana-* and *prasādaṇa puṭapāka* for 100, 200 and 300 *mātra*, respectively. *Snēhana-* and *lēkhana puṭapāka* are to be performed with lukewarm medicinal fluid and *prasādaṇa puṭapāka* with cold fluid. *Dhūmapāna* is to be carried out on completion of *snēhana-* and *lēkhana puṭapāka.* This step is not necessary in the case of *prasādaṇa puṭapāka.* Signs of optimum, improper and excessive performance of *puṭapāka* are the same as in the case of *tarpaṇa. Tarpaṇa* and *puṭapāka* are not to be performed in patients in whom *nasya* is contra-indicated. Dietary and behavioral restrictions are to be observed for double the number of days of *puṭapāka.* Flowers of *Jasminum multiflorum* and *Jasminum sambac* are to be tied over the eyes at night. Table 3.17 lists the diseases and conditions in which *tarpaṇa* and *puṭapāka* are indicated. A blind man cannot enjoy the diversity and beauty of this world. The miserable predicament of such a person can be understood by spending a whole day closing both eyes. Therefore, Vāgbhaṭa advises everyone to keep eyes in good health by the judicious application of *nasya, tarpaṇa* and *puṭapāka* (*Sū.* 24: 17–22).

TABLE 3.17
Diseases and Conditions in Which *tarpaṇa* and *puṭapāka* Are Indicated

	Diseases and Conditions		
Sl. No.	Sanskrit Term	English Equivalent	Reference
1	Abhiṣyanda	Conjunctivitis	*Sū.* 24: 2
2	Adhimandha	Glaucoma	*Sū.* 24: 2
3	Anyatōvāta	Neuralgia of the fifth cranial nerve	*Sū.* 24: 2
4	Arjuna	Phlyctenular conjunctivitis	*Sū.* 24: 2
5	Kṛchrōnmīla	Eye disease characterized by stiff and painful eyelids, sensation of eyes filled with sand, difficulty in opening eyes and moving the lid upward	*Sū.* 24: 2
6	Kumbhikā vartma	Meibomian cyst	*Ut.* 9: 11
7	Mandanidra	Sleeplessness	*Sū.* 7: 66
8	Pūyālasa	Inflammation of inner canthus of eye	*Ut.* 11: 4–5
9	Sirāharṣa	Redness and inflammation of eye, leading to impaired vision	*Sū.* 24: 2
10	Śirōgatavāta	*Vāta* enragement in head	*Ci.* 21: 17
11	Sirōtpāta	Redness and inflammation of eye	*Sū.* 24: 2
12	Śukḷarōga	Diseases of sclera	*Sū.* 24: 2, *Ut.* 11: 42
13	Timira	Cataract	*Sū.* 24: 2, *Ut.* 13: 47, 61
14	Vātaparyaya	Ophthalmic neuralgia	*Sū.* 24: 2

3.14 CONCLUSION

According to Ayurveda, all life processes are regulated by *vāta*, *pitta* and *kapha*, collectively known as *tridōṣa*. The human body is made up of seven tissue elements (*dhātu*), which undergo sequential transformation and produce waste products (*mala*) as well as an end product called *ōjas* that circulates in body imparting vigor and vitality. *Tridōṣa* exist in *dhātu* and *mala*. Diseases are caused when the *tridōṣa* are destabilized and health results when they return to steady state. Due to dietary and behavioral indiscretions and effects of season, the *tridōṣa* increase or decrease, generating many diseases. All forms of matter (*dravya*) are made up of varying combinations of the five primordial elements such as earth, water, fire, air and sky, collectively called *pañcamahābhūta*. *Dravya* possess six basic tastes such as sweet, sour, salty, pungent, bitter and astringent having ability to influence *tridōṣa*. Increase or decrease of *tridōṣa*, *dhātu* and *mala* produce specific signs and symptoms, which help the physician to understand the nature of pathology, following which medicines made up of appropriate *dravya*, food, beverages and measures, having ability to modulate *tridōṣa* in the desired way are administered for the cure of the disease in question.

The *Sūtrasthānam* explains in nutshell all the medical information presented in the other five sections of AH. It also includes a chapter on herbs (*Śōdhanādigaṇasaṅgrahādhyāyaḥ*) that can be used in the treatment of almost all the disease conditions mentioned in the text (*Sū.* 15: 1–47). Therefore, *Sūtrasthānam* can be considered to be a summary of AH. Interpreters of AH comment that the *Sūtrasthānam* is analogous to the thread that holds all the pearls in a garland. It is traditionally believed that a very intelligent person can learn all aspects of Ayurveda and become an accomplished physician just by studying diligently the *Sūtrasthānam* alone.

REFERENCES

Mooss, N.S. 1983a. *Kāyasēka* or *Piẓiccil*. In *Ayurvedic Treatments of Kerala*, 18–28. Kottayam: Vaidyasarathy Press (P) Ltd.

Mooss, N.S. 1983b. *Śirasseka or dhāra*. In *Ayurvedic Treatments of Kerala*, 37–46. Kottayam: Vaidyasarathy Press (P) Ltd.

Mooss, N.S. 1983c. *Śirōvasti*. In *Ayurvedic Treatments of Kerala*, 29–36. Kottayam: Vaidyasarathy Press (P) Ltd.

4 Surgical Armamentarium

4.1 INTRODUCTION

AH marks a milestone in Ayurveda. It is the text that brought together for the first time on one platform, in a concise format, the diagnosis and treatment of diseases narrated in the eight branches of Ayurveda namely *kāyacikitsa* (internal medicine), *śalya* (surgery), *śālākya* (diseases of supraclavicular region), *kaumārabhṛtya* (pediatrics including obstetrics and gynecology), *agadatantra* (toxicology), *bhūtavidya* (treatment of spiritual afflictions), *rasāyana* (rejuvenation therapy) and *vājīkaraṇa* (virilification). General medicine (*kāyacikitsa*) is described in *Cikitsāsthānam*, and the remaining seven branches form the theme of *Uttarasthānam*. All diseases and conditions that afflict human beings cannot be cured by the administration of medicines alone. Conditions such as intestinal obstruction (*baddhōdara*), intestinal perforation (*chidrōdara*), ascites (*jalōdara*), anal fistula (*bhagandara*) and urinary stone (*aśmari*) are managed only by resorting to surgery. Similarly, surgery is the only recourse to remove sharp objects, arrows, thorns, pebbles and other foreign bodies that get lodged inside the body.

4.2 SURGICAL INSTRUMENTS

4.2.1 Blunt Instruments

Surgical instruments are broadly divided into two types: *yantra* (blunt instruments) and *śastra* (sharp instruments). *Yantra* are made use of in the extrication of arrows and sharp objects struck inside the body. They are also useful to observe hemorrhoids and anal fistula. *Yantra* also protect the body parts surrounding the areas subjected to surgery, chemical cautery (*kṣārakarma*) and thermal cautery (*agnikarma*). Elimination measures such as *vasti* and *dhūmapāna* are carried out with *vastiyantra* and *dhūmapānayantra*, respectively. In addition to these, tubular instruments such as *ghaṭika*, *śṛṅgam* and *jāmbavouṣtha* are also known as *yantra*. *Yantra* are of various shapes and are intended for different purposes. They are further subdivided into groups such as *svastikayantra*, *sandaṃśa yantra*, *tālayantra*, *nāḍīyantra* and *śalāka*. Therefore, a physician should have possession of various *yantra* (*Sū.* 25: 1–3).

4.2.1.1 Cruciform Instruments

The word *svastika* is a technical term signifying 24 sacred symbols of the Jains. It is represented by two lines crossing each other, the arms of the cross being bent at their extremities, toward the same direction. Therefore, these instruments are described as cruciform. As a rule, they have a length of 18 *aṅgula* (one *aṅgula* is equal to a finger's breadth). Their tips are shaped like the heads of birds and ferocious beasts such as lion, bear, crow, osprey and so on. Made of steel or iron, these instruments are named after such beings. They bear names like *kaṅkamukha yantra* (heron-faced pincer), *siṃhamukha yantra* (lion-faced pincer), *ṛkṣamukha yantra* (bear-faced pincer), *kākamukha yantra* (crow-faced pincer), *sarpamukha yantra* (serpent-faced pincer) and *kuraramukha yantra* (osprey-faced pincer) (Figures 4.1–4.4). Cruciform instruments (*svastika yantra*) are used in the extrication of sharp objects struck into bones (*Sū.* 25: 3–7). Of all the varieties of *svastika yantra*, heron-faced pincer is the best, as it can be easily introduced and turned in all directions. It also grasps firmly and extracts foreign bodies with ease and without harming any part of body (Mukhopadhyaya, 1913a).

DOI: 10.1201/9781003148296-4

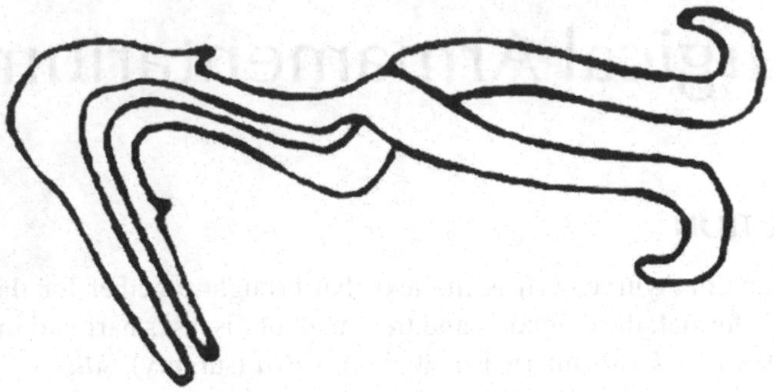

FIGURE 4.1 Heron-faced pincer.

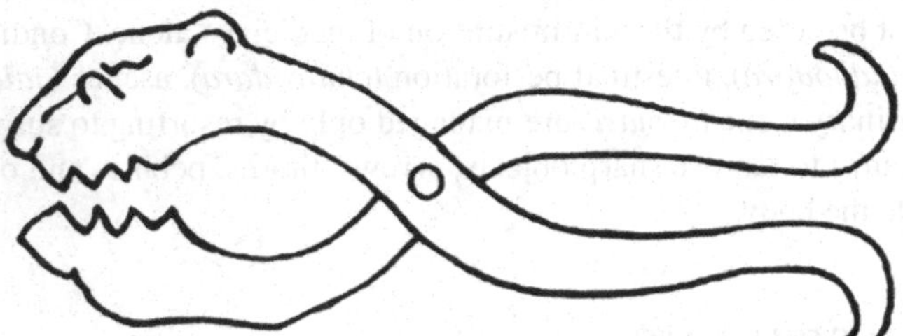

FIGURE 4.2 Lion-faced pincer.

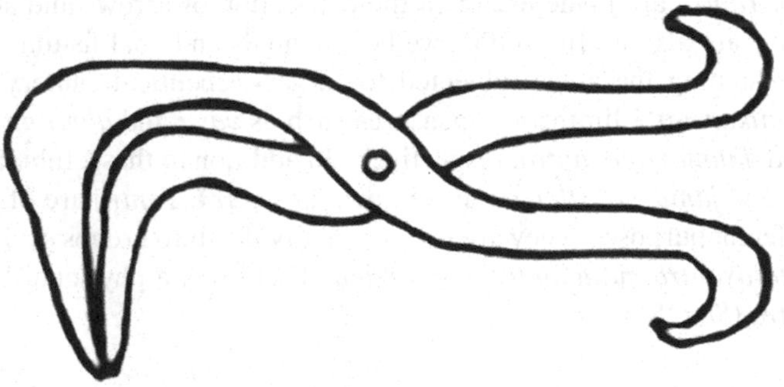

FIGURE 4.3 Crow-faced pincer.

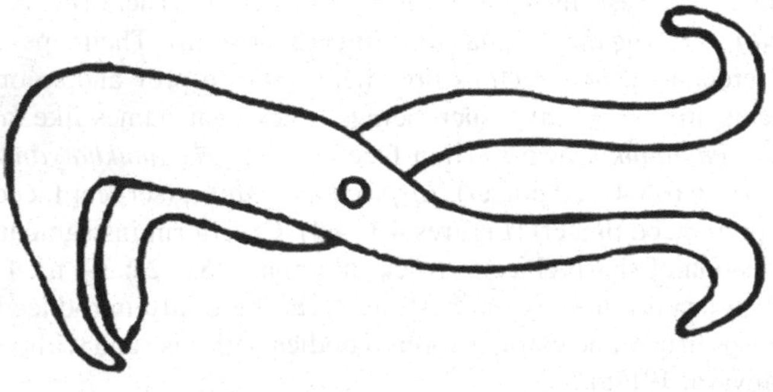

FIGURE 4.4 Osprey-faced pincer.

FIGURE 4.5 *Mucuṇḍi or* teethed forceps.

4.2.1.2 Pincher-Like Forceps

Two 16 finger-breadth long pincher-like forceps (*sandaṃśa yantra*) with or without screws attached to the tip are used in the extraction of sharp objects struck inside skin, blood vessels, nerves, tendons and muscles. Very small sharp objects and overgrown eyelashes are plucked with another pincher-like forceps that are six finger-breadth long (*Sū.* 25: 7–8). *Yantra* called *mucuṇḍi,* which has very small teeth and is straight without any curve, having a ring at the base is employed to excise tissues overgrown inside sinuses and ulcers (Figure 4.5). It is also useful in extracting without surgery pterygium (*arma*) remaining on the sclera (*Sū.* 25: 9). Pincher-like forceps are comparable to the modern dressing forceps (Mukhopadhyaya, 1913a).

4.2.1.3 Picklock-Like Instruments

Sharp objects pierced into the external auditory canal are removed with two, 12 finger-breadth long picklock-like instruments (*tālayantra*). They have either one disc (*tāla*) or two discs, resembling the operculum that covers the gills of a fish. The one with single disc is called *ēkatāla yantra*, and the one with two discs is known as *dvitāla yantra* (*Sū.* 25: 10).

4.2.1.4 Tubular Instruments

Nāḍīyantra or tubular instruments are described to be of several kinds and serve many purposes. They are open either at one or both ends. They are used for the extrication of foreign bodies from outlets of the body and are recommended as a diagnostic tool for observing stages of pathological conditions in several diseases. They are also the means of sucking out fluid discharges from cavities and they facilitate the performance of other operations. They vary in length and diameter in proportion to the sizes of the outer canals of body or according to the several purposes served by them. *Nāḍīyantra* are used in anal fistula, hemorrhoids, tumors and abscesses. They are used for introducing medicaments into rectum, vagina, urethra and in inhalation of medicinal smoke (Mukhopadhyaya, 1913a).

Tubular instrument with length of either ten finger-breadth or five finger-breadth is used for observing sharp objects struck inside throat. It is known as *kaṇṭhāvalōkini* or *kaṇṭhaśalyāvalōkini*. These and several other *nāḍīyantra* should be of suitable length, so that they match the shape, length and thickness of the foreign body. *Nāḍīyantra* with an upper part shaped like the carpellary receptacle of lotus flower is known as *śalyanirghātini*. One-fourth of this 12 finger-breadth long *yantra* is hollow (*Sū.* 25: 11–16).

Arśōyantra shaped like the teat of cow is used to view the inside of hemorrhoids. It is open at both ends and four finger-breadths long. *Arśōyantra* for men should have diameter of five finger-breadth, and the one for women must have a diameter of six finger-breadth. The *arśōyantra* intended for viewing hemorrhoids is to be open at both, and the one for chemical cautery (*kṣārakarma*) should be closed at one end and open at the other. There should be a three finger-breadth long aperture at its middle, with a half finger-breadth long and wide limiting ridge (*Sū.* 25: 16–18).

The *yantra*, known as *śami*, is similar in shape to *arśōyantra*. But it will not have aperture, as it is intended to crush the hemorrhoid. *Bhagandara yantra* is also similar in shape. But it will not have limiting ridge. *Nāḍīyantra* to be used in nasal tumor and nasal polyp is similar in shape to *bhagandara yantra*. However, it will have only one aperture. It is to have length of two finger-breadth and thickness of index finger (*Sū.* 25: 19–20).

Aṅgulītrāṇaka yantra is to be fabricated with ivory or heartwood. It is four finger-breadth long and open at both ends. Shaped like the teat of a cow, it is employed to open the mouth of patient, as in fainting and *vāta* disorders. Mouth of the patient can be opened by wearing this *yantra* on finger. As it is made of ivory or wood, finger is protected from bite of the patient. *Yōnivraṇēkṣaṇa yantra* is hollow and 16 finger-breadth long. It is shaped like a lotus flower, split into four equal parts and fitted with four thin rods. On pressing the ring at the base, the tip of the *yantra* opens up. It is used for viewing the inside of vagina (*Sū.* 25: 21–23).

When a *śōpha* in the ripe stage is not incised or a burst *śōpha* (*vraṇa*) is not properly managed, the pus inside the lesion goes into deeper structures and traverses longer distances across the tissues. This condition is known as *nāḍīvraṇa*, as it resembles a tube. Two instruments are recommended to wash the sinus with decoctions of herbs and apply medicinal oil. These instruments are also shaped like a *vastiyantra*. The difference is that the tube requires only a length of six finger-breadth. The base should be hollow enough to insert thumb finger; tip is to be without limiting ridge and sufficient to hold a pea. The bladder is made of soft leather. *Nāḍīyantra* open at both ends is used for draining fluid from *jalōdara*. The straw-shaped shaft of peacock feather may also be used for this purpose (*Sū.* 25: 23–25).

Vastiyantra (rectal clyster) has two parts: one balloon-shaped bladder and a tube (*vastinētra*) connected to it. The tube required for *vastiyantra* needs to be fabricated with gold, silver, wood, bone or bamboo stem. It should be shaped like the switch of cow's tail, with no perforations. It should be smooth, linear and round at the tip. Tube should be of six types with a length of 5-, 6-, 7-, 8-, 9- or 12 finger-breadth. Five finger-breadth long tube is suitable for an infant less than a year old. From one year to seven-year-old children a tube of six finger-breadth, seven years to twelve years seven finger-breadth, twelve year onward eight finger-breadth, sixteen year onward nine finger-breadth and twenty year onward 12 finger-breadth are recommended. The base of the tube should be as thick as the subject's thumb finger and the tip as thin as the subject's little finger. Bladder of the *yantra* is to be made with urinary bladder of goat, sheep or buffalo, processed to acquire brown or red color, with no perforations or odor. The bladder is to be tied firmly to the tube, so as to hold medicinal fluids. Waxed cloth can be used if urinary bladder of these animals is unavailable (*Sū.* 19: 9–17) (Figures 4.6 and 4.7).

Tube (*dhūmanētra*) of the *dhūmayantra* used in inhalation of medicinal smoke is to be made with gold, silver, wood, bone or bamboo stem. It should be linear and divided into three chambers. The base of the *yantra* should be hollow, so as to pass thumb finger and the tip sufficient to hold the seed of *Ziziphus mauritiana* fruit (*kōlāsthi*). There are three categories of smoke: *tikṣṇa dhūma*,

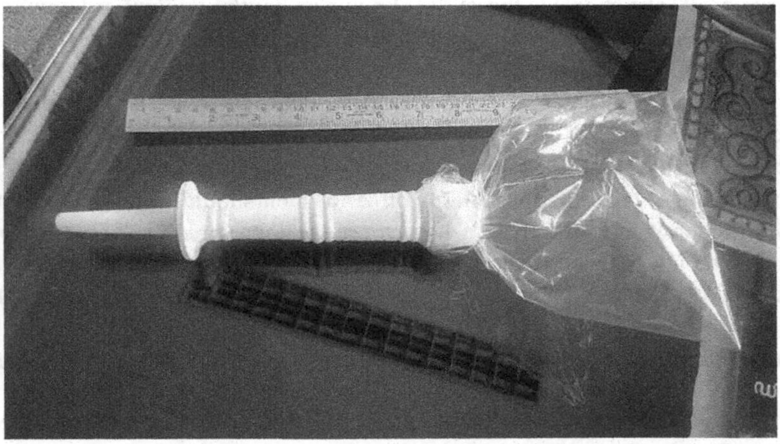

FIGURE 4.6 Instrument for medicinal enema using decoctions. (Photo courtesy: Divakara Concepts, Coimbatore.)

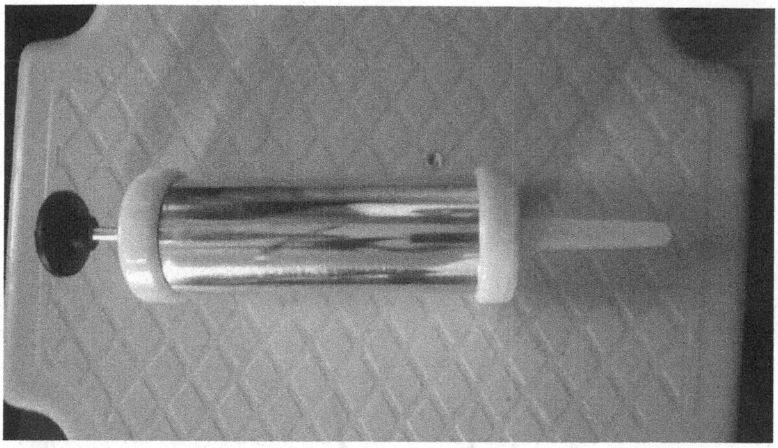

FIGURE 4.7 Instrument for medicinal enema using oleaginous substances. (Photo courtesy: Divakara Concepts, Coimbatore.)

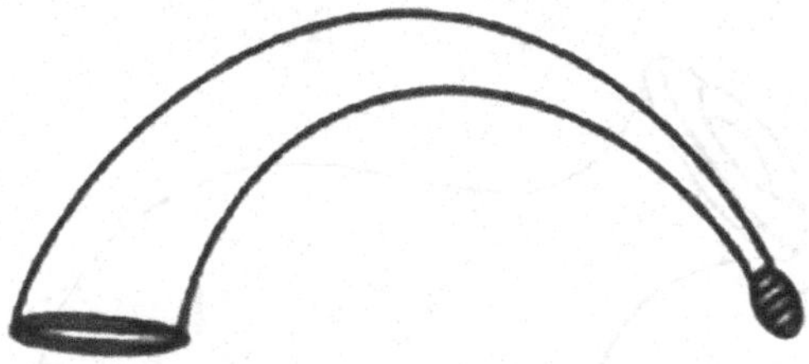

FIGURE 4.8 Horn-shaped instrument for drawing vitiated blood.

snēha dhūma and *madhyama dhūma*. The length of tubes for these will be 24, 32 and 40 finger-breadth, respectively (*Sū*. 21: 7–9).

Śṛṅgayantra should be eight or ten finger-breadth long, with a base that is three finger-breadth wide. Its tip should be hollow enough to hold a mustard seed and have appearance of nipple. *Śṛṅgayantra* is intended to remove vitiated blood and breast milk. The *yantra* is named so because of its similarity to horn of ox (Figure 4.8). The mouth of the *yantra* should be pressed firmly on the part from where the fluid is to be removed. The physician sucks through the tip as much air as possible and then seals the aperture at tip with beeswax. The vacuum created inside the *yantra* draws the fluid, which is discarded thereafter (*Sū*. 25: 26).

Alābuyantra is made out of *Alābu* fruit (*Lagenaria siceraria*). A gourd that is 12 finger-breadth long and 18 finger-breadth in circumference is to be selected. Its mouth should be circular and have a diameter of three or four finger-breadth. The pulp is scraped away, and the hollowed-out fruit is allowed to dry up and is used as *alābuyantra* (Figure 4.9). The body part is cut with a sharp instrument used in *sirāvēdha*. A fire is lit inside the pot by burning a strip of dry cloth, so as to create a vacuum. The *yantra* is immediately placed firmly over the body part cut. Blood flows into the *yantra* because of the pressure difference. *Kapha*-vitiated blood is drawn thus (*Sū*. 25: 27, Mukhopadhyaya, 1913a).

Ghaṭīyantra is similar to *alābuyantra* and is used in the disintegration of *gulma* (Figure 4.10). The *yantra* consists of a brass pot. A fire is lit inside as before and the pot applied to the surface of the body, covered with a cloth. It soon gets fixed to the body and the *gulma* is slowly raised, so that it can be crushed. *Alābuyantra* and *ghaṭīyantra* are similar to the modern-day cupping glasses (*Sū*. 25: 27–28, Mukhopadhyaya, 1913a).

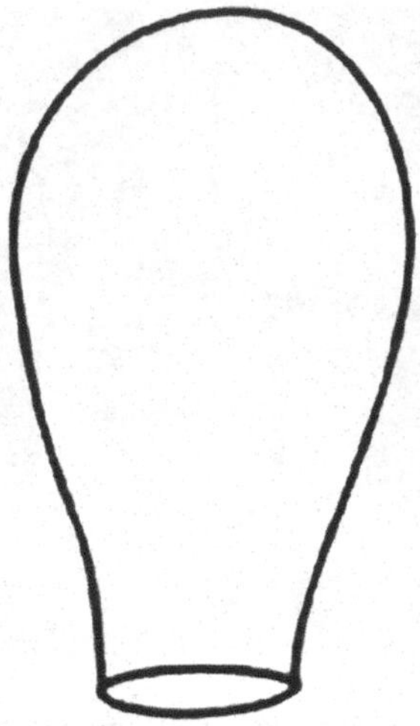

FIGURE 4.9 Gourd-shaped instrument for drawing vitiated blood.

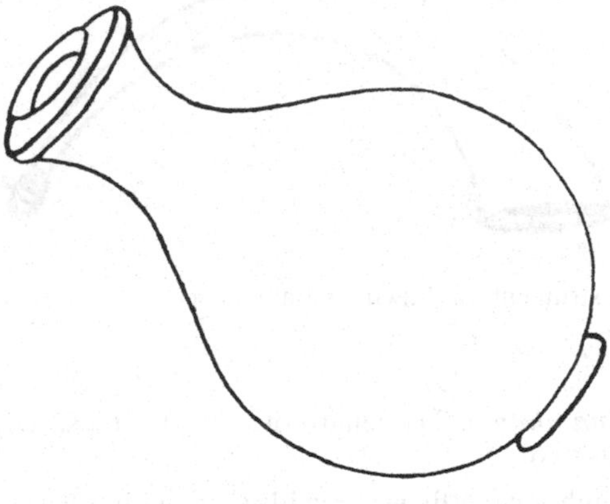

FIGURE 4.10 Brass pot for crushing phantom tumors.

4.2.1.5 Rods

Śalāka (rods or probes) are described to be of various kinds and serve several purposes. Therefore, their length and circumference vary according to some special uses required of them. Among them, two rods having heads resembling that of an earthworm are used in probing. Two rods with heads having shape similar to the half cotyledon of lentil (*Lens culinaris*) and length of eight or nine finger-breadth are intended to extricate sharp objects from body channels (*Sū*. 25: 28–30).

Six rods having their heads covered with cotton are used in the cleaning of sores and ulcers. Among them two, with length of ten and twelve finger-breadth are to be introduced into anal orifice. Another two which are six and seven finger-breadth long are used for cleaning nasal passages. External auditory canal is cleaned with two rods that are eight and nine finger-breadth long. Rod named *karṇaśōdhanaṃ* and which is intended for cleaning external auditory canal has tip similar to the leaf of *Aśvattha* (*Ficus religiosa*) and a head that resembles the bowl of a ladle (*Sū*. 25: 34–35).

Śalāka and *jāmbavouṣṭha* are used in chemical and thermal cautery. Each one of them is of three varieties: *sthūla*, *aṇu* and *dīrgha*. Large one is *sthūla*, long one is *dīrgha* and the one which is not very big and long is called *aṇu*. *Śalāka* with a half-moon-shaped blade is ideal to perform thermal cautery in *antarvidradhi*. Thermal cautery in nasal polyp and nasal tumor is performed with *śalāka* having a blade resembling one half of cotyledon of *Ziziphus mauritiana* (*Sū*. 25: 36–37). *Śalāka* used in chemical cautery are eight finger-breadth long and with tips having depression. The blades resemble the nails on little finger, middle finger and ring finger (*Sū*. 25: 38–39).

4.2.1.6 Hooks

There are six hooks (*śaṅku*). Among them, 2 which are 16 and 12 finger-breadth long and have a head resembling the hood of cobra are intended for raising the operable part into proper position. Another two *śaṅku* having length of ten and twelve finger-breadth and having a head similar to the fletch of an arrow are used for transferring or transporting. Two more hooks are useful in extraction. Hook that is eight finger-breadth long and with a bent tip is called *garbha śaṅku*. It is used for pulling out stillbirth (*mūḍhagarbha*). *Śaṅku* with head similar to that of a cobra is intended to be used in the surgical removal of urinary stone. It is known as *sarpāsya* or *sarpaphaṇāsya śaṅku*. *Śaṅku* with head similar to the fletch of an arrow is intended for extraction of teeth. It is known as *dantapātanam* (*Sū*. 25: 30–33).

4.2.1.7 Minor Instruments

Lodestone, rope, cloth, stone, hammer, leather strap, intestines of animals, tongue, hair, branches of trees, fingernail, mouth, teeth, time span, ripening of *dōṣas*, hands and feet, fear and cheerfulness are minor instruments (*anuyantra*). With intelligence and skill, a physician should identify the actions to be carried out with minor instruments (*Sū*. 25: 39–40).

The various actions to be performed by *yantra* include extraction after crushing (*nirghātana*), extraction after twisting or shaking (*unmathana*), filling up cavities (*pūraṇa*), cleansing the passage (*mārgaśuddhi*), bringing together (*vyūhana*), extraction (*āharaṇa*), bandaging (*bandhana*), rubbing or pressing (*pīḍana*), sucking or suction (*ācūṣaṇa*), lifting up (*unnamana*), pushing down (*nāmana*), loosening by repeated shaking (*cālana*), breaking (*bhaṅga*), rolling over or turning (*vyāvartana*) and straightening (*ṛjukaraṇa*) (*Sū*. 25: 41–42). Table 4.1 lists the various blunt instruments.

4.2.2 Sharp Instruments

Twenty-six sharp instruments (*śastra*) are used in surgical measures. A physician should get the *śastra* made by skilled blacksmiths. The *śastra* are to be of good shape, not frightening in appearance and having a bluish tinge. Among them, the *śastra* named *maṇḍalāgram* should be flat and curved like the nail on index finger (Figure 4.11). It is used for scraping, excising and cutting, in the eye disease *pōthaki* and *gaḷaśuṇṭhika*, a disease of throat. The *śastra* known as *vṛddhipatra* is shaped like a traditional shaving razor and is used for cutting, splitting, tearing and incising (Figure 4.12). Its tip has to be very thin for use in boil (*śōpha*) that is raised or deeply suppurated. These *śastra* have tips bent downward. *Utpalapatra* should have long blade, and *adhyardhadhāra* should be with short blade. They are used for excising and incising (Figure 4.12). *Sarpāsya* has a blade that is half finger-breadth long only. Polyps in nose and ear can be ablated with it. *Ēṣiṇi* is of two types. The one that is smooth and having a blade similar to the head of an earthworm is called *gaṇḍupadamukha*. It is employed to gauge the depth of sinuses. Another *ēṣiṇi* named *sūcimukha* is for splitting. Its tip is fine like a needle, and the base has a hole, through which a thread can be inserted. This thread is usually impregnated with alkali (*Sū*. 26: 1–9).

Vētasapatra is for puncturing. *Śarārimukha* and *trikūrcaka* are meant for multiple puncturing (*pracchāna*). *Kuśapatramukha* (Figure 4.12) and *āṭāvadana* are also used in *pracchāna*. These four instruments have two finger-breadth long blades. *Antarmukha* is used for *pracchāna*. Its blade should be one and half finger-breadth long and shaped like a half-moon. *Vrīhivaktraṃ* is used in venesection and abdominal tapping. Venesection is a measure carried out by cutting blood vessels, and abdominal tapping is resorted to in *mahōdara*, for draining fluid by making an incision on abdominal wall. *Kuṭhāri* (axe) is flat and shaped like the tooth of cow. It has a length of half finger-breadth and is used in venesection (*Sū*. 26: 9–12).

TABLE 4.1
Blunt Instruments Described in AH (*Sū.* 25: 1–42, 23: 12–13, Mukhopadhyaya, 1913a)

Serial No.	Name of *yantra*: Sanskrit Name	English Equivalent	Intended Function
I	***Svastika yantra***	**Cruciform Pincers**	Removal of sharp objects struck into bones
1	*Kaṅkamukha yantra*	Heron-faced pincer	
2	*Siṃhamukha yantra*	Lion-faced pincer	
3	*Ṛkṣamukha yantra*	Bear-faced pincer	
4	*Kākamukha yantra*	Crow-faced pincer	
5	*Sarpamukha yantra*	Serpent-faced pincer	
6	*Kuraramukha yantra*	Osprey-faced pincer	
II	***Sandaṃśa yantra***	**Pincher-Like Forceps**	
1	*Sandaṃśa yantra* 16 finger-breadth long (2)		Extrication of sharp objects struck inside skin, blood vessels, nerves, tendons and muscles
2	*Sandaṃśa yantra* six finger-breadth long (1)		Plucking out very sharp objects and overgrown eyelashes, extrication of very minute foreign objects
3	*Mucuṇḍi*	Teethed-forceps	Removal of tissue overgrown inside sinuses and pterygium remaining on sclera after surgery
III	***Tālayantra***	**Picklock-Like *yantra***	
1	*Ēkatāla*	Single-blade	Removal of sharp objects lodged inside external auditory canal
2	*Dvitāla*	Double-blade	
IV	***Nāḍīyantra***	**Tubular Instruments**	
1	*Kaṇṭhāvalōkini*	Throat speculum	Viewing and removal of sharp objects struck inside
2	*Śalyanirghātini*	The impellent	Removal of sharp objects struck inside throat
3	*Gōstanākāra arśōyantra*	Cow teat-shaped tubular instrument for hemorrhoids	Viewing the inside of hemorrhoid. In surgery and application of alkali
4	***Śami yantra***	Hemorrhoid crusher	Ablation of hemorrhoid
5	*Nāsādvārēṣaṇa yantra*	Rudimentary rhinoscope	Ablation of nasal tumor and nasal polyp
6	*Aṅgulītrāṇaka yantra*	Finger guard	To be used in trismus (lockjaw)
7	*Yōnivraṇēkṣaṇa yantra*	Vaginal speculum	Viewing the inside of vagina
8	*Nāḍīvraṇaśōdhana yantra*	Sinus-cleaning tube	Washing sinus and application of *taila*
9	*Picchānaḷika nāḍīyantra*	Canula	Draining fluid in *jalodara*
10	*Vastiyantra*	Rectal clyster	Performing *vasti*
11	*Dhūmayantra*	Smoking tube	Inhalation of medicinal smoke
12	*Śṛṅgayantra*	Horn instrument	Sucking of vitiated blood and breast milk
13	*Alābuyantra*	Gourd instrument	Drawing *kapha*-vitiated blood
14	*Ghaṭīyantra*	Brass pot	Pushing *gulma* outward
V	***Śalāka***	**Rod**	
1	*Gaṇḍūpadamukha* (2)	Earthworm probe	Searching (*anvēṣaṇa*)
2	*Masūradalavktra* (2)	Lentil-faced rod	Extrication of sharp objects from body channels
3	*Kārpāsavihitōṣṇiṣa* (6)	Swab probes	Removal of pus from anal orifice, nasal passages and external auditory canal
4	*Śalāka* (3)	Rod	*Agnikarma* and *kṣārakarma*
5	*Jāmbavouṣṭha* of small and large diameter (3)		*Agnikarma* and *kṣārakarma*
6	*Āścyōtana śalāka*	Collyrium probe	Application of collyrium (*Sū.* 23: 12–13)

(*Continued*)

TABLE 4.1 (*Continued*)
Blunt Instruments Described in AH (*Sū.* 25: 1–42, 23: 12–13, Mukhopadhyaya, 1913a)

Serial No.	Name of *yantra* Sanskrit Name	English Equivalent	Intended Function
VI	***Śaṅku***	**Hook**	
1	*Ahiphaṇavaktra* (2)	Serpent-hooded hook	Raising the operable part into proper position (*vyūhana*), removal of urinary stone
2	*Śarapuṅkhāsya* (2)	Fletch-shaped hook	Transferring (*cālana*), extraction of teeth
3	*Baḍiśākāra* (2)	Hook-shaped rod	Pulling out objects, removal of stillbirth

Śalāka is a rod-shaped *śastra* made of copper. It has blades at both sides, shaped like the bud of

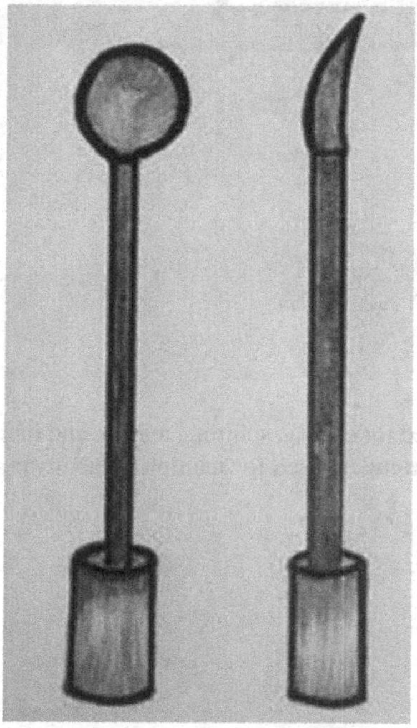

FIGURE 4.11 *Maṇḍalāgram* used for scraping, excising and cutting.

Barleria cristata (*Kuravaka*). It is useful in surgery of the eye disease *liṅganāśa* (near blindness). *Aṅgulīśastra* has shape similar to that of *vṛddhipatra* and *maṇḍalāgram*. It is hollow at the base and has a thread attached to it. It is used in excision and incision of diseases of throat. *Aṅgulīśastra* is worn on the index finger, tied with the thread and introduced into throat. *Aṅgulīśastra* is also known as *mudrikāśastra* (Figure 4.13). *Baḍiśa* is a sharp hook with a bent blade, meant for holding the operable part in diseases such as *gaḷaśuṇṭhika* and *arma* (pterygium) (Figure 4.14). *Karapatra* is used for cutting bones. It has a serrated cutting edge like a hacksaw. It has a handle for holding comfortably (Figure 4.15). *Kartari* is utilized for cutting sinews and ropes and is shaped like a scissors. *Nakhaśastra*, having curved and straight cutting edge, has two blades and is nine finger-breadth long. It is handy in extrication of small sharp objects and for performing excision, incision, multiple puncturing and scraping or scarification (*Sū.* 26: 13–18).

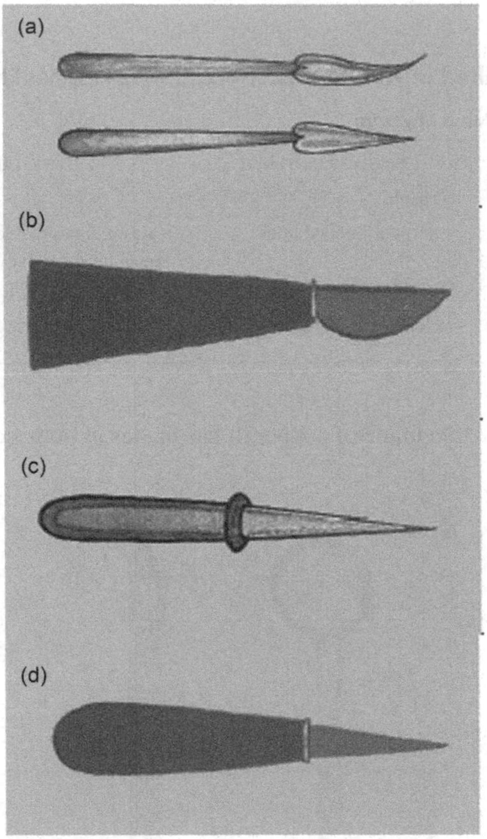

FIGURE 4.12 (a) *Vṛddhipatra* used for cutting, splitting, tearing and incising, (b) *Adhyardhadhāra* used for excising and incising, (c) *Kuśapatramukha* used for multiple puncturing, (d) *Utpalapatra* used for excising and incising.

FIGURE 4.13 *Mudrikāśastra* used for excision and incision.

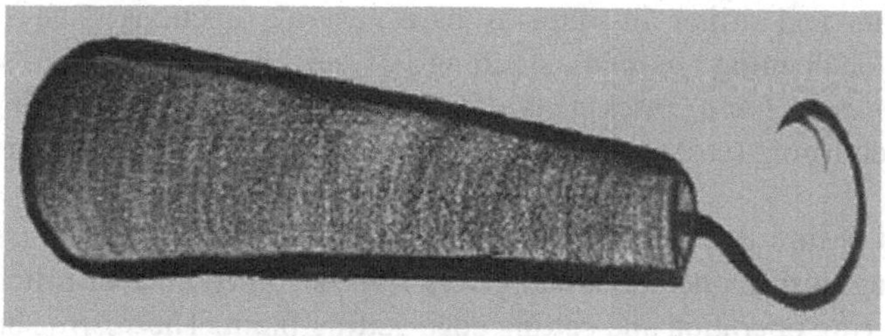

FIGURE 4.14 *Baḍiśa* meant for holding the operable part.

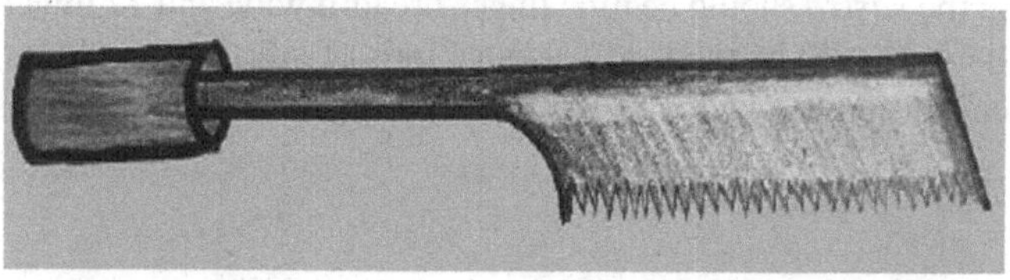

FIGURE 4.15 *Karapatra* used for cutting bones.

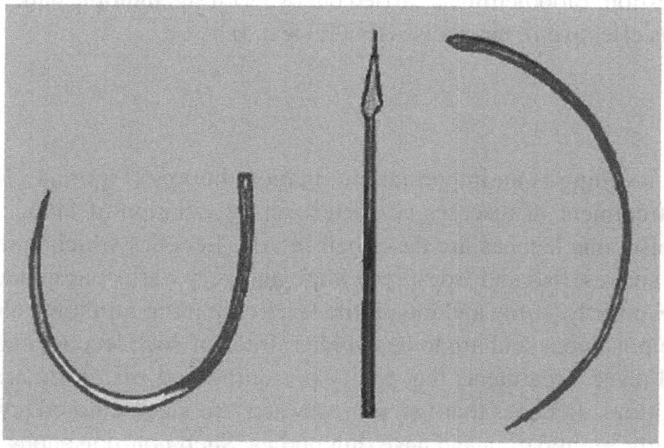

FIGURE 4.16 Needles meant for suturing and pricking body parts.

Dantalēkhana is a four-angled lancet, with cutting edge on one side and having a folded appearance. Dental plaques are scraped off with this *śastra*. Needles (*sūci*) are meant for suturing and pricking body parts (Figure 4.16). They are of three types. Three finger-breadth long *sūci* is for suturing skin in fleshy parts like thigh. Two finger-breadth long *sūci* is used for suturing wounds in parts having not much muscle and joints. *Sūci* having blade shaped like leaf of paddy and curved like a bow is suitable for suturing wounds in intestine, stomach and vital points. *Kūrca* or *sūcikūrca* is a set of seven or eight *sūci* fixed on a round piece of wood, with a handle. It is used for pricking in naevi, black pigmentation on face and hair loss. *Khaja* is a round-shaped *śastra* with eight pins projecting from it. Blood can be removed from nasal passages, using this *śastra*. *Karṇavēdhini* is a *sūci* with which ear lobule is pierced, so that gold ornaments can be worn on the ear lobe. *Āra* is a *śastra* with which boils are punctured, to judge the stage of suppuration. Leech, alkali, fire, glass, stone and fingernail are minor instruments (*anuśastra*) (*Sū.* 26: 19–28).

Sharp instruments and minor instruments have several functions. They include extraction (*utpāṭana*), tearing or opening (*pāṭana*), stitching (*sīvana*), probing (*ēṣaṇa*), scraping (*lēkhana*), multiple puncturing (*pracchāna*), pricking (*kuṭṭana*), excising (*chēdana*), incising (*bhēdana*), puncturing (*vyadhana*), churning (*mathana*), holding or grabbing (*grahaṇa*) and burning (*dahana*) (*Sū*. 26: 28–29).

Physician should ensure that *śastra* in his possession do not have defects (*śastradōṣa*) such as being blunt, breaking during operation, being very thin or too rotund, insufficient length, assuming curved shape during operation and cutting edge getting blunt. The *śastra* should be held firmly between the handle and the blade with index finger, middle finger and thumb while performing excision, incision and scraping. While doing puncturing, the *śastra* should be held at the end of the handle, with index finger and thumb. In the case of extraction, the handle should be held at its base. The wallet for storing the *śastra* should be nine finger-breadth wide and 12 finger-breadth long, with several compartments. It should be made of various kinds of silk, cotton cloth or thin leather. The *śastra* are to be placed inside, packed with wool (*Sū*. 26: 29–34). Table 4.2 lists the various sharp instruments.

4.3 BLOODLETTING

Raktamōkṣa or bloodletting is an important measure included in *pañcakarma*. It is carefully controlled removal of small volumes of impure blood, so as to detoxify the body, thereby hastening cure of the disease in question. Bloodletting is carried out by leeching, multiple puncturing and venesection. This measure is effective in many diseases (Table 4.3).

4.3.1 Leeching

Jaḷūkāvacaraṇa or leeching is an important elimination therapy (Figure 4.17). Leeching is recommended in the treatment of diseases characterized by vitiation of blood. Characteristics of several kinds of poisonous leeches are described in AH. Leeches which emerge from polluted water, carcasses of snakes, fish and frogs; red, white and very dark colored leeches are not to be used for leeching. Similarly, slimy and too motile leeches, having rainbow-colored lines on body and hairy ones, are poisonous and are to be avoided. Bites of such leeches cause itching, suppuration, fainting and fever. Treatments that pacify poisoning and *raktapitta* are recommended to manage these conditions. Leeches that live in freshwater are algae-colored, cylindrical in shape, have bluish lines in the frontal area and have thin bodies. Such non-poisonous leeches are dipped in buttermilk, rice washing or water mixed with turmeric paste. Thereafter, they are washed in fresh water. After rubbing the patient's body part with sand, clarified butter, breast milk or blood, the sucker at the anterior part of the leech is placed on the part from where blood is to be drawn. As the leech draws blood, it starts swelling up. Then the leech is covered with a wet cloth. Leeches have the ability to draw impure blood from a mixture of pure and vitiated blood. The leech should be disengaged if the subject feels pain or itching at the bitten site. Rock-salt mixed with sesame oil is smeared on the anterior sucker and fine bran dusted on the body to disengage the leech. Following this, the leech should be made to regurgitate the sucked blood and stored in a clean pot of water with sand at bottom. It is to be used again for leeching only after seven days. If impure blood is found to remain at the spot even after leeching, the wound is made to bleed by smearing a paste of turmeric, jaggery and honey. On complete removal of vitiated blood, a piece of cloth soaked in *Śatadhouta ghṛta* or paste of cold-potency herbs is to be applied on the wound (*Sū*. 26: 35–48). *Alābu* and *ghaṭīyantra* are not to be used for removing blood vitiated with *pitta*. Nevertheless, they can be used in *kapha-vāta*-vitiated blood. Blood vitiated with *kapha* is not to be removed with *śṛṅgayantra*. *Śṛṅgayantra* is ideal for removal of *vāta-pitta*-vitiated blood (*Sū*. 26: 49–50).

TABLE 4.2
Sharp Instruments Described in AH (*Sū*. 26: 1–34, Mukhopadhyaya, 1913b)

	Name of *śastra*		Intended Function	
Sl. No.	Sanskrit Term	English Equivalent	Sanskrit Term	English Equivalent
1	*Adhyardhadhāra*	Single-edged knife	*Lēkhana*	Scraping
			Bhēdana	Incising
			Pāṭana	Opening
2	*Aṅgulīśāstra*	Finger knife	*Chēdana*	Excising
			Bhēdana	Incising
3	*Antarmukha*	Scissors with half-moon-shaped blades	*Pracchāna*	Multiple puncturing
4	*Āra*	Awl	*Vyadhana*	Puncturing
5	*Āṭāvadana*	*Śāstra* with tip resembling beak of the bird *Āṭi*	*Pracchāna*	Multiple puncturing
6	*Baḍiśa*	Sharp hook	*Grahaṇa*	Holding
7	*Dantalēkhana*	Tooth scaler	*Lēkhana*	Scraping
8	*Ēṣiṇi*	Sharp probe	*Bhēdana*	Incising
			Ēṣaṇa	Probing
9	*Karapatra*	Saw	*Chēdana*	Excising
10	*Karṇavēdhini*	Ear lobule piercer	*Vyadhana*	Puncturing
11	*Kartarimaṇḍalāgram*	Scissor-shaped *śāstra*	*Chēdana*	Excising
12	*Khaja*	Churner	*Mathana*	Churning
13	*Kūrca*	Brush-like *śāstra*	*Kuṭṭana*	Pricking
14	*Kuśapatramukha*	*Śāstra* with *Kuśa* grass-like tip	*Pracchāna*	Multiple puncturing
15	*Kuṭhāri*	Axe-shaped *śāstra*	*Vyadhana*	Puncturing
16	*Maṇḍalāgram*	Round-headed knife	*Lēkhana*	Scraping
			Chēdana	Excising
17	*Nakhaśastra*	Nail parer	*Lēkhana*	Scraping
			Chēdana	Excising
			Bhēdana	Incising
			Pracchāna	Multiple puncturing
			Uddharaṇa	Raising
18	*Śalāka*	Rod	*Vyadhana*	Puncturing
19	*Śarārimukha*	Heron-faced *śāstra*	*Pracchāna*	Multiple puncturing
20	*Sarpāsya*	Serpent-faced *śāstra*	*Chēdana*	Excising
21	*Sūci*	Needle	*Sīvana*	Suturing
22	*Trikūrcaka*	Thin-edged sharp *śāstra*	*Pracchāna*	Multiple puncturing
23	*Utpalapatra*	Shaped like petal of blue lotus *Utpala*	*Chēdana*	Excising
			Bhēdana	Incising
			Pāṭana	Opening
24	*Vētasapatra*	Shaped like *Vētasa* leaf	*Vyadhana*	Puncturing
25	*Vṛddhipatra*	Shaped like leaf of the herb *Vṛddhi*	*Chēdana*	Excising
			Bhēdana	Incising
			Pāṭana	Opening
26	*Vrīhivaktraṃ*	Grain-shaped trocar	*Vyadhana*	Puncturing

TABLE 4.3
Diseases and Conditions in Which Bloodletting Is Indicated

	Diseases and Conditions		
S. No.	**Sanskrit Name**	**English Equivalent**	**References**
1	*Abhighātaja ōṣṭhakōpa*	Inflammation of lips caused by physical trauma	*Ut.* 22: 5–6
2	*Abhighāta jvara*	Fever due to physical trauma	*Ci.* 1: 166–167
3	*Ajaka*	Coppery red eruptions on cornea, with slimy surface and bloody discharge, resembling droppings of goat in color and appearance	*Ut.* 11: 51–52
4	*Apaci*	Cervical glandular swelling	*Ut.* 30: 14
5	*Apakva vidradhi*	Unripe abscess	*Ci.* 13: 16–18
6	*Ardita*	Facial palsy	*Ci.* 21: 43
7	*Arśas*	Hemorrhoids	*Ci.* 8: 28–29
8	*Aṣṭhīlika*	Small and hard swelling on the phallus	*Ut.* 34: 14
9	*Avamantha*	Large plaques with pain, on the middle part of the phallus	*Ut.* 34: 9
10	*Carmadala*	A skin disease characterized by itching, redness, pimples and elevated round areas caused by enragement of *pitta* and *kapha*	*Ci.* 19: 90
11	*Chardi vēgavidhāraṇāt*	Disorders arising from restrainment of urge to vomit	*Sū.* 4: 18
12	*Gaṇḍa*	Swelling on feet	*Ut.* 30: 17
13	*Granthi visarpa*	A type of skin disease with red and painful blisters arranged in the form of beads	*Ci.* 18: 34, 36–37
14	*Jālakagardabha*	Lymphangitis	*Ut.* 32: 6
15	*Kāca*	Advanced stage of cataract	*Ut.* 13: 80–82
16	*Kaṇṭharōga*	Diseases of throat (18)	*Ut.* 22: 54
17	*Kaphajābhiṣyanda*	Conjunctivitis due to enragement of *kapha*	*Ut.* 16: 18–19
18	*Kaphaja ślīpada*	Elephantiasis due to enragement of *kapha*	*Ut.* 30: 11
19	*Kēvala vātajvara*	Fever due to enragement of *vāta*	*Ci.* 1: 166–167
20	*Kṛṣṇasarpa daṃśanam*	Bite of black lance-hooded cobra	*Ut.* 36: 58
21	*Kṣataja śōpha*	Edema due to physical trauma	*Ci.* 17: 41
22	*Kṣataśukḷa*	Small growth on cornea turning the sclera red, with pricking pain and lachrymation	*Ut.* 11: 30
23	*Kuṣṭha*	Leprosy and other skin diseases	*Ci.* 19: 15–17, 55–56
24	*Lāñchana*	Congenital hyperpigmentation	*Ut.* 32: 15
25	*Liṅganāśam*	Blurred vision (near blindness)	*Ut.* 14: 29
26	*Lūtāviṣa*	Spider bite	*Ut.* 37: 69
27	*Madātyaya*	Alcoholism	*Ci.* 7: 107–108
28	*Mukhadūṣika*	Pimples	*Ut.* 32: 4
29	*Mukharōga*	Diseases of mouth	*Ut.* 22: 108
30	*Mūṣikaviṣa*	Rat bite	*Ut.* 38: 17–20
31	*Nīlika*	Naevi	*Ut.* 32: 15
32	*Ōṣṭhakōpa*	Inflammation of lips	*Ut.* 22: 108
33	*Pakṣmarōdha*	Trichiasis	*Ut.* 9: 34–36
34	*Pittajābhiṣyanda*	Conjunctivitis due to enragement of *pitta*	*Ut.* 16: 18–19

(*Continued*)

TABLE 4.3 (*Continued*)
Diseases and Conditions in Which Bloodletting Is Indicated

S. No.	Diseases and Conditions: Sanskrit Name	Diseases and Conditions: English Equivalent	References
35	*Pillarōga*	Group of 18 eye diseases that stay for a long time	*Ut.* 16: 47–48
36	*Pittaja granthi*	Glandular swelling caused by aggravation of *pitta*	*Ut.* 30: 3
37	*Pittaja gulma*	*Gulma* caused by enragement of *pitta*. Gulma is abdominal mass characterized by tumor-like hard, round mass, unstable in size and consistency, moving or immobile, situated in the bowel	*Ci.* 14: 70–71
38	*Pittaja ōṣṭhakōpa*	Inflammation of lips caused by aggravation of *pitta*	*Ut.* 22: 5–6
39	*Pittaja ślīpada*	Elephantiasis due to enragement of *pitta*	*Ut.* 30: 11
40	*Pittaja vṛddhi*	Inguinoscrotal swelling due to enragement of *pitta*	*Ci.* 13: 32
41	*Pittōnmāda*	Insanity due to enragement of *pitta*	*Ut.* 6: 46
42	*Pittōtkḷiṣṭa*	Inflammation of eyelids, with redness, burning sensation, exudation and pain	*Ut.* 9: 16–17
43	*Plīhōdara*	Enlargement of spleen	*Ci.* 15: 85, 97–98
44	*Pūyālasa*	Inflammation of inner canthus of eye	*Ut.* 11: 4
45	*Rājayakṣma*	Kingly consumption	*Ci.* 5: 66–70
46	*Rājilasarpa daṃśanam*	Bite of krait	*Ut.* 36: 80
47	*Raktaja granthi*	Glandular swelling caused by aggravation of *rakta*	*Ut.* 30: 3
48	*Raktaja karṇaśūla*	Pricking pain in ear due to enragement of *rakta*	*Ut.* 18: 16
49	*Raktaja vṛddhi*	Inguinoscrotal swelling due to enragement of *rakta*	*Ci.* 13: 32
50	*Raktōtkḷiṣṭa*	Inflammation of eyelids due to enragement of *rakta* and *tridōṣa*	*Ut.* 9: 16–17
51	*Rāsayanavidhi*	Rejuvenation therapy	*Ut.* 39: 3
52	*Sannipāta jvara*	Fever due to enragement of *tridōṣa*	*Ci.* 1: 149–151
53	*Śaradcarya*	Part of regimen for autumn season	*Sū.* 3: 50
54	*Sarṣapika*	Mustard-like pustules appearing inside or outside vagina and phallus	*Ut.* 34: 8
55	*Śuddhaśukḷa*	Eye disease characterized by white cornea and slight pain	*Ut.* 11: 42
56	*Sūryāvartam*	Migraine	*Ut.* 24: 11
57	*Śvitra*	Vitiligo	*Ci.* 20: 18
58	*Timira*	Cataract	*Ut.* 13: 47
59	*Vātajābhiṣyanda*	Conjunctivitis due to enragement of *vāta*	*Ut.* 16: 18–19
60	*Vātaja kuṣṭha*	Skin disease due to enragement of *vāta*	*Ci.* 19: 92–94
61	*Vātasōṇita*	*Vāta*-induced blood disorders	*Ci.* 22: 1–3
62	*Vidārika*	Hard tumor appearing in axilla and groins, resembling the root of the herb *Vidāri*	*Ut.* 32: 7
63	*Viṣamajvara*	Irregular fever	*Ci.* 1: 152
64	*Visarpa*	Cellulitis	*Ci.* 18: 1, 8, 10
65	*Visarpa jvara*	Fever due to cellulitis	*Ci.* 1: 166–167
66	*Visphōṭa jvara*	Fever due to eruptions on body	*Ci.* 1: 166–167
67	*Visphōṭaka kuṣṭha*	Skin disease characterized by white and bluish-black eruptions	*Ci.* 19: 90
68	*Vraṇa*	Wound, sore, ulcer	*Ut.* 25: 25–26
69	*Vṛścika daṃśanam*	Scorpion sting	*Ut.* 37: 31, 39
70	*Vyaṅga*	Black pigmentation on face	*Ut.* 32: 15

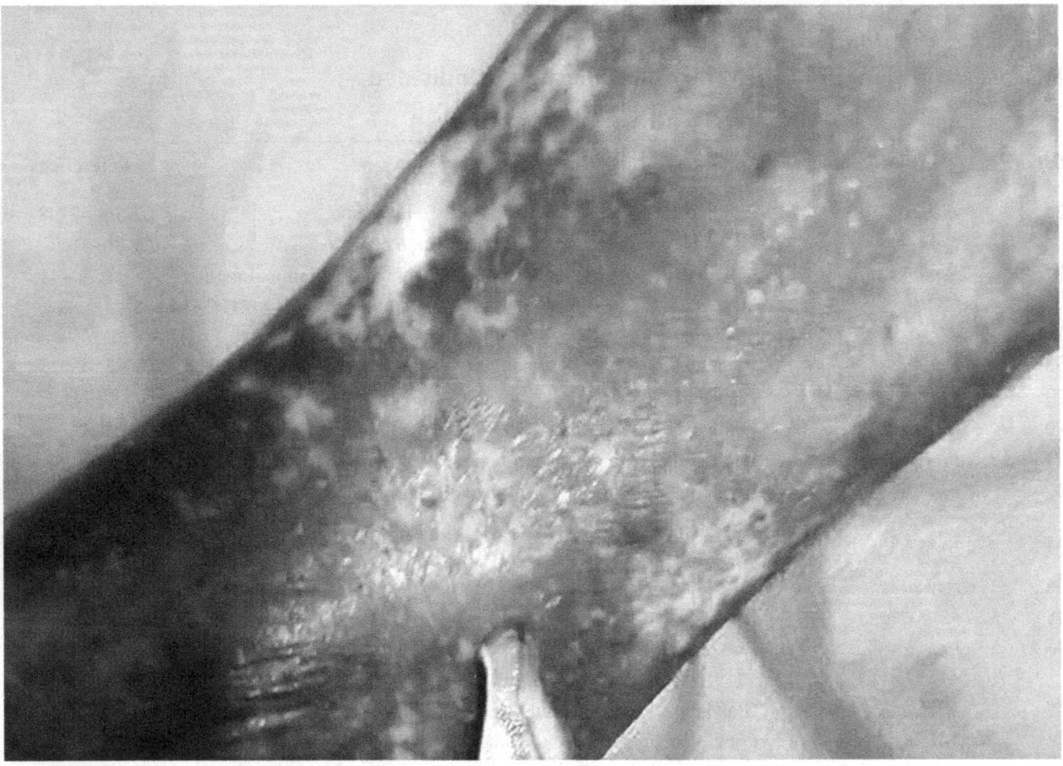

FIGURE 4.17 Bloodletting using leech. (Photo courtesy: Ayursoukhya Ayurveda, Kottayam).

4.3.2 Multiple Puncturing

Multiple puncturing (*pracchāna*) is suitable for removing localized or solidified blood. It is carried out using the instruments *śarārimukha* and *trikūrcaka* (*Sū.* 26: 9–10) (Figures 4.18 and 4.19). The area above the part intended to be subjected to multiple puncturing is bound firmly with rope or cloth. Thereupon, *pracchāna* is carried out from down to upward, without injuring nerves, ligaments, joints and bones. Pricked spot is not to be pricked again. Vitiated blood lodged at a particular site is removed by *pracchāna*. Leeching is effective in the removal of clotted blood, and *śṛṅgayantra* or *Alābuyantra* are suitable for removal of impure blood affecting tactile sensation. Vitiated blood circulating in the systemic circulation needs to be removed by venesection. *Śṛṅgayantra*, leech and *alābu* can be employed at sites of *vāta*, *pitta* and *kapha*, respectively. As cold-potency pastes are applied on patients subjected to bleeding, *vāta* may get enraged, resulting in edema with pricking pain and itching. These discomforts can be pacified by pouring warm clarified butter in a thin stream (*sēka*) (*Sū.* 26: 51–55).

4.3.3 Venesection

Pure blood has sweet and slightly salty tastes. It is cold in nature, mildly hot as well, and remains in fluid state. Color is that of red lotus flower, velvet mite (*indragōpa*) and blood of goat and rabbit. Blood maintains the human body. It gets vitiated by food that increases *pitta* and *kapha*. Such vitiation causes the appearance of cellulitis, abscesses, splenomegaly, *gulma*, lowering of abdominal fire; fever; diseases of mouth, eyes and head; stupor; polydipsia; salty taste in mouth; leprosy and other skin diseases; hemorrhages of obscure origin; acid eructation; and giddiness. Certain diseases that are curable do not subside even after treating them intelligently with opposite qualities such as coolness, warmth, unctuousness and dryness. These diseases have their origin in enragement of

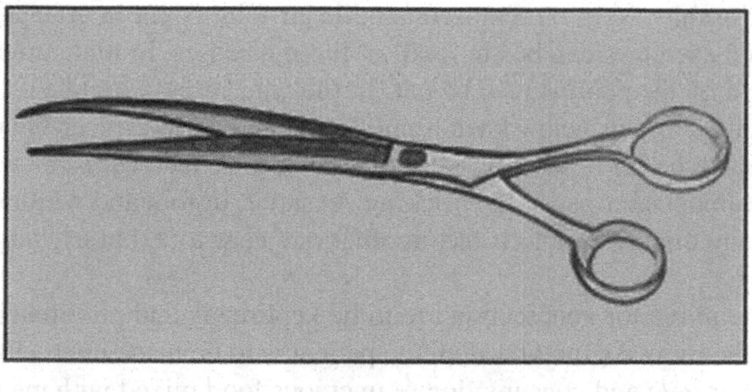

FIGURE 4.18 *Śarārimukha* used in multiple puncturing.

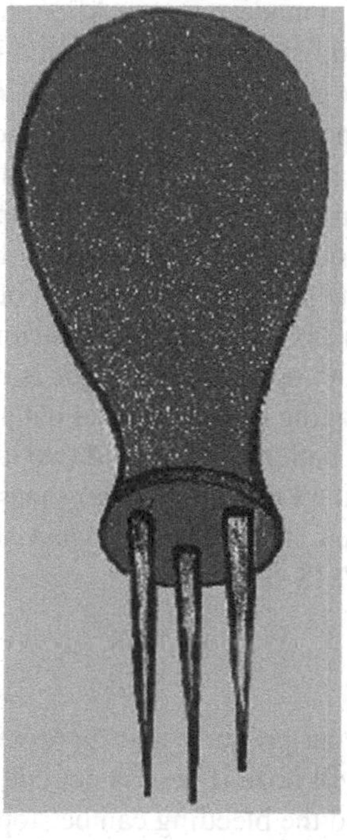

FIGURE 4.19 *Trikūrcaka* used in multiple puncturing.

blood. Veins are to be cut and impure blood drained in such conditions with venesection (*sirāvēdha*) (*Sū*. 27: 1–5).

Venesection is contra-indicated in children who have not crossed sixteen years of age and those above seventy years of age, persons who have not undergone oleation or sudation, those suffering from *vāta* diseases, pregnant and post-partum mothers, patients suffering from indigestion, hemorrhages of obscure origin, respiratory distress, cough, enlargement of abdomen, dysentery, vomiting, morbid pallor and edema all over body (*sarvāṅgaśōpha*). Vein that is not visible should not be cut. Venesection should not be performed on very cold, hot or windy days (*Sū*. 27: 6–9).

Vein on the forehead, outer canthus of eye or near nose are to be cut in diseases of head and eyes. In ear diseases, the one near ear is recommended. In diseases of nose, veins at tip of nose and in chronic rhinitis, those on nose and forehead are to be cut. Venesection can be carried out in diseases of mouth by cutting veins on tongue, lips, jaw and palate. Veins on chest, outer canthus of eye and forehead are cut in insanity. Vein on temporomandibular joint is cut in epilepsy. Vein that passes through area between eyebrows can be cut in all of these diseases. In male infertility and diseases of male genitals, vein on the phallus is to be cut. In the *vāta* disease known as *gṛdhrasi*, vein that lies four finger-breadth above or below knee is cut. Vein that lies two finger-breadth above the vital point *kṣipramarma* is to be cut in burning sensation in soles of feet, *vāta*-induced blood disorders, numbness of feet, palmoplantar psoriasis, pricking sensation in foot and whitlow. If vein specified in the treatment of any disease is not visible, another one near to it, but excluding vital points can be cut (*Sū*. 27: 9–18).

All instruments required for venesection are to be kept ready and propitiatory rites performed, before starting this measure. As the first step, the patient is to be made unctuous by administration of appropriate *ghṛta* or *taila* and consumption of unctuous food mixed with meat soup. The patient is thereafter subjected to sudation by staying in the proximity of a fireplace or getting exposed to hot sun. Following this, he sits on a knee-high stool and the hair on his head caught together and tied with fine cloth. This was the practice in ancient days, when having long hair was a fashion for men. However, this step is redundant in modern times. After placing elbow joints on knee and holding small bundles of cloth in clenched fists, the patient should rub firmly the vein on back of neck. The patient should strain himself by grinding teeth, breathe out strongly against a closed glottis (Valsalva's maneuver) and puff up cheeks. Another person should stand at the back of the patient and bandage the neck with a long strip of cloth, keeping his left index finger between the cloth and neck. This bandaging enables the vein to become visible. This procedure is applicable in cutting the vein on head, except the mouth. Following the bandaging with cloth, the physician taps firmly on the vein that is to be cut. After making sure that the vein has become visible, an incision is made at the center of the vein. Vein on nose is cut with the sharp instrument called *vrīhivaktraṃ*, raising the tip of nose with thumb. With the mouth open, tip of tongue is raised and the vein below the tongue is cut. For cutting vein on neck, bandaging is to be carried out above breast. Before cutting the veins on thoracic and abdominal areas, the patient should hold two large pieces of stone in clenched fists, both hands extended and placed on the knee joint. After massaging the area firmly, the physician cuts the vein on thorax and abdomen. Thus, the physician should identify the appropriate sites where cloth band is to be tied (*Sū*. 27: 18–33).

4.3.4 Management of Mishaps

Blood will flow profusely if vein is cut properly and there will not be bleeding on removing the bandage. A small volume of blood will ooze if vein is not cut sufficiently well. Vein that has been cut excessively will bleed heavily, and the bleeding can be stopped only with difficulty. Fear, fainting, looseness of the bandage, blunt instruments, eating heavily before the measure, remaining in empty stomach, urge for voiding stool and urine and failure to subject the patient to sudation are the causes of poor bleeding, even if the vein is cut properly. In the event of poor bleeding, powder of *Embelia ribes*, *Piper longum*, *Piper nigrum*, *Zingiber officinale*, *Curcuma longa* and *Valeriana wallichii* mixed with chimney soot, rock salt and sesame oil is to be applied on the site of venesection. Slightly warm mixture of sesame oil and rock salt is to be applied if bleeding is satisfactory (*Sū*. 27: 34–38).

The patient may sometimes faint during venesection. In such condition, the bandage is to be untied and cool air blown on body. Venesection can be attempted again after consoling the patient. If the patient faints again, the measure is to be carried out on the next day or three days later. Blood vitiated with *vāta* has bluish red color and flows fast. *Pitta*-dominant blood will be yellowish black and with bad odor. Because of the inherent heat, it will not clot and will be differently colored like a peacock feather. Blood vitiated with *kapha* will be oily, pale yellowish white in color, slimy and with thread-like lines. Blood vitiated with two *dōṣa*s will have signs of these *dōṣa*s, and *tridōṣa*-vitiated blood will be dirt-like and turbid. More than one *prastha* (768 ml) of blood should not be drained, however strong the patient may be. Excessive bleeding can be fatal or can cause severe *vāta*-induced discomforts like various kinds of pain and muscular contractions. In such condition, the patient may be smeared with medicinal oil all over the body (*abhyaṅga*) or served with meat soup and milk. Freshly drawn blood from a goat or rabbit can also be consumed (*Sū*. 27: 39–43).

The bandage is to be untied slowly and the wound cleaned with cold water, if blood has been removed to the satisfaction of the physician. The wound is to be bandaged with strip of cloth dipped in *taila*. Venesection can be carried out in the evening or next day if impure blood remains to be removed. If blood is excessively vitiated, *snēha* is applied topically and internally. Venesection can be performed after fifteen days. Disease will not progress even if a small volume of impure blood remains in body. Excessive removal of impure blood is not to be attempted. Impure blood that remains even after venesection is to be removed with *śṛṅgayantra*. Application of cold-potency medicines, treatment for *raktapitta*, various elimination measures and fasting are also recommended (*Sū*. 27: 43–47).

Hemostatic measures for arresting hemorrhage are to be resorted to, if bleeding continues for long time. Powder of *Symplocos spicata* bark, *Callicarpa macrophylla* flower, *Phaseolus mungo* seeds, *Glycyrrhiza glabra* root, red ochre, galena, black sand, pieces of earthen pot, ash of silk, barks and apical buds of *Ficus glomerata*, *Ficus retusa*, *Ficus benghalensis* and *Ficus religiosa* is to be applied on the wound. Cold infusion of *Padmakādi gaṇa* can be administered. The wound may be cauterized with a heated *śalāka* as well. The patient should observe strictly dietary regimen till all discomforts disappear. Food and beverages are to be moderately hot, cold, light and carminative. Pleasing appearance, good vision and auditory perception, appetite, mental and bodily well-being and satisfactory stamina are the signs of a person endowed with unvitiated blood (*Sū*. 27: 48–53).

4.4 EXTRACTION OF SHARP FOREIGN BODIES

Extraction of sharp foreign bodies (*śalyāharaṇa*) is the branch of surgery that relates to the removal of thorns, arrows and splinters from body. These sharp objects get lodged in the body in five directions: bent, straight, horizontal, upward and downward. Wound caused by a *śalya* (sharp object) is edematous, painful, bluish red in color, exuding blood, raised like a bubble, surrounded by small eruptions and soft in texture. When *śalya* is struck underneath skin, the wound will be discolored, hard, elongated and edematous. In the case of *śalya* struck inside muscle, the edema enlarges with burning sensation and suppuration. If pressed, the pain will be excruciating and the wound remains unhealed. *Śalya* struck in between muscles shows all the signs except edema. *Śalya* lodged in tendons causes convulsions, stiffness, loss of movement and many kinds of pain. It can be extricated only with difficulty. Veins will swell when struck with *śalya*. Channels of body will lose their ability to function properly when struck with sharp objects. Injuries to arteries (*dhamani*) result in hemorrhage with froth. There will be pain in limbs and feeling of nausea. Heavy friction and feeling of bones getting filled up result, if the joint of bone is affected. When struck in the bone, there will be edema and various kinds of pain. Rumbling sounds and bloating of abdomen result when alimentary canal is struck with *śalya*. Food, stool and urine may exude (*Sū*. 28: 1–9).

When a *śalya* strikes a healthy person straight into the body, the wound heals quickly. However, with *dōṣakōpa* and physical trauma, the wound flares up and becomes a sore. If a person harboring such a *śalya* is subjected to *abhyaṅga*, sudation and rubbing, the body part with concealed *śalya* experiences redness, pain, warmth and stiffness. Clarified butter placed on that spot will melt, and a paste applied will dry up fast. In such instances, it is to be inferred that the spot has a *śalya* hidden underneath. When a person with *śalya* hidden in the muscle, but with a healed wound is subjected to elimination measures like emesis and purgation, the body leans. The minor oscillations of such a body generate signs like redness, pain, warmth and stiffness, enabling the physician to locate the spot. In this way, spots with hidden *śalya* in muscles, bones, joints and alimentary canal can be recognized. Especially, *śalya* hidden inside bones can be recognized from the pain that is felt on applying measures like *abhyaṅga*, sudation, bandaging, pressing and rubbing. Pain felt in the joints on extending or folding limbs suggests the location of the hidden *śalya* (*Sū*.28: 10–15).

Subjects with such *śalya* hidden inside tendons, veins, body channels and arteries are to be seated in horse-drawn carriages having wheels with several broken spokes. The carriage should then be driven very fast along uneven paths with potholes. The spot harboring the *śalya* can be identified from the discomfort and pricking pain in the parts of body caused by the jerking movements of the carriage. The round, flat, square and triangular shape of the *śalya* can be inferred from the shape of the sore (*Sū*. 28: 15–18).

Śalya that strike the body from above and below are to be pulled out in the opposite directions. Those which hit horizontally should be removed by incising that part conveniently. *Śalya* that is lodged in an accessible spot is to be extricated by incising the area. However, *śalya* that have pierced into chest, armpits, groin and flanks are not to be removed by incising the area, as vital points exist there. The *śalya* are to be allowed to remain as such in those areas so that they dislodge themselves after the area gets suppurated. *Śalya* that can be removed by pulling with hand may be extricated in that way. *Śalya* which are visible but cannot be pulled out with hand are removed using blunt instruments such as *Siṃhamukha*, *Sarpamukha* and so on. Invisible *śalya* are recognized by the shape of the sore and extricated with blunt instruments like *Kaṅkamukha*, *Kākamukha* and *Kuraramukha*. *Śalya* lodged inside skin, veins, tendons and muscles are to be removed with two *sandaṃśa yantra* fitted with and without screws. Two *tālayantra* are required for removing *śalya* which are hollow. *Śalya* that are lodged in hollow spaces are to be removed with tubular instruments, and different kinds of *śalya* embedded in varied sites are to be extricated with appropriate *yantra* (*Sū*. 28: 19–24).

The spot where the *śalya* remains needs to be cut, even when *yantra* are used. After cessation of bleeding, warm clarified butter is to be poured over the area and then bandaged. *Śalya* remaining inside veins and tendons are to be removed with *śalāka*, by repeated shaking. Sometimes *śalya* may be firmly struck inside bones. In such cases, the physician should hold the subject firmly with both legs and pull it out. If the base of the *śalya* is not very strong, it should be pulled out by holding with blades of grass. *Śalyanirghātini nāḍīyantra* is to be used to pull out *śalya* that has pierced the organ and protrudes to the other side. *Śalya* that stays inside intestine is to be eliminated with purgation. Air, poison, breast milk, blood or water that gets trapped inside the body are to be sucked out using *śṛṅgayantra*. Lotus leaf stalk with a thread tied at one end is introduced into the gullet to extricate *śalya* lodged there. The leaf stalk is swirled, and in that process, the thread gets entangled with the *śalya*. The *śalya* and the leaf stalk are pulled out slowly. Fish bone struck inside throat is removed using a small bunch of human hair tied to a thread. The subject drinks an emetic medicine with the bunch of hair suspended in it. The thread should be pulled along with the bunch of hair, as soon as the subject vomits. During the process, the bone gets entangled in the hair. Powder of *Triphala* mixed with honey, sugar and clarified butter is to be consumed to heal the abrasion caused by the pulling out of the fish bone (*Sū*. 28: 25–38).

Śalya that cannot be pulled out through mouth or nose should be brought to alimentary canal and removed. A bolus of food may at times get stuck in the throat and acts as a *śalya*. The bolus is to be passed into the alimentary canal by making the subject drink water and hitting forcefully on his shoulders. Very small *śalya* that remain attached to the wound in eye may be removed using silk thread, human hair or by sprinkling water. A drowned person is to be shaken vigorously holding him upside down. Otherwise, he should be made to vomit. Ear canal is filled with brine or warm vinegar if small insect gets trapped there. In case the insect dies, treatment recommended for *karṇasrāva* is to be carried out. *Śalya* made of lac, gold or silver will get softer due to body heat over a period of time. *Śalya* such as bamboo splinters, sharp pieces of wood, animal horn, bone, tooth, hair, stone and iron do not disintegrate inside body. Such *śalya* vitiate quickly muscles and blood. If *śalya* is embedded deep inside muscle, the area surrounding it does not suppurate. Therefore, measures such as massaging, sudation, elimination and smoothening are to be performed. The area surrounding the *śalya* is to be subjected to *upanāha* using *tīkṣṇa* medicines, food, beverages and wounding with sharp instruments. Once the spot gets suppurated, the *śalya* can be removed by incising that area. *Śalya* are of numerous shapes and similarly *yantra* are also of different types. Therefore, any *śalya* needs to be removed with a *yantra* suitable to it. The sites that hold *śalya* are also innumerable. Certain *śalya* are not to be removed through the sites where they are lodged and they need to be moved to another place. *Śalya* are not removable with certain incisions. Such incisions need to be made bigger. *Śalya* lodged in some sites are not to be removed at all. They are left to remain there and get dislodged when the area gets suppurated. Therefore, after understanding the varied nature of *śalya*, the affected body parts and the *yantra* required to extricate them, an intelligent physician needs to know the proper shape of the *śalya*. He should then choose the appropriate *yantra* for its removal (*Sū*. 28: 38–47).

4.5 SURGERY

4.5.1 Stages of Edema

Sores and ulcers are formed from edema (*śōpha*). Therefore, a physician should strive to prevent edema from transforming to that stage, which necessitates the performance of surgery. Topical application of cold-potency medicines, bloodletting, emesis, purgation and such elimination measures are to be carried out for its prevention. Edema that remains in the unripe stage (*āma* stage) will be hard and firm. It will have color of the body, very little heat and pain. Edema undergoing suppuration will turn red and swollen (*pacyamāna* stage). There will be pricking pain and body ache. The patient experiences stiffness, anorexia, burning sensation, thirst, fever and sleeplessness. Semi-solid clarified butter placed over the edema melts and intense pain will be felt when touched. Edema that is ripe shows lessening of pain. Skin wrinkles with the central area protruding out. Itching and edema reduce and when pressed, the edema feels like water-filled bladder (*pakva* stage). Pain does not occur without enragement of *vāta*. Similarly, heat is not generated without the involvement of *pitta*, edema does not arise without increase in *kapha* and redness does not appear without the involvement of *rakta* (blood). Therefore, suppuration takes place due to *vāta-pitta-kapha* associated with *rakta*. Edema that has attained excessive transformation (*atipakva* stage) develops perforations and skin wrinkles significantly. Color changes into bluish red, and the edema turns hairless. Therefore, it is to be understood that edema has four stages: *āma*, *pacyamāna*, *pakva* and *atipakva* (*Sū*. 29: 1–7). It is notable that Vāgbhaṭa places equal importance to *rakta* and *tridōṣa* in the causation of suppuration, the healing of which calls for adoption of surgical measures. He adopts a centrist attitude, as AH is an attempt to synthesize the Ātreya and Dhanvantari schools of Ayurveda.

In edema due to *kapha*, *rakta* stays deep and undergoes suppuration. Therefore, signs of *pakva* stage may not be evident. Edema that feels cold, painful, hard and devoid of discoloration is termed *raktapāka*. Excessively suppurated edema and edema appearing in vital points and joints are not to be subjected to surgery. In these cases, in weak-willed individuals, those with poor stamina and in children, the edema is to be treated with alkalis. This procedure is called surgical rupturing (*dāraṇa*). Surgery is suitable for all others. Veins and tendons get damaged, and bleeding, pain and ruptures may occur when *śōpha* is cut in the *āma* stage. The edema may turn into *kṣatavisarpa* (skin eruption occurring due to physical trauma) also. If the edema is still left unoperated, pus will increase causing damage to veins, tendons, blood and muscles. A physician should neither operate upon an edema that has not reached ripe (*pakva*) stage nor leave a suppurated (*pakva*) edema unoperated (*Sū*. 29: 8–14).

4.5.2 Procedure of Surgery

In AH, there is no mention of a general anesthetic. So, it can be inferred that anesthesia was unknown in ancient days. Nevertheless, Vāgbhaṭa felt the necessity of such an agent to cause insensibility to pain. Before starting surgery, the patient should consume suitable food. Patients who drink *madya* (wine) regularly and those who cannot bear the pain are to consume strong *madya*. Fainting does not happen, as the patient has already consumed food. One who is stupefied with *madya* does not feel the pain caused by surgery. (This is the procedure except in stillbirth, urinary stone, diseases of mouth and enlargement of abdomen. In these cases, the patient is not to be fed or served *madya*.) Thereafter, the physician seats the patient facing east. He sits opposite to the patient and with a sharp surgical instrument cuts the edema quickly without injuring vital points and joints. After cutting the edema, the instrument is pulled out quickly. The incision is not to exceed a length of two finger-breadth. If all the pus is not drained completely, the physician should probe with *ēṣiṇiśastra*, finger or hair from pig tail and find out where the undrained pus remains. Incisions may be made in those areas also at two or three finger-breadth interspaces. The depth of sinus is to be gauged, and several incisions made if essential. The sinus is to be freed of pus and cleaned. Fearlessness, swiftness in action, sharp instruments, presence of mind and the ability to do the right thing at the right time are the qualities required in surgical operations. Incisions are to be made horizontally curved on forehead, eyebrows, gum of teeth, shoulders, abdomen, armpit, eyeball, lip, cheek, neck and groin. There is chance of vein and tendons getting cut, if at other sites incisions are made horizontally curved (*Sū*. 29: 14–23).

4.5.3 Suturing of Wounds

Fresh wounds caused by sword thrusts and stabbing are deep in nature. The edges of such a wound need to be sutured immediately. Suturing is unnecessary if the edges are not wide. Lumps that arise from excessive growth of adipose tissue are to be scraped with sharp instruments and their edges sutured. Lobules of ear cut by sharp weapons are also to be sutured without any delay. Suturing is recommended in wounds caused on head, eye sockets, nose, lip, cheek, anus and abdomen which are fleshier and do not move. Wounds caused on groin, armpits and body parts having less muscles and which move are not to be sutured. Similarly, wounds caused by alkalis, poisons and fire are also not to be sutured (*Sū*. 29: 49–52).

Before suturing the wound, the physician removes from it pieces of broken bone, clotted blood, grass and hair. The cut flesh, dislocated joint and bone are placed back in their original position, when bleeding ceases. Suturing is carried out using tendons, thread or hair. Sutures should not be made too far or too close. Very little or too large areas are not to be covered in a suture. The patient is consoled after completing the suturing. Paste of *Symplocos spicata*, *Glycyrrhiza glabra*, antimony sulphide (*añjana*), ash of burnt silk, *Callicarpa macrophylla* fruit and *Phoenix dactylifera* fruit is applied and bandaged as before. Wound having dry edges devoid of much blood should be made to bleed by scraping and then sutured (*Sū*. 29: 52–56).

4.5.4 Dressing of Wounds

After putting down the surgical instruments, the physician should console the patient with cheerful words and by sprinkling cold water on face. The area surrounding the operated edema is to be pressed slowly and all the pus removed. The wound is then cleaned using water boiled with suitable herbs. Following that, the wound is fumigated with powder of guggul gum, *Aquilaria agallocha* heartwood, *Brassica juncea*, *Ferula foetida*, gum of *Shorea robusta*, rock salt, *Acorus calamus* roots and neem leaves, mixed with clarified butter. Subsequently, a cylindrical pill (*varti*), smeared with paste of sesame seeds, clarified butter and honey, is to be introduced into the wound. The same paste is to be applied heavily on the wound as well, so as to cover it completely. Paste of powder of parched rice (*lāja*), mixed with clarified butter, is also applied on the site. A round piece of soft cloth is placed over the dressed wound and bandaged. Clean, thin, strong and fumigated strips of cloth and bandage are desirable in dressing of operated edema. Propitiatory rites are then performed to protect the patient from evil spirits. The patient is expected to wear always on his head herbs such as *Nervilia aragoana*, *Desmodium gangeticum*, *Pseudarthria viscida*, *Nardostachys jatamansi*, *Tricholepis glaberrima*, *Acorus calamus*, *Foeniculum vulgare*, *Desmostachya bipinnata* and *Brassica juncea* (*Sū*. 29: 24–31).

4.5.5 Bandaging

The material to be used for bandaging is selected according to the region and season. Hide and wool of sheep and silk fiber are of hot potency. Flax is of cold potency. Various kinds of cotton, ligaments and bark of trees are of cold-hot potency. Hot-potency bandages are appropriate in cold regions and in cold season. On the contrary, cold-potency bandages are for use in hot regions and in summer. Cold-hot potency bandages are suitable when heat and cold are not very severe. Copper, iron, lead and tin are to be used in wounds with high content of adipose tissue and phlegm. Broken bones are to be bandaged with flat pieces of wood, hide, bark of trees and stems of *Desmostachya bipinnata* (*Sū*. 29: 57–59).

Vraṇa or an ulcer can be bandaged in 15 different ways. They are *kōśa* (fingers, joints of fingers), *svastika* (ears, axillae, breast, joints), *muttōli* (penis, neck), *cīna* (outer canthus of eye), *dāma* (groins), *anuvēllita* (extremities), *khaṭvā* (cheek, lower jaw, temples), *vibandhaṃ* (back, abdomen, thighs), *sthagikā* (fingers, tip of penis, inguinoscrotal selling), *vitānaṃ* (organs that are thick, head), *utsaṅga* (hanging parts), *gōṣphaṇa* (nose, lips, joints), *yamakaṃ* (parts having two adjacent ulcers), *maṇḍala* (parts that are round) and *pañcāṅgi* (parts above shoulders). A wise physician chooses the one appropriate for the situation. Bandages on thigh, buttocks, armpit, groin and head are to be bound firmly. The bandage is tied neither firmly nor loosely in limbs, face, ear, chest, back, flanks, neck, abdomen, penis and scrotum. The bandage is to be loose in eyes and joints. However, in the predominance of *vāta* and *kapha*, the bandage over eyes and joints has to be tied neither tightly nor loosely. Bandage needs to be moderate over ulcers caused by *vāta* and *kapha*. Bandage should be renewed once in three days during pre-winter, winter and spring. Bandages should be bound moderately tight when the ulcer is caused by *pitta-rakta* vitiation. In summer and autumn, the bandage is to be renewed in the morning and evening. In the absence of bandage, the wound tends to get vitiated by mosquitoes, flies, cold and wind. Oily medicines will not stay on the wound for a long period. Healing takes place slowly and the healed spot may have discoloration. When bandaged, the wound heals very fast even if tissue is split or tendons and veins are cut. The wound will not be pressed even when the patient stands or lies on bed. The paste applied on wounds that heal slowly is to be covered with clean leaves of *Ficus glomerata*, *Ficus retusa*, *Ficus benghalensis*, *Ficus religiosa*, *Betula edulis*, *Terminalia arjuna* and *Neolamarckia cadamba* (*Sū*. 29: 59–72).

Sores of lepers, burn injuries, diabetic carbuncles, wounds caused by alkalis or poisons, edematous wounds, skin eruptions and those with pain and burning sensation are not to be bandaged. Flies lay eggs on sores and ulcers when they are not taken care of properly. The emerging worms feed upon the tissues causing pain, edema and bleeding. Such sores and ulcers are to be washed with decoction of *Surasādi gaṇa*. The decoction may be poured into sinuses as well. Paste of barks of *Alstonia scholaris*, *Pongamia glabra*, *Calotropis gigantea*, *Azadirachta indica* and *Manilkara hexandra* ground with cow urine can also be applied. The patient should stay without indigestion, exercise, sexual intercourse, excitement, anger and fear till the wound is healed completely and the affected part of body regains its health. These restrictions are to be observed religiously for six to seven months (*Sū*. 29: 72–79).

4.5.6 Dietary Regimen

The physician should advise the patient on the food he should consume and those which are to be avoided. Guideline to lead a morally sound life is also to be recommended. Sleeping during day time causes itching, redness, pain, edema and accumulation of pus. Thought, touch and sight of women are also to be avoided. Cooked barley, wheat and *Ṣaṣṭika* rice are ideal for convalescing patients. Broth (*yūṣa*) of *Lens culinaris*, *Cajanus cajan* and *Vigna radiata* can be consumed. So are curries prepared with *Holostemma annulare*, *Marsilea quadrifolia*, tender *Raphanus sativus*, *Solanum melongena*, *Amaranthus spinosus*, *Chenopodium murale*, *Momordica charantia*, *Trichosanthes cucumerina*, *Lagenaria siceraria*, *Punica granatum*, *Emblica officinalis*, clarified butter and rock salt. Boiled and cooled water can be drunk. Consumption of aged *Śāli* rice, cooked and mixed with clarified butter and *jāṅgala* meat, heals sores and ulcers very quickly. The patient should consume the recommended food in small measures at the appropriate time. Food will be easily digested in this way. However, *tridōṣa* get enraged if food remains undigested. Consequently, edema, pain, pus, burning sensation and bloating of abdomen may occur (*Sū*. 29: 32–38).

Contra-indicated food includes freshly harvested grains, sesame seeds, black gram, *madya*, meats other than *jāṅgala* variety, food prepared with milk or sugarcane juice; food predominantly acidic, salty or pungent; cold food and any food that is heavy and causes constipation and burning sensation. Their consumption causes many disorders. *Madya* is penetrating, hot, dry and acidic. Therefore, sores and ulcers get enraged by its use. Cold air is to be directed at the operated part, which should not be pricked or scratched. The part of body which underwent surgery should be moved frequently, but protecting it from any physical trauma. The patient who wishes to be healed quickly should spend his time listening to stories about kind persons, venerable elders and philosophers (*Sū*. 29: 39–43).

4.5.7 Medicated Wick

The bandage is to be removed three days after keeping the patient on controlled diet. The operated edema is to be washed, cleaned and fumigated as before. However, these measures should not be carried out on the second day, as doing do would harden the wound and generate intense pain. Wick of cotton threads smeared with paste of herbs (*vikēśika*) is to be inserted in the operated wound properly. It should not be too unctuous, dry or thin. The paste applied also should not have these characteristics. Too much unctuousness causes exudation of fluid and dryness causes tissues to break, produce cracks, bleeding and generate pain. Lightness or hardness of the medicated wick or its improper insertion causes friction in the edges of the operated edema. Medicated wick inserted inside cleanses putrefying, swollen, pus-filled sinus very quickly. Therefore, insertion of medicated wick is essential. At times an edema that has not reached ripe stage may get operated. In such cases, the edema is induced to undergo suppuration by adopting food that encourages suppuration, but which is acceptable in the treatment of edema. *Upanāha* is also to be carried out (*Sū*. 29: 43–48).

4.6 CHEMICAL CAUTERY

Alkali (*kṣāra*) is the best one among all sharp instruments and accessory instruments. Operations such as excision and incision can be carried out with it. Alkali can be applied in inaccessible wounds, sores and ulcers in which surgical measures cannot be performed, an example being nasal polyp. Additionally, alkali can be administered internally in conditions such as *gulma*, enlargement of abdomen, hemorrhoids, lowering of abdominal fire and poisoning. *Kalyāṇaka kṣāra* (*Ci*. 8: 140–144), *Pūtika kṣāra* (*Ci*. 14: 38), *Kṣārāgada* (*Ci*. 14: 103–107) and several unnamed *kṣāra* (*Ci*. 10: 53–56, 56–58, 58–60) are administered internally in their treatment. Alkali needs to be applied topically in moles, vitiligo, external piles, leprosy and other skin diseases, numbness, anal fistula, tumor, glandular swelling, obstinate ulcers and sinuses. Alkali should not be applied either internally or topically in many diseases including those due to vitiation of *pitta*, *rakta* and *vata*; fever; dysentery; heart disease; gynecological diseases; morbid pallor; distaste for food; and cataract. Application of alkali is contra-indicated in children and the aged. This measure is not to be carried out in winter, summer and rainy seasons (*Sū*. 30: 1–8).

4.6.1 Preparation of Alkali

Method of preparation on caustic alkali is described in AH. Roots, fruits, leaves, stems and branches of 19 climbers, shrubs and trees are the major ingredients of this formula. They include *Schrebera swietenoides* (*Kāḷamuṣkaka*), *Cassia fistula* (*Śamyāka*), *Musa paradisiaca* (*Kadaḷi*), *Erythrina variegata* (*Pāribhadraka*), *Terminalia paniculata* (*Aśvakarṇa*), *Euphorbia neriifolia* (*Mahāvṛkṣa*), *Trichosanthes tricuspidata* (*Kākajaṅgha*), *Butea monosperma* (*Palāśa*), *Vallaris solanacea* (*Asphōṭa*), *Holarrhena antidysenterica* (*Vṛkṣaka*), *Toona ciliata* (*Indravṛkṣa*), *Calotropis gigantea* (*Arka*), *Holoptelia integrifolia* (*Pūtika*), *Pongamia glabra* (*Naktamāla*), *Nerium indicum* (*Aśvamāra*), *Achyranthes aspera* (*Apamārga*), *Premna serratifolia* (*Agnimantha*), *Plumbago zeylanica* (*Agni*) and *Symplocos racemosa* (*Tilvaka*). These 19 herbs are cut into small pieces, dried under shade and heaped separately on clean stone slabs. Four *Luffa* species such as *Luffa echinata* (*Dēvadāḷi*), *Luffa cylindrica* (*Dhamārgava*), *Luffa acutangula* (*Kṣvēḍa*) and *Luffa acutangula* var. *amara* (*Kṛtavēdhana*) are also chopped, dried under shade and heaped separately. Spikes and reeds of *Hordeum vulgare* are also heaped similarly. Blocks of limestone are buried in the heap of *Kāḷamuṣkaka*. All the heaps are set on fire using dried chaff of *Sesamum indicum*. When the burnt ash has cooled down, one *drōṇa* (12,288 g) of ash of limestone with ash of *Kāḷamuṣkaka* and ¼ *drōṇa* (3,072 g) of ash of other herbs are mixed well in ½ *bhāra* (48,000 g) each of water and cow urine. The mixture is filtered through a thick cloth to collect a slimy, reddish, clear, alkaline liquid. The filtrate is poured into an iron cauldron and heated over fire, stirring frequently with a ladle. To this is added one *kuḍava* (192 g) each of pastes of limestone ash, calcined pearl, clay and conch shell; the uric acid-containing portion of excreta of rooster, peacock, falcon, heron and pigeon; bile of quadrupeds and birds; orpiment, realgar and five salts. The mass is heated with continuous stirring. Heating is stopped when the mass acquires thick consistency. On cooling, the caustic alkali is stored in an iron container and buried in a heap of barley. This is called medium alkali (*madhyama kṣāra*) (*Sū*. 30: 8–19).

For preparing mild alkali (*mṛdu kṣāra*), ash of limestone, excreta, bile and others are not to be made into a paste. They are added to the liquid filtrate, filtered and then discarded. Very caustic alkali (*tīkṣṇa kṣāra*) is prepared with more herbs. To the ingredients of mild alkali in process, paste of *Gloriosa superba* rhizome, *Baliospermum montanum*, *Plumbago zeylanica*, *Aconitum heterophyllum*, *Acorus calamus*, *Ferula foetida*, tender leaves of *Holoptelia integrifolia*, *Curculigo orchioides*, sonchal salt and *viḍa* salt is to be added. The prepared *tīkṣṇa kṣāra* needs to be buried in a heap of barley for seven days (*Sū*. 30: 20–22).

Tīkṣṇa kṣāra is applied in tumor and hemorrhoids which are caused by *vāta*, *kapha* and *mēdas*. *Madhyama kṣāra* is applied in these diseases with medium virulence. *Mṛdu kṣāra* is indicated in hemorrhoids due to enragement of *pitta* and *rakta*. On storage, the alkali may lose moisture and turn hard. It may be reconstituted with sufficient water. Moderate pungency, smoothness, sliminess, spreadability, white color, easy removability, causing moderate exudation of fluid and causing less pain are the desirable characteristics of an alkali (*Sū*. 30: 24–25).

4.6.2 Application of Alkali

The spot for applying alkali is made to bleed by cutting or scraping. Alkali picked on a cotton-covered *śalāka* is placed on the spot for a duration of 100 *mātra*. If pile mass is concealed in hemorrhoids, it is manipulated by finger and alkali smeared inside. In diseases of eyelids, the eyelid is to be kept everted, and with cornea covered by cotton swab, a small quantity of alkali is applied on the inner side of eyelid. In nasal tumor, the patient is seated facing the sun, tip of nose is raised and alkali applied inside the nostril for a duration of 50 *mātra*. The same procedure is applicable in aural polyp also. After the stipulated duration, the alkali is wiped away with a cotton-covered *śalāka*. The cauterized spot is then smeared with clarified butter-honey mixture and milk, whey or fermented rice water poured in a thin stream. Finally, a paste of cold-potency herbs mixed with clarified butter is applied. The patient should also consume food that causes the dissolution of *kapha* (*Sū*. 30: 27–34).

4.6.3 Management of Mishaps

Black color like that of ripe *Jambu* fruit (*Syzygium cumini*) and depression of the cauterized site are signs of optimum cauterization. Copper color, pricking pain and itching are signs of improper cauterization. Bleeding occurs at over-cauterized site. Fainting, burning sensation and fever may also appear. Excessive cauterization in anus causes in addition to these signs, obstruction or excessive voiding of stool and urine. Sexual vigor may be reduced also. Over-cauterization in nostrils causes contraction of nasal septum, burning sensation and anosmia. Ear and eye also will be affected similarly. In such adverse conditions, pouring of acidic liquids is helpful. Application of paste of sesame seeds mixed with honey and clarified butter and all measures that pacify *vāta-pitta* vitiation are recommended. If the cauterized part does not get torn from an ulcer because of being deep-rooted, paste of residue in the bottom of the tank that holds *Dhānyāmḷa* (rice gruel fermented with grains and herbs), *Glycyrrhiza glabra* and *Sesamum indicum* seeds is to be smeared. Thereafter, paste of *Sesamum indicum* seeds and *Glycyrrhiza glabra* mixed with clarified butter is to be applied for facilitating healing (*Sū*. 30: 34–39).

4.7 THERMAL CAUTERY

Thermal cautery (*agnikarma*) is performed in diseases that are not cured with medicines, chemical cautery and surgery. It is superior to chemical cautery and is applied on skin, muscle, veins, tendons, joints and bones. Thermal cautery is carried out with a lighted wick, tooth of cow, rock crystal and arrow head in conditions such as moles, tiredness, headache, glaucoma, wart and non-elevated mole. Honey, clarified butter, the blunt instrument *jāmbavouṣṭha* and jaggery are used in thermal cautery of hemorrhoids, anal fistula, glandular swelling, sores, sinus and ulcers (Figure 4.20). These ingredients and blunt instrument are also used for cauterizing veins, tendons, joints and bones in eye diseases, excessive bleeding, black mole and improper venesection. Thermal cautery is contra-indicated in those who are unfit for chemical cautery, wounds which carry foreign body or accumulated blood inside, persons with perforated abdomen and those suffering from severe sores. Emergence of crackling sound, appearance of lymph, color of palmyra palm fruit or neck of pigeon,

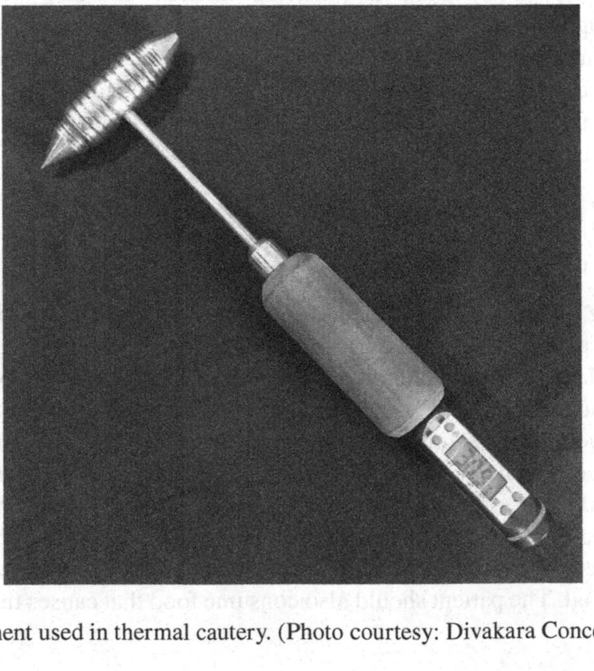

FIGURE 4.20 Instrument used in thermal cautery. (Photo courtesy: Divakara Concepts, Coimbatore).

quick healing and less pain are signs of properly performed thermal cautery (*sudagdha*). *Tuccha* is improperly conducted thermal cautery characterized by discoloration and burning sensation. In this mode, body part is slightly exposed to heat. A blister appears in *durdagdha* (improperly cauterized) type along with burning sensation all over the body. *Atidagdha* (excessively cauterized) is characterized by sagging of muscles, constriction, burning sensation, feeling of hot fume emanating from body, pain, damage to veins and tendons, thirst, fainting, worsening of the sore and death. *Tuccha* is pacified with warming of the spot, application of heat and hot potency medicines. Warming is ideal to pacify the immediate effects of a subclinical burn, rather than applying a cold potency medicine that may result in coagulation of blood. *Durdagdha* is treated at first with alternating cold and hot medicines, followed by cold alone. Paste of *Tinospora cordifolia*, *Maranta arundinacea* starch, *Ficus retusa* bark, *Santalum album* heartwood, red ochre and clarified butter is to be applied in *sudagdha*. It is to be followed by treatment for *pitta vidradhi*. All treatments for *pitta visarpa* are to be carried out fast in *atidagdha*. Skin blistered by hot clarified butter is to be pacified quickly with herbs which are very dry in nature (*Sū*.30: 40–52).

4.8 CONCLUSION

As AH is a text that integrates Ātrēya and Dhanvantari schools of Ayurveda, surgery is given a prominent position in the text. *Sūtrasthānam* devotes six chapters to describe venesection, extrication of foreign bodies, surgery, cauterization and the instruments required for performing these measures. Chapter 25 describes the 34 blunt instruments (*yantra*) that are used in surgery. These *yantra* are also useful in viewing the inside of affected body parts. Chapter 26 deals with 26 sharp instruments which can be used in performing 12 actions related to surgery. Leeching and multiple puncturing are also described in this chapter. Chapter 27 is devoted to venesection. Arrows, sharp objects and pebbles that get lodged inside the body can also be considered as *dōṣa* in one way, as they vitiate the body. Nevertheless, they cannot be extricated with administration of medicines. The removal of these foreign bodies is described in Chapter 28. Information on qualities of a surgeon, surgical procedure, suturing of operated parts, bandaging and dressing of wounds, wound healing formulations, foods that enhance healing of wounds and behavioral restrictions are detailed in Chapter 29. Finally, various aspects of cauterization such as chemical cautery, preparation of caustic alkalis, thermal cautery and treatment of cauterized parts are described in Chapter 30.

REFERENCES

Mukhopadhyaya, G. 1913a. Description of the blunt instruments. In *The Surgical Instruments of the Hindus*, 100–224. Calcutta: Calcutta University.

Mukhopadhyaya, G. 1913b. The *śastra* or the sharp instruments. In *The Surgical Instruments of the Hindus*, 225–281. Calcutta: Calcutta University.

5 Diagnosis and Treatment of Diseases

5.1 INTRODUCTION

The Sanskrit terms that denote disease are *rōgaḥ*, *pāpmā*, *jvara*, *vyādhi*, *vikāra*, *duḥkham*, *āmayaḥ*, *yakṣmā*, *ātaṅka*, *gada* and *ābādhaḥ*. Comprehensive knowledge of disease is gained using five parameters, namely *nidāna*, *pūrvarūpa*, *rūpa*, *upaśaya* and *samprāpti*. *Nidāna* means cause of the disease. The signs and symptoms that herald the appearance of a disease are called *pūrvarūpa*. They give vague information about the nature of the disease in the making, while its relationship with the *dōṣa* is not evident. Slowly the disease progresses and produces distinct signs and symptoms. This is called *rūpa* (form). Temporary adoption of medicines, food and measures that act against the cause and the disease itself is indicated by the term *upaśaya*. For example, a physician is confronted by a patient suffering from skin disease. However, he is unable to identify clearly the predominant *dōṣa* involved – whether it is *kapha*-dominant or *pitta*-dominant. Therefore, a *pitta*-pacifying cold-potency medicine is prescribed. However, instead of evoking a positive effect, the medicine causes discomfort to the patient, indicating that the treatment is *anupaśaya* (incompatible) to the patient. So, the physician changes the strategy and advises to consume a *kapha*-pacifying medicine with *tīkṣṇa* and *uṣṇa* qualities. The changed medication promptly causes a beneficial effect, suggesting that *kapha*-lowering treatment is *upaśaya* (compatible) to the patient. Definite treatment strategy is thereafter planned on the basis of information gained from *upaśaya*, which can also be studied in relation to food, beverages and measures. As the disease progresses, *dōṣa*s get vitiated and *dhātu*s also undergo pathological changes. These developments are denoted by the term *samprāpti*, which is inferred by careful observation of the symptoms and signs during treatment (*Ni.* 1: 1–10).

5.2 BASIC CONCEPTS

These five-fold characteristics can be understood better using *jvara* (fever) as an example. Due to effects of dietary or behavioral indiscretions or season of the year, the *tridōṣa* get enraged and reach *āmāśaya* (stomach). There they interact with *āma* (toxins) derived from improper digestion of food and obstruct body channels (*srōtāmsi*). Thereafter, they displace *agni* from its preferred site, move along with it and spread in the entire body, raising temperature. The *dōṣa* vitiation is unclear at this stage and prodromes (*pūrvarūpa*) appear. The *pūrvarūpa* of *jvara* are lassitude, disinclination to indulge in any activity, heaviness of body, inability to discriminate tastes; anorexia; yawning; lachrymation; pressing pain all over body; indigestion; slight breathing; hypersomnia; horripilation; cramps in calf muscles; tiredness; interest in food having sour, salty and pungent tastes; aversion toward sweetness; excessive thirst; and irritability (*Ni.* 1: 3–5, *Ni.* 2: 3–6).

From the *pūrvarūpa* stage, the disease progresses further and exhibits characteristic symptoms of *jvara* (*rūpa*). If left untreated, the disease advances and results in further vitiation of *dōṣa*s and pathological changes in *dhātu*s. These developments are denoted by the term *samprāpti*. On the basis of *samprāpti*, the disease entity *jvara* is subdivided into eight types, namely *vāta jvara*, *pitta jvara*, *kapha jvara*, *vāta-pitta jvara*, *kapha-vāta jvara*, *kapha-pitta jvara*, *vāta-pitta-kapha jvara* and *āgantu jvara*. *Samprāpti* is studied from five angles (*Ni.* 1: 5–8).

Assorting diseases on the basis of their characteristics into groups like 8 *jvara*, 5 *kāsa* (cough), 18 *kuṣṭha* (skin diseases) and so on is known as *saṅkhyāsamprāpti*. Measuring the intensity of vitiation of *vāta*, *pitta* and *kapha* in a disease is known as *vikalpasamprāpti*. The physician is expected to

DOI: 10.1201/9781003148296-5

quantify the destabilization of each *dōṣa* to diagnose the disease accurately. *Prādhānyasamprāpti* means ascertaining the independent nature of the disease (*svatantra rōgam*) or its association with other diseases (*paratantra rōgam*). For example, a patient presents himself with *pramēha* and *pramēha piṭaka* (diabetic carbuncles). In this case, *pramēha* is the independent disease (*svatantra rōgam*) and *pramēha piṭaka* is the associated disease (*paratantra rōgam*). Studying the aspects of *prādhānyasamprāpti* helps the physician to decide the line of treatment – whether to treat the independent disease first or the associated disease. The strength (*bala*) and weakness (*balahīnata*) of a disease are judged by observing meticulously how many components of causative factors (*nidāna*), prodromes (*pūrvarūpa*) and characteristic symptoms (*rūpa*) are manifested in the patient. Thus, *balasamprāpti* of a disease is a measure of its intensity. *Kālasamprāpti* is an estimate of the influence of day, night, season and consumption of food on the increase or decrease of the *dōṣa*(s) in the disease in question. For example, the components of *tridōṣa are* dominant in various phases of day, night, season and the period between consumption of food and its digestion (*Sū.* 1: 8, *Ni.* 1: 9–11).

The cause of all diseases is the destabilization of *dōṣa*s resulting from dietary or behavioral indiscretions and effects of season. For example, *vāta* gets deranged by consumption of food having bitter, pungent, astringent tastes, consumption of small quantities of food, oil-free food, having food at odd times, suppression of natural urges, keeping awake at night, speaking loudly, excessive adoption of measures like *vamana* and *virēcana*, fear, grief, worry, overdoing exercises and too much indulgence in sexual activity. *Vāta* also increases in rainy season and in the last one-third of day, night and period between consumption of food and completion of its digestion (*Ni.* 1: 14–15). *Pitta* gets destabilized by consumption of food having pungent, sour, salty, hot and penetrating attributes and having the ability to cause *vidāha* (food turning acidic in stomach). *Pitta* also gets enraged by anger and during autumn, midday, midnight and the midpoint between consumption of food and completion of its digestion (*Ni.* 1: 16). *Kapha* increases by consumption of unctuous, heavy and cold food having sweet, sour and salty tastes, sedentary life style, indigestion, sleeping during day time, excessive use of roborant food, medicines and improper performance of *vamana* and *virēcana* measures. *Kapha* increases soon after consumption of food, spring season and beginning of day and night (*Ni.* 1: 17–18).

Tridōṣa increase if causes for destabilization of all the three *dōṣa*s are involved such as consumption of food having complex and incompatible combination of qualities, overeating and indigestion. Other causative factors of vitiation of *tridōṣa* are stale-fermented beverages, polluted water, decomposed or desiccated meat, poisoning, overeating, staying in mountainous areas without sufficient protection, sinful actions and spiritual influences (*Ni.* 1: 19–23). The section *Nidānasthānam* describes the causative factors, prodromes, characteristic symptoms, nosology and prognosis of 40 major groups of diseases belonging to the branch of *kāyacikitsa* (Table 5.1). The etiology, prodromes, symptoms, nosology and prognosis of diseases of the other seven limbs of *Aṣṭāṅga Ayurveda* (*śalya*, *śālākya*, *kaumārabhṛtya*, *agadatantra*, *bhūtavidya*, *rasāyana*, *vājīkaraṇa*) are described in *Uttarasthānam*. In addition to these, treatment of poisoning and bites of poisonous snakes, spiders, insects, rats and rabid dogs is described in *Uttarasthānam* (Table 5.2). These conditions belong to the *āgantuja* type of diseases – diseases caused by exogenous factors. However, their effects are mediated by *tridōṣa*.

5.3 DIAGNOSIS OF DISEASES

AH teaches that symptoms and signs (*lakṣaṇa*) are the only tools for gauging the underlying pathology of a disease state. Therefore, a quick-witted physician should strive to collect as many symptoms or prodromes as possible, using the principles of *darśana* (observation), *sparśana* (palpation) and *praśna* (interrogation) (*Sū.* 1: 22, *Sū.* 12: 54–55). Observation of the patient reveals information on complexion of body, color and shape of the affected part, whether the patient is emaciated, of unsound mind, weak or fearful, and severity of the disease. Hardness or softness of the affected

TABLE 5.1
Diseases Described in *Nidānasthānam*

Sl. No.	Name of Disease	Subgroup	Reference to Text
1	*Jvara* (8)[a]	Vāta jvara	*Ni.* 2: 10–79
		Pitta jvara	
		Kapha jvara	
		Vāta-kapha jvara	
		Vāta-pitta jvara	
		Pitta-kapha jvara	
		Sannipāta jvara	
		Āgantu jvara	
2	Raktapitta (3)	Ūrdhvaga raktapitta	*Ni.* 3: 4–16
		Adhōga raktapitta	
		Ūrdhvaga-adhōga raktapitta	
3	Kāsa (5)	Vāta kāsa	*Ni.* 3: 17–38
		Pitta kāsa	
		Kapha kāsa	
		Kṣata kāsa	
		Kṣaya kāsa	
4	Śvāsa (5)	Kṣudra śvāsa	*Ni.* 4: 1–18
		Tamaka śvāsa	
		Chinna śvāsa	
		Mahā śvāsa	
		Ūrdhva śvāsa	
5	Hidhmā (5)	Bhaktōdbhava	*Ni.* 4: 18–31
		Kṣudra	
		Yamaḷa	
		Mahati	
		Gambhīra	
6	Rājayakṣma (3)	Vātaja rājayakṣma	*Ni.* 5: 1–23
		Pittaja rājayakṣma	
		Kaphaja rājayakṣma	
7	Svarasādam (6)	Vātaja svarasādam	*Ni.* 5: 24–27
		Pittaja svarasādam	
		Kaphaja svarasādam	
		Tridōṣaja svarasādam	
		Kṣayaja svarasādam	
		Medōja svarasādam	
8	Aruci (3)	Vātāruci	*Ni.* 5: 28–29
		Pittāruci	
		Kaphāruci	
9	Chardi (5)	Vātaja chardi	*Ni.* 5: 30–37
		Pittaja chardi	
		Kaphaja chardi	

(Continued)

TABLE 5.1 (*Continued*)
Diseases Described in *Nidānasthānam*

Sl. No.	Name of Disease	Subgroup	Reference to Text
		Tridōṣaja chardi	
		Dviṣṭārthayōgaja chardi	
10	Hṛdrōga (5)	Vātaja hṛdrōga	*Ni.* 5: 37–45
		Pittaja hṛdrōga	
		Kaphaja hṛdrōga	
		Tridōṣaja hṛdrōga	
		Kṛmi hṛdrōga	
11	Tṛṣṇa (6)	Vātaja tṛṣṇa	*Ni.* 5: 45–57
		Pittaja tṛṣṇa	
		Kaphaja tṛṣṇa	
		Tridōṣaja tṛṣṇa	
		Rasakṣayaja tṛṣṇa	
		Upasarga tṛṣṇa	
12	Madātyaya (4)	Vātaja madātyaya	*Ni.* 6: 14–23
		Pittaja madātyaya	
		Kaphaja madātyaya	
		Tridōṣaja madātyaya	
13	Mada (7)	Vātaja mada	*Ni.* 6: 24–30
		Pittaja mada	
		Kaphaja mada	
		Tridōṣaja mada	
		Raktaja mada	
		Madyaja mada	
		Viṣaja mada	
14	Mūrcha (4)	Vātaja mūrcha	*Ni.* 6: 30–35
		Pittaja mūrcha	
		Kaphaja mūrcha	
		Tridōṣaja mūrcha	
15	Sannyāsa		*Ni.* 6: 36–39
16	Arśas (6)	Vātārśas	*Ni.* 7: 8–59
		Pittārśas	
		Kaphārśas	
		Samsargārśas	
		Sannipātārśas	
		Raktārśas	
17	Udāvarta		*Ni.* 7: 46–52
18	Atisāraḥ (6)	Vātātisāraḥ	*Ni.* 8: 1–15
		Pittātisāraḥ	
		Kaphātisāraḥ	
		Sannipātātisāraḥ	
		Bhayātisāraḥ	

(*Continued*)

TABLE 5.1 (*Continued*)
Diseases Described in *Nidānasthānam*

Sl. No.	Name of Disease	Subgroup	Reference to Text
		Śōkātisāraḥ	
19	Grahaṇi (4)	Vāta grahaṇi	*Ni.* 8: 15–30
		Pitta grahaṇi	
		Kapha grahaṇi	
		Sannipāta grahaṇi	
20	Mūtrāghāta (20)	Vātaja mūtrakṛchra	*Ni.* 9: 1–40
		Pittaja mūtrakṛchra	
		Kaphaja mūtrakṛchra	
		Sannipātaja mūtrakṛchra	
		Vātāśmari	
		Pittāśmari	
		Kaphāśmari	
		Śukḷāśmari	
		Vātavasti	
		Vātāṣṭhīla	
		Vātakuṇḍalika	
		Mūtrātītaṃ	
		Mūtrajaṭharaṃ	
		Mūtrōtsaṅgaṃ`	
		Mūtragranthi	
		Mūtraśukḷaṃ	
		Viḍvighātaṃ	
		Uṣṇavātaṃ	
		Mūtrakṣayaṃ	
		Mūtrasādaṃ	
21	Pramēha (20)	Udakamēha	*Ni.* 10: 1–23
		Ikṣumēha	
		Sāndramēha	
		Surāmēha	
		Piṣṭamēha	
		Śukḷamēha	
		Sikatāmēha	
		Śītamēha	
		Śanairmēha	
		Lālāmēha	
		Kṣāramēha	
		Nīlamēha	
		Kāḷamēha	
		Hāridramēha	
		Māñjiṣṭhamēha	
		Raktamēha	

(*Continued*)

TABLE 5.1 (*Continued*)
Diseases Described in *Nidānasthānam*

Sl. No.	Name of Disease	Subgroup	Reference to Text
		Vasāmēha	
		Majjamēha	
		Hastimēha	
		Madhumēha	
22	Pramēhapiṭaka (10)	Śarāvika	*Ni.* 10: 25–36
		Kachapika	
		Jālini	
		Vinata	
		Alaji	
		Masūrika	
		Sarṣapika	
		Putriṇi	
		Vidārika	
		Vidradhi	
23	Vidradhi (6)	Vātaja vidradhi	*Ni.* 11: 1–21
		Pittaja vidradhi	
		Kaphaja vidradhi	
		Sannipāta vidradhi	
		Raktaja vidradhi	
		Kṣata vidradhi	
24	Vṛddhi (7)	Vāta vṛddhi	*Ni.* 11: 21–32
		Pitta vṛddhi	
		Kapha vṛddhi	
		Rakta vṛddhi	
		Mēdōvṛddhi	
		Mūtravṛddhi	
		Āntravṛddhi	
25	Gulma (8)	Vāta gulma	*Ni.* 11: 32–59
		Pitta gulma	
		Kapha gulma	
		Vāta-Pitta gulma	
		Vāta-kapha gulma	
		Pitta-kapha gulma	
		Sannipāta gulma	
		Raktagulma	
26	Ānāham		*Ni.* 11: 60
27	Aṣṭhīla		*Ni.* 11: 61
28	Pratyaṣṭhīla		*Ni.* 11: 61
29	Pratūni		*Ni.* 11: 62–63
30	Udaram (8)	Vātōdaram	*Ni.* 12: 1–46
		Pittōdaram	

(*Continued*)

TABLE 5.1 (*Continued*)
Diseases Described in *Nidānasthānam*

Sl. No.	Name of Disease	Subgroup	Reference to Text
		Kaphōdaram	
		Sannipatōdaram	
		Pḷīhōdaram/yakṛdōdaram	
		Baddhōdaram	
		Kṣatōdaram	
		Jalōdaram	
31	Pāṇḍu (5)	Vāta pāṇḍu	*Ni.* 13: 1–15
		Pitta pāṇḍu	
		Kapha pāṇḍu	
		Sannipāta pāṇḍu	
		Mṛtjanya pāṇḍu	
32	Kāmala		*Ni.* 13: 15–17
33	Kumbhakāmala		*Ni.* 13: 17–18
34	Halīmakam		*Ni.* 13: 18–20
35	Śōpha (9)	Vāta śōpha	*Ni.* 13: 21–42
		Pitta śōpha	
		Kapha śōpha	
		Vāta-pitta śōpha	
		Vāta-kapha śōpha	
		Pitta-kapha śōpha	
		Sannipāta śōpha	
		Abhighāta śōpha	
		Viṣa śōpha	
36	Visarpam (8)	Vāta visarpam	*Ni.* 13: 43–67
		Pitta visarpam	
		Kapha visarpam	
		Vāta-pitta visarpam	
		Vāta-kapha visarpam	
		Kapha-pitta visarpam	
		Sannipāta visarpam	
		Abhighāta visarpam	
37	Kuṣṭha (18)	Kapālam	*Ni.* 14: 1–36
		Oudumbaram	
		Maṇḍalam	
		Vicarci	
		Ṛkṣajihva	
		Carmakuṣṭham	
		Ēkakuṣṭham	
		Kiṭibham	
		Sidhma	
		Alasam	

(*Continued*)

TABLE 5.1 (*Continued*)
Diseases Described in *Nidānasthānam*

Sl. No.	Name of Disease	Subgroup	Reference to Text
		Vipādika	
		Dadru	
		Śatārūṣi	
		Puṇḍarīkam	
		Visphōṭam	
		Pāmā	
		Carmadalam	
		Kākaṇam	
38	Śvitram (3)	Vāta śvitram	*Ni.* 14: 37–41
		Pitta śvitram	
		Kapha śvitram	
39	Vātavyādhi (21)	Ākṣēpakam	*Ni.* 15: 25–56
		Apatantrakam	
		Antarāyāmam	
		Bāhyāyāmam	
		Hanusramsam	
		Jihvāstambham	
		Arditam	
		Sirāgraham	
		Pakṣavadham	
		Daṇḍakam	
		Apabāhukam	
		Viśvāci	
		Khañjam	
		Paṅgu	
		Kaḷāyakhañjam	
		Ūrustambham	
		Krōṣṭukaśīrṣam	
		Vātakaṇṭakam	
		Gṛdhrasi	
		Pādaharṣam	
		Pādadāham	
40	Vātaśōṇitam (6)	Uttānam	*Ni.* 16: 1–17
		Gambhīra	
		Vātādhika vātaśōṇitam	
		Raktādhika vātaśōṇitam	
		Pittādhika vātaśōṇitam	
		Kaphādhika vātaśōṇitam	

[a] Figures in parentheses indicate the number of entities in the disease group.

TABLE 5.2
Diseases Described in *Uttarasthānam*

Sl. No.	Name of Disease	Subgroup	Reference to Text
1	Stanyadōṣam (5)[a]	Vātaja stanyadōṣam	*Ut*. 2: 2–4
		Pittaja stanyadōṣam	
		Kaphaja stanyadōṣam	
		Samsargaja stanyadōṣam	
		Sannipāta stanyadōṣam	
2	Kṣīrālasakam		*Ut*. 2: 20–23
3	Tālukaṇṭakam		*Ut*. 2: 63–65
4	Gudakuṭṭam		*Ut*. 2: 69–70
5	Mṛttikōdbhava rōgam		*Ut*. 2: 76–77
6	Bālagrahabādha (12)	Skanda grahabādha	*Ut*. 3: 6–30
		Viśākha grahabādha	
		Naigamēṣa grahabādha	
		Śvagrahabādha	
		Pitṛgrahabādha	
		Śakuni grahabādha	
		Pūtana grahabādha	
		Śītapūtana grahabādha	
		Andhapūtana grahabādha	
		Mukhamaṇḍitika grahabādha	
		Rēvati grahabādha	
		Śuṣkarēvati grahabādha	
7	Grahabādha (18)	Dēva grahabādha	*Ut*. 4: 13–43
		Asura grahabādha	
		Gandharva grahabādha	
		Sarpa grahabādha	
		Yakṣa grahabādha	
		Brahmarākṣasa grahabādha	
		Rākṣasa grahabādha	
		Piśāca grahabādha	
		Prēta grahabādha	
		Kūśmāṇḍaka grahabādha	
		Niṣāda grahabādha	
		Aukiraṇa grahabādha	
		Vētāḷa grahabādha	
		Pitṛgrahabādha	
		Gurugrahabādha	
		Vṛddhagrahabādha	
		Ṛṣigrahabādha	
		Siddhagrahabādha	
8	Unmādam (6)	Vātōnmādam	*Ut*. 6: 6–10
		Pittōnmādam	*Ut*. 6: 10–11

(*Continued*)

TABLE 5.2 (*Continued*)
Diseases Described in *Uttarasthānam*

Sl. No.	Name of Disease	Subgroup	Reference to Text
		Kaphōnmādam	*Ut.* 6: 12–13
		Sannipātōnmādam	*Ut.* 6: 14
		Ādhijōnmādam	*Ut.* 6: 15–16
		Viṣajōnmādam	*Ut.* 6: 17
9	Apasmāraḥ (4)	Vātāpasmāraḥ	*Ut.* 7: 9–12
		Pittāpasmāraḥ	*Ut.* 7: 12–13
		Kaphāpasmāraḥ	*Ut.* 7: 14–15
		Sannipātāpasmāraḥ	*Ut.* 7: 15
10	Vartmarōga (24)	Kṛchrōnmīlam	*Ut.* 8: 1–24
		Nimēṣam	
		Vātahata	
		Kumbhi	
		Pittōtkḷiṣṭam	
		Pakṣmaśātam	
		Pōthaki	
		Kaphōtkḷiṣṭam	
		Lagaṇa	
		Utsaṅga	
		Raktōtkḷiṣṭam	
		Arśaḥ	
		Añjana	
		Visavartma	
		Utkḷiṣṭavartma	
		Śyāvavartma	
		Śḷiṣṭavartma	
		Sikatāvartma	
		Kardamavartma	
		Bahaḷavartma	
		Kukūṇakaḥ	
		Pakṣmōparōdha	
		Alajivartma	
		Arbuda	
11	Sandhigata rōga (9)	Jalasrāvam	*Ut.* 10: 1–9
		Kaphasrāvam	
		Upanāham	
		Raktāsrāvam	
		Parvaṇi	
		Pūyasrāvam	
		Pūyālasam	
		Alaji	
		Kṛmigranthi	

(*Continued*)

TABLE 5.2 (*Continued*)
Diseases Described in *Uttarasthānam*

Sl. No.	Name of Disease	Subgroup	Reference to Text
12	Śukḷagata rōga (13)	Śuktika	*Ut.* 10: 10–19
		Śukḷārmam	
		Valāsagranthitaḥ	
		Piṣṭaka	
		Sirōtpātam	
		Sirāharṣam	
		Sirājālam	
		Śōṇitārmam	
		Arjunam	
		Prastāryarmam	
		Snāvārmam	
		Adhimāmsārmam	
		Sirāpiṭaka	
13	Kṛṣṇagata rōga (5)	Kṣataśukḷam	*Ut.* 10: 22–31
		Śuddhaśukḷakam	
		Ajaka	
		Sirāśukḷam	
		Pākātyayaśukḷam	
14	Dṛṣṭi rōga (27)	Vāta timiram	*Ut.* 12: 1–32
		Vāta kācam	
		Vāta liṅganāśam	
		Gambhīra	
		Pittaja timiram	
		Pitta kācam	
		Pitta liṅganāśam	
		Hrasva rōga	
		Pitta vidagdham	
		Kapha timiram	
		Kapha kācam	
		Kapha liṅganāśam	
		Raktaja timiram	
		Rakta kācam	
		Raktaja liṅganāśam	
		Samsarga timiram	
		Samsarga kācam	
		Samsarga liṅganāśam	
		Sannipāta timiram	
		Sannipāta kācam	
		Sannipāta liṅganāśam	
		Nakulāndha	
		Dōṣāndha	
		Uṣṇavidagdha	

(*Continued*)

TABLE 5.2 (*Continued*)
Diseases Described in *Uttarasthānam*

Sl. No.	Name of Disease	Subgroup	Reference to Text
		Amḷavidagdha	
		Dhūmaraḥ	
		Oupasarga liṅganāśam	
15	Sarvākṣirōga (16)	Vātābhiṣyandam	*Ut.* 15: 1–23
		Vātādhimantham	
		Hatādhimantham	
		Anyatōvātam	
		Vātaparyayam	
		Pittābhiṣyandam	
		Pittādhimantham	
		Kaphābhiṣyandam	
		Kaphādhimantham	
		Raktābhiṣyandam	
		Raktādhimantham	
		Śukḷākṣipākam	
		Saśōpham	
		Alpaśōpham	
		Akṣipākātyayam	
		Amḷōṣitam	
16	Karṇarōga (25)	Vātaja karṇaśūla	*Ut.* 17: 1–26
		Pittaja karṇaśūla	
		Kaphaja karṇaśūla	
		Raktaja karṇaśūla	
		Sannipātaja karṇaśūla	
		Karṇanādam	
		Badhiratvam	
		Pratīnāham	
		Karṇakaṇḍu	
		Karṇaśōpham	
		Pūtikarṇakam	
		Kṛmikarṇakam	
		Karṇavidradhi	
		Karṇārśas	
		Karṇārbudam	
		Kūcikarṇakam	
		Pippali	
		Vidārika	
		Pāḷīśōṣam	
		Tantrika	
		Paripōṭaḥ	
		Utpātakaḥ	

(*Continued*)

TABLE 5.2 (*Continued*)
Diseases Described in *Uttarasthānam*

Sl. No.	Name of Disease	Subgroup	Reference to Text
		Unmanthaḥ	
		Duḥkhavardhanaḥ	
		Lihya	
17	Nāsārōga (18)	Vātaja pratiśyāya	*Ut.* 19: 1–27
		Pittaja pratiśyāya	
		Kaphaja pratiśyāya	
		Sannipātaja pratiśyāya	
		Raktaja pratiśyāya	
		Duṣṭa pratiśyāya	
		Pakva pratiśyāya	
		Bhṛśakṣavaḥ	
		Nāsikāśōṣam	
		Nāsānāham	
		Ghrāṇapākam	
		Nāsāsrāvam	
		Apīnasam	
		Dīpti	
		Pūtināsam	
		Pūyaraktam	
		Puṭakam	
		Nāsārśas/Nāsārbudam	
18	Ōṣṭharōga (11)	Khaṇḍōṣṭham	*Ut.* 21: 3–10
		Vātōṣṭhakōpam	
		Pittōṣṭhakōpam	
		Kaphōṣṭhakōpam	
		Sannipātōṣṭhakōpam	
		Raktōṣṭhakōpam	
		Raktārbudam	
		Māmsōṣṭhakōpam	
		Mēdōjōṣṭhakōpam	
		Kṣatōṣṭhakōpam	
		Jalārbudam	
19	Gaṇḍālaji		*Ut.* 21: 11
20	Dantarōga (10)	Śītadantam	*Ut.* 21: 11–20
		Dantaharṣam	
		Dantabhēdam	
		Dantacālam	
		Karāḷaladantam	
		Adhidantam	
		Śarkarādantam	
		Dantakapālikā	
		Śyāvadantam	
		Kṛmidantakaḥ	

(*Continued*)

TABLE 5.2 (*Continued*)
Diseases Described in *Uttarasthānam*

Sl. No.	Name of Disease	Subgroup	Reference to Text
21	Dantamūlarōga (13)	Śītāda	*Ut.* 21: 20–31
		Upakuśaḥ	
		Dantapupputaḥ	
		Dantavidradhi	
		Suṣiradantam	
		Mahāsuṣiram	
		Adhimāmsam	
		Vidarbhadantam	
		Dantanāḷi (5)	
22	Jihvārōga (6)	Vātadūṣitajihva	*Ut.* 21: 31–35
		Pittadūṣitajihva	
		Kaphadūṣitajihva	
		Kaphapittajajihva	
		Adhijihvaḥ	
		Upajihva	
23	Tālurōga (8)	Tālupiṭaka	*Ut.* 21: 36–41
		Gaḷaśuṇḍika	
		Tālusamhati	
		Tālvarbudam	
		Kaccapatālu	
		Pupputatālu	
		Pākatālu	
		Tāluśōṣam	
24	Kaṇṭharōga (18)	Vātajarōhiṇi	*Ut.* 21: 41–57
		Pittajarōhiṇi	
		Kaphajarōhiṇi	
		Raktarōhiṇi	
		Sannipātarōhiṇi	
		Śālūkakaṇṭharōga	
		Vṛndakaṇṭharōga	
		Tuṇḍikērika	
		Gaḷaugham	
		Valayagaḷam	
		Gilāyukaḥ	
		Śataghnī	
		Gaḷavidradhiḥ	
		Gaḷārbudam	
		Vātagaḷagaṇḍam	
		Kaphajagaḷagaṇḍam	
		Mēdōjatagaḷagaṇḍam	
		Svaraghnakaṇṭharōga	
25	Sarvāsyagatarōga (8)	Vātaja mukhapāka	*Ut.* 21: 58–64
		Pittaja mukhapāka	

(*Continued*)

TABLE 5.2 (*Continued*)
Diseases Described in *Uttarasthānam*

Sl. No.	Name of Disease	Subgroup	Reference to Text
		Kaphaja mukhapāka	
		Sannipātaja mukhapāka	
		Raktaja mukhapāka	
		Ūrdhvagudam	
		Kapōlārbudam	
		Pūtyāsyata	
26	Śirōrōga (10)	Vātaja śiraḥstāpam	*Ut.* 23: 1–20
		Ardhāvabhēdakaḥ	
		Pittaja śiraḥstāpam	
		Kaphaja śiraḥstāpam	
		Raktaja śiraḥstāpam	
		Tridōṣaja śiraḥstāpam	
		Kṛmija śiraḥstāpam	
		Śiraḥkampam	
		Śaṅkhakaḥ	
		Sūryāvartaḥ	
27	Kapālarōga (12)	Upaśīrṣakam	*Ut.* 23: 21–31
		Arūmṣikāḥ	
		Dāruṇakam	
		Indraluptam	
		Vātaja khalati	
		Pittaja khalati	
		Kaphaja khalati	
		Sannipāta khalati	
		Vātaja palitam	
		Pittaja palitam	
		Kaphaja palitam	
		Tridōṣaja palitam	
28	Vraṇam (16)	Vātaja vraṇam	*Ut.* 25: 1–12
		Pittaja vraṇam	
		Kaphaja vraṇam	
		Vātaraktaja vraṇam	
		Pittaraktaja vraṇam	
		Kapharaktaja vraṇam	
		Vātapittaja vraṇam	
		Vātakaphaja vraṇam	
		Pittakaphaja vraṇam	
		Vātapittaraktaja vraṇam	
		Vātakapharaktaja vraṇam	
		Pittakapharaktaja vraṇam	
		Vātapittakaphaja vraṇam	
		Vātapittakapharaktaja vraṇam	

(*Continued*)

TABLE 5.2 (*Continued*)
Diseases Described in *Uttarasthānam*

Sl. No.	Name of Disease	Subgroup	Reference to Text
		Raktaja vraṇam	
		Śuddha vraṇam	
29	Sadyōvraṇam (8)	Ghṛṣṭasadyōvraṇam	*Ut*. 26: 1–5
		Avagāḍhasadyōvraṇam	
		Vicchinnasadyōvraṇam	
		Pravilambisadyōvraṇam	
		Pātitasadyōvraṇam	
		Viddhasadyōvraṇam	
		Bhinnasadyōvraṇam	
		Vidalitasadyōvraṇam	
30	Bhaṅgam (2)	Sandhibhaṅgam	*Ut*. 27: 1–2
		Asandhibhaṅgam	
31	Bhagandaram (8)	Vātaja bhagandaram	*Ut*. 28: 1–20
		Pittaja bhagandaram	
		Kaphaja bhagandaram	
		Vātapitta bhagandaram	
		Vātakapha bhagandaram	
		Pittakapha bhagandaram	
		Sannipātaja bhagandaram	
		Āgantuja bhagandaram	
32	Granthi (9)	Vāta granthi	*Ut*. 29: 1–13
		Pitta granthi	
		Kapha granthi	
		Rakta granthi	
		Māmsa granthi	
		Mēdōgranthi	
		Asthi granthi	
		Sirā granthi	
		Vraṇa granthi	
33	Arbudam (6)	Vātaja arbudam	*Ut*. 29: 14–17
		Pittaja arbudam	
		Kaphaja arbudam	
		Raktaja arbudam	
		Māmsaja arbudam	
		Mēdōja arbudam	
34	Śḷīpadam (3)	Vātaja śḷīpadam	*Ut*. 29: 18–21
		Pittaja śḷīpadam	
		Kaphaja śḷīpadam	
35	Apaci		*Ut*. 29: 23–25
36	Nāḷīvraṇam (5)	Vātaja nāḷīvraṇam	*Ut*. 29: 26–31
		Pittaja nāḷīvraṇam	
		Kaphaja nāḷīvraṇam	

(*Continued*)

TABLE 5.2 (*Continued*)
Diseases Described in *Uttarasthānam*

Sl. No.	Name of Disease	Subgroup	Reference to Text
		Sannipātaja nāḷīvraṇam	
		Śalyaja nāḷīvraṇam	
37	Kṣudrarōga (36)	Ajagallikā	*Ut.* 31: 1–33
		Yavaprakhyā	
		Alajī	
		Kacchapī	
		Panasikā	
		Pāṣāṇagardabhaḥ	
		Mukhadūṣikāḥ	
		Padmakaṇṭakā	
		Vivṛtā	
		Masūrapiṭakā	
		Visphōṭā	
		Viddhā	
		Gardabhī	
		Kakṣya	
		Gandhanāma	
		Rājikā	
		Jālakagardabhaḥ	
		Agnirōhiṇi	
		Irigallikā	
		Vidāri	
		Sarkārarbudam	
		Valmīika	
		Kandara	
		Ruddhagudam	
		Cippa	
		Kunakha	
		Alasam	
		Tilakāḷakam	
		Maśa	
		Carmkīlam	
		Jatumaṇiḥ	
		Lāñchanam	
		Vyaṅgam	
		Prasuptiḥ	
		Utkōṭhaḥ	
		Kōṭha	
38	Puruṣa guhyarōga (23)	Upadamśam	*Ut.* 33: 1–26
		Guhyārśas	
		Sarṣapikāḥ	
		Avamanthaḥ	

(*Continued*)

TABLE 5.2 (*Continued*)
Diseases Described in *Uttarasthānam*

Sl. No.	Name of Disease	Subgroup	Reference to Text
		Kumbhikā	
		Alajī	
		Uttamām	
		Puṣkarika	
		Samvyūḍha piṭakā	
		Mṛditam	
		Aṣṭhīlikā	
		Nivṛtta	
		Avapāṭikā	
		Niruddhamaṇi	
		Grathitā	
		Sparśahāni	
		Śatapōnakam	
		Tvakpākam	
		Māmsapākam	
		Asṛgarbudam	
		Māmsārbudam	
		Vidradhi	
		Tilakāḷakam	
39	Strīguhyarōga (20)	Vātikī	*Ut*. 33: 27–52
		Aticaraṇā	
		Prākcaraṇā	
		Udāvṛttā	
		Jātaghnī	
		Antarmukhī	
		Sūcimukhī	
		Śuṣkā	
		Vāmini	
		Ṣaṇḍha	
		Mahāyōni	
		Paittikī	
		Raktayōni	
		Śḷaiṣmiki	
		Lōhitakṣayā	
		Paripḷutā	
		Upapḷutā	
		Vipḷutā	
		Karṇinī	
		Sānnipātikī	

[a] Figures in parentheses indicate the number of entities in the disease group.

part, sensation of pain, nature of swelling, ability to perceive tactile sensation, warm or cold nature of body and affected part can be found out by palpation. Interrogation of the patient provides information on the symptoms and signs, feelings, fears, nature and severity of pain or the disease, time of day at which the discomforts are severe, appetite, digestive efficiency, voiding of urine and stool, ability to perceive tastes, gustatory sensations, nature of sleep, addictions, likes, dislikes and dreams.

Dreams have diagnostic and prognostic value. The three constitutional types of individuals can be identified by specific signs, some of which relate to dreams. For example, a *vātaja*-type person often dreams of flying in the air and around hills and trees, indicating the similarity between atmospheric air and the "air" (*vāta*) operating inside the human body. Dreams have diagnostic value. Months before kingly consumption (*rājayakṣma*) sets in, the person dreams of being attacked by flying termites, lizard, snake, monkey, dog and birds. He also gets visions of walking into heaps of hair, bones, rice husk and ash. As a symbolic prelude to the emaciation that he is bound to suffer from, he dreams of completely deserted villages, dried-up lakes and burning trees. A person who has ingested poison often starts having weird dreams, which help the physician to arrive at diagnosis. Persons suffering from alcoholism experiences nightmares. Dreams also have prognostic value. Many of them are harbingers of imminent death. Some such dreams having diagnostic and prognostic value are listed in Table 5.3.

The physician should then study and evaluate these symptoms at the levels of increase or decrease of *tridōṣa* (*Sū*. 11: 6–8, *Sū*. 11: 15–16); *dhātus* (*Sū*. 11: 8–12, *Sū*. 11: 17–20); *mala* (*Sū*. 11: 13–14, *Sū*. 11: 21–23); *pūrvarūpa* of diseases, symptoms of diseases (*rūpa*); *āvaraṇa* (*Ni*. 16: 31–58); and *saṃprāpti* (*Ni*. 1: 8–11) (Figure 5.1). This exercise helps the physician to pinpoint the disease entity. Once the disease is identified, the physician follows the line of treatment recommended in its treatment. AH provides lines of treatment for all diseases mentioned in it.

5.4 LINES OF TREATMENT

5.4.1 FEVER

Laṅghana therapy is to be adopted to stimulate abdominal fire and clear obstructions in the body channels. Fasting (*upavāsa*) is to be carried out with emetic therapy, using appropriate emetic herbs, in *āma* related *jvara* and when there is predominance of *kapha*, nausea, exudation of mucus and anorexia. Even if *vamana* is carried out or not, the patient should adopt fasting to disintegrate and eliminate the *dōṣas*. *Upavāsa* is to be continued till abdominal fire is stimulated, and *dōṣas* undergo *pācana* (transformation). In *vāta-kapha jvara*, when the patient feels thirst while fasting, he should drink frequently small measures of warm water. Warm water is not to be administered in *pitta jvara*, *pitta*-dominant *saṃsarga jvara*, *jvara* with feeling of warmth in eyes, burning sensation, fainting and loose bowels; *jvara* due to poisoning and alcoholism; *jvara* appearing in summer season; and *jvara* manifested in patients with injury to thorax and hemorrhages of obscure origin. To such patients, cooled decoction of *Cyperus rotundus*, *Santalum album*, *Zingiber officinale*, *Coleus vettiveroides*, *Mollugo cerviana* and *Vetiveria zizanoides* is to be administered. It will cause transformation of *dōṣas*, quench thirst and reduce *jvara*. In *pitta*-dominant *jvara*, the patient should abstain from *pitta*-increasing food and measures (*Ci*. 1: 1–17).

Pain-relieving decoctions are not to be administered in *jvara* associated with dominance of *āma*. Sudation is beneficial in *vāta-kapha jvara* with association of respiratory distress and joint pain. Fasting, sudation, porridge (*yavāgu*) and decoctions with bitter taste can be applied in this case, as they pacify *āmadōṣa* by digesting the wastes. Fasting is contra-indicated in *vāta jvara*, *kṣaya jvara*, *āgantu jvara* and *jīrṇa jvara* (chronic fever). Incomplete fasting can be ascertained by symptoms of fever with *āma* (*sāma jvara*). For identifying optimum and excessive *laṅghana*, the physician should look for symptoms described in the *Dvidhōpakramaṇīyam* chapter of *Sūtrasthānam*. On achieving optimum *laṅghana*, the patient is to be fed with medicated porridge for six days or till the *jvara* is pacified. Various medicated porridges are recommended in the case of *jvara* associated with loose

TABLE 5.3

Importance of Dreams in Diagnosis and Prognosis

Sl. No.	Description of the Dream	Significance	Reference
1	Images of men	The pregnant mother will be get a boy child	Śā. 1: 70
2	Flying in the air, around hills and trees	Characteristic of *vātaja* constitution	Śā. 3: 88
3	Images of *Erythrina variegata* and *Butea monosperma* trees in bloom bearing red flowers, fire and lightning	Characteristic of *pittaja* constitution	Śā. 3: 93
4	Lakes with lotus in bloom surrounded by birds and clouds	Characteristic of *kaphaja* constitution	Śā. 3: 102
5	Objects with blue, red or yellow color	Prodrome of *raktapitta*	Ni. 3: 6–7
6	Visions of being attacked by flying termites, lizard, snake, monkey, dog and birds; visions of walking into heaps of hair; bones, rice husk and ash; deserted villages and completely dried-up lakes; burning trees; visions of moon, stars and mountains falling down	Prodromes of *rājayakṣma*	Ni. 5: 11–13
7	Person dreams of dancing and singing	Prodrome of epilepsy	Ut. 7: 8
8	Nightmares	Common feature of all types of *madātyaya* (alcoholism)	Ni. 6: 17
9	Visions being thrown around and speaking to spirits	Characteristic of *vātaja madātyaya*	Ni. 6: 18
10	Visions of fox, cat, mongoose, snake and monkey	Effect of poisoning	Ut. 35: 52
11	Visions of dried-up vegetation and completely dried-up lakes	Effect of poisoning	Ut. 35: 53
12	A fair person sees himself in dream as a dark-skinned person. Conversely, a swarthy person sees himself as of fair complexion.	Effect of poisoning	Ut. 35: 53
13	Person dreams of drinking spirituous beverage in the company of persons dead long ago, then gets bitten and dragged away by a dog.	He will die very soon afflicted with fever.	Śā. 6: 40–41
14	The subject sees himself in dream as a ruddy-complexioned person, wearing red clothes and red garland. While sitting in a happy mood laughing, he is abducted by a woman.	He will die soon of *raktapitta*.	Śā. 6: 41–42
15	Person journeys southward, riding a buffalo, dog, pig, camel or donkey	He will die of *rājayakṣma*.	Śā. 6: 42
16	Person dreams of spiny creeper, bamboo or palmyra palm springing from his heart.	He will die soon of *gulma*.	Śā. 6: 43
17	Person has visions of himself sitting naked, body smeared with clarified butter and performing fire ritual, with no fire in the fire pit. Thereupon, a lotus springs from his heart.	He will die of skin disease.	Śā. 6: 43–44
18	Person dreams of drinking various fatty liquids in the company of garbage collectors.	He will die afflicted with polyuric disease.	Śā. 6: 44
19	Person dreams of drowning in water, while dancing in the company of devils	He will die of insanity.	Śā. 6: 45
20	Person dreams of being abducted by a ghost, while dancing	He will die of epilepsy.	Śā. 6: 45
21	Person dreams of riding donkey, camel, cat, monkey, tiger, pig, fox or a ghost	His death is imminent.	Śā. 6: 46–47
22	Person wakes up dreaming of having eaten various cakes and then vomiting, the vomitus resembling these foods	He will die very soon.	Śā. 6: 47
23	Person dreams of witnessing solar and lunar eclipses	He will soon be afflicted with eye disease.	Śā. 6: 48
24	Person dreams of sun and moon falling down	He will soon become blind.	Śā. 6: 48
25	Person dreams of himself surrounded by ghosts, devils, women, crows and vultures	Death is imminent.	Śā. 6: 50

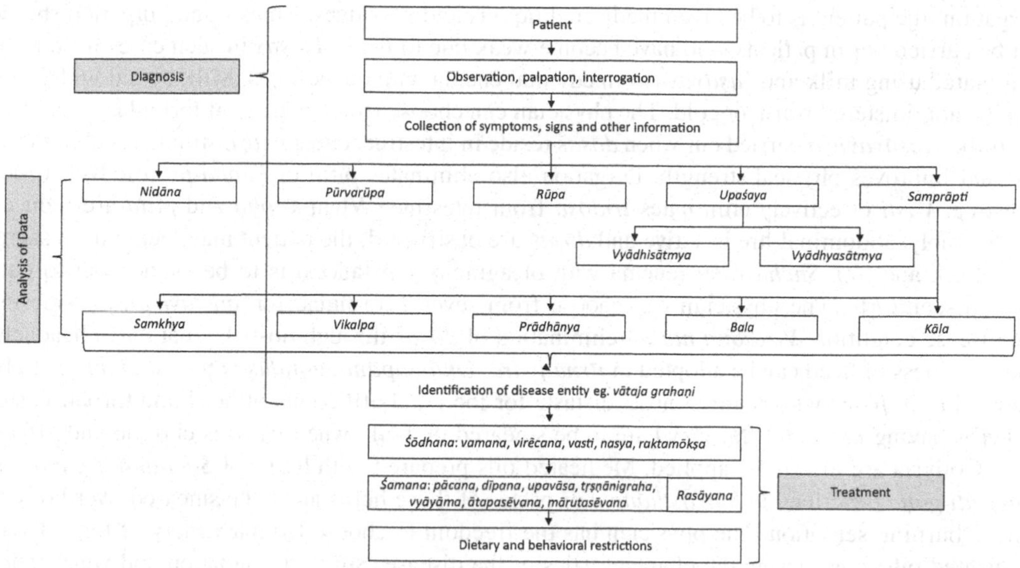

FIGURE 5.1 Protocol for diagnosis and treatment of diseases.

bowels, constipation, *pitta* predominance, pain in urinary bladder and flanks, hiccup, body pain, respiratory distress, cough, excruciating pain in abdomen, hyperhidrosis, excessive thirst, insomnia and vomiting. Soups fortified with ingredient herbs of these medicated porridges can also be administered. Medicated porridges are contra-indicated in *jvara* arising from alcoholism, *jvara* due to *kapha* enraged in site of *pitta*, *jvara* manifesting in summer season, *jvara* with excessive thirst, vomiting and burning sensation and *ūrdhvaga raktapitta*. In such conditions, *mantha* (a liquid obtained by churning of any grain, millet or parched rice, after adding 14 times of water) along with sugar and honey should be administered with juice of *jvara*-pacifying fruits or water. When this *mantha* is digested, the patient should consume for six days rice prepared from fried rice grains mixed with green gram and soup of meat of *Lāva* bird (Bustard quail, *Turnix suscitatus*). This treatment known as *ṣaḍaha cikitsa* prepares the patient for treatment with *kvātha*. *Jvara* caused by minor causes will be cured by this treatment alone (*Ci.* 1: 18–38).

To eliminate the *dōṣas* remaining after *laṅghana* and other treatments, the patient may be treated with appropriate *kvātha*. Formulae of many such *kvātha* useful in the treatment of different kinds of *jvara* are described in Chapter 1 of *Cikitsāsthānam*. Some of them also cure other conditions like vomiting, fainting, tiredness, jaundice, respiratory distress, cough and so on. The patient may consume light food like porridges, on digestion of the *kvātha*. Patients with *kapha* dominance should not consume porridges. Instead of that they should consume before meals, soups prepared with *Dolichos biflorus* seeds and vegetables having bitter taste. Various grains and vegetables having *jvara*-lowering properties are also recommended as food. Patients suffering from *jvara* and those cured of *jvara* should consume light food at dinner. Consumption of heavy food is prohibited as they impair digestion and cause indigestion, worsening the disease. After spending ten days consuming *kvātha* and medicinal food, the patient can start consuming *ghṛta*, which possesses ambrosia-like qualities. The *kvātha* described earlier can also be administered along with *ghṛta*. Formulae of several medicated *ghṛta* are also described. On digestion of the *ghṛta*, the patient can consume food along with meat soup. This helps in improving stamina (*Ci.* 1: 39–97).

Elimination therapy is to be carried out if *jvara* is not pacified by these measures. If the *dōṣa* remains in the stomach and if the patient has strength, emetic therapy is to be carried out using the emetic herbs mentioned earlier. If *dōṣa* remains in the intestine and in *jvara* due to poisoning or alcoholism, purgation therapy needs to be carried out using appropriate herbs. After emesis and purgation, the patient is to be given medicated liquids and porridges. Emesis and purgation should not be carried out in patients who have become weak due to *jvara*. *Dōṣas* in such cases need to be eliminated using milk and *kaṣāyavasti* (medicinal enema with decoctions). Milk boiled with herbs is to be administered warm or cold. The physician can choose from an array of formulae of medicinal milk. *Kaṣāyavasti* carried out when *dōṣa*s reside in intestine, cures *jvara*, stimulates abdominal fire and improves physical strength. Purgation also eliminates *pitta* or *kapha-pitta* lodged inside intestine. *Vasti* effectively eliminates *tridōṣa* from intestine. When *kapha* and *pitta* are reduced considerably, abdominal fire is active and *dōṣa*s are obstructed, the patient may feel pain in sacral area, back and hip. *Snēha vasti* (enema with oleaginous substances) is to be carried out to cure these discomforts. The physician can choose from several formulae for *snēhavasti*, considering the disease condition. *Virēcana nasya* (elimination of *dōṣa*s through nostrils) that cures headache and heaviness of head can be adopted in *jīrṇa jvara*. *Dhūmapāna*, *gaṇḍūṣa* and *kabaḷam* are to be resorted to in *jvara* with running nose, distaste for food and afflictions of head and throat. Pastes of herbs having *uṣṇa* and *śīta vīrya* are to be smeared on body when *jvara* is chronic and affects skin. Collyria are also to be applied. Medicated oils prepared with herbs of *Śōdhanādi gaṇa* and *Dūrvādi gaṇa* described in *Sūtrasthānam* or pastes of those herbs are to be smeared over body to relieve burning sensation. The physician has the freedom to choose from a variety of formulae of medicated oils, considering the characteristics of the disease. Similarly, sudation and venesection are to be performed in different types of *jvara* like *viṣama jvara*, *śīta jvara* and *sannipāta jvara* (*Ci.* 1: 97–152).

All these treatments can be adopted in *viṣama jvara* after considering in detail the conditions of *vāta*, *pitta* and *kapha*. Several *kvātha*, single herb remedies and *ghṛta* are recommended in its treatment. Medicinal smokes (*dhūpa*) like *Aparājitaḥ*, all collyria and formulations for *nasya* indicated in the treatment of *unmāda* can be applied in *viṣama jvara* as well. *Ghṛta*, medicinal pastes for topical application, meat soups and venesection (*sirāvēdham*) can be of value in the treatment of *vāta jvara*, *visarpa jvara*, *visphōṭa jvara* and *abhighāta jvara* (*Ci.* 1: 153–167).

To treat *jvara* caused by spiritual affliction, sacrificial rites and *mantrā*s described in *bhūtapratiṣēdha* chapter of *Uttarasthānam* are to be applied. *Pitta*-pacifying treatment is to be adopted in the case of *jvara* caused by irritant odors. *Jvara* caused by poisons is to be treated with medicines that cure poisoning. By the judicious application of measures that impart happiness to mind and with medicines appropriate for stabilizing *tridōṣa*, *jvara* caused by anger, grief and uncontrolled desires can be pacified. In *jvara* caused by curse (magical spells) and witchcraft, spiritual healing (*daivavyāpāśraya cikitsa*) should be adopted. Patient should refrain from exercise, bath, libido, heavy and incompatible food and all factors that cause *jvara*, till the body and mind are completely free from the grip of the disease. He should not eat all kinds of food immediately after the treatment. Dietary and behavioral indiscretions can cause the return of the disease when the body is weak (*Ci.* 1: 168–177).

5.4.2 Hemorrhages of Obscure Origin

Raktapitta (hemorrhages of obscure origin) that works upward (*ūrdhvaga*), with less intensity, involving only one *dōṣa*, of acute nature and devoid of secondary complications, can be cured if the patient is not weak. *Raktapitta* that works downward (*adhōga*) and is associated with two *dōṣa*s can be controlled with treatment. However, *raktapitta* that reappears even after treatment, that which works both upward and downward, involving *tridōṣa* and with weak abdominal fire is difficult to cure (*Ci.* 2: 1–3).

First of all, purgation is to be achieved using appropriate *lēhya* and *mōdaka*. Emesis can also be adopted by adding powder of pericarp of *Randia dumetorum* (*Madanaphala*), to *mantha* containing sugar and honey, sugary water and milk or sugarcane juice. On elimination of *dōṣa*s through emesis and purgation, medicinal substances are to be administered as *manthādi krama* (beginning with *mantha* and gradually ending in solid food) in *raktapitta* that works upward (*ūrdhvaga*) and as *pēyādi krama* (beginning with *pēya* or gruel with more fluid, less grains and gradually ending in solid food) in *raktapitta* that works downward (*adhōga*). *Drākṣādi kvātha* mentioned in *jvara cikitsa* can be given as *mantha*. *Mantha* prepared with *pitta*-lowering fruits or with honey, dates, grapes, *Parūṣakam* (*Phoenix pusilla*) and sugary water can also be used. *Mantha* prepared with clarified butter and powder of *lāja* is also equally useful. Porridge prepared with powder of specific herbs, rice, honey and sugar can be administered. All *laghu* and *śīta* grains, millets, green herbs and vegetables described in chapter *annasvarūpa vijñānīya* (Chapter 6 of *Sūtrasthānam*) can be included in the patient's diet. Water boiled with *laghu pañcamūla* herbs or boiled and cooled water or cold water with honey can be given for drinking frequently. The physician can administer a variety of *kvātha*, *ghṛta*, *lēhya* and medicated beverages. *Nasya* can be performed using *kvātha*, milk, sugarcane juice and juice of several fresh herbs. Pastes of *śīta vīrya* herbs can be applied topically all over the body. The patient should avoid all factors that cause *raktapitta* (*Ci.* 2: 5–50).

5.4.3 Cough

In *vāta kāsa* (without the involvement of *pitta* or *kapha*), the patient should at first be treated with *vāta*-lowering *ghṛta*, *pēya*, *yūṣa*, meat soups, *lēhya*, *dhūma* (medicinal smokes), *abhyaṅga* with medicinal oils, *svēdana*, *pariṣēka* (continuous pouring in a stream, of oil, warm water, *kvātha*, *taila* or *ghṛta* on any part of body) and *avagāha* (sitting immersed in warm medicinal fluid). *Vasti* is to be applied if there is obstruction of stool. In condition with involvement of *pitta*, *ghṛta* and milk are to be administered post-prandially. Condition with involvement of *kapha* is to be pacified with *snēhavirēcana* (purgation with unctuous ingredients). The physician can choose appropriate medicines from a variety of *ghṛta*, *lēhya*, medicated porridges and beverages (*Ci.* 3: 1–25).

In *pitta kāsa* with involvement of *kapha*, emesis is to be carried out with emetic medicine containing *ghṛta*. *Kvātha* can also be used in emesis. When *kapha* becomes lighter, purgation is to be carried out with herbs having sweet taste. In cases where *kapha* is thick, emetic medicine should contain bitter herbs. When *dōṣa*s have been eliminated, *pēya* of cold, sweet and unctuous nature is to be administered. In cases where *kapha* is thick, the *pēya* is to be prepared with herbs having cold, dry and bitter qualities. *Yūṣa* prepared with *jāṅgala* meat having sweet taste and green gram can be used along with food prepared from barley and millets. Dishes can be prepared with green vegetables having bitter taste. Rice cooked using *Śāli* and *Ṣaṣṭika* varieties of grains, meat soups and dishes prepared from ingredients having sweet taste are recommended as food in *pitta kāsa*. Grape juice and sugarcane juice can be used as beverages. *Lēhya* can be administered in *pitta kāsa*. They are to be prepared with bitter herbs and administered along with honey (*Ci.* 3: 27–40).

Kapha kāsa patient should, first of all, consume oil of *Dēvadāru* (*Cedrus deodara*) mixed with powder of *Trikaṭu* and *yavakṣāra*. Thereafter, he should undergo emesis, purgation or nasal instillation, according to the condition of the disease. After that he should be subjected to purgation with *tīkṣṇa* drugs, followed by consumption of liquid food in *pēyadikrama*. Thereafter, he should be fed with barley, green gram and horse gram cooked with herbs that are of pungent taste and soup made out of meat of *jāṅgala* and *vilēśaya* types of animals. The food should be mixed with small quantities of sesame oil, mustard oil and neem oil. Water boiled with *Daśamūla* group of herbs or warm water or wine or honey with water can be served as beverage. *Kvātha*, *cūrṇa*, *ghṛta* and *lēhya* appropriate for the condition can also be administered. This is to be followed by *dhūmapāna*. *Śōdhana dhūmapāna* is advised if *kapha* is very thick. *Dhūmapāna* is to be followed by drinking of warm milk. This *dhūmapāna* cures chronic cough of *vāta-kapha* dominance (*Ci.* 3: 41–69).

Tamaka śvāsa may develop, if *kapha kāsa* is associated with *pitta*. In such cases, treatment for *pitta kāsa* is recommended, depending on the condition. *Kapha*-pacifying treatment is to be adopted in *vāta kāsa* associated with *kapha*. *Pitta*-lowering treatment should be resorted to in the case of *vāta-kapha kāsa* with involvement of *pitta*. Treatment with *snigdha* drugs is required in dry cough of *vāta-kapha* nature. Similarly, treatment with *rūkṣa* drugs is needed in *ārdra kāsa* or *kāsa* with predominance of watery element. In *kapha kāsa* with association of *pitta*, treatment with *tikta* drugs is ideal (*Ci.* 3: 70–72).

Patients of *kṣata kāsa* or cough arising from internal injury to thorax should drink milk containing honey and powder of lac. When the milk is digested, he should consume cooked *Śāli* rice mixed with sugar and milk. Those suffering from pain in flanks and urinary bladder and those with weak abdominal fire should consume lac powder suspended in *Sura* wine. Patients with loose stools should, however, consume it with powder of *Cyperus rotundus*, *Aconitum heterophyllum*, *Cissampelos pareira* and *Holarrhena antidysenterica*. Several medicated milk, *ghṛta* and pills are also indicated in *kṣata kāsa*. Treatment for *raktapitta* is to be adopted, if bleeding occurs through nostrils. Body pain can be alleviated by *abhyaṅga* with *ghṛta* and medicated oils. All medicines and measures advised in treatment of *rājayakṣma* can be applied in *kṣata kāsa* as well, depending upon the dominance of *dōṣas*. When injury to thorax is healed and *kapha* increases, the patient may experience excruciating pain in thorax and head. This can be pacified by *dhūmapāna* (*Ci*. 3: 73–151).

A person suffering from *kṣaya kāsa* should, at first, be treated to improve abdominal fire and physique. Individuals with high titer of *dōṣa* should be subjected to *mṛduvirēcana* (mild purgation) with unctuous ingredients. This should be followed by administration of *ghṛta* to nourish the body. In case of difficult micturition and urine with unusual color, specially processed *ghṛta* and milk are to be administered. *Snēha vasti* with *Miśraka snēha* should be adopted, if there is heaviness and pain in phallus, anus, hip and inguinal area. After that he should consume food with cooked *jāṅgala* meat. Slowly he should consume moderately meat of animals of *prasaha* type. On account of their *uṣṇa vīrya* nature, these meats clear the obstructions in body channels, cause proper flow of *dhātus* and improve health. *Ghṛta* and *lēhya* can also be administered. All *dhūma* advised in the treatment of *kṣata kāsa* and *rājayakṣma* can be applied in the treatment of *kṣaya kāsa* also. Medicines that are roborant, stimulate abdominal fire and clear blockages in body channels are suitable in the treatment of *kṣaya kāsa* (*Ci*. 3: 151–180).

5.4.4 Respiratory Disorders and Hiccup

As the etiology and characteristics of respiratory disorders and hiccup (*śvāsa* and *hidhma*) are similar, their treatments are also similar. *Śvāsa-hidhma* patient should first of all be subjected to *abhyaṅga* with sesame oil containing salt. Thereafter, he should be subjected to sudation of the unctuous variety. This treatment liquefies hardened *kapha* lodged in the body channels and brings it to the alimentary canal for easy elimination. With that the body channels get softened and *vāta* starts moving in its natural direction (*vātānulōmana*). After that he should consume unctuous food along with curd and fish curry. Following this, mild emetic medicine is to be administered. This emesis measure is essential if the patient has associated problems like cough, vomiting, catching pain in precordial area and *svarasāda* (lassitude of voice). The emetic medicine should contain *Piper longum* and rock salt. It should not vitiate *vāta* in the patient. The patient is relieved of *śvāsa-hidhma* when the *kapha* in body channels get eliminated following emesis. As the respiratory tract is cleansed and softened, *vāta* also starts moving without any hindrance. If the patient complains of bloating of abdomen, retention of feces and *tamaka śvāsa*, he should be fed with cooked rice containing appropriate herbs. Alternatively, he can be given moderately warm purgative medicine, containing juices of acidic fruits and rock salt. Elimination of *kapha* with emesis and purgation generally pacifies *śvāsa-hidhma* (*Ci*. 4: 1–7).

If these treatments do not pacify the disease, *dōṣa* remaining in the stomach and body channels is to be eliminated by *dhūmapāna*. The physician can select appropriate *dhūma* from the several formulae available. Sudation of thorax and neck is a prerequisite in the treatment of *śvāsa-hidhma*. Therefore, patients in whom sudation is essential and others in whom this measure is contra-indicated are to be subjected to sudation in a unique way. Sweetened milk, warm water, several medicated oils or *ghṛta* may be poured in a stream over thorax and neck of such patients. *Piṇḍasvēdam* with *Utkārika* (balls of paste of rice hardened in boiling water), *upanāha* (application of herbal paste on any part of body, covering it up with leaves or cloth and keeping for a specific period of time) and medicines described in the chapter on sudation in *Sūtrasthānam* can also be utilized for performing sudation. Treatment for *āma* should be initiated, if signs of *āma* are evident. Meat soups, medicinal milk and *vata*-lowering *ghṛta* are to be administered, if *vāta* gets enraged because of excessive sudation. *Vāta* can also be pacified by *abhyaṅga* with *taila* that are not very *uṣṇa vīrya* in nature (*Ci.* 4: 8–17).

Dōṣas of patients who are unfit for sudation and those in whom sudation has been performed should be pacified with *kvātha, lēhya* and *ghṛta*. *Śvāsa-hidhma* patients who suffer from leaning of *dhātus* (*dhātukṣaya*), *uraḥkṣataḥ* (internal injury to thorax), dysentery, *raktapitta* and *sarvāṅga santāpa* (burning sensation all over body) are to be treated with mild medicines that are sweet, unctuous and cold in nature. Several *pēya, yūṣa, kvātha, cūrṇa,* medicines for nasal instillation and *jāṅgala* meat processed with *kvātha* of curative herbs are also recommended in the treatment of *śvāsa-hidhma*. The food recommended for the patients are rice cooked from *Śāli, Ṣaṣṭika* grains, wheat, barley, green gram and horse gram. For quenching thirst, patients can drink *kvātha* of *Daśamūla* or *Dēvadāru* or *Vāruṇi* wine. Several formulations are recommended for use in the treatment of *śvāsa-hidhma* with involvement of *kapha, pitta-kapha* and *pitta-vāta* (*Ci.* 4: 17–59).

5.4.5 Kingly Consumption

Rājayakṣma or "kingly consumption" is called so, as this disease is followed by several other diseases. It is basically a disease that needs to be treated with roborant medicines. Nevertheless, as the body is vitiated with *dōṣas*, elimination therapy is essential to eliminate the *dōṣas*. *Rājayakṣma* patient who is not weak should be made unctuous at first, by administering orally, suitable *ghṛta*. Thereafter, he should be subjected to sudation. Following that, depending on the *dōṣas* to be eliminated, proper emetic or purgative medicine mixed with *ghṛta* or oil should be administered. Emetic medicine may be mixed with milk, fruit juice or meat soup with sweet taste. Purgative medicine may be added to a mixture of sugar, honey, clarified butter and milk or porridge or *mantha* (*Ci.* 5: 1–4).

When the alimentary canal is cleaned of *dōṣas*, *bṛṃhaṇa* and *dīpana* treatment is to be initiated. All food and beverages given to the patient should be pleasant, light in nature and having *vāta*-lowering properties. Food prepared with one-year-old *Śāli, Ṣaṣṭika* grains, wheat, barley, green gram, goat milk and its butter and meat of carnivorous animals have the ability to increase strength of body and cure *rājayakṣma*. Meat of crow, owl, wolf, tiger, cow, horse, mongoose, snake, vulture, fish-eating kite (*Bhāsa, Gypaetus barbatus*), donkey and camel have the ability to strengthen the body. Food prepared with any of these meats has to be served "covered with lie". To ensure the acceptability of the food, the patient should be told that it is prepared from his most favorite meat (*Ci.* 5: 4–7).

Patients of *pitta-kapha rājayakṣma* should consume meat of *mṛga* or *viṣkira* type of animals (vide *Annasvarūpavijñānīyādhyāyaḥ* chapter of *Sūtrasthānam*). Those who suffer from *vāta*-dominant *rājayakṣma* are to eat meat of *prasaha* and *mahāmṛga* animals. These meats should be consumed as meat soups. Alternately, they can be served fried in mustard oil or clarified butter. Consumption of goat meat cooked along with powder of *Piper longum*, barley, horse gram, dry ginger; juices of pomegranate and gooseberry and clarified butter cures rhinitis and associated afflictions of *rājayakṣma*. Aged wine drunk along with this food clears the body channels. Several *ghṛta* and *lēhya* are prescribed in the treatment of *rājayakṣma* (*Ci.* 5: 8–34).

Svarasāda or "lassitude of voice" is another secondary complication generally identified in *rājayakṣma* patients. Nasal instillation and medicinal smoke are to be applied in these conditions. *Ghṛta* is to be consumed post-prandially, especially in *vātaja svarasāda*. *Pittaja svarasāda* patients should consume *ghṛta* prepared from the buds of the four *Kṣīrī vṛkṣa* (*Ficus benghalensis, Ficus religiosa, Ficus glomerata, Ficus retusa*), followed by drinking of boiled and cooled milk. Rice cooked in *kvātha* of *Yaṣṭimadhu* (*Glycyrrhiza glabra*) and milk can also be consumed, mixed with clarified butter. Consumption of *lēhya* and nasal instillation are also indicated. Powder of herbs with pungent taste is to be suspended in cow urine and drunk in *kaphaja svarasāda*. This should be followed by oil-free food. After consuming porridge made of barley, the patient should chew either dry ginger or *Piper longum* fruits. Emesis with *tīkṣṇa* emetic medicine is also advised (*Ci*. 5: 35–46).

Aruci or *arōcakam* (distaste for food) is another disease accompanying *rājayakṣma*. Cleanliness of body and mind is essential to ward off this disease. Teeth should be brushed in the morning as well as in the evening. Mouth is to be rinsed with *kvātha* of cleansing herbs. *Tālīsapatrādi cūrṇa* has the ability to improve gustatory sensation. In *vātaja arōcakam*, suspension of a specific formulation of herbs is to be drunk, mixed with clarified butter. Thereafter, emesis is to be carried out by drinking water mixed with paste of *Acorus calamus*. In *pittaja arōcakam*, emesis is to be carried out after drinking jaggery water. Emesis in *kapha*-dominant *arōcakam* is to be carried out after drinking the *kvātha* of *Azadirachta indica* bark. Several powders are also recommended in the treatment of *arōcakam* (*Ci*. 5: 47–60).

Those suffering from exudation of mucus (*kaphaprasēkam*) should be subjected to emesis with medicines having pungent and bitter tastes. Emesis is to be followed by consumption of *jāṅgala* meat cooked with little fat and light food. *Kaphaprasēkam* is caused by *vāta* activity resulting in expulsion of *kapha*. Therefore, this condition is to be treated with medicines having *snigdha* and *uṣṇa* qualities. These treatments can be adopted in the case of rhinitis and vomiting. *Abhyaṅga* and consumption of *ghṛta* are recommended in rhinitis. Unctuous sudation is to be carried out on head, flanks and neck using lumps of *Utkārika*. Meat soups with salty, sour, pungent, hot and unctuous qualities are to be consumed regularly (*Ci*. 5: 61–65).

Headache (*śiraḥśūla*), pricking pain in flanks (*pārśvaśūla*) and pricking pain in arm (*āṃsaśūla*) are to be treated based on the involvement of *tridōṣa*. Topical application of herbal pastes (*pralēpaḥ*) and *upanāha* with *taila* (medicated oils) made with meat soups are useful in these conditions. Whenever involvement of more than one *dōṣa* is evident, the patient is to be treated with *nasya*, *dhūmapāna*, *abhyaṅga*, *vasti* and post-prandial consumption of *ghṛta*. Vitiated blood of patients suffering from these three secondary complications is to be removed using appropriate venesection techniques. Because of dull abdominal fire, *rājayakṣma* patents show a tendency to void slimy and loose stool. Such patients are to be treated with medicines recommended in the treatment of *atisāra* and *grahaṇi*. The physician should endeavor to control this digestive disorder, as conservation of stool is essential in the case of the already emaciated patient. Appropriate fatty substances, milk and water are to be boiled in large cylindrical vessels. The patient should be smeared with *taila* all over the body and made to sit immersed in the tolerably warm liquid for a stipulated period of time. On completion of this measure, known as *avagāha*, he should undergo *abhyaṅga* with *Miśraka snēha*, followed by massage. Massage with dry herbal powders (*udvartana*) is also advised. *Rājayakṣma* patient should spend time among kind-hearted friends and relatives. He should listen to pleasant music and always try to be cheerful. *Vasti*, moderate use of wine, meat and clarified butter are permitted in the case of the patient. *Daivavyāpāśraya cikitsa* (spiritual healing) hastens healing of the *rājayakṣma* patient (*Ci*. 5: 66–83).

5.4.6 Vomiting

Even though *chardi* (vomiting) can be due to several reasons, the contents of stomach can be vomited out only if they are of watery nature. The stomach and buccal cavity secrete water and hasten vomiting. Therefore, stopping the accumulation of water (*utkḷēśa*) in stomach is the first step in treatment of vomiting. *Laṅghana* is adopted in this regard. *Upavāsa* is recommended in all forms of *chardi*, except *vātaja chardi*. After that, in slow degrees purgation is carried out by administering purgative medicines mixed with palatable wine, fruit juices or milk. This induces purgation which brings down the *dōṣa*s that tend to go upward. After purgation, medicines that induce pacification of *dōṣa*s (*śamana*) are to be administered. *Śamana* therapy is of choice, if the patient is weak and of dry nature. Pleasing *yūṣa*, meat soups and light and dry food are to be served to the patient.

Good fragrance and washing of face with cold water after food also reduce the tendency to vomit. Several medicinal formulations for internal administration, *snēha virēcana* and liquid food are recommended in the treatment of *vātaja chardi* (*Ci.* 6: 1–10).

Purgation is to be carried out in *pittaja chardi* by administering powder of *Operculina turpethum* suspended in grape juice or sugarcane juice. *Pitta* staying in the stomach is to be eliminated through emesis, using herbs having sweet and bitter tastes. *Pittaja chardi* patient cleansed of *dōṣas* through emesis and purgation is to be fed with *mantha*, *yavāgu* (porridge), mixture of honey and sugar or rice cooked from *Śāli*, *Ṣaṣṭika* grains accompanied by *jāṅgala* meat soup and *yūṣa* of green gram. Emesis is to be induced in patients of *kaphaja chardi* by administering paste of *Azadirachta indica* bark, *Piper longum* fruits, *Randia dumetorum* pericarp and *Brassica juncea* (mustard) seeds, suspended in lukewarm water. *Kvātha* of *Āragvadhādi gaṇa* described in *Sūtrasthānam* can also be used mixed with honey. Weak patients should, on the other hand, be subjected to fasting. *Mantha* with *kapha*-pacifying herbs and *lēhya* are also recommended as medicines to be used after emesis. *Dviṣṭārthayōgaja chardi* (*chardi* caused by the sight or smell of nauseating substances) is to be cured by adopting measures that are pleasing to the patient. *Chardi* caused by *kṛmi* can be cured by treatment recommended for *kṛmi hṛdrōga*. If *chardi* becomes very frequent, treatment advised for excessive emesis (*atiyōga* of *vamana*) should be applied. Several *ghṛta* and medicated milk are indicated in its treatment (*Ci.* 6: 10–24).

5.4.7 Heart Disease

Patients suffering from *vātaja hṛdrōga* (heart disease with *vāta* predominance) should consume along with *viḍ lavaṇa* (black salt), *taila* prepared with whey, rice washing and buttermilk. Another *taila* prepared from *Aegle marmelos* (root), *Alpinia galanga* (rhizome), *Hordeum vulgare* (grains), *Ziziphus mauritiana* (seed kernel), *Cedrus deodara* (heartwood), *Boerhaavia diffusa* (root), *Dolichos biflorus* (seed), *Gmelina arborea* (root), *Premna serratifolia* (root), *Steroespermum suaveolens* (root) and *Oroxylum indicum* (root) is to be used for internal administration, *nasya* and *vasti*. Several *ghṛta*, *taila* and medicinal pastes are recommended for the treatment of *vātaja hṛdrōga*. Treatment with *rūkṣa* and *uṣṇa* herbs is ideal for *vātaja hṛdrōga* with association of *kapha*. *Pittaja hṛdrōga* patients should be subjected to purgation. The purgative medicine should be a mixture of purgative herbs, grape juice, sugarcane juice, honey and *Phoenix pusilla* fruits. After purgation the patient should consume *pēya* having *pitta*-lowering properties. He should also adopt measures for internal and external cleanliness recommended in treatment of *uraḥkṣataḥ* and *pitta jvara*. He should also drink a paste of *Picrorhiza kurroa* and *Glycyrrhiza glabra* suspended in sugar water. The physician can choose from a few *ghṛta* described. *Taila* prepared with *kvātha* of *Glycyrrhiza glabra* is to be mixed with honey and administered as *vasti* (*Ci.* 6: 25–49).

Kapha hṛdrōga patients should at first be subjected to sudation, followed by emesis by drinking *kvātha* of *Azadirachta indica* bark and *Acorus calamus* rhizome. After that the patient should consume cooked barley along with *yūṣa* prepared with horse gram, soup of *jāṅgala* meat and wine with *tīkṣṇa* quality. Several *cūrṇa*, *kvātha*, *lēhya*, *ghṛta* and *kṣāra* (alkalis) are recommended as *śamana* medicine. *Kṛmi hṛdrōga* should be treated with all medicines that destroy *kṛmi* (*Ci.* 6: 49–59).

5.4.8 Excessive Thirst

Vāta-pitta lowering treatment is essential in all forms of *tṛṣṇa* disease (excessive thirst). Medicines of *śīta* nature are to be applied topically and internally. *Śōdhana* and *śamana* therapy are also ideal. *Kvātha* of *Tṛṇapañcamūla* with sugar and *mantha* prepared with powder of parched rice can be consumed. Rice cooked from aged *Śāli* grains and *yavāgu* made of *Kōdrava* (*Paspalum scrobiculatum*) are recommended food for *tṛṣṇa* patients. *Nasya* with *śīta vīrya* herbs is to be carried out. All *pitta*-lowering *gaṇḍūṣa* described in *Sūtrasthānam* and *kalka* (medicinal pastes) indicated in treatment of *dāha jvara* can be applied in *tṛṣṇa* as well. Patient should take complete rest and maintain a positive mental attitude. *Vātaja tṛṣṇa* patients are to be fed with jaggery mixed with curd and meat soups that are roborant and of *śīta vīrya*. *Kvātha* of *Vidāryādi gaṇa* described in *Sūtrasthānam* is to be administered. In *pittaja tṛṣṇa*, the fruit of *Ficus glomerata* or its *kvātha* should be mixed with sugar and administered. *Kvātha* of *Śāribādi gaṇa* can also be administered mixed with sugar and honey. A few other *kvātha* are also indicated in *pittaja tṛṣṇa*. *Kaphaja tṛṣṇa* patients are to be subjected to emesis using *kvātha* of flowers of *Azadirachta indica*. *Kvātha* prepared from *Aegle marmelos* root, *Cajanus cajan* seeds, *Pañcakōla* (*Piper longum* fruit, *Zingiber officinale* rhizome, *Piper chaba* root, *Piper longum* root, *Plumbago zeylanica* root) and *Tṛṇapañcamūla* (roots of *Imperata cylindrica, Desmostachya bipinnata, Saccharum officinarum, Saccharum arundinaceum, Oryza sativa*) is to be drunk with sugar and honey. Similarly, *kvātha* of *Curcuma longa* rhizome is to be administered with sugar and honey. *Yūṣa* of green gram fortified with *Trikaṭu* (*Piper longum* fruits, *Piper nigrum* fruits, *Zingiber officinale* rhizome), *Trichosanthes cucumerina* root and tender leaves of *Azadirachta indica* is also ideal for internal administration. Recommended food is boiled barley. *Kabaḷa, nasya* and administration of *lēhya* made of herbs with *tīkṣṇa* quality are also recommended for the treatment of *kaphaja tṛṣṇa*. *Tridōṣaja tṛṣṇa* and *tṛṣṇa* due to *āma* are to be treated respectively with *tridōṣa*-lowering and *āma*-eliminating medicines and measures that remove the *dōṣa*s and *āma*. Patients suffering from *tṛṣṇa* due to starvation should be fed with warm *maṇḍa* (clear and thick fluid filtered from *yavāgu*) and cold *mantha*. *Tṛṣṇa* due to exertion is to be treated with meat soup or wine, mixed with sugar. Those suffering from sun stroke should drink *mantha* made of powder of barley and *kōlamajja* (seed kernels of *Zizyphus mauritiana*) mixed with sugar. Paste of sesame seed oil cake and *kāñjika* (fermented rice-water) is to be smeared all over the body (*Ci.* 6: 60–77).

Those suffering from *tṛṣṇa* caused by bathing in very cold water is to be fed with either dilute wine or jaggery water. Patients of *tṛṣṇa* due to alcoholism should take bath and drink diluted wine mixed with acidic fruit juice and salt. Those who experience sharp abdominal fire (*tīkṣṇāgni*) due to consumption of clarified butter should drink cold water. *Tṛṣṇa* arising from indigestion due to excessive consumption of *ghṛta* should drink warm water, followed by *maṇḍa*. Those who suffer from *tṛṣṇa* due to unctuous food should drink cold jaggery water. Patients of *tṛṣṇa* due to consumption of heavy, acidic and salty food should vomit out after drinking warm water. *Tṛṣṇa* due to *kṣaya* (emaciation) should be treated with roborant medicines indicated in treatment of *rājayakṣma*. *Tṛṣṇa* manifesting in lean, weak and dry-bodied individuals should be cured by drinking of milk and meat soups. *Tṛṣṇa* accompanied by belching needs to be pacified by drinking of milk boiled with herbs that cure *kṣaya kāsa*. Drinking of water boiled with green coriander, mixed with honey and sugar, cures *upasargaja tṛṣṇa* or *tṛṣṇa* appearing as a secondary complication of chronic diseases like consumption, diabetes and fever. The disease that generated *tṛṣṇa* is to be treated with medicines and measures, depending on the dominance of *dōṣa*s. When *tṛṣṇa* is cured, it is easier to control the disease that caused it (*Ci.* 6: 78–84).

5.4.9 Alcoholism

Madātyaya (alcoholism) is essentially of *sannipāta* nature or due to vitiation of all the three *dōṣa*s. *Sannipāta* can be either *samasannipāta* or *viṣamasannipāta*. When the *tridōṣa* are increased in equal degree, it is called *samasannipāta*, and when they are increased unequally, it is called *viṣamasannipāta*. Treatment of *viṣamasannipāta* begins with stabilizing of the *dōṣa* which is maximally destabilized among the *tridōṣa*. *Samasannipāta*, on the other hand, is to be treated in relation to *kapha* or the dependent site. In *madātyaya*, the first *dōṣa* to be destabilized is *kapha*, followed by *pitta* and *vāta*. In the site-dependent approach also, *kapha* is to be pacified first, as *madya* (intoxicating beverage, wine) damages stomach, which also happens to be the site of *kapha*. On account of its *tīkṣṇa* and *uṣṇa* nature, *madya* causes *utkḷēśa* of *kapha*, which gradually results in vitiation of *pitta* and *vāta*. Therefore, it is logical to pacify *kapha* first and restore the health of stomach. *Pitta* and *vāta* can be gradually brought back to normal state using roborant medicines (*Ci.* 7: 1–4).

Treatment of *madātyaya* is to be carried out for seven to eight days. By this time, the toxins generated by *madya* will have been transformed and eliminated from body. If any disease continues after seven days, that particular disease should be treated, while continuing the treatment for *madātyaya*. Patients of *vātaja madātyaya* should be administered *Piṣṭamadya* (wine brewed from rice powder), mixed with *Citrus medica* (fruit juice), *Garcinia gummi-gutta* (fruit juice), *Zizyphus mauritiana* (seed kernel), *Punica granatum* (fruit juice), *Trachyspermum ammi* (seeds), *Sphaeranthus indicus* (root), *Cuminum cyminum* (seed), *Trikaṭu*, rock salt, *sauvarcala* salt (sodium sulphate mixed with sodium chloride) and black salt. Barbecued meat, vegetables, and *lāja* powder mixed with clarified butter should be served. Several kinds of meat, beverages prepared from acidic fruits, delicacies made of wheat and black gram, spiced *vāruṇi* wine and spiced whey are also recommended as food and beverage. *Abhyaṅga* followed by bathing, fumigation with fragrant herbs and smearing of fragrant pastes on body are also useful in relieving the ill effects of *madātyaya* (*Ci.* 7: 10–19).

Those suffering from *pittaja madātyaya* are to be served sweet and diluted wine, mixed with juices of fruits such as pomegranate, dates, grapes, *Parūṣaka* (*Phoenix pusilla*) and so on. Cold and sweetened water mixed with powder of *lāja* can also be served. Meat of rabbit, goat and deer; green gram; gooseberry; pomegranate; fruit of *Paṭōla* (*Trichosanthes cucumerina*); and *Śāli* and *Ṣaṣṭika* grains are recommended food ingredients. *Pittaja madātyaya* patients suffering from *tṛṣna* and burning sensation in stomach are to be subjected to emesis after drinking copious volumes of sugarcane juice, cold water, grape juice or wine. After emesis, he should be served at first liquid food and gradually change over to solid food. This measure helps in stimulating *jaṭharāgni* that can digest food and the residual *dōṣa*s remaining in the stomach. A few herbal juices and soups are indicated, if patient experiences pain in flanks and thorax or spits blood-stained phlegm. Several *kvātha* are indicated in *tṛṣṇa* due to increased *vāta* and *pitta* vitiation. Cold measures are to be adopted in cases of excessive burning sensation (*Ci.* 7: 19–33).

Kaphaja madātyaya needs to be pacified with emesis and fasting. Patient should drink cooled decoction prepared from any of the following herbs: *Zingiber officinale* (rhizome), *Pseudarthria viscida* (root), *Coleus vettiveroides* (root) and *Tragia involucrata* (root). On disappearance of *āmadōṣa* and improvement in appetite, the patient can consume any wine mixed with *lāja* powder, *Trachyspermum ammi* and ginger powder. Thereafter, he may be fed with cooked barley and wheat, accompanied by thin soup of horse gram, meat soup or medicated soup of dried root of *Raphanus sativus*. Several medicated foods and beverages, a medicinal salt formula named *Aṣṭāṅga lavaṇa*, dry massage with powder of herbs having *rūkṣa* and *uṣṇa* qualities, fasting and keeping awake at night in the company of entertainers are also beneficial in *kaphaja madātyaya*. The treatment for *madātyaya* due to increased *vāta*, *pitta* or *kapha* is to be carefully adapted for application in the ten types of *sannipātaja madātyaya* (*Ci.* 7: 33–44).

5.4.10 Stupor

Śōdhana and *śamana* measures are to be carried out first in patients of *mada* (stupor). If *mada* continues even after these therapies, the patient should drink milk regularly. As body gains strength, drinking of milk can be discontinued. Instead, he should drink wine in very small measures. *Vāta-pitta*-pacifying treatment is ideal in *mada* and *mūrcha* (faining). However, special care needs to be taken to control *pitta*. Cold-natured pastes for topical application, juices of sweet fruits, meat soups, *Kalyāṇakaṃ ghṛta*, *Mahātiktakaṃ ghṛta*, *Ṣaṭpalaṃ ghṛta*, *vamana*, *virēcana*, *vasti*, *nasya*, *dhūma* and venesection are useful in the treatment of *mada* and *mūrcha*. Treatment recommended for *sannyāsa* (coma) is to be adopted if *mada* and *mūrcha* get aggravated. Emesis, purgation, nasal instillation, collyria, venesection and medicinal smokes with *tīkṣṇa* medicines are to be applied in *sannyāsa* (*Ci.* 7: 47–52; 100–115).

5.4.11 Hemorrhoids

Arśas (hemorrhoids) is protrusion of the lining of rectum, caused by dietary indiscretions. Removal of *arśas* is the main treatment for this disease. After performing *śōdhana* measures, the patient is to be fed light food. After that he is seated in a comfortable posture, and clarified butter is smeared inside and outside rectum. Thereafter, a *yantra* (blunt instrument) is inserted slowly into the rectum. With a *śalāka* (spatula) covered at the tip with cotton, physician should ablate the *arśas* if it is large, or cauterize with chemical or thermal cautery if it is small. Only one *arśas* should be ablated or cauterized at a time. Afterward, the patient is to adopt appropriate food and measures. After seven days, the procedure is to be repeated. Thus, over a span of several weeks all the *arśas* will have been removed. One whose *arśas* has been removed in this manner will have appetite and interest to consume food. Paste of *Boerhaavia diffusa* (root), *Sassurea lappa* (root), *Selinum tenuifolium* (root), *Anethum sowa* (seed), *Aquilaria agallocha* (heartwood) and *Cedrus deodara* (heartwood) is to be applied below the navel if pain is experienced in lower abdomen. In case of obstruction of stool and urine, *kvātha* of some specific herbs mixed with sesame oil should be poured in a thin stream (*sēka*) around the anus. *Vāta*-lowering *taila*, medicated milk and food that causes laxation are to be adopted (*Ci.* 8: 1–14).

A different treatment is advised in patients in whom thermal cautery is contra-indicated or when *arśas* is caused by *kapha-vāta* vitiation and vexed by secondary problems like feeling of heaviness, itching, pain and edema. *Sēka* is to be performed using *taila* prepared from *Aegle marmelos* (root), *Plumbago zeylanica* (root), *yavakṣāra* (alkali from barley straw) and *Sassurea lappa* (root). Alternatively, *sēka* can also be performed with several animal fats. After *sēka*, the rectal area is to be subjected to sudation using *piṇḍasvēda*, *dravasvēda* or boluses of herbal powders tightly covered in cloth. Fumigation can also be carried out with mixture of powders of *Withania sonifera* (root), *Ocimum sanctum* (herb), *Solanum xanthocarpum* (root), *Piper longum* (fruit) and clarified butter. *Varti* made of herbal powders and *taila* can also be applied. Because of these treatments, the vitiated blood in the cauterized area drains out and the patient feels relief (*Ci.* 8: 14–28).

Blood contained in hard or swollen *arśas* is to be drained continuously with leeches and surgical instruments. Buttermilk (*takra*) prepared from curd containing *Plumbago zeylanica* also cures *arśas*. Several *cūrṇa*, *mantha*, *kvātha*, *ghṛta* and a *lēhya*, known as *Guḍaharītaki*, are also recommended for internal administration. Buttermilk is recommended as a beverage that has *arśas*-pacifying properties. By the regular consumption of *takra*, body channels get cleared of obstructions and *dhātu* cycle functions optimally. Majority of *vāta-kapha arśas* get cured by the consumption of medicated *takra*. All these treatments are for *arśas* patients who void loose stool (*Ci.* 8: 28–52).

Arśas patients who suffer from constipation are advised to drink *Vāruṇi* wine with powder of *lāja* and rock salt. They can also drink salted buttermilk, *Sīdhu* wine or *Dhānyāmḷa* (rice gruel fermented with grains and herbs). Several formulations including the famous *Abhayāriṣṭam* and *Durālabhāriṣṭam* are recommended for the treatment of such patients. *Snēhavasti* (*vasti* using *taila* or *ghṛta*) and *nirūhavasti* (*vasti* performed with decoction of herbs) are advised for *arśas* patients suffering from *udāvarta* and stomach pain (*Ci.* 8: 52–94).

Treatment of *raktārśas* (bleeding piles) is to be initiated after assessing its association with *vāta* or *kapha*. Association with *vāta* is inferred if stool is of reddish black color, hard and if pain is experienced in hip, thigh and anus. *Kapha* involvement is inferred if stool is loose, pale or yellow in color, heavy and slimy. Several *cūrṇa*, *ghṛta*, medicated milk and *lēhya* including the reputed *Kuṭajāvalēha* are recommended in the treatment of *raktārśas*. All *raktārśas* which bleed frequently turn invariably into *vātaja arśas*. Therefore, it is vital to pacify *vāta*. To control *pitta* and *rakta* involved in *raktārśas*, the physician should strive to conserve *vāta* and *kapha*. If bleeding does not stop by these treatments, *ghṛta* should be administered to the patient. *Sēka* should be carried out on the anal area with warm sesame oil, milk and clarified butter. *Picchāvasi* (mucilaginous enema) with *kvātha* is also recommended (*Ci.* 8: 94–134).

Udāvarta is a complication arising from *arśas*. *Udāvarta* patient should first of all be subjected to *abhyaṅga* with *taila* that cures *śītajvara*. After that sudation should be carried out using *piṇḍasvēda* techniques. A medicated suppository prepared with *kvātha* of a specific combination of herbs, rock salt, *souvarcala* salt and jaggery is to be introduced into the rectum. Powder of these herbs should be administered rectally also. If this treatment fails, *snēhavasti* with *tīkṣṇa* herbs is to be carried out. If *udāvarta* still persists, purgation is to be carried out with *vāta*-lowering *ghṛta*. *Snēhavasti* also should be performed. A calcined herbal preparation known as *Kalyāṇaka kṣāra* is to be administered mixed with clarified butter. It cures 18 diseases including *arśas*. Many *lēhya*, *guḷika* (pills), *cūrṇa*, *ghṛta* and *kvātha* are indicated in treatment of *arśas*. *Arśas*, *atisāra* and *grahaṇi* can be causes for the appearance of each other, when *jaṭharāgni* is dull. They do not appear when *jaṭharāgni* is optimal. Therefore, conservation of *jaṭharāgni*, without any fluctuation, is vital in the treatment of these diseases (*Ci*. 8: 135–164).

5.4.12 Dysentery

Atisāra (dysentery) originates in stomach, due to lowering of *jaṭharāgni*. The first step in the treatment of all forms of *atisāra* is *laṅghana*. Patients suffering from stomach pain, bloating of stomach and *prasēkam* (excessive salivation) are to be subjected to emesis. Those suffering from *āmātisāra* are to be administered powder or *kvātha* of *Terminalia chebula*, which helps in eliminating the *āma* through purgation. Patients with medium *āmadōṣa* and scanty *āmadōṣa* should adopt fasting. Thereafter they should consume appropriate *kvātha*. Many *cūrṇa*, *taila* or *ghṛta* are prescribed to be administered once the *atisāra* is brought under control. Consumption of medicated food and beverages is also encouraged to strengthen the body (*Ci*. 9: 1–54).

Laṅghana is to be performed in *pittātisāra* with association of *āma*. *Ṣaḍaṅga kvātha* is to be used for quenching thirst. If dysentery is not cured with these treatments, the patient should be administered *kvātha* of specific herbs. *Kvātha* of *Holarrhena antidysenterica* (seeds) should be mixed with meat soup and consumed along with food. *Kvātha* of *Cyperus rotundus* also can be consumed mixed with honey. *Kvātha* of *Asparagus racemosus* or *ghṛta* prepared with it is to be administered in *raktātisāra*. Several other formulations are also indicated in *raktātisāra*. *Picchāvasi* with a *kvātha*, mixed with clarified butter and milk, is prescribed in treatment of *gudabhraṃśa* (rectal prolapse) and abdominal pain. Several *ghṛta* are also effective in curing *raktātisāra* (*Ci*. 9: 55–102).

Laṅghana is to be performed in *kaphātisara* to eliminate *āma*, followed by administration of suitable *cūrṇa*, *kvātha* or *lēhya*. *Picchāvasi* is to be carried out if there is abdominal pain and voiding of *kapha*. *Snēhavasti* is recommended in curing *kapha-vāta* conditions. Special care is to be taken to pacify *vāta*, which may get enraged due to diminution of *kapha*. After that *pitta* and *kapha* are to be pacified. *Vāta* gets enraged due to fear and grief. Therefore, *vāta*-pacifying treatment is advised in *bhayātisara* (dysentery due to fear) and *śōkātisara* (dysentery due to grief). The patient should be consoled to overcome grief and persuaded to be cheerful (*Ci*. 9: 103–124).

5.4.13 Grahaṇi

Grahaṇi is very similar to *ajīrṇa* (indigestion). The *āmadōṣa* of *grahaṇi* is to be eliminated, as instructed in treatment of *atisāra*. Patient should consume *yavāgu* processed with *Pañcakōla*. After food, he should consume medicines having properties similar to that of *Pañcakōla*. Many formulations are to be used in the elimination of *āma* and to treat associated complications like abdominal pain and vomiting. While consuming these, the patient may use *madya*, *yūṣa*, meat soup, whey, *pēya* or milk post-prandially (*Ci*. 10: 1–21).

After elimination of *āma*, the *vāta grahaṇi* patient should be administered in small measures, *ghṛta* prepared with *dīpana* herbs. *Jaṭharāgni* will get improved with administration of *ghṛta*. If there is obstruction of stool or urine, the patient is to be subjected to sudation for two or three days, followed by *nirūhavasti*. When *vāta* gets subsided, the patient is to be subjected to purgation, using castor oil, *Tilvaka ghṛta* or *yavakṣāra*. As alimentary tract gets cleared by purgation, *ghṛta* prepared with *vāta*-lowering herbs should be administered as *snēhavasti*. After these measures, the patient can get used to light food, along with small quantities of clarified butter. *Abhyaṅga* may also be performed with *vāta*-lowering *taila* (*Ci*. 10: 22–32).

Elevated *pitta* present in *pitta grahaṇi* patient is to be eliminated with purgation or emesis. Thereafter, he may be served food that is light, *grāhi* (constipating) and *dīpana* in nature. His *Jaṭharāgni* is to be improved by use of *cūrṇa* and *ghṛta* prepared with herbs having bitter taste. The physician can choose appropriate medicines from the several formulae presented (*Ci*. 10: 32–44).

Vitiated *kapha* which is the underlying cause of *kapha grahaṇi* needs to be eliminated through emesis, using herbs having *tīkṣṇa* quality. Thereafter, *Jaṭharāgni* is to be improved gradually with food and medicines fortified with *kṣāra* (alkalis) and herbs possessing pungent, sour and salty tastes. *Pēya* blended with *Pañcakōla*, *Terminalia chebula* (pericarp), *Coriandrum sativum* (herb), *Cissampelos pareira* (root), *Kaempferia galanga* (tuber) and *Citrus medica* (tender leaves) can be consumed regularly for improving *Jaṭharāgni*. Several *ghṛta* (*Dhānvantara-*, *Ṣaḍpala-*, *Bhallātaka-*, *Abhayā ghṛta*), *kṣāra*, *cūrṇa* and *āsava* including the famous *Madhūkāsava* are known to cure *grahaṇi* and impart strength (*Ci*. 10: 45–65).

Dōṣa present in *sannipāta grahaṇi* is to be eliminated using appropriate *śōdhana* measures such as emesis, purgation, medicated enema, nasal instillation and inhalation of medicated smoke. Following *śōdhana*, the various treatments and measures recommended earlier may be adapted judiciously, considering the dominance of *dōṣa*s. Treatment of *agnimāndya* (lowering of *Jaṭharāgni*) is described as an annexure to treatment of *grahaṇi*. *Agnimāndya* patient with excess of *kapha* should use medicines of dry and bitter nature to improve digestion. When increase of *kapha* is evident in a lean person, he should use *snigdha* and *rūkṣa* medicines alternatingly. A weak person needs to use *dīpana* drugs, mixed with clarified butter. Patient with increased *pitta* should take bitter-tasting medicines, mixed with sweet ingredients. Persons with *vāta* predominance should be administered sour-salty drugs mixed with clarified butter. Clarified butter has the ability to improve *Jaṭharāgni* of a weak-bodied persons. *Jaṭharāgni* improved by clarified butter cannot be lowered even by very heavy food. Various modes of administering clarified butter are also described. Proper use of several *taila*, *ghṛta*, *āsava*, *sura*, *ariṣṭa*, *cūrṇa*, *kvātha* and desirable foods improve *Jaṭharāgni* and impart strength to body. Proper digestion is pivotal to healthy life. Therefore, one should conserve *Jaṭharāgni* and maintain it in steady state, not allowing to fluctuate (*Ci*. 10: 65–93).

5.4.14 Dysuria

Vātaja mūtrakṛchra (dysuria with enragement of *vāta*) patient is to be subjected to *abhyaṅga* with *vāta*-pacifying *taila*. Thereafter, sudation with *piṇḍasvēda*, *parisēka* and *avagāha* is to be applied on the lower abdominal area. *Parisēka*, *lēpa* (smearing of medicinal paste) and *avagāha* of cold nature should be applied in *pittaja mūtrakṛchra*. After that the patient should consume *kvātha* of specific herbs. *Kaphaja mūtrakṛchra* patient should be subjected to emesis and sudation. He should regularly use food of *tīkṣṇa*, hot and pungent nature, *kṣāra*, different kinds of food prepared from barley and drink buttermilk. Many medicated foods, *kṣāra* and *kvātha* are recommended. All these treatments and measures are to be applied judiciously in *sannipāta mūtrakṛchra*, newly diagnosed *aśmari* (urinary stone) and *vātavasti*, after due consideration of *dōṣa*s involved (*Ci*. 11: 1–16).

5.4.15 Urinary Stone

Aśmari (urinary stone) is a disease that is very painful. It can be cured with medicines if diagnosed at an early stage. When fully developed, the urinary stone can be removed only by surgery. Therefore, *aśmari* is to be treated in its prodromal stage, using sudation and administration of *ghṛta*. Several *ghṛta* are recommended in the treatment of *vātaja*, *pittaja* and *kaphaja aśmari*. The ingredient herbs of these formulations can be processed into *kvātha* and medicated milk. They can be added to *yavāgu* and consumed as food as well. *Tailvaka ghṛta* can be used for purgation. *Uttaravasti* can also be applied. The urinary bladder of *śukḷāśmari* patient needs to be cleansed at first with *uttaravasti*. Thereafter, he should consume cooked chicken meat and indulge in sexual intercourse (*Ci*. 11: 16–43).

If *aśmari* is not cured by these treatments, the alternative is surgical removal of the urinary stone. Detailed procedure of surgical removal of the urinary stone, post-operative care and dressing of wound is described (*Ci*. 11: 43–63). Etiology and nosology of 12 other urinary diseases such as *vātavasti, vātāṣṭhīla, vātakuṇḍalika, mūtrātītam, mūtrajaṭharaṃ, mūtrōtsaṅgaṃ, mūtragranthi, mūtraśukḷaṃ, viḍvighātaṃ, uṣṇavātaṃ, mūtrakṣayaṃ* and *mūtrasādaṃ* are described in *Nidānasthānam* (*Ni*. 9: 20–39). All protocols described in the treatment of *mūtrāghāta* and *aśmari* can be extended to the treatment of diseases beginning with *vātavasti* (*Ci*. 11: 7; 15–16; 32–34; 35).

5.4.16 Polyureic Diseases

Physically strong patient of *pramēha* (polyuria) should be made unctuous, by administering internally mustard oil, neem oil, *Nagādanti* (*Baliospermum montanum*) oil, *Vibhītaki* (*Terminalia belerica*) oil, *Karañja* (*Pongamia glabra*) oil, *Trikaṇṭakādi snēhaṃ* or *taila* appropriate for the *dōṣas* of the patient. Thereafter, he should be subjected to emesis and purgation, followed by *nirūhavasti*. After eliminating *dōṣas* from body, he should consume meat soup. Curative (*śamana*) medicines are to be administered, thereafter, to prevent recurrence of *pramēha*. Those patients in whom *śōdhana* therapy is contra-indicated are to be administered with curative medicines. Formulae of many *svarasa* (juice of fresh herbs) and *kvātha* are suggested for the treatment of *kaphaja* and *pittaja pramēha*. Food and beverages are to be fortified with these herbs, in consideration of the dominance of *dōṣas*. *Snēha* (*ghṛta, taila*) are to be administered in *pramēha* with *vāta* dominance. Many grains, millets, meat and wines are recommended for preparing food and beverages for *pramēha* patients. Medicines like *Lōdhrāsavaṃ, Ayaskṛti* and several *ghṛta* are useful in the treatment of *pramēha*. Massaging with powders, exercise, keeping awake at night and all food, beverages and measures that reduce *kapha* and *mēdas* can be applied in treatment of *pramēha* (*Ci*. 12: 1–38).

Śarāvika, kacchapika and similar *pramēha piṭaka* (diabetic carbuncles) appear in *pramēha* patients. In the initial stage, these are to be treated as *śōpha* (edema) and as *vraṇa* (wounds, sores, ulcers), when they suppurate. In the early stage, it is advisable to administer *kvātha* of *Kṣīrī vṛkṣa* and goat urine. Purgation is also desirable, as *pramēha* patients have a tendency to constipate. *Taila* prepared with herbs of *Ēlādi gaṇa* can be applied topically on the carbuncles. *Kvātha* prepared with *Āragvadhādi gaṇa* can be used for powder massage. *Pariṣēka* can be carried out with *kvātha* of *Asanādi gaṇa*. Food and beverages can be fortified with herbs of *Vatsakādi gaṇa*. Powders like *Navāysam cūrṇa* can be administered mixed with honey. *Śilājit* (mineral bitumen) is a miracle medicine in the treatment of *pramēha*. On consuming *Śilājit* in the prescribed manner, even a chronic *pramēha* patient disowned by physician will get a new lease of life, with a disease-free body (*Ci*. 12: 38–43).

5.4.17 Abscess

In their early stage, all *vidradhi* (abscess) should be treated as a *śōpha* (edema). Bloodletting needs to be carried out frequently. When fully developed, they should be treated as *vraṇa*. *Vāta vidradhi, pitta vidradhi, kapha vidradhi, rakta vidradhi* and *āgantuja vidradhi* should be washed with *kvātha* of specific herbs and medicinal pastes smeared upon them. In the early stage of *antarvidradhi* (internal abscess), patient should drink in the morning, *kvātha* of *Varuṇādi gaṇa,* mixed with powder of *Ūṣakādi gaṇa*. *Ghṛta* prepared with herbs having ability to cause purgation or herbs of *Varuṇādi* and *Ūṣakādi gaṇas* should also be administered. *Nirūhavasti* and *snēhavasti* should be performed using herbs of these two *gaṇas*. Pastes for topical application and *kvātha* and *ghṛta* for internal administration are also recommended (*Ci*. 13: 1–18).

Vidradhi formed inside the alimentary tract may protrude out as a swelling. In such cases, *upanāha* should be carried out. Burning sensation will subside if the sides of the *vidradhi* are pressed, indicating that the *vidradhi* has ripened. The ripened *vidradhi* should be cut with sharp instruments and treated like a *vraṇa*. Ripened *antarvidradhi* will rupture and its contents will get drained. The patient should consume recommended medicinal food and take rest for ten to twelve days. The physician should take care of the patient and facilitate the drainage of pus. Powder of herbs of *Varuṇādi gaṇa* or *Madhuśigru* (*Moringa concanensis*) bark should be suspended in warm water and administered to the patient. *Yavāgu* fortified with powder of *Madhuśigru* bark should be served as food. *Yūṣa* of barley, *Zizyphus mauritiana* seed kernels and *Dolichos biflorus* seeds can be used as beverage (*Ci.* 13: 18–23).

Śōdhana is to be carried out after ten days, with *Tailvaka ghṛta* or *ghṛta* prepared with *Brahmi* (*Bacopa monnieri*). After *sōdhana*, the patient should consume *Tiktaka ghṛta* with honey. Guggulu and *śilājit* mixed with *kvātha* appropriate for the *dōṣa* state are to be administered in all types of *vidradhi*. Physician should strive to prevent the ripening of *vidradhi*. If *vidradhi* develops in one with *pramēha*, treatment for *pramēha* also needs to be initiated. *Stanavidradhi* (*vidradhi* appearing in breasts) should be treated as *vraṇa*. However, *upanāha* is contra-indicated in this case (*Ci.* 13: 24–29).

5.4.18 Inguinoscrotal Swelling

In *vātaja vṛddhi* (inguinoscrotal swelling with enragement of *vāta*), oil of *Kōśāmra* (*Schleichera oleosa*), *Symplocos racemosa* (*Tilvaka*), castor oil or *Sukumāra ghṛta* should be administered internally and purgation initiated by powder of *Operculina turpethum*. After that *nirūhavasti* is to be carried out using a mixture of *kvātha*, *kalka* (paste) and *ghṛta* of *vāta*-pacifying herbs. This should be followed by consumption of food along with meat soup. *Snēhavasti* is to be performed after that, using *taila* prepared with *Glycyrrhiza glabra* root, followed by *lēpa* and sudation with *vāta*-pacifying herbs. *Vṛddhi* in the ripe stage is to be operated and treated as a *vraṇa*. *Pittaja vṛddhi* and *raktaja vṛddhi* should be treated as *śōpha* in early stage and as *vraṇa* in advanced stage. Vitiated blood needs to be drained out with venesection. *Kaphaja vṛddhi* patient should drink cow urine, mixed with paste of *Pītadāru* (*Coscinium fenestratum*). Treatment advised for *śḷēṣmagranthi* (glandular swelling with predominance of *kapha*), excluding *vimḷāpana* measure (mutilation of swelling by massaging), is also advised. Ripe *kaphaja vṛddhi* is to be operated upon and dressed with a wound-healing *ghṛta*. Patient suffering from *mēdōjavṛddhi* is to be subjected to sudation, with paste of *Surasādi gaṇa* or *Vēllāpamārgādi gaṇa* ground with cow urine. Thereafter, the *vṛddhi* is to be operated upon, using an instrument called *vṛddhipatram*, and the *mēdas* removed. This is followed by dressing of the operated wound with a wound-healing *taila*. Appropriate *ghṛta* can be administered internally as well (*Ci.* 13: 29–38).

Mūtraja vṛddhi patient is to be subjected to sudation with unctuous herbs and urine drained by surgical intervention. The wound should be bandaged with a bandage called *sthagikā*. *Āntravṛddhi* (hernia) which has not descended to the testes needs to be treated as *vātavṛddhi*. *Sukumāra ghṛta* is a very important medicine that cures *vṛddhi*, *vidradhi*, *gulma*, *arśas*, *śōpha* and many other diseases. Its regular use is recommended in the treatment of all these diseases. Thermal cautery is indicated if *vṛddhi* is not cured by oleation, purgation and *snēhavasti* (*Ci.* 13: 39–51).

5.4.19 Phantom Tumor

Gulma (phantom tumor) characterized by pain, constipation and caused by dry and cold qualities is to be treated with *taila* recommended in *vāta* diseases. The patient is to be made unctuous by application of *taila* through food and beverages, *abhyaṅga* and *snēhavasti*. Thereafter, he should be subjected to sudation. Sudation is all the more essential, if there is bloating of abdomen, abdominal pain, feeling of heaviness and constipation. Sudation reduces *vāta*, clears obstructions in body channels and pacifies *gulma*. Internal administration of *taila* is beneficial in *gulma* that appears above the navel (in stomach). *Vasti* is better in intestine-dependent *gulma*. Internal administration of *taila* and *vasti* are essential in stomach-dependent *gulma*. *Bṛṃhaṇa* and *snigdha* food and beverages are to be served to the *vāta gulma* patient, if he has good appetite and suffering from constipation. Internal administration of *taila* is to be continued along with *snēhavasti* and *nirūhavasti*, without disturbing *kapha* and *pitta* in body. As *vasti* pacifies *vāta* at its own site, regularly conducted *snēhavasti* and *nirūhavasti* pacify *vāta-*, *pitta-* and *kapha gulma*. *Vāta gulma* patient is to be subjected to emesis, if *kapha* increases, lowering *jaṭharāgni* and causing distaste for food, heaviness of body and drowsiness. As a result of *kapha* increase, stomach pain, bloating of abdomen and constipation may also be experienced. In that case, ingredient herbs of a specific *ghṛta* are to be powdered and consumed as *kvātha* or *cūrṇa*. If *kapha* is associated with *vāta* in *vāta gulma* patient, purgation should be carried out by drinking castor oil mixed with wine. In case of *pitta* association, castor oil is to be drunk mixed with milk. If *pitta* increases in *vāta gulma* patient causing burning sensation, purgation is to be induced by administering along with *snēha*, herbs that normalize the *dōṣa*. Venesection is to be carried out if burning sensation persists (*Ci*. 14: 1–60).

Purgation is to be carried out, if *pitta gulma* is due to causative agents having *snigdha* and *uṣṇa* qualities. Purgation is to be induced by powder of *Terminalia chebula* mixed with grape juice and honey. *Pitta*-pacifying purgation described in *Kalpasthānam* and chapter on treatment of *raktapitta* can also be applied. *Śamana* therapy is better in *pitta gulma* caused by *rūkṣa* and *uṣṇa* causes. Several formulations including *Tiktaka ghṛta* and *Vāśa ghṛta* can be used for *śamana*. If the physician feels that the disease can take a serious turn if purgation is not carried out, purgation can be induced by *ghṛta* prepared with purgative herbs or plain milk. *Abhyaṅga* with *ghṛta* made up of *śīta vīrya* herbs is advised in case patient feels burning sensation. Paste of *śīta vīrya* herbs ground with clarified butter can be applied all over the body. Venesection is essential if patient feels stomach pain and indigestion. If tiredness is experienced due to elimination measures, the patient may consume soup of *jāṅgala* meat. After feeling relief, he should be administered *ghṛta* for pacifying the residual *dōṣa*. If the *gulma* starts to ripen due to increase of *rakta* and *pitta* or treatment not being carried out properly, all the treatments advised in *pitta vidradhi* are to be initiated (*Ci*. 14: 61–75).

Kapha gulma patient is to be subjected first to emesis. Those in whom emesis is contra-indicated are to be subjected to fasting. Thereafter, *jaṭharāgni* is to be stimulated by adopting *pēya* having bitter and pungent tastes or *Hiṅgvādi cūrṇa*. *Pēya* should be gradually changed to solid food. In *gulma* that is shaped like a tumor, protruding out, deep-seated, hard and with associated problem of bloating of abdomen, the patient should be subjected to appropriate *śōdhana*, followed by pressing of the *gulma* firmly. After that, the patient should consume *ghṛta* processed with *kṣāra* and *Trikaṭu*. This is to be followed by sudation of the entire body, including the area where the *gulma* is located. When the *gulma* is softened by these measures, it should be made to protrude out using *ghaṭīyantra* (small brass pot). Following this the physician should cover the *gulma* with a sheet of cloth and press it repeatedly with any handy tool such as *vimārga*, *ajapada* or *ādarśa*, taking care not to press precordial area or the area where intestine is located. According to *Caraka Samhita*, *kapha gulma* is to be compressed with *vimārga*, *ajapada* (pieces of hides of deer or goat) or *ādarśa* (glass piece) (Sharma, 1983). After disintegrating the *gulma*, the area should be smeared with the paste of seeds of sesame, castor, *Atasi* (*Linum usitatissimum*) and mustard. The area should be warmed by placing small heated metallic bowls. Thereupon the patient is to be subjected to purgation using purgative medicines containing *snēha*, like *Miśraka snēham*. *Vasti* measure called *Dāśamūlikaḥ*, with *Daśamūla* herbs as ingredients, is also to be carried out. Regular consumption of several *ghṛta* such as *Nīlini ghṛta*, *Sukumāra ghṛta*, *Dantīharītaki lēhya* and various *āsava* and *ariṣṭa* cures all kinds of *gulma*. The treatment protocols described for individual *dōṣa*s are to be applied judiciously to their combinations, after careful review of the *dōṣa* states (*Ci*. 14: 76–119).

Rakta gulma is an ailment that appears only in women, mimicking pregnancy. On completing twelve months from the appearance of *rakta gulma*, the patient is to be subjected to *snehavirēcana*, after conditioning the body with oleation and sudation. In case of frequent bleeding, treatments recommended for *raktapitta* are ideal. Pain due to enraged *vāta* is to be pacified by *vāta*-lowering treatment. Treatments that pacify *udāvarta* and *kapha* enragement are to be initiated in bloating of abdomen. Several *kvātha*, *cūrṇa* and *lēhya* cure *rakta gulma* (*Ci.* 14: 119–129).

5.4.20 Enlargement of Abdomen

Udara (enlargement of abdomen) is caused by the movement of *dōṣa*s in high magnitude and obstruction of body channels. Therefore, *udara* patient is to be subjected to purgation regularly. Castor oil is to be administered with cow urine or milk. For one or two months, the patient should drink cow urine or buffalo urine. Food is to be consumed along with cow milk. Patient suffering from burning sensation, bloating of abdomen, excessive thirst and bouts of fainting should survive only on camel milk. *Vāta*-dominant *udara* patients requiring *śōdhana* and those with dry body should consume *ghṛta* that improves unctuousness in body and cures *udara*. On consuming these *ghṛta*, purgative medicines recommended in *Kalpasthānam* are to be administered. On achieving satisfactory purgation, the patient should consume *pēya* followed by cooked rice and *jāṅgala* meat. Thereafter, for six days consecutively, he should drink milk mixed with powder of *Operculina turpethum* stem. The treatment cures all *udara*, including the one that has turned into *jalōdara* (ascites) (*Ci.* 15: 1–53).

Patient suffering from *vātōdara* should consume *ghṛta* prepared with herbs of *Vidāryādi gana*, followed by sudation. Thereafter, purgation is to be induced many times using *Tailvaka ghṛta* or *Miśraka snēham*. After purgation, the patient should start consuming *pēya* and in graded stages switch over to solid food. After that, he should regularly consume milk for improving stamina. Drinking of milk is to be slowly discontinued when stamina is regained (*Ci.* 15: 53–59).

Pittōdara patient needs to be made unctuous by consuming *ghṛta* prepared with herbs having sweet taste. After that he should be subjected to purgation using appropriate purgatives. *Nirūhavasti* should then be performed using *kvātha* of herbs of *Nyagrōdhādi gana*, mixed with sugar, honey and clarified butter. *Snēhavasti* using *ghṛta* made with *Nyagrōdhādi gana* herbs is also to be carried out (*Ci.* 15: 59–65).

Kaphōdara patient is to be administered *ghṛta* prepared with *Vatsakādi gana*. Thereafter, he should be subjected to sudation. After that, he should undergo purgation induced by *ghṛta* prepared with latex of *Euphorbia neriifolia*. He should, thereafter, start consuming *pēya* containing *kvātha* of herbs with pungent taste and in graded stages switch over to solid food; *Nirūhavasti* should also be carried out using *kvātha* of *Muṣkakādi gana*, mixed with cow urine, *Trikaṭu* and sesame oil. *Snēhavasti* is also to be carried out using *ghṛta* prepared with herbs of *Muṣkakādi gana*. Food can be consumed with soup of *Dolichos biflorus* seeds or milk boiled with powder of *Trikaṭu*. Cow urine, various *ariṣṭa*, *cūrṇa* and *Ayaskṛti* can be used regularly (*Ci.* 15: 66–75).

Sannipātōdara is a very serious condition. Therefore, before starting treatment, the physician should seek permission from patient's relatives, as sometimes the patient may die during the treatment. Roots of *Abrus precatorius* and *Nerium indicum* are to be ground into a paste, mixed with wine and administered to the patient. Mineral and vegetable poisons can also be administered, mixed with food and beverages. This treatment causes either recovery from disease or death of patient. If patient survives, he should be bathed in cold water and *pēya* or milk served to him. Thereafter, for one month he should eat only parboiled leaves of *Operculina turpethum*, *Centella asiatica*, *Chenopodium album*, *Murraya koenigii* or barley, without adding tamarind, salt and clarified butter. Juices of these leaves can be used as beverage. When *dōṣa*s have been eliminated after a month, patient should drink regularly camel milk (*Ci.* 15: 76–84).

Plīhōdara patient is to be made *snigdha* based on *dōṣa* state and subjected to sudation. Venesection should be performed on the left hand, after consuming boiled rice with curd. On recovering from the tiredness caused by venesection, he should be administered *ghṛta* and subjected to purgation. The *kṣāra* prepared from pearl oyster shells is to be administered suspended in milk. *Hiṅgvādi cūrṇa* mixed with *ghṛta* made from *kṣāra* can also be administered (*Ci.* 15: 85–98).

Baddhōdara patient should be subjected to sudation and then *snēhavasti*. After that *nirūhavasti* is mandatory, using cow urine mixed with *kalka* of *tīkṣṇa* herbs, sesame oil and rock salt. Foods that cause laxation and purgation with *tīkṣṇa* herbs are ideal for him. Measures that pacify *vāta* and *udāvarta* also need to be performed. All treatments except sudation recommended for *kaphōdara* are to be carried out in *chidrōdara*. The fluid generated is also to be drained out. *Jalōdara* patient should be administered cow urine containing *kṣāra* and various *tīkṣṇa* herbs. He should consume food that increases *jaṭharāgni*, *ōjas* and lowers *kapha* (*Ci.* 15: 99–131).

5.4.21 Morbid Pallor

Pāṇḍu (morbid pallor) is caused by the decline in quality of *rakta dhātu*, which is closely associated with *pitta*. Therefore, normalizing *pitta* and improving the properties of *rakta* are the two ways in which *pāṇḍu* is pacified. Patient should start consuming *ghṛta* such as *Kalyāṇaka*, *Mahātiktaka*, *Pañcagavya* or *ghṛta* prepared with herbs of *Āragvadhādi gaṇa*. One who has become unctuous with consumption of *ghṛta* is to be subjected to emesis using *tīkṣṇa* medicines. He should again be made unctuous with consumption of *ghṛta*. Thereafter, he is to be subjected to purgation several times, induced by a mixture of cow urine and milk or milk alone. After successful elimination of *dōṣa* from body, *śamana* therapy is to be initiated (*Ci.* 16: 1–33).

Vāta pāṇḍu is pacified mostly by *snēha* like *Kalyāṇaka ghṛta*. Medicines with bitter and cold qualities are used for pacifying *pitta pāṇḍu*. An example is *Drākṣādi lēhya*. Medicines with pungent, dry and hot qualities pacify *kapha pāṇḍu*. *Viśālādi cūrṇa* and *Vyōṣādi cūrṇa* are commonly employed. As *sannipāta pāṇḍu* is a combination of *vāta-*, *pitta-* and *kapha pāṇḍu*, medicines that pacify *tridōṣa* are used, *Maṇḍūra vaṭakam* being an example. Patient suffering from *Mṛtjanya pāṇḍu* (*pāṇḍu* caused by ingestion of sand) is first subjected to *śōdhana* with *tīkṣṇa* medicines. This brings out the sand that is deposited inside the alimentary tract. Thereafter, *ghṛta* that pacify *pāṇḍu* and improve stamina are administered (*Ci.* 16: 34–39).

Pitta-pacifying treatment is required in *kāmala* (jaundice). For example, *ghṛta* prepared with decoction and powder of *Terminalia chebula* pericarp cures *kāmala*. Consumption of decoction of *Triphala* (pericarps of *Emblica officinalis*, *Terminalia belerica* and *Terminalia chebula*), *Tinospora cordifolia* stem, *Coscinium fenestratum* bark or *Azadirachta indica* bark mixed with honey cures *kāmala*. Administration of *Śilājit* for one month mixed with cow urine cures *kumbhakāmala* (advanced stage of *kāmala*). *Halīmaka* patient is to be made unctuous by consumption of *ghṛta* prepared with milk, clarified butter and juice of *Tinospora cordifolia*. After that he should drink juice of gooseberry, mixed with powder of *Operculina turptheum*. Following purgation, he should consume food that is *pitta-vāta*-pacifying and sweet in taste. *Drākṣādi lēhya*, *ghṛta* prepared with herbs having sweet taste, *snēhavasti* and *kṣīravasti* (*vasti* with milk) are also recommended in the treatment of *halīmaka*. Treatment and measures recommended for *śōpha* are to be adapted by an intelligent physician in the treatment of *pāṇḍu* (*Ci.* 16: 40–57).

5.4.22 Edema

In *śvayathu* (edema spread all over the body), associated with *āma* and not of exogenous causes, the patient should be subjected to fasting, followed by consumption of light food. After that he should consume with warm water, *Navāyasa cūrṇa* or powder of herbs having *kaṭu* and *tīkṣṇa* qualities. Those who are affected by high degree of *dōṣa* derangement should consume *Terminalia chebula* powder mixed with cow urine or the formulation *Gōmūtra harītaki*. One who has dull *jaṭharāgni*, voids stool that is heavy and with signs of *āma*, should drink buttermilk containing *sauvarcala* salt, powder of *trikaṭu* and honey (*Ci.* 17: 1–27).

Vātaja śvayathu patient should consume for fifteen days *Trivṛt snēha* or castor oil. He should also be subjected to sudation and *abhyaṅga* with *vāta*-pacifying herbs and *taila*. *Pitta śvayathu* patient should first of all consume *Tiktaka ghṛta*. *Ghṛta* or milk processed with herbs of *Nyagrōdhādi gaṇa* should be administered, if patient complains of excessive thirst, burning sensation and fainting. *Lēpa* prepared with *śīta vīrya* herbs is to be applied on body. Patient suffering from *kaphaja śvayathu* should consume *taila* prepared with herbs of *Āragvadhādi gaṇa*. *Kṣāra*, *cūrṇa*, *āsava*, *ariṣṭa*, cow urine and buttermilk are to be administered to *kaphaja śvayathu* patient having dull *jaṭharāgni*, obstruction of body channels and distaste for food. In combinations of *dōṣas*, as in *samsargaja* and *sannipātaja śvayathu*, the dominant *dōṣa* is to be pacified first, followed by the rest. These are the treatment for *nijaśvayathu* or *śvayathu* arising from endogenous causes. *Āgantuja śvayathu* or *śvayathu* arising from exogenous causes, as in *abhighātaja śvayathu*, is to be treated with bloodletting, *ghṛta*, *lēpa* and wine and subjected to purgation with *śīta vīrya* medicines. Anti-poisonous medicines are desirable in *viṣaja śvayathu* (edema caused by poison). *Śvayathu* patient is permitted to include in food meat of domestic animals, *ānūpa* meat, meat of weak-bodied creatures, dried leafy vegetables, rice cooked with sesame seeds, rice cooked with jaggery, curd, salt, preparations with rice flour, wine, sour substances and dried meat. However, he should abstain from heavy and incompatible food, food that causes *vidāha*, sleeping during day time and sexual intercourse (*Ci*. 17: 28–42).

5.4.23 Cellulitis

Treatment of *visarpa* (cellulitis) begins with *laṅghana*, *rūkṣa* measures, bloodletting, emesis and purgation. *Randia dumetorum* pericarp, *Glycyrrhiza glabra* root and *Holarrhena antidysenterica* seeds form an ideal combination to induce emesis. Purgation can be induced by administering powder of *Operculina turpethum* in juice of *Bacopa monnieri*, grape juice, *kvātha* of *Triphala*, milk or clarified butter. *Śamana* therapy is to be applied in the case of patient with less derangement of *dōṣas* and those in whom *śōdhana* is contra-indicated. Venesection is to be carried out if blood is found to be vitiated. *Visarpa* becomes *vāta-pitta*-dominant when *āma* is eliminated and *kapha* is reduced. In such cases *Tiktaka ghṛta*, *Mahātiktaka ghṛta* or *Trāyamāṇa ghṛta* is to be administered. When venesection is performed, the alimentary tract is cleansed and the *dōṣa* that caused *visarpa* will get transferred to skin, muscles and joints. Topical application of *lēpa* quickly pacifies the *visarpa*. The physician can choose an appropriate *lēpa* from the several formulae suggested. Ingredient herbs of those formulae can be processed into *kvātha* for performing *sēka* on body, *ghṛta* and *cūrṇa*. *Lēpa* to be applied on *vāta visarpa* should contain more of clarified butter. Patient suffering from *agni visarpa* is to be smeared with clarified butter. *Sēka* is also to be performed with cooled *kvātha* of *Glycyrrhiza glabra*, sugared water, cooled *kvātha* of *Cyperus rotundus*, milk or sugarcane juice. *Mahātiktaka ghṛta* is ideal for internal administration, topical application and *sēka*. Treatment recommended for *raktapitta* is to be applied in *granthi visarpa*, followed by *piṇḍasvēda* and *upanāha*, which pacify *kapha-vāta* conditions. *Sēka* with cow urine or plain water pacifies pain experienced in *granthi visarpa*. Venesection is to be carried out in *granthi visarpa* following *vāta-kapha* treatment. Internal and topical application of medicine is required in *granthi visarpa* characterized by burning sensation or exudation of fluid, as seen in ulcers. *Ghṛta* prepared with juice of *Cynodon dactytlon* is to be applied internally and topically in *kardama visarpa*, characterized by *kapha-pitta* dominance (*Ci*. 18: 1–38).

5.4.24 Leprosy and Other Skin Diseases

Ghṛta or *taila* are to be administered to all patients suffering from *kuṣṭha* (leprosy and other skin diseases). Specific *ghṛta* are recommended in *vāta*-, *pitta*- and *kapha kuṣṭha*. *Abhyaṅga* is also to be performed with these *snēha*. *Śōdhana* medicines recommended in treatment of *visarpa* are to be administered, when the inside and outside of the *kuṣṭha* patient have become unctuous. Venesection is to be carried out by cutting veins on forehead, hands and legs. This is to be followed by administration of *kuṣṭha*-pacifying *snēha*. Patient suffering from *kuṣṭha* can consume food prepared with

Śāli rice, barley, wheat, millets, pulses, bitter leafy vegetables, medicated wines and *jāṅgala* meat processed with medicinal herbs. Buttermilk containing powder of *Psoralea corylifolia* seeds can be used as beverage. He should not use food and beverages that are too sour and salty. Curd, milk, jaggery, *ānupa* meat, sesame seeds and black gram are to be avoided completely. Many *kvātha*, *ghṛta*, *cūrṇa*, *vaṭaka*, *guḷika* (pills) and *lēhya* can be used as *śamana* medicines (*Ci*. 19: 1–53).

After pacifying the *dōṣa* inside the body, treatment should focus on *dōṣa*s located in the skin. Various *lēpa* are to be applied for that purpose. Hard and round *kuṣṭha* are to be softened by application of *pōṭaḷa svēda*, followed by surgical ablation and application of *lēpa*. Venesection should be performed and *kṣāra* applied on those *kuṣṭha* which cannot be removed surgically and those with loss of tactile sensation. Poisons should be administered first to patients suffering from hard, rough, chromic and insensitive *kuṣṭha*. Thereafter, paste of poisonous herbs should be applied as *lēpa*, accompanied by chanting of *mantra*. *Kuṣṭha* which do not perspire and have itching are to be scraped off with a rough-surfaced object like cuttlefish bone and *lēpa* applied. *Kuṣṭha* which looks like scalded skin are to be smeared with *ghṛta*, which can also be used for *abhyaṅga*. *Vāta*-dominant *kuṣṭha* is to be treated with *ghṛta*, *kapha*-dominant with emesis and *pitta*-dominant with venesection and purgation. *Kuṣṭha* patient should apply *nasya* once in three days, emesis once in fifteen days, purgation once in a month and venesection once in six months. This protocol is intended to eliminate effectively, all *dōṣa*s residing in the patient. As *kuṣṭha* is said to have some connection with sinful acts, the patient should lead a virtuous life, on recovering from this malady (*Ci*. 19: 54–98).

5.4.25 Vitiligo

Śvitra (vitiligo) is more disgusting than *kuṣṭha* and very quickly it reaches the incurable stage. Therefore, a physician should work hard to cure this disease, as if he were dousing a house fire. As first step in treatment of *śvitra*, the *dōṣa*s are to be eliminated through purgation using appropriate purgatives. Juice of *Ficus hispida* fruits can be administered, mixed with jaggery. After consuming this purgative, the patient should undergo *abhyaṅga* with a *taila* and stand in the sun for some time, till he feels an urge to void his bowels. After purgation he should consume only *pēya* for three days, avoiding any solid food (*Ci*. 20: 1–3).

The eruptions that appear on body are to be punctured with needles. When the fluid in all these eruptions is drained completely, the patient should drink in the morning for three days *kvātha* of *Ficus hispida* root, *Pterocarpus marsupium* heartwood, *Callicarpa macrophylla* flowers and *Anethum sowa* seed. The *kṣāra* prepared from *Butea monosperma* bark, suspended in water, can also be consumed in moderate volume. Thereafter, several *lēpa* can be applied topically on the vitiligo patches. *Śōdhana* measure, bloodletting and consumption of *saktu* are recommended for the cure of *śvitra* (*Ci*. 20: 4–18).

5.4.26 Worm Infestation

One who is infested with worms (*kṛmi*) should undergo oleation and sudation. He should then consume jaggery, milk and fish. After spending one night comfortably, he should be subjected to *vasti* in the next morning. The *vasti* medicine should consist of herbs of *Surasādi gaṇa*, made into *kvātha* with diluted cow urine, mixed with sesame oil, *souvarcala* salt and powders of *Piper longum*, *Randia dumetorum* and *Embelia ribes* seeds. After *vasti* he should be subjected to purgation, induced by paste of *Operculina turpethum*, suspended in *kvātha* of *Randia dumetorum* and *Piper longum*. This is to be followed by consumption of *pēya*, mixed with powder of *Pañcakōla* herbs. He should then be bathed in water boiled with herbs having pungent, bitter and astringent tastes. On achieving good appetite and stamina, he should be subjected to *snēhavasti* with *ghṛta* made of *Embelia ribes* seeds (*Ci*. 20: 19–23).

For freeing the patient from the grip of *kṛmi* residing in head, he should be treated according to the protocol recommended in the treatment of diseases of head (*Śirōrōga pratiṣēdham*) in *Uttarasthānam*. He should consume food having bitter, pungent tastes and less of oil. He should also consume *yavāgu* prepared with buttermilk and powders of *Embelia ribes*, *Piper longum*, *Piper nigrum*, *Phyla nodiflora* root and *Moringa oleifera* bark. *Souvarcala* salt can be used instead of common salt. Several medicines, medicated food and beverages are recommended in the treatment of *kṛmi* (*Ci*. 20: 24–32).

Vasti and purgation are to be employed more in the case of *purīṣōttha kṛmi*. Nasal instillation, emesis and *śamana* therapy are recommended in treatment of *kaphaja kṛmi*. Treatment of *kuṣṭha* is to be adapted in the case of *raktaja kṛmi*. *Kṛmi* that cause loss of hair are to be treated according to the treatment of *indralupta* (alopecia). Cow milk, buffalo milk, various kinds of meat, clarified butter, jaggery, curd, leafy vegetables and all sour and sweet substances are to be avoided by patients suffering from infestation of *kṛmi* (*Ci*. 20: 33–35).

5.4.27 Vāta Diseases

Patients suffering from *vātavyādhi*, which are essentially *vāta*-dominant and devoid of involvement with other *dōṣas* (*kēvalavāta*), are to be administered at first *snēha* such as *ghṛta*, animal fat, bone marrow and *taila*. Thereafter, *snēha*na should be stopped, and he should be fed with milk, *yūṣa* mixed with *snēha*, meat, pudding, rice cooked with sesame seeds and food that are *bṛṃhaṇa* as well as *vāta* -pacifying. He should also be subjected to *snēhavasti*, using *snēha* mixed with ingredients having sour and salty tastes. On achieving optimum unctuousness, he should be subjected to *abhyaṅga* and then repeatedly to sudation, with various sudation measures. Patients affected with neurological diseases will feel light and flexible by these measures. Numbness, pricking pain, pressing pain, catching pain and edema will vanish after the sudation measures. When properly administered, *snēha* nourish the body and impart stamina. Therefore, *vātavyādhi* patient is to be repeatedly subjected to oleation and sudation. If the condition is still not improved, he should be subjected to purgation with oily purgative medicines. In *vātavyādhi* dependent on stomach, the patient should undergo emetic therapy, followed by consumption of *pēya* according to *pēyādikrama*. *Ṣaḍdharaṇaṃ cūrṇa* or *Vacādi cūrṇa* should be administered in luke warm water. After stimulating *jaṭharāgni*, treatment for essentially *vāta*-dominant disease (*kēvalavāta*) is to be carried out (*Ci*. 21: 1–15).

Enragement of *vāta* on navel area is to be treated by feeding fish cooked with pulp of unripe fruits of *Aegle marmelos*. *Vasti* and pressing with hands are desirable in *vāta* enraged below the navel. Carminative *kṣāra* and *cūrṇa* are to be administered in *vāta* enraged in the alimentary tract (*kōṣṭha*). Consumption of milk boiled with roots of *Pseudarthria viscida* pacifies *vāta* enraged in heart. *Vāta* enraged in head needs to be pacified by *śirōvasti*, *nasya*, *dhūmapāna* and *tarpaṇa* measures in eyes and ears. Sudation, *abhyaṅga*, staying indoors and consumption of pleasant food normalise *vāta* is deranged in skin. *Lēpa* with *śīta vīrya* herbs, purgation and bloodletting are recommended in the case of *vāta* deranged in *rakta*. *Vāta* derangement in bone and bone-marrow are pacified by oleation performed internally and topically. Food that improves vigor and vitality cures *vāta* enraged in *śukra*. Obstruction of *śukra* is cleared by purgation and adoption of *pēyādikrama*, followed by general measures for pacifying *vāta*. Purgation, *nirūhavasti* and *śamana* medicines are recommended in the event of *vāta* deranged in *māmsa* and *mēdas*. Administration of internal medicine to the pregnant mother is advised in the case of emaciation of fetus *in-utero*. Enragement of *vāta* in sinews, joints and blood vessels can be corrected by oleation, thermal cautery and *upanāha*. *Abhyaṅga* using *taila* prepared with black gram and rock salt is ideal in spasm of limbs. *Lēpa* prepared with chimney soot, rock salt and sesame oil is useful in bleeding due to *vāta* enragement. Numbness of organs can be pacified by tight bandaging and *upanāha*. Patient suffering from *apatānaka* (Stokes-Adams syndrome) is to be subjected to oleation and sudation, followed by *nasya* of *tīkṣṇa* medicines. On coming back to sense, he is to be administered with *ghṛta* prepared with herbs of *Vidāryādi gaṇa*, curd, milk and meat soup. Treatment protocol recommended for *ardita* is to be followed for patients suffering from *bāhyāyāma* and *antarayāma*. *Taila* is to be applied on jaw, followed by sudation in the case of *hanusraṃsa* disease (locked jaw). Physician should fix the jaw properly. Physical manipulations are required when mouth remains open or shut continuously. These are to be followed by treatment recommended for *ardita*. *Nasya*, application of *taila* on apex of head and keeping *taila* in eyes and ears are recommended in the treatment of *ardita*. Oleation and purgation with *snēha* are to be carried out in *pakṣāghāta*. *Nasya* and post-prandial administration of *snēha* are desirable in *apabāhuka* (paralysis of arms) (*Ci*. 21: 15–44).

Oleation is not desirable in *ūrustambha* (stiffness of thighs), which is characterized by increased *āma* and *mēdas*. *Śōdhana* in the form of emesis, purgation and medicinal enema is also contra-indicated. Therefore, *kapha* and *mēdas* are to be reduced, without allowing *vāta* to get enraged. This is achieved by administration of various internal medicines such as *kvātha*, *taila* and *ghṛta*. Light exercise is also recommended (*Ci*. 21: 45–83).

5.4.28 Vāta-Induced Blood Disorders

Patient suffering from *vātaśōṇita* (*vāta*-induced blood disorders) is to be made unctuous by internal and topical application of *sneha*. Thereafter, considering *dōṣa*, stamina and without allowing *vāta* to get enraged, he should be subjected to bloodletting several times. If there is need for purgation, the patient should be subjected to oleation and then to purgation, using medicines mixed with *snēha*. *Śamana* medicines are to be applied after elimination of *dōṣa* and in consideration of the *samprāpti*. Aged clarified butter is to be administered to patient suffering from *vātaja vātaśōṇita*. *Ghṛta* and medicated milk can also be administered. *Kvātha* of *Asparagus racemosus* root, *Picrorhiza kurroa* root, *Trichosanthes cucumerina* root, *Triphala* pericarp and *Tinospora cordifolia* stem can be administered for pacifying *pittaja vātaśōṇita*. *Ghṛta* or milk processed with herbs having sweet and bitter tastes can also be administered. *Kvātha* of *Triphala* or *kvātha of Cyperus rotundus* tuber, *Vitis vinifera* fruit and *Berberis aristata* bark can be administered with honey in *kaphaja vātaśōṇita*. Mild *rūkṣaṇa* (the process of inducing dryness) can also be applied, after consumption of appropriate *snēha*, followed by emesis (*Ci*. 22: 1–15).

Topical application of medicines is advised in the case of pain and burning sensation associated with *vātaśōṇita*. They include *lēpa*, *ghṛta* and *taila* such as *Piṇḍa taila*. They are to be used for topical application and *sēka*. Bloodletting is recommended in burning sensation accompanied by redness and pain. *Upanāha* using *vēśavāram* prepared from *ānūpa* and *prasaha* kind of meat cures *stambhanam* (stiffness or loss of movement), pricking pain, contraction, spasm and temporary stiffness of limbs. *Lēpa*, *abhyaṅga*, *sēka* and *avagāha* are recommended in the treatment of *vātaśōṇita* called *uttānaṃ*. *Vātaśōṇita* of *gambhīra* variety is to be pacified by purgation, *nirūhavasti* and *snēhapāna*. *Abhyaṅga* and *sēka* with *ghṛta* are recommended in *kapha*-dominant *vātaśōṇita*. *Lēpa* is advised in pain associated with *vāta-kapha*-dominant *vātaśōṇita*. However, the *lēpa* should be applied slightly warm. Cold *lēpa* are recommended in *pitta-rakta*-dominant *vātaśōṇita* (*Ci*. 22: 21–40).

Snēhana and *bṛṃhaṇa* treatment are contra-indicated in *vāta* enraged due to *kapha* and *mēdas*. Treatment recommended for *āḍhyavāta* (rheumatic palsy on the loins) is to be applied in such cases, followed by treatment recommended for *vātaśōṇita*. In *vātaśōṇita* characterized by enragement of the five subtypes of *vāta*, treatment should be based on their enragement, localization and intensity. *Āma* associated with *vātaśōṇita* is to be eliminated completely with sudation, *laṅghana*, *pācana* measures, coupled with *lēpa* and *sēka* of *rūkṣa* nature. This is to be followed by *vāta*-pacifying therapy. *Śīta* and *uṣṇa* treatments are to be applied alternatively in *vāta* enveloped by *pitta* (*āvaraṇa*). Sudation, *nirūhavasti* with *tīkṣṇa* medicines, emesis, purgation, consumption of *jāṅgala* meat, aged clarified butter and mustard oil are desirable in envelopment of *vāta* by *kapha*. *Pitta* is to be pacified first in envelopment of *vāta* by *kapha* and *pitta*. Treatment protocol recommended for *vātaśōṇita* is advised in *vāta* enveloped by *rakta*. Sudation, *abhyaṅga*, meat soups, milk and *snēha* are desirable in *vāta* enveloped by *māṃsa*. Treatments that pacify *pramēha*, *mēdas* and *vāta* are advised in *āḍhyavāta* or envelopment of *vāta* by *mēdas*. *Mahāsnēha* is to be applied in *vāta* enveloped by *asthi* and *majja*. Treatment advised for *vāta* associated with *śukra* described in chapter on *vātavyādhi cikitsa* needs to be applied in envelopment of *vāta* by *śukra*. Emesis, medicines and food having *pācana and dīpana* properties are recommended in *annāvṛta vāta*. Diuretic medicines, various sudation measures and *uttaravasti* pacify *vāta* enveloped by *mūtra*. Castor oil, *vasti* and *snēha* are advised in *vāta* enveloped by *mala*. Medicines that do not provoke *kapha* and *pitta* and which encourage *vāta* to move in its natural course (*anulōmana*) are to be applied in *vāta* envelopment by all *dhātu*. Unctuous medicines that clear body channels and do not cause exudation of fluids (*abhiṣyanda*) are desirable in all envelopments of *vāta*. Carminative, constipation-inducing medicines that cleanse the urinary bladder and cause *anulōmana* of *vāta* are advised in envelopment of *apāna vāta*. Various kinds of *yāpana vasti*, herbs with sweet taste, *snēhavasti*, *mṛduvirēcana*, *rasāyana* medicines, especially *Cyavanaprāśa*, *Śilājit*, *guggulu* and *Brāhma rasāyana* are invaluable in the treatment of envelopment of *vāta* (*Ci*. 22: 47–74).

Cikitsāsthānam of AH describes *kāyacikitsa* or treatment of diseases usually assigned to the modern category of general medicine. The other seven limbs of *Aṣṭāṅga Ayurveda* such as *śalya*, *śālākya*, *kaumārabhṛtya*, *agadatantra*, *bhūtavidya*, *rasāyana* and *vājīkaraṇa* are described in *Uttarasthānam*. *Uttarasthānam* essentially deals with the diagnosis and treatment of pediatric diseases, mental disorders, epilepsy, diseases of eye, ear, nose, mouth and head; wounds, fractures, anal fistula; venereal diseases, bites of snakes, rats, rabid dogs, insects; rejuvenation therapy and virilification.

5.4.29 Diseases of Infants

Infants are usually affected by vitiated breast milk. The physician is expected to know the intensity of the affliction from the crying of the infant. Infants cry loudly when pain is intense and cry softly when the pain is mild. As communicating with the infant is impossible, the physician should understand more about the ailment by careful observation of the infant's behavior and body movements. Thereafter, the breast-feeding mother should be treated according to her *dōṣa* state. Her vitiated *dōṣa*s should be eliminated using appropriate elimination measures like mild purgation and medicinal enema, followed by administration of pacifying medicines such as *kvātha*, *cūrṇa*, *ghṛta* and so on (*Ut*. 2: 1–27).

Pediatric diseases (*bālāmaya*) that manifest after the appearance of teeth are to be treated in consideration of *dōṣa* and age of the baby. Only mild medicines are to be employed. Infants are not to be troubled by dietary restrictions, as these diseases disappear once the teeth are fully exposed. *Cūrṇa* mixed with honey and clarified butter pacify most of these conditions. *Cūrṇa* recommended in diseases of infants are to be ground to fine paste and applied on the nipples of the breast-feeding mother. After two hours, the dried paste is to be washed away and the infant suckled. Diseases of infants are cured in this way (*Ut*. 2: 28–77).

5.4.30 Spiritual Afflictions of Children

Children are prone to possession by 12 evil spirits (*bālagraha bādha*). These afflictions can be identified by definite signs. Some of them are incurable or difficult to cure. The rest of these possessions can be exorcised with magico-religious practices, bath in medicated water and fumigation with such substances like hides of leopard, tiger, bear and monkey. Fumigating with many strong-smelling herbs like mustard seeds, *Acorus calamus* rhizomes, *Trachyspermum ammi* seeds and neem leaves is said to dispel all afflictions by spirits. Internal administration of several *ghṛta* is also recommended (*Ut*. 3: 1–60).

5.4.31 Spiritual Afflictions of Adults

Adult humans are prone to possession by 18 spirit entities causing insane behavior (*bhūtagraha bādha*) (*Ut.* 4: 1–44). *Graha* (spirit entities) that do not cause violent behavior are to be pacified with magical incantations, sacrificial fire, offerings and kindness. Certain *graha* affect the victim on specific days. Therefore, these propitiatory measures are to be carried out on those days. The possessed person is also to be subjected to *abhyaṅga* with specific *ghṛta* and *taila*. These medicines may be administered orally and as *nasya*. The ingredients of these medicines can be used for fumigating the patient. Many herbs and animal products can be ground to a fine paste, which should be used for *nasya*, topical application and to be suspended in water used for *sēka*. The paste can also be used as collyrium. Sacrificial rites for exorcising spirits of angels (*Dēvagraha*), sages (*Ṛṣigraha, Siddhagraha*) and aged persons (*Vṛddhagraha*) are to be performed in temples. For other spirits, these rites may be carried out at specific sites such as woods, banks of rivers and uninhabited or deserted houses. Offerings for these spirits include various flowers, lemon, rice cooked with jaggery, sweet puddings, honey, parched rice, rice cakes, cooked or uncooked meat and so on. Exorcision of spirits is to be carried out with great caution, as they can cause the death of the healer as well as the patient, when displeased (*Ut.* 5: 1–53).

5.4.32 Insanity

Six types of *unmāda* (insanity) are described in AH. *Vātōnmāda* patient is to be subjected at first to *snēhana* using appropriate *ghṛta* or *taila*. If *vāta* is obstructed in such patients, emesis or purgation needs to be carried out before *snēhapāna*. Patients of *pittaja* and *kaphaja unmāda* are to be administered internally *snēha*, followed by emesis, purgation and nasal instillation. On elimination of *dōṣa*, the patient is to be administered *śamana* medicines, which include various *ghṛta* such as *Kalyāṇaka*, *Mahākalyāṇaka* and *Mahāpaiśācika*. *Vāta-kapha unmāda* patient is to be fumigated with foul-smelling substances such as fish and canine or bovine flesh. *Pittōnmāda* patient is to be administered with *Tiktaka ghṛta* and *ghṛta* prepared with herbs of *Jīvanīya gaṇa*. *Unmāda* patient is to be housed in a clean and quiet place. His relatives should console him with kind words. If *unmāda* patients show signs of spiritual possession, treatment for exorcising those entities also needs to be carried out. He who abstains from consumption of meat and wine, remains patient and maintains purity of mind will never fall a prey to *unmāda* caused by *tridōṣa* and spirit entities (*Ut.* 6: 1–60).

5.4.33 Epilepsy

Patients suffering from *apasmāra* (epilepsy) are to be subjected to elimination measures such as emesis, purgation and nasal instillation using medicines having *tīkṣṇa* quality, so that channels carrying intellect and mind are cleared of obstructions. *Vātāpasmāra* patient is to be subjected to medicinal enema, *pittāpasmāra* patient to purgation and the one suffering from *kaphāpasmāra* to emesis. Patients freed of *dōṣa* are to be administered *śamana* medicines such as *Pañcagavya ghṛta*, *Mahatpañcagavya ghṛta*, *Vacādi ghṛta*, *Jīvanīyataila ghṛta*, *Kṣīrēkṣurasa ghṛta* and *Kūśmāṇḍasvarasa ghṛta*. *Nasya* with *Triphalādi taila*, *Triphalā cūrṇa* and fumigation are also recommended. *Sannipātāpasmāra* is to be cured with *rasāyana* medicines. *Apasmāra* patient must be kept away from fire, water bodies and uneven sites. He should always be consoled with kind words (*Ut.* 7: 15–37).

5.4.34 Diseases of Eyelids

Twenty-four diseases (*vartmarōga*) affect the eyelids. Among them, *arjuna* is removed surgically. *Pakṣmasadanaṃ* is pricked with needles and the blood removed using leeches. This is followed by emesis, consumption of *ghṛta*, application of collyria (*añjana*) and *nasya*. *Lagaṇa, kumbhīkā, bisavartma, utsaṅga, añjana* and *alaji* are to be cut using surgical instruments. After draining the fluid and blood, *lēpa* prepared with herbs is applied. *Kapha*-pacifying emesis, *nasya* and collyria are also recommended. *Kardama vartma, bahaḷa vartma, kukūṇaka vartma, pōthakī, śyāva vartma, pittōtkḷiṣṭa, kaphōtkḷiṣṭa* and *utkḷiṣṭa vartma* are to be subjected to scraping using cuttlefish bone or rough-surfaced leaves. After scraping, the wound is washed with *kvātha* of herbs and *lēpa* applied. Pediatric eye diseases are of *kapha* predominance. Therefore, emesis is the best remedial measure. Following emesis, the alimentary tract is cleansed by purgation, which can be induced by *Brahmī ghṛta* or *varti* (medicated rod or stick). Non-surgical measures are also adopted in the treatment of *vartmarōga* such as *kṛcchrōnmīla*. Patient is administered *ghṛta* prepared from grape juice and aged clarified butter. The *ghṛta* is administered mixed with sugar. *Nasya, dhūmapāna* and *añjana* using medicines mixed with *ghṛta* are also recommended (*Ut*. 9: 1–41).

5.4.35 Diseases of Sclera and Cornea

Twenty-seven diseases affect lachrymal apparatus, sclera, bulbar conjunctiva and cornea (*sandhi sitāsitarōga*). These diseases are treated with surgical and non-surgical measures. For example, diseases like *upanāha, parvaṇī, pūyālasa, kṛmigranthi, arma* and so on are corrected by surgery. Before surgery, the eyelid is subjected to sudation by pressing small balls of cloth dipped in hot water. After that surgery is performed. *Sēka* with *kvātha* of herbs is carried out, and it is followed by the application of *lēpa* and *añjana*. Many formulations such as *Triphalādi maṣi, Jātimukuḷādi varti, Dantavarti, Tamālapatrādi guḷika, Mahānīla guḷika, Mudgāñjana* and *Madhukādyañjana* are recommended to be applied in eyes. Non-surgical measures are adopted in treating diseases affecting the sclera. In such cases, *Triphala* is applied internally and externally, with or without *snēha*, depending on the *dōṣa* states. *Tiktaka ghṛta* is also recommended. Purgation, *sēka* with *kvātha* of herbs and application of *añjana* are also ideal. *Sēka* with *kvātha* made from *Curcuma longa* rhizome, *Glycyrrhiza glabra* root, *Hemidesmus indicus* root and *Symplocos racemosa* bark is advised in the treatment of *śuddhaśukra*. *Sēka* can also be carried out by squeezing and pressing cloth bag filled with powder of *Symplocos racemosa* bark, after dipping it in hot water. *Ghṛta* can be administered in all eye diseases, after carefully reviewing the *dōṣa* states. Such *ghṛta* can also be used for performing *nasya* daily. Eyes strengthened by such measures can never be affected by sharp or pungent substances (*Ut*. 11: 1–58).

5.4.36 Cataract

If left untreated, *timira* (cataract) progresses into *kāca* and then into blindness. Therefore, *timira* is to be treated expeditiously. *Timira* can be cured by internal administration of *Jīvantyādi ghṛta, Drākṣādi ghṛta, Traiphala ghṛta* and *Mahātraiphala ghṛta*. Consumption of *Triphala* powder mixed with honey and clarified butter in unequal proportions cures all diseases including *timira*. Application of collyria such as *Māṃsyādi cūrṇāñjana, Maricādi cūrṇa, Bhāskara cūrṇa* and *Amalāñjana* is also reputed to cure *timira*. Similar effect is attributed to *nasya* with *Jīvantyādi taila* and *Sitairaṇḍādi taila*. *Tarpaṇa* with *Śatāhvādi ghṛta* or an appropriate *ghṛta* should be performed, if all other treatments fail to cure *timira* (*Ut*. 13: 1–62).

Vātaja timira should be treated as detailed in treatment of *vātaja pīnasa*, along with *snēhavasti* and *nirūhavasti*. Patient suffering from *pittaja timira* should be administered *ghṛta* prepared with *Jīvanīya gaṇa* herbs, mixed with *Triphala*, followed by cutting of the vein on the forehead. *Lēpa* and *sēka* should be performed in eyes, face and head, using *śīta vīrya* herbs. Venesection should be performed on patients suffering from *kaphaja timira*, after administering *ghṛta* prepared with *Tinospora cordifolia* stem, *Triphala* pericarp and *Piper longum* fruit. *Raktaja timira* is to be treated according to the protocol advised for *pittaja timira*. *Śīta vīrya* medicines are to be administered internally. *Timira* that has progressed into *kāca* is to be treated according to the protocol for treatment of *sannipātaja timira*, excluding venesection. Patients suffering from *dhūmara*, *amḷavidagdha*, *pittavidagdha* and *uṣṇavidagdha* should be made unctuous by administering aged clarified butter, followed by purgation. Paste of *śīta vīrya* herbs is to be applied all over the body. This is to be followed by application of *añjana*, *nasya* and *tarpaṇa* (*Ut*. 13: 62–100).

5.4.37 Near Blindness

Liṅganāśaṃ (blurred vision, near blindness) is treated surgically. However, surgery is contra-indicated in persons in whom venesection is not permitted, persons suffering from thirst, rhinitis, cough, indigestion, fever, vomiting, headache, earache and pain in eyes. Surgery is to be carried out using a surgical instrument called *śalāka*. After surgery, the physician should console the patient and do *sēka* in the operated eye using breast milk. Thereafter, the *kapha* that has been removed from eye should be expelled by the patient, spitting forcefully. After removal of *śalāka* from eye, the eye should be bandaged with long strip of cloth dipped in clarified butter and the eye covered with bandage. After serving good food and smearing *taila* on head and legs, the patient should be moved to clean room and laid properly on bed, ensuring that the operated eye is not pressed, while lying on bed. For seven days from the day of surgery, the patient should not sneeze, cough, belch, spit, keep face downward, bathe or brush teeth. After surgery, the patient should undergo fasting according to strength of his body. *Sēka* with lukewarm clarified butter is advised in case of pain in eyes. He should consume *vāṭya* (gruel made of roasted and powdered barley), spiced with *Trikaṭu* powder. He can also consume gruel. Bandage over eyes can be removed on third day. *Sēka* with *vāta*-pacifying *kvātha* is recommended, and eyes can be covered again. However, bandage should not remain over eyes beyond seven days. For a few days, the patient should not look at very minute and bright objects. Treatment for *śōpha* should be initiated in case redness, edema and pain are experienced in the operated eye. *Sarṣapādi lēpa*, *Payasyādi lēpa*, *Āḍhakīmūlādyañjana* and *Piṇḍāñjana* can be applied in eyes (*Ut*. 14: 1–32).

5.4.38 Global Diseases of the Eye

Upavāsa, *gaṇḍūṣa* and *nasya* of *tīkṣṇa* nature are advised in *abhiṣyanda* (conjunctivitis) caused by *pitta-kapha* and *rakta*. These measures are contra-indicated in *vātaja abhiṣyanda*. *Lepa* is to be applied on the eyelids for pacifying heat, redness, lacrimation and edema in all *abhiṣyanda*. *Lēpa* mixed with clarified butter is to be applied over eyelids in *vātaja abhiṣyanda*. Paste of specific herbs ground along with water is ideal in *pttaja* and *raktaja abhiṣyanda*. *Lēpa* mixed with honey is to be applied in *kaphaja abhiṣyanda*. In *sannipātaja abhiṣyanda*, *lēpa* of ingredients recommended for *vāta-*, *pitta-*, *kapha-* and *raktaja abhiṣyanda* is to be applied. Many *cūrṇa* are recommended for topical application. Powders of medicinal herbs packed in a small cloth bag is soaked in *Dhānyāmḷa* (rice gruel fermented with grains and herbs). Thereafter, the bag is squeezed and the dripping extract poured into eyes. This measure is recommended in all *abhiṣyanda*. *Lēpa*, *sēka* and *cūrṇa* recommended in the treatment of *abhiṣyanda* can be extended to conditions like *mantha*, *atimantha* and *anyatōvāta*. Many *varti* are also recommended for topical application in *adhimantha*. Patients suffering from *śuṣkākṣipāka* are to be administered clarified butter. *Ghṛta* prepared with *Jīvanīya gaṇa* herbs should be poured in a thin stream into the eyes, and *nasya* is to be carried out using *Aṇutaila*. *Sēka* with medicated milk and *añjana* are also recommended. Those suffering from *akṣipāka*, with or without edema, should at first be subjected to venesection, after making them unctuous, using appropriate *snēha*. Thereafter, they should be subjected to purgation using *kvātha* or *ghṛta*. Application of *añjana*, *guḷika* and *sēka* with *kvātha* or juice of herbs is also advised. Treatment recommended for *pittābhiṣyanda* is to be extended to *amḷōṣita*. Eighteen eye diseases such as *kaphōtkḷiṣṭa*, *pittōtkḷiṣṭa* and so on are known as *pilla*, as they can continue to trouble the patient for a long time. Patient suffering from *pillarōga* is to be made unctuous applying *snēhapāna* and then subjected to emesis, followed by venesection. Thereafter, he should undergo purgation and then be subjected to scraping. Various *añjana* can be applied thereafter (*Ut*. 16: 1–67).

5.4.39 Diseases of Ear

Consumption of *ghṛta* is advised at night in diseases that manifest above shoulder. Therefore, patients suffering from *vātaja karṇaśūla* should have dinner with meat soup, followed by consumption of *ghṛta* prepared with *vāta*-pacifying herbs. After performing sudation of the ear according to *nāḷīsvēda* procedure, the juice of specific herbs processed according to *puṭapāka* procedure should be poured into the ear and retained for some time. This measure is called *karṇapūraṇam*. *Karṇapūraṇam* performed using *ghṛta* prepared with *vāta*-pacifying herbs and substances having sour taste pacifies quickly even very excruciating *karṇaśūla*. *Karṇapūraṇam* can also be carried out with *Mahatpañcamūla taila* and *Bhadrakāṣṭhādi taila*. Drinking of cold water, sleeping during day time and washing of head are prohibited to *vātaja karṇaśūla* patients (*Ut*. 18: 1–6).

Pittaja karṇaśūla patient should be subjected to purgation after consuming clarified butter with sugar. Thereafter, *karṇapūraṇam* should be performed using milk processed with grapes and *Glycyrrhiza glabra* root. Those suffering from *kaphaja karṇaśūla* should be subjected to emesis, after administering *ghṛta* prepared with *Piper longum* fruits. *Dhūmapāna*, *nasya*, *gaṇḍūṣa* and sudation can also be applied. *Karṇapūraṇam* with juices of several herbs and medicated mustard oil is also recommended. *Raktaja karṇaśūla* is to be treated just as *pittaja karṇaśūla*. Venesection should also be performed in the ear. Suppuration in ear can be pacified by *dhūmapāna*, *gaṇḍūṣa*, *nasya*, *nāḷīsvēda*, *piṇḍasvēda* and treatment recommended for *duṣṭavraṇa*. The treatment recommended for *vātaja karṇaśūla* should be extended to *karṇanāda* (tinnitus) and *bādhirya* (deafness). *Nasya*, *abhyaṅga* and *karṇapūraṇam* with *taila* also pacify these conditions. *Pratināha* can be pacified by cleaning of ear, followed by *karṇapūraṇam*. *Kapha*-pacifying *nasya* is advised in *karṇakaṇḍu* and *karṇaśōpha*. Treatment recommended for *karṇasrāva* should be extended to *pūtikarṇam* and *kṛmikarṇam*. *Karṇavidradhi* patient should be treated according to the protocol for treatment of *vidradhi*, after subjecting him to emesis. Treatment recommended for *pittaja karṇavidradhi* is to be extended to *kṣataja karṇavidradhi*. *Karṇārśas* and *karṇārbuda* should be treated according to the protocol advised for *nāsārśas* and *nāsārbuda* (*Ut*. 18: 7–65).

5.4.40 Diseases of Nose

In the early stages of all nasal diseases (*nāsārōga*), the patient should be housed in a clean and wind-free room. Thereafter, he should be subjected to oleation, sudation, emesis, *dhūmapāna* and *gaṇḍūṣa*. He should wear warm and thick clothes. Head should be covered with thick headgear. He should consume food made of barley and wheat, which is light, sour, unctuous, hot and salty. The food should include *jāṅgala* meat, jaggery, milk, ginger, black pepper and fruits of *Piper longum*. Curd, juice of pomegranate fruit and soup of *Dolichos biflorus* seeds are ideal in nasal diseases. Patient suffering from *vātaja pīnasa* should consume *ghṛta* prepared with herbs of *Vidāryādi gaṇa* or five kinds of salts. Sudation and *nasya* recommended in treatment of *ardita* can also be applied. Patients of *pittaja* and *raktaja pīnasa* should consume *ghṛta* prepared with sweet and *śīta vīrya* herbs. *Lēpa* and *sēka* of *śīta vīrya* nature can also be employed. *Kaphaja pīnasa* patient should undergo *upavāsa* and apply paste of mustard on head. After consuming *ghṛta* mixed with *yavakṣāra*, he should undergo emesis. *Nasya* can also be applied. *Sannipātaja pīnasa* can be cured by *nasya* and *kabaḷam*, using *ghṛta* prepared with herbs that are pungent and *tīkṣṇa*. *Duṣṭapīnasa* should be pacified by applying measures like *abhyaṅga*, *snēhapāna* and *uṣṇasvēda* recommended in treatment of *rājayakṣma*. Treatment prescribed for *kṛmi hṛdrōga* and *śiraḥstāpa* caused by *kṛmi* can also be made use of. *Dhūmapāna* with *Vyōṣādi dhūmavarti* is also effective. *Nāsāpāka* and *dīpti* can be cured by *pitta*-pacifying treatment. *Nasya* and *dhūmapāna* with *tīkṣṇa* herbs are also advised in *nāsāsrāva*. Treatment recommended for *kaphaja pīnasa* should be applied in *pūtināsa* and *apīnasa*. In its early stage, *pūyarakta* can be cured by treatment recommended for *raktaja pīnasa*. Chronic *pūyarakta* needs to be treated like *nāḷīvraṇa*. *Nāsārśas* and *nāsārbuda* are to be treated surgically (*Ut*. 20: 1–25).

5.4.41 Diseases of Mouth

Khaṇḍōṣṭham (cleft lip) is to be treated surgically, followed by application of *taila* like *Yaṣṭhyādi taila* and *nasya* with *vāta*-pacifying *taila*. *Vātōṣṭhakōpam* is to be treated by topical application of *taila*. Leeching is advised in *pittaja ōṣṭhakōpam* and *abhighātaja ōṣṭhakōpam*, followed by topical application of *lēpa* and *Guḍūcyādi ghṛta*. Treatment advised for *pittaja vidradhi* can also be applied. Treatment advised for *pittaja vidradhi* is suitable for acute *raktaja ōṣṭhakōpam* as well. Vitiated blood is to be drained in *kaphōṣṭhakōpam*, followed by application of *lēpa*, *dhūmapāna*, *nasya* and *gaṇḍūṣa*. *Jalārbudam* should be incised and the fluid drained. After that, powder of *tīkṣṇa* herbs is to be applied, mixed with honey and the wound subjected to *gharṣanaṃ*. *Kṣāra* and *agnikarma* may be applied if the wound has uneven surface. Treatment recommended for *śōpha* is to be applied in *gaṇḍālaji*. *Śītadantam* is to be treated surgically, followed by *kabaḷam*. *Gaṇḍūṣa* using milk boiled with sesame seeds and *Glycyrrhia glabra* root is ideal in *dantaharṣam* and *dantabhēdam*. *Gaṇḍūṣa* with *taila* is advised in *dantacāla*. Teeth are to be extracted and treatment for *kṛmidantam* applied in the case of *adhidantakam*. *Dantaśarkara* is to be removed using sharp instruments and *yavakṣāra* applied. *Kapālika* is to be treated like *dantaharṣam*. *Kṛmidantam* is to be pacified by sudation, application of *lēpa*, *gaṇḍūṣa*, *nasya* and appropriate food. Holes caused by *kṛmidantam* are to be filled with latex of *Alstonia scholaris* and *Calotropis procera*. This measure pacifies pricking pain (*kṛmiśūla*) associated with *kṛmidantam*. *Śītāda* is to be treated with removal of vitiated blood, followed by *lēpa*, *gaṇḍūṣa* and *nasyaṃ*. *Upakuśa* is to be treated with sudation using warm water, scraping, application of *lēpa* and *gaṇḍūṣa*. *Gaṇḍūṣa* with herbs having pungent, *tīkṣṇa*, *uṣṇa* qualities and application of *lēpa* are advised in *dantavidradhi*. Treatment for *vātōṣṭhakōpam* is to be applied to *vātajihvākaṇṭakam*. *Kaphajihvākaṇṭakam* is to be treated with *gaṇḍūṣa*, using herbs having *tīkṣṇa* and *uṣṇa* qualities. *Gaḷaśuṇṭhika* is to be pacified by *kapha*-pacifying *gaṇḍūṣa* and *nasyaṃ*. *Tālupākam* in early stage can be cured by scraping and *gaṇḍūṣa*. *Lēpa* is to be applied in *kapharōhiṇi* and the fluid drained. *Nasya* and *gaṇḍūṣa* with *taila* are also recommended. *Vṛnda*, *śālūka*, *tuṇḍikēri* and *gilāyu* are to be treated just like *kapharōhiṇi*. Five types of *mukhapākam* (inflammation of mouth) are to be treated by *gaṇḍūṣa*, using *kvātha* mixed with honey. Patient suffering from *pūyāsyata* is to be subjected to emesis, *nasya*, *dhūmapāna* and *gaṇḍūṣa*. *Dantapuppuṭa*, *suṣiraṃ*, *adhimāṃsam*, *vidarbha*, *dantanāḷi*, *pittajihvākaṇṭakam*, *jihvālasa*, *adhijihva*, *upajihva*, chronic *tālupākam*, *vātarōhiṇi*, *pittarōhiṇi*, *raktaja rōhiṇi*, *gaḷavidradhi*, *vātaja gaḷagaṇḍam*, *vāta-pitta-kaphaja gaḷagaṇḍam*, *mēdōbhava gaḷagaṇḍam* and *nāsārbudam* are to be treated surgically (*Ut*. 22: 1–111).

5.4.42 Diseases of Head

Snēhapāna measure described in treatment of *vātavyādhi* is to be adopted in *vātaja śiraḥstāpa*. The patient should apply *vāta*-pacifying *taila* on head and consume *ghṛta* at night, followed by drinking of warm milk. He should also consume at night food prepared with black gram, horse gram or green gram. *Piṇḍasvēda* and *upanāha* are also recommended. *Sēka* of head with medicated milk, *bṛṃhaṇa nasya*, *śiraḥtarpaṇam* and *śravaṇatarpaṇam* also pacify *vātaja śiraḥstāpa*. These treatments are also recommended in *ardhāvabhēdaka*, after reviewing the *dōṣa* state. *Sūryāvarta* also needs to be treated similarly. However, venesection is to be performed. Patient suffering from *pittaja śiraḥstāpa* is to be made unctuous using *snēha* and then venesection should be performed. After that paste of *śīta vīrya*, herbs should be applied on head and face. *Sēka* and *śōdhana vasti* can also be applied. *Ghṛta* prepared with herbs of *Jīvantyādi gaṇa* is to be used for internal administration and *nasya*. *Raktaja śiraḥstāpa* should be treated just like *pittaja śiraḥstāpa*. *Kaphaja śiraḥstāpa* patient should be administered aged clarified butter and then subjected to emesis. *Rūkṣa* and *tīkṣṇa* herbs are to be applied as sudation, *lēpa* and *nasya*. *Upanāha* is also beneficial. Treatments advised for *vātaja*-, *pittaja*- and *kaphaja śiraḥstāpa* are to be adapted judiciously in the case of *sannipātaja śiraḥstāpa*. *Nasya* with blood is recommended in *kṛmija śiraḥstāpa*. When *kṛmi* are expelled by this measure, *nasya* and *dhūmapāna* with *tīkṣṇa* medicines are to be performed. *Śiraḥkampa* is to be treated just like *vātaja śiraḥstāpa*. Newly formed and non-suppurating *upaśīrṣaka* are to be treated just like a *vātavyādhi*. Suppurating *upaśīrṣaka* needs to be treated as *vidradhi*. Suppurating and non-suppurating *śirōvidradhi*, *śirōpiṭaka* and *śirārbudam* are to be treated in consideration of the *āma* and *pakva* states of the disease condition. Vitiated blood in *arūṃṣika* is to be drained by leeching, washed with *kvātha* of neem bark and *lēpa* applied. *Indralupta* is to be treated with venesection, followed by pricking of the area and applying *lēpa*. Several *nasya* are also recommended. *Palitaṃ* (greying) can be cured by *nasya*, *lēpa* and application of *taila* (*Ut*. 24: 1–59).

5.4.43 Wounds and Ulcers

Elimination of *dōṣa* is to be carried out in early stage of *vraṇa* (wounds, ulcers), characterized by edema. Emesis is to be performed if *vraṇa* is in the upper part of body, and purgation is appropriate if *vraṇa* presents in the lower part of body. *Sēka* with *śīta vīrya* herbs is required when the *vraṇa* is associated with unsuppurating edema. Removal of vitiated blood using leeching is advised in hard, discolored and painful edema and *vraṇa*. After removal of vitiated blood, paste of *śīta vīrya* herbs, mixed with milk, sugarcane juice and *Śatadhouta ghṛta* should be applied frequently on the *vraṇa*. *Sēka* and *abhyaṅga* with these medicines are also beneficial. *Vraṇa* which are hard, painful and previously subjected to bloodletting are qualified for sudation with *Utkārika* and *Vēśavāra* prepared from *ānūpa* meat. Hard and less painful *vraṇa* should be subjected to sudation using *snēha* having *vāta-kapha*-pacifying properties. Following this, the *vraṇa* should be pressed softly with fingers to loosen the edema. If this measure does not loosen the edema, the *vraṇa* should be subjected to *upanāha*. Suppurating *vraṇa* should be smeared with *lēpa* prepared with guggul gum, *Linum usitatissimum* seeds, *Garcinia morella* latex, teeth of cow and droppings of pigeon. *Vraṇa* filled with pus are to be smeared with *lēpa* made of barley, wheat, green gram, black gram and pressed frequently. Obstinate ulcers (*duṣṭavraṇa*), diabetic carbuncles and wounds on lepers are to be smeared with *lēpa* prepared from herbs of *Surasādi gaṇa* and *Āragvadhādi gaṇa*. Obstinate ulcers which exude pus are to be fumigated with mixture of barley, clarified butter, *Betula edulis* leaf, beeswax, *Pinus roxburghii* gum resin and *Cedrus deodara* heartwood. *Vraṇa* with high content of *pitta*, *rakta* and toxins are to be healed by application of *lēpa* and *sēka* with *śīta vīrya* herbs. *Vraṇa* which have become sinus are to be smeared with *lēpa* made of *Ficus benghalensis* bark, *Prunus cerasoides* root, *Sida rhombifolia* root, *Withania somnifera* root and sesame seeds. Sores which are hard, chronic, accompanied by itching and difficult to clean are to be treated with *kṣāra*, followed by *lēpa* of wound healing medicines. Thermal cautery is recommended in the treatment of *vraṇa* which are unresponsive to treatment. *Lēpa* are to be applied to impart normal color to skin of wounds that have healed. *Vātaja*-, *pittaja*-, *kaphaja*- and *sannipātaja vraṇa* are to be treated with specific herbs. These herbs can be used for washing the *vraṇa* and for preparing *lēpa*, *ghṛta*, *taila*, *cūrṇa* and *varti* for applying on the *vraṇa* (*Ut*. 25: 1–67).

5.4.44 Traumatic Wounds

Wounds that appear due to exogenous causes like injuries from weapons, thorns and accidents are called *sadyōvraṇa* (traumatic wounds). Freshly formed wound should be subjected to *sēka* using *ghṛta* prepared with *Glycyrrhiza glabra* root. *Sēka* with the same *ghṛta* warmed or *Balā taila* is to be performed in wounds with intense pain. *Lēpa* prepared with herbs having *śīta* and *snigdha* qualities should be applied on the wound to reduce the heat of the wound. Clarified butter mixed with honey should be applied on large wounds to hasten their healing. *Pitta*-pacifying measures are also ideal. If the wound is bleeding profusely, the patient should be subjected to *snēhapāna*, *sēka*, *upanāha*, sudation and medicinal enema. Other *vāta*-pacifying treatments are also to be initiated. These treatments are to be continued for seven days. *Kalka*, *kvātha*, *ghṛta* and *taila* are to be administered and wound dressed with *lēpa* and *snēhavarti* in all types of *sadyōvraṇa*. Pressing and bandaging are also to be done on all of them. Surgical measures are advised in large wounds, dislocated limbs and for removing sharp objects like arrows. Special treatments are to be applied on injuries to vital points in body.

Patients suffering from any kind of wound should consume oil-free food. They should eat rice mixed with soup of barley, *Ziziphus mauritiana* seed kernel, and horse gram. Consumption of *yavāgu* with a pinch of rock-salt is also recommended to such patients (*Ut.* 26: 6–57).

5.4.45 Fractures

As first step in setting broken bones (*bhaṅga*), the bone that has gone down should be raised and the bone that has gone up should be brought down by soft massage. Following this, the broken bone should be bandaged with long strip of cloth soaked in *ghṛta*. After tying the bandage firmly, split pieces of bamboo stem or barks of large trees should be placed around, so as to offer protection. The bandage is to be removed and the fracture tied firmly again with fresh cloth at regular intervals. *Sēka* with *kvātha* of herbs of *Nyagrōdhādi gaṇa* should be performed to relieve the pain and edema of the fractured part of body. *Lēpa* and *sēka* should always be carried out with herbs having *śīta vīrya*. The patient should drink in the morning milk boiled with *śīta vīrya* herbs and mixed with powder of lac. Powder of wound healing herbs mixed with honey and clarified butter should be applied on wounds caused by the fracture injury. The edema can be reduced by this measure. Powder of herbs should be sprinkled on the wounds when they are healed and showing neither depressions nor elevations. Patients suffering from fracture injury should lie on flat and firm wooden platform, with provision to prevent movement of body. Treatments described for *vātavyādhi* can be applied to fracture also. The patient should consume food and beverages that improve stamina. *Śāli* rice, clarified butter, meat soups and milk should be consumed in limited measures. *Gandha taila* can be used for internal administration, *abhyaṅga*, *nasya* and *snēhavasti*. It imparts strength to bones and cures all diseases (*Ut.* 27: 11–41).

5.4.46 Anal Fistula

Non-suppurating *bhagandara* (anal fistula) is to be subjected to emesis, purgation, bloodletting, *sēka* and *lēpa*. Following this, pacifying treatment is to be initiated for the healing of the fistula. Suppurating *bhagandara* is to be subjected to *avagāha svēda*. Thereafter, it is to be treated surgically. *Piṇḍasvēda* and *nāḷīsvēda* are effective in pacifying pain associated with *bhagandara*. Several *lēpa*, *taila* and *cūrṇa* are recommended for healing the operated *bhagandara*. Patient who has recovered from surgery should avoid for one year or more riding on horse, camel, mule, elephant and bull; consumption of wine, sexual intercourse, unfamiliar food, food that causes indigestion and various adventurous acts (*Ut.* 28: 22–44).

5.4.47 Glandular Swelling, Tumor

Non-suppurating *granthi* (glandular swelling) is to be treated like *śōpha*. Patient is to be made unctuous by internal administration of *ghṛta*. Thereafter, his *dōṣa*s should be eliminated through emesis, purgation and medicinal enema. Following these measures, *lēpa* of *tīkṣṇa* herbs is to be applied on the *granthi*. After subjecting to sudation many times, the *granthi* is to be pressed repeatedly. *Vātaja granthi* is treated in this way. Bloodletting is to be carried out in *pittaja granthi* followed by application of *lēpa* of *śīta vīrya* herbs. Treatment advised for *vātaja granthi* is to be applied to *kaphaja granthi*. Surgery and thermal cautery are also advised. *Māṃsagranthi* and *vraṇagranthi* are also treated in this way. *Mēdōjagranthi* is treated like *kaphaja granthi*. Patient having *sirāgranthi* (varicose vein) which is not one-year-old should be administered *Sahacarādi taila* advised in the treatment of *vātavyyādhi*. After applying *upanāha* and various kinds of *vasti*, he should be subjected to venesection. Treatment protocols advised for *granthi* are to be adapted in the treatment of *arbuda* (tumor) (*Ut.* 30: 1–8).

Patient suffering from *vātaja śḷīpada* should be made *snigdha*, followed by *upanāha* and sudation. Thereafter, the *śḷīpada* is to be subjected to surgery. *Pittaja*- and *kaphaja śḷīpada* are also treated surgically. He who suffers from *apaci* should consume specific *ghṛta* for performing emesis and purgation. Appropriate *dhūmapāna*, *gaṇḍūṣa* and *nasya* should be applied everyday. Venesection is to be conducted after these measures. Surgery is advised in the case of *kaṭhināpaci*, non-suppurating *gaṇḍa*, *amarmaja granthi*, *vātaja nāḷīvraṇa* and *śalyaja nāḷīvraṇa*. *Taila*, *lēpa* and *varti* are to be applied in all *nāḷīvraṇa* (*Ut*. 30: 8–40).

5.4.48 Minor Diseases

This is a group of 36 diseases (*kṣudrarōga*), which are difficult to cure and causing much misery to the patients. Many of them are treated surgically and with thermal or chemical cautery. Emesis, purgation and venesection are adopted in the treatment of some *kṣudrarōga*. Suppurating *ajagallika* is treated according to the protocol advised for treatment of *vraṇa*. *Irigallika*, *vivṛta* and *jālakagardabha* are treated as *pittaja visarpa*. *Prasupti*, *kōṭha* and *utkōṭha* share some similarities with skin diseases, and therefore, they are treated according to the protocol for *kuṣṭha*. Various *lēpa*, *taila* and *ghṛta* are also used in the treatment of *kṣudrarōga* (*Ut*. 32: 1–33).

5.4.49 Diseases of the Genitals

Diseases that affect the genitals of men and women are detailed in *Guhyarōgapratiṣēdhādhyāyaḥ* of *Uttarasthānam*. Such diseases of men are treated by several means like surgery, scraping, chemical and thermal cautery, physical manipulations and sudation. However, some of them are treated with emesis, purgation, *lēpa*, washing with *kvātha*, *sēka* with *taila*, sprinkling with *cūrṇa*, *upanāha* and *upanāha svēda* (*Ut*. 34: 1–21).

Vāta-pacifying measures like oleation, sudation and medicinal enema are of importance in treatment of gynecological disorders. A physician should first of all pacify *vāta*, followed by the others. *Balā taila*, *Miśraka snēha* and *Sukumāra ghṛta* can be administered internally. Patients of gynecological diseases are to be subjected to emesis using medicines which are not very *tīkṣṇa* in nature. After purifying the body, the patient should be subjected to medicinal enema, followed by *abhyaṅga*, *sēka* and *picu*. *Kvātha* and *ghṛta* can be used as pacifying medicines. *Vāta*-dominant gynecological diseases are treated in this way (*Ut*. 34: 22–34).

Patients of *pittaja yōnirōga* are to be subjected to *sēka*, *abhyaṅga* and *picu* using medicines of *śīta vīrya* nature. The disease known as *raktayōni* is to be treated with medicines that arrest bleeding. *Puṣyānuga cūrṇa* mixed with honey can be administered internally. All medicines having *rūkṣa* and *uṣṇa* qualities can be administered to pacify *kaphaja yōnirōga*. *Abhyaṅga*, *picu* and *vasti* with *Dhātakyādi taila* cure all gynecological diseases. Women suffering from *yōnirōga* should consume cooked barley, *Abhayāriṣṭa* and *Sīdhu* wine. *Sannipātaja yōnirōga* is to be treated with medicines and measures that pacify *tridōṣa*. Powder of *Terminalia chebula* pericarp and *Piper longum* fruits mixed with honey wards off all gynecological diseases (*Ut*. 34: 35–67).

5.4.50 Poisoning

Mineral and vegetable poisons are collectively known as *sthāvara viṣa* or immovable poisons. Poisons of animals such as snakes, insects and rabid dogs are known as *jaṅgama viṣa* (movable poisons). *Sthāvara viṣa* acts in seven stages (*vēga*). Patient in the first stage of poisoning should be subjected to emesis and *sēka* with cold water. Medicine that nullifies the effects of poisons should be administered, mixed with honey and clarified butter. Patient in the second stage should be subjected to emesis and purgation followed by the treatment advised for stage one poisoning. Internal administration of medicine, *nasya* and *añjana* are recommended in the third stage of poisoning. Internal medicine mixed with *ghṛta* or *taila* is to be administered in fourth stage. Medicines that counter the effects of poisons are to be administered with honey and *kvātha* of *Glycyrrhiza glabra* root to patient in fifth stage of poisoning. Treatment for *atisāra* is essential in sixth stage of poisoning. In the seventh and final stage, *avapīḍaka ghṛta* should be administered followed by *avapīḍaka nasya*.

A wound resembling the digit of crow should be inflicted on the apex of head and flesh of an animal or bird kept over that wound. Poison in the patient's body will slowly get transferred to that piece of flesh, thereby saving his life. Two medicated porridges and a pill named *Candrōdaya guḷika* can be used in all stages of poisoning (*Ut*. 35: 1–32).

5.4.51 Latent Poisons

Ayurveda describes the remedies to counteract the effects of poisons formed naturally due to transformation of metals, certain grains and herbs. These poisons are collectively called *dūṣīviṣa*, which can remain dormant for several years. *Dūṣīviṣāri guḷika* is generally used in the treatment of *dūṣīviṣa*. Suppurating wounds caused by poisoned weapons are to be treated according to the protocol for treatment of *pittaja visarpa* (*Ut*. 35: 33–48).

5.4.52 Formulated Poisons

Formulated poisons are sometimes administered surreptitiously through food and beverages for influencing individuals subliminally or for homicide. They produce distinct symptoms and are known as *garaviṣa*. If left untreated, *garaviṣa* will result in the death of the victim. *Tāpyādi lēhya*, *Mūrvādi cūrṇa* and *Pāravatādi kvātha* are to be administered to *garaviṣa* patients. *Garaviṣa* with predominance of *kapha* is to be subjected to emesis, followed by administration of internal medicines and *lēpa* having *uṣṇa*, *rūkṣa* and *tīkṣṇa* qualities. He should consume food having astringent, pungent and bitter tastes. Patient suffering from *pitta*-dominant *garaviṣa* is to be treated with medicines, *sēka* and *lēpa* having *śīta vīrya*. He should consume food having astringent, bitter and sweet tastes. *Garaviṣa* with dominance of *vāta* should be treated with medicines and food having sweet, sour and salty tastes, mixed with clarified butter. Medicines and food used in treatment of *viṣa* should be mixed with clarified butter (*ghṛta*), which is of utmost importance in the treatment of all kinds of *viṣa* (*Ut*. 35: 48–70).

5.4.53 Snake Bite

As soon as the snake bitten person is brought to the physician, he should administer *śīta vīrya* medicine and apply *śīta vīrya lēpa* on the bitten part. Cold air should be blown on his body using a small fan till he experiences horripilation. After that medicines that neutralize toxins should be administered, mixed with honey and clarified butter. As a result of consuming clarified butter, the patient may feel discomfort in chest. To overcome this, appropriate medicines may be administered, mixed with fermented rice and water mixture, sesame oil, soup of horse gram and wine. Emesis will follow, and thereafter, venom will not spread in body. Appropriate treatment may be initiated thereafter, considering *dōṣa* and *prakṛti*. Treatment varies with the type of snake that bit the person. The type of snake can be ascertained from the characteristics of the wound and symptoms manifested in the patient. *Candanādi lēpa*, *Himavān guḷika*, *Kāśmaryādi lēpa* and *Vilvādi guḷika* are commonly used in the treatment of snake bite. *Vāta*, *pitta* and *kapha* that remain enraged even after neutralizing the venom are to be normalized by internal medicines and measures (*Ut*. 36: 1–93).

5.4.54 Stings of Poisonous Pests

Stings of scorpions, centipedes, spiders and poisonous insects (*kīṭalūtādi viṣa*) produce symptoms related to the *tridōṣa*. Therefore, medicines having qualities opposite to those of the *dōṣa* (s) are to be used. In *vāta*-dominant bite, aqueous paste of oil cake is to be applied on the wound caused by the sting. This is to be followed by *abhyaṅga*, *nāḍīsvēda* and consumption of *bṛṃhaṇa* medicines. *Pitta*-dominant sting is to be pacified by *sēka* and *lēpa* of *śīta vīrya* nature. *Kapha*-dominant stings can be cured by emesis, scraping and sudation. *Tridōṣa*-pacifying treatment is to be initiated in *sannipātaja* stings. *Taṇḍulīyakādi cūrṇa, Kṣīrīvṛkṣatvaga lēpa, Daśāṅga guḷika, Śaivalādi lēpa* and *Hiṅgvādi guḷika* pacify problems arising from poisonous stings (*Ut.* 37: 16–86).

5.4.55 Bites of Rats and Rabid Dogs

A *lēpa* is to be applied on the wound caused by the bite of rats (*mūṣikaviṣa*), followed by washing with sour liquids and venesection. Emesis, administration of *ghṛta, cūrṇa* and application of *añjana* are also effective in pacifying the effects of rat bite. Rat bite, if left untreated, will cause the toxins to remain dormant and create more problems later. Treatment of *dūṣiviṣa* can be carried out at that time (*Ut.* 38: 16–34).

The wound caused by the bite of rabid dog (*alarkaviṣa*) is to be subjected to thermal cautery, using hot clarified butter. Following that, a *lēpa* should be applied on the wound and patient should consume aged clarified butter. The patient should also be subjected to purgation using medicines prepared with the latex of *Calotropis procera*. Consumption of *kvātha* of *Alangium salvifolium* root mixed with clarified butter neutralizes the poison of rabid dogs (*Ut.* 38: 35–40).

5.4.56 Rejuvenation Therapy

The term *rasāyana* means "movement of *rasa*". *Rasa* can be the taste of any form of matter, deciding its biological activity or the first tissue element *rasa dhātu*, from which the other six *dhātus* and finally *ōjas* evolve. The seven *dhātus* should transform sequentially with an optimum velocity for maintaining the steady state of *tridōṣa*. *Rasāyana vidhi* is a guideline for regulating the transformation of matter in a human body. A substance that facilitates the movement of *dhātu* cycle in the desired way is called *rasāyana*. Consumption of *rasāyana* improves memory, comprehension, overall health, youthfulness, voice quality, strength of body and sense organs, conversational skills, virility and longevity. *Rasāyana* therapy is to be applied on a young person, having control over sense organs and whose body is cleansed using various elimination measures. *Rasāyana* therapy is of two types: *kuṭīpravēśika* and *vātātapika*. In *kuṭīpravēśika* method, the subject is lodged in a house that is impervious to wind and sunlight. These restrictions are not observed in *vātātapika* method. *Kuṭīpravēśika* is the ideal system for conducting *rasāyana* therapy. After performing purificatory measures like emesis, purgation, medicinal enema, nasal instillation and inhalation of medicinal smoke, the subject should begin therapy on an auspicious day. The *rasāyana* is selected on the basis of the disease and various characteristics of the subject, who will remain inside the *kuṭi* (hut) during the entire period of treatment. Food, beverages and behavior of the patient are to be controlled strictly during the course of the therapy. *Śilājit* is one of the celebrated *rasāyanas*. Its consumption according to *rasāyana vidhi* banishes all curable diseases (*Ut.* 39: 1–181). The *rasāyana* medicines described in AH are listed in Table 5.4.

5.4.57 Virilification

Vāgbhaṭa remarks that a childless person is like a solitary tree that casts no shade, has only one branch, bears foul-smelling flowers and produces no fruits (*Ut.* 40: 9). *Vājīkaraṇa* (virilification) is, therefore, prescribed to childless persons to improve their fertility (*Su.* 1: 48). *Vājīkaraṇa* makes a man strong and virile like a horse (*vāji*=horse). *Vājīkaraṇa* medicines stimulate the *dhātu* cycle, resulting in the circulation of more *ōjas* in body. The person who is to be subjected to *vājīkaraṇa* should consume *ghṛta* in the stipulated manner. Thereafter, his alimentary canal should be cleansed by applying emesis and purgation. After these measures, he should consume milk, meat and meat soups. Following this, he should be subjected to *nirūhavasti* and *snēhavasti* using *kvātha, ghṛta, taila,* meat soups, milk, sugar and honey. Thereafter, the physician should administer appropriate *vājīkaraṇa* medicine, depending on the digestive efficiency of the subject (*Ut.* 40: 10–89). The *vājīkaraṇa* medicines described in AH are presented in Table 5.5.

TABLE 5.4
***Rasāyana* Medicines Described in AH**

Sl. No.	Name of *rasāyana*	Mode of Preparation and Administration	Reference
1	Abhayāmalakādi rasāyana	500 fruits of *Terminalia chebula*, 500 fruits of *Emblica officinalis* and 1,000 fruits of *Piper longum* to be immersed in alkali prepared from bark of a young *Butea monosperma* tree. The fruits are dried when all the liquid is absorbed, powdered, mixed with sugar, honey, clarified butter and stored in a pot that is buried in sand. The *rasāyana* can be consumed after six months.	*Ut*. 39: 24–27, 60, 146–147
2	Agastya rasāyana	Attributed to sage Agastya, this *rasāyana* cures many diseases, enhances complexion and increases longevity.	*Ci*. 3: 127–132
3	Aguru rasāyana	Paste of heartwood of *Aquilaria agallocha* to be smeared inside an iron vessel at night. The dried-up paste is to be washed out with water in the morning and consumed.	*Ut*. 39: 104–105
4	Āmalaka rasāyana	Fruits of *Emblica officinalis* are put inside the hollowed-out base of a *Butea monosperma* tree. The heap is packed with grass and set on fire. The cooked fruits are to be eaten *ad libitum*, along with honey and clarified butter. Milk is post-prandial drink.	*Ut*. 39: 28–32, 60, 148–149
5	Amṛtaprāśa rasāyana	This *ghṛta* prepared with 35 herbs, milk, juices of *Emblica officinalis*, *Ipomoea digitata*, sugarcane, honey, sugar and meat soup is an excellent remedy for emaciation and 12 diseases. It causes the birth of a male child as well.	*Ci*. 3: 94–101
6	Aṃśumatī rasāyana	Paste of root of *Pseudarthria viscida* is to be suspended in milk and consumed for fourteen days, two months, six months or a year.	*Ut*. 39: 155
7	Asana rasāyana	Paste of heartwood of *Pterocarpus marsupium* to be smeared inside an iron vessel at night. The dried-up paste is to be washed out with water in the morning and consumed.	*Ut*. 39: 104–105, 153, 155
8	Aśvakarṇa rasāyana	Paste of bark of *Terminalia paniculata* to be smeared inside an iron vessel at night. The dried-up paste is to be washed out with water in the morning and consumed.	*Ut*. 39: 104–105
9	Atibalā rasāyana	Root of *Abutilon indicum* ground to paste, suspended in milk and consumed for one month	*Ut*. 39: 60, 104–105
10	Balā rasāyana	Root of *Sida rhombifolia* ssp. *retusa* ground to paste, suspended in milk and consumed for one month	*Ut*. 39: 60, 104–105, 155
11	Bhallātaka rasāyana	Powder or decoction of seeds of *Semecarpus anacardium* is consumed along with any of the following adjuvants such as milk, honey, clarified butter, gooseberry powder, sesame oil, sesame seeds, jaggery, barley powder, meat soup or lentil soup.	*Ut*. 39: 66–83
12	Brahma rasāyana	3,000 fruits of *Emblica officinalis*, 1,000 fruits of *Terminalia chebula*, 39 herbs, sugar, clarified butter, sesame oil and honey are ingredients. *Ṣaṣṭika* rice cooked in milk is the only food to be consumed.	*Ut*. 39: 15–23
13	Brahmī rasāyana	*Bacopa monnieri*, eight other herbs, gold leaf and clarified butter to be ground to a paste and consumed for one year. Food mixed with clarified butter and honey to be consumed on complete digestion of the *rasāyana*.	*Ut*. 39: 50–53

(*Continued*)

TABLE 5.4 (*Continued*)
Rasāyana Medicines Described in AH

Sl. No.	Name of *rasāyana*	Mode of Preparation and Administration	Reference
14	Bṛhatī rasāyana	Paste of root of *Solanum indicum* is to be suspended in milk and consumed for fourteen days, two months, six months or a year.	*Ut*. 39: 155
15	Candana rasāyana	Paste of heartwood of *Santalum album* is to be suspended in milk and consumed for fourteen days, two months, six months or a year.	*Ut*. 39: 155
16	Catuṣkavalaya rasāyana	*Ghṛta* prepared with stem, rhizome, leaf, stamens of *Monochoria vaginalis* and gold leaf.	*Ut*. 39: 49, 104–105
17	Citraka rasāyana	Root powder of *Plumbago zeylanica* to be consumed with clarified butter, honey-clarified butter mixture, milk, water, sesame oil, cow urine or buttermilk.	*Ut*. 39: 62–65, 104–105
18	Cyavanaprāśa rasāyana	500 fruits of *Emblica officinalis*, 40 herbs, sesame oil, clarified butter and jaggery processed to produce this *rasāyana*. It is to be consumed according to the *kuṭīpravēśika* mode.	*Ut*. 39: 33–41
19	Dhavā rasāyana	Paste of root of *Anogeissus latifolia* to be smeared inside an iron vessel at night. The dried-up paste is to be washed out with water in the morning and consumed.	*Ut*. 39: 104–105
20	Ēlājamōdādi rasāyana	This *rasāyana* improves mental faculties, vision and longevity. It cures polyuric diseases, *gulma*, kingly consumption and morbid pallor. Its consumption does not call for any dietary or behavioral restrictions.	*Ci*. 5: 28–32
21	Gōkṣuraka rasāyana	Powder of whole *Tribulus terrestris* herb, subjected to *bhāvana* with its own juice, is to be suspended in milk and consumed. Cooked *Śāli* rice is to be eaten along with milk.	*Ut*. 39: 56–57, 159
22	Guḍūcī rasāyana	Juice of *Tinospora cordifolia* stem	*Ut*. 39: 44, 60, 104–105
23	Haridrā rasāyana	Paste of rhizome of *Curcuma longa* to be smeared inside an iron vessel at night. The dried-up paste is to be washed out with water in the morning and consumed.	*Ut*. 39: 104–105
24	Jaladādi rasayana	*Ghṛta* prepared with *Cyperus rotundus*, 17 other ingredients, clarified butter and milk.	*Ut*. 39: 46–47
25	Jīvantī rasāyana	Root of *Holostemma annulare* ground to paste, suspended in milk and consumed for one month.	*Ut*. 39: 60
26	Kāḷānusāri rasāyana	Paste of seeds of *Trigonella foenum-graecum* is to be suspended in milk and consumed for fourteen days, two months, six months or a year.	*Ut*. 39: 155
27	Kṛṣṇatila rasāyana	Seeds of *Sesamum indicum* are to be chewed in the morning, followed by drinking of cold water.	*Ut*. 39: 158
28	Kūśmāṇḍa rasāyana	This *rasāyana* prepared with juice of *Benincasa hispida* cures cough, hiccup, fever, respiratory distress, hemorrhages of obscure origin, trauma to thorax and kingly consumption.	*Ci*. 114–118
29	Lāṅgalyādi rasāyana	Rhizome of *Gloriosa superba*, *Triphala* powder and calcined iron are to be ground in juice of *Eclipta alba*. Pills rolled from the resultant mass are to be consumed for a month, following strict dietary and behavioral restrictions.	*Ut*. 39: 165–168

(*Continued*)

TABLE 5.4 (*Continued*)
***Rasāyana* Medicines Described in AH**

Sl. No.	Name of *rasāyana*	Mode of Preparation and Administration	Reference
30	Laśūna rasāyana	Paste or juice of *Allium sativum* can be administered along with sesame oil, clarified butter, animal fat or meat soup.	*Ut*. 39: 110–129
31	Maṇḍūkaparṇī rasāyana	Juice of *Centella asiatica*	*Ut*. 39: 44, 61
32	Mūrvā rasāyana	Paste of root of *Marsdenia volubilis* is to be suspended in milk and consumed for fourteen days, two months, six months or a year.	*Ut*. 39: 155
33	Mustā rasāyana	Paste of tubers of *Cyperus rotundus* to be smeared inside an iron vessel at night. The dried-up paste is to be washed out with water in the morning and consumed.	*Ut*. 39: 104–105
34	Nāgabalā rasāyana	Roots of *Sida vernonicaefolia* collected on the first day with *Puṣya* star in autumn season are to be powdered. This powder is to be consumed for one year, suspended in milk and followed by ingestion of honey-clarified butter mixture. Milk alone is the food for the entire period of treatment.	*Ut*. 39: 54–55 *Ci*. 3: 118–125
35	Nārasiṃha rasāyana	*Ghṛta* prepared with eight herbs, milk, juice of *Eclipta alba*, decoction of *Triphala*, butter, iron powder, candied sugar, sugar and 35 honey is to be consumed along with tasty food and beverages.	*Ut*. 39: 170–176
36	Pañcāravinda rasāyana	*Ghṛta* prepared with leaf stalk, rhizome, stamens, leaf, seed of *Nelumbo nucifera*, gold leaf and milk.	*Ut*. 39: 48, 104–105
37	Pāribhadrasāra rasāyana	Paste of heartwood of *Azadirachta indica* to be smeared inside an iron vessel at night. The dried-up paste is to be washed out with water in the morning and consumed.	*Ut*. 39: 104–105
38	Pāṭhā rasāyana	Paste of root of *Cyclea peltata* is to be suspended in milk and consumed for fourteen days, two months, six months or a year.	*Ut*. 39: 155
39	Pippalī rasāyana	Powder of fruits of *Piper longum* to be consumed along with honey or clarified butter.	*Ut*. 39: 96–103
40	Punarnava rasāyana	Paste of freshly collected root of *Boerhaavia diffusa* is to be suspended in milk and consumed for fourteen days, two months, six months or a year.	*Ut*. 39: 154
41	Saptasama guḷika	This *rasāyana* cures skin diseases and improves virility.	*Ci*. 19: 43
42	Śaṅkhupuṣpi rasāyana	Root and flower of *Clitoria ternatea* ground to paste	*Ut*. 39: 44, 61
43	Śāribā rasāyana	Paste of root of *Hemidesmus indicus* is to be suspended in milk and consumed for fourteen days, two months, six months or a year.	*Ut*. 39: 155
44	Śatāvarī rasāyana	Root of *Asparagus racemosus* ground to paste, suspended in milk and consumed for one month.	*Ut*. 39: 61, 156
45	Śilājatu rasāyana	Mineral bitumen is administered in different ways as a *rasāyana*.	*Ut*. 39: 130–142, 161
46	Sōmarājī rasāyana	Powder of seeds of *Psoralea corylifolia* can be administered in different ways.	*Ut*. 39: 107–110
47	Śrēyasī rasāyana	Root of *Piper chaba* ground to paste, suspended in milk and consumed for one month.	*Ut*. 39: 60
48	Sthirā rasāyana	Root of *Desmodium gangeticum* ground to paste, suspended in milk and consumed for one month.	*Ut*. 39: 60

(*Continued*)

TABLE 5.4 (*Continued*)
Rasāyana Medicines Described in AH

Sl. No.	Name of *rasāyana*	Mode of Preparation and Administration	Reference
49	Sukumāraṃ rasāyana	Prepared with *Boerhaavia diffusa* as the major ingredient, this *rasāyana* cures several diseases including abscess, *gulma*, *vātavyādhi* and *vāta*-induced blood disorders. It can be consumed without complying with any restrictions.	*Ci*. 13: 41–47
50	Śuṇṭhī rasāyana	Paste of rhizome of *Zingiber officinale* to be smeared inside an iron vessel at night. The dried-up paste is to be washed out with water in the morning and consumed.	*Ut*. 39: 104–105
51	Surāhvā rasāyana	Paste of heartwood of *Cedrus deodara* to be smeared inside an iron vessel at night. The dried-up paste is to be washed out with water in the morning and consumed.	*Ut*. 39: 104–105
52	Tēkarāja rasāyana	He who chews leaves of *Eclipta alba* and sucks its juice for one month will live with stamina and valor for hundred years.	*Ut*. 39: 162
53	Triphalā rasāyana	*Triphala* powder is to be consumed for one year, with sugar or honey-clarified butter mixture.	*Ut*. 39: 42–43, 104–105, 152, 160
54	Tuvaraka rasāyana	Oil extracted from seeds of *Hydnocarpus laurifolia* to be consumed along with honey, clarified butter or decoction of *Acacia catechu*. Nasal instillation of the oil is also recommended.	*Ut*. 39: 84–95
55	Uśīra rasāyana	Paste of root of *Vetiveria zizanioides* is to be suspended in milk and consumed for fourteen days, two months, six months or a year.	*Ut*. 39: 155
56	Vacā rasāyana	Powder of rhizome of *Acorus calamus* is to be consumed for one month mixed with milk, sesame oil or clarified butter.	*Ut*. 39: 163
57	Vājīgandha rasāyana	Roots of *Withania somnifera* ground to paste, suspended in milk and consumed for one month.	*Ut*. 39: 61, 157
58	Vārāhīkanda rasāyana	Roots of *Curculigo orchioides* ground to paste, suspended in milk and consumed for one month.	*Ut*. 39: 58–59
59	Vasiṣṭha rasāyana	Named after sage Vasiṣṭha, this *rasāyana* can be consumed in all seasons, without dietary or behavioral restrictions.	*Ci*. 3: 133–141
60	Vāyasī rasāyana	Root of *Fritillaria roylei* ground to a paste, suspended in milk and consumed for one month.	*Ut*. 39: 60
61	Viḍaṅga rasāyana	Paste of seeds of *Embelia ribes* to be smeared inside an iron vessel at night. The dried-up paste is to be washed out with water in the morning and consumed.	*Ut*. 39: 104–105, 149–151
62	Vidārī rasāyana	Tuber of *Ipomoea digitata* ground to paste, suspended in milk and consumed for one month.	*Ut*. 39: 60
63	Yaṣṭimadhuka rasāyana	Powder of *Glycyrrhiza glabra* roots suspended in milk	*Ut*. 39: 44, 60, 104–105
64	Yuktā rasāyana	Rhizome of *Alpinia galanga* ground to paste, suspended in milk and consumed for one month.	*Ut*. 39: 60

TABLE 5.5
***Vājīkaraṇa* Medicines Described in AH**

Sl. No.	Name of Formula	Mode of Preparation and Administration	Reference
1	Āmalaka rasāyana	Consumption of this medicine prepared from *Emblica officinalis* fruits improves vigor and vitality.	*Ut*. 39: 28–32
2	Amṛtaprāśa rasāyana	Regular consumption of this *rasāyana* pacifies sexual debility and improves fertility.	*Ci*. 3: 94–101
3	Cyavanaprāśa rasāyana	The celestial physicians *Aśvinīkumāra* formulated this famous medicine for improving vigor and vitality.	*Ut*. 39: 33–41
4	Dadhisāra yōga	Cream isolated from curd is to be mixed with sugar and cooked *Ṣaṣṭika* rice. On consumption of this medicated rice, even an old man will entertain women, as if he were a young man.	*Ut*. 40: 33
5	Dhātrīphalādi ghṛta	Consumption of this *ghṛta* prepared with juices of *Emblica officinalis*, *Ipomoea digitata* and sugarcane improves stamina and virility.	*Ci*. 3: 108–110
6	Gōdhūmādi vājīkaraṇayōga	Powders of wheat and *Mucuna pruriens* beans are to be cooked in milk. On cooling, it is to be mixed with honey, clarified butter and consumed.	*Ut*. 40: 23–24
7	Gōkṣuraka rasāyana	This formula prepared with *Tribulus terrestris* is reputed to improve libido.	*Ut*. 39: 56–57
8	Kṛṣṇādi vājīkaraṇayōga	Mixture of powders of *Piper longum* and *Emblica officinalis* is to be subjected to *bhavana* with their own juices. This powder is to be mixed with sugar, honey, clarified butter and consumed. Milk is post-prandial drink.	*Ut*. 40: 27–28
9	Kṣīrabhāvitatila yōga	Sesame seeds are subjected to *bhāvana* using milk boiled with beans of *Mucuna pruriens*. Consumption of these seeds with sugar improves sexual vigor.	*Ut*. 40: 25
10	Kulīraśṛṅgi vājīkaraṇayōga	Galls of *Pistacia integerrima* are to be ground, suspended in milk and consumed. Sugar, milk and clarified butter are to be added to food.	*Ut*. 40: 29–30.
11	Madhukacūrṇa yōga	Consumption of powder of *Glycyrrhiza glabra* mixed with honey and clarified butter improves sexual vigor.	*Ut*. 40: 28–29
12	Māṣacūrṇa prayōga	Powder of black gram is cooked in milk. On cooling, it is mixed with honey, clarified butter and consumed.	*Ut*. 40: 24
13	Nārasiṃha rasāyana	Consumption of this *rasāyana* improves sexual vigor as well.	*Ut*. 39: 169–176
14	Payasyā vājīkaraṇayōga	Roots of *Holostemma annulare* are to be cooked in milk, ground to paste and mixed with honey and clarified butter. Milk is to be drunk after consuming this paste.	*Ut*. 40: 30–31
15	Saptasama guḷika	This *rasayana* improves virility.	*Ci*. 19: 43
16	Śarādi vājīkaraṇayōga	This compound formulation is made of 47 herbs, juices of *Emblica officinalis*, *Ipomoea digitata*, *Saccharum officinarum*, clarified butter and honey. One *pala* (48 g) is the dose. Meat soup and milk are post-prandial drinks.	*Ut*. 40: 12–21
17	Sarpirguḍa	Consumption of sweet balls prepared with *Maranta arundinacea* starch, *Piper longum* and parched rice improves stamina, sexual vigor and vitality.	*Ci*. 3: 112–113
18	Śatāvarīcūrṇa prayōga	Roots of *Asparagus racemosus* are to be powdered and suspended in sweetened milk. This medicated milk enhances vigor and vitality.	*Ut*. 40: 32
19	Śvadaṃṣṭrādi yōga	Powders of *Tribulus terrestris* fruits, *Hygrophila auriculata* seeds, *Phaseolus mungo* beans, *Mucuna pruriens* beans and *Asparagus racemosus* root are to be cooked in milk and consumed.	*Ut*. 40: 34
20	Svayaṅguptādi yōga	Beans of *Mucuna pruriens* and seeds of *Hygrophila auriculata* are to be powdered and suspended in sweetened milk. He who consumes this medicated milk will copulate like a donkey.	*Ut*. 40: 31–32
21	Uccaṭācūrṇa prayōga	Roots of *Curculigo orchioides* are to be powdered and suspended in sweetened milk. This medicated milk improves sexual vigor.	*Ut*. 40: 32
22	Vidāryādi vājīkaraṇayōga	*Ipomoea digitata* and five other herbs are processed with honey, clarified butter and sugar. Consumption of this medicine bestows strength to indulge in sexual intercourse with 100 women.	*Ut*. 40: 21–23
23	Vidārīcūrṇa yōga	Powder of *Ipomoea digitata* tubers subjected to *bhāvana* with its own juice is to be consumed, mixed with honey and clarified butter.	*Ut*. 40: 26

5.5 CONSUMPTION OF INCOMPATIBLE FOOD

AH states that certain food ingredients are incompatible to each other. Food prepared with such ingredients turn to be toxic and cause serious health problems or even death. For example, milk is incompatible with sour ingredients and all sour fruits. Similarly, one should avoid drinking milk after eating radish and vegetables. Curd is incompatible with meat of deer and chicken (*Sū.* 7: 30–33). Combining in equal parts any two or three from honey, clarified butter, visceral fat, sesame oil and water turns to be toxic. If honey and clarified butter are to be consumed together, they should be mixed in unequal proportions (*Sū.* 7: 39–40). Consumption of meat of yellow-footed green pigeon (*Hārīta, Treron phoenicoptera*) mixed with honey causes instant death (*Sū.* 7: 45). Elimination measure such as emesis or purgation is to be applied in the treatment of discomforts arising from consumption of incompatible food. *Śamana* therapy with medicines antagonistic to the incompatible food may be applied if the discomforts are not severe. Incompatible food does not harm youth, unctuous-bodied persons, those who exercise regularly and those with good stamina and digestive efficiency. Conditioning the body with protective properties is, therefore, the best way to combat the adverse effects of incompatible food (*Sū.* 7: 45–47).

5.6 PROGNOSIS OF DISEASES

Prognosis of diseases can be of four types: *sādhya* (curable), *kṛcchraḥsādhya* (difficult to cure), *yāpya* (manageable with medicines) and *asādhya* (incurable) (*Sū.* 1: 30–33). However, a prognostic technique, incorporating some elements of astrology, is also recommended. Known as *dūtavijñānīyam* or knowledge gleaned from the messenger who arrives to accost the physician to the patient, this technique provides additional information on the prognosis of the disease in question. The physician should not accompany the messenger if he seems to be frightened, in great haste, speaks impolitely, carries weapons or is shabbily dressed. Similarly, if the messenger arrives when the physician is sleeping, is in the midst of prayer or worship of deity, doing destructive actions like cutting, crushing or splitting materials, or engaged in discussion of topics related to violence and such inauspicious matters, the physician must understand that accepting the message brought will be in vain. Contrarily, if the physician obtains easily the resources and helpers, it should be understood that the treatment will be successful. *Dūtavijñānīyam* helps the physician to have prior knowledge about the outcome of the treatment (*Śā.* 6: 1–73).

5.7 TREATMENT PROTOCOL

In addition to elimination measures collectively called *pañcakarma*, several other measures such as *upavāsa* (fasting), *vyāyāma* (exercise), *abhyaṅga* (applying medicinal oil all over body), *sēka* (pouring of medicinal oils, decoction or other fluids in a thin stream), *avagāha* (keeping the body immersed in warm water or medicinal fluids), *lēpa* (application of medicinal pastes), *dhūpana* (fumigation) and *udvartana* (powder massage) are also applied in treatment of diseases. Tables 5.6–5.13 list the diseases in which these measures are indicated. Compatible medicines are administered after conditioning the patient with these measures. Suitable post-prandial drinks (*anupāna*), desirable diets and behavioral restrictions are also recommended.

5.8 ADMINISTRATION OF MEDICINES

A wide range of traditional dosage forms is recommended in treatment of diseases. They include decoctions (*kvātha*), pastes (*kalka*), juices (*svarasa*) and powders of herbs (*cūrṇa*), medicinal lipids (*taila, ghṛta*), fermented liquids (*ariṣṭa, āsava*), electuaries (*lēhya*), pills (*guḷika*), collyria (*añjana*) and medicinal smoke (*dhūma*). In addition to these, medicated porridges, soups and beverages are also used. At the end of the chapter on *vājīkaraṇa*, Vāgbhaṭa mentions many excellent remedies

TABLE 5.6
Diseases and Conditions in Which *upavāsa* Is Indicated

	Name of Disease or Condition		
Sl. No.	Sanskrit Term	English Equivalent	Reference
1	Abhiṣyanda	Conjunctivitis	*Ut.* 16: 1
2	Akālanidra	Discomforts arising from sleeping at odd times	*Sū.* 7: 62
3	Atisāra	Dysentery	*Ci.* 9: 5–9
4	Atinidra	Discomforts arising from sleeping excessively	*Sū.* 7: 63
5	Chardi	Vomiting	*Ci.* 6: 4
6	Chardi vēgavidhāraṇāt	Disorders arising from restrainment of urge to vomit	*Sū.* 4: 18
7	Gulma	Phantom tumor caused by enragement of *kapha*	*Ci.* 14: 76
8	Jālakagardabha	Lymphangitis	*Ut.* 32: 6
9	Jvara	Fever	*Ci.* 1: 9
10	Kaphājīrṇa	Indigestion due to enragement of *kapha*	*Sū.* 8: 27
11	Liṅganāśa	Blurred vision (near blindness)	*Ut.* 14: 20
12	Madātyaya	Alcoholism	*Ci.* 7: 33, 42
13	Pratiśyāya	Common cold	*Ut.* 20: 13
14	Raktapitta	Hemorrhages of obscure origin	*Ci.* 2: 6–7
15	Śūla	Pricking pain in stomach	*Ci.* 6: 58–59
16	Udara	Enlargement of abdomen	*Ci.* 15: 117
17	Viṣūci	Cholera	*Sū.* 8: 17

TABLE 5.7
Diseases and Conditions in Which *vyāyāma* Is Indicated

	Name of Disease or Condition		
Sl. No.	Sanskrit Term	English Equivalent	Reference
1	Chardi vēgavidhāraṇāt	Disorders arising from restrainment of urge to vomit	*Sū.* 4: 18
2	Dinacarya	Exercise as a promoter of health	*Sū.* 2: 10–14
3	Dōṣa śamanam	Pacification of *dōṣas*	*Sū.* 14: 7
4	Hēmanta carya	Regimen for pre-winter season	*Sū.* 3: 10–17
5	Kaphavṛddhi	Treatment for enragement of *kapha*	*Sū.* 13: 11–12
6	Pramēha	Polyuria	*Ci.* 12: 33, 36–37
7	Laṅghana	Measure to induce leaning	*Sū.* 14: 14–15
8	Mēdaḥ kaphāvṛta vātōpacāra	Measure to pacify *vāta* enveloped by *mēdas* and *kapha*	*Sū.* 17: 28
9	Śiśira carya	Regimen for winter season	*Sū.* 3: 10–17
10	Ūrustambha	Stiffness of thighs	*Ci.* 21: 53–55
11	Vasanta carya	Regimen for spring season	*Sū.* 3: 19

(*agrouṣadham*) for pacifying disease conditions (Table 5.14). These medicines and measures can be used for effective management of diseases. The herbs mentioned are to be processed into various dosage forms like *kvātha*, *cūrṇa*, *taila*, *ghṛta* and so on. They can be administered after considering factors such as age, stamina and *dōṣa* status of the patient (*Ut.* 40: 48–58).

The time of administration of these medicines varies with the *dōṣa* and disease. Generally, medicines are administered at the beginning, middle and end of meals. Sometimes they are administered with a morsel of food. In *kapha*-dominant diseases and in physically strong patients, the medicine

TABLE 5.8
Diseases and Conditions in Which *abhyaṅga* Is Recommended

Sl. No.	Name of Disease or Condition: Sanskrit Term	English Equivalent	Reference
1	Apaci	Cervical glandular swelling	*Ut.* 30: 24
2	Aparā pātanam	Expulsion of placenta	*Śa.* 1: 91
3	Apasmāra	Epilepsy	*Ut.* 7: 30
4	Arśas	Hemorrhoids	*Ci.* 8: 14–15, 26–27, 100, 135–137
5	Bādhirya	Deafness	*Ut.* 18: 24
6	Bhaṅga	Fracture	*Ut.* 27: 33
7	Bhēṣajakṣaya	Mishaps in treatment	*Sū.* 4: 29
8	Chardivēga vidhāraṇāt	Diseases arising from restrainment of urge to vomit	*Sū.* 4: 19
9	Garbhakāla upadravaḥ	Discomforts during pregnancy	*Śa.* 1: 60
10	Garbha vyāpat	Mishaps in pregnancy	*Śa.* 2: 19
11	Garbhiṇī paricarya	Preparing pregnant mother for parturition	*Śa.* 1: 77
12	Grahabādha	Spiritual afflictions	*Ut.* 5: 6
13	Gudabhramśa	Prolapse of rectum	*Ci.* 9: 53–54
14	Gulma	Phantom tumor	*Ci.* 14: 2
15	Hēmanta carya	Regimen for pre-winter season	*Sū.* 3: 10
16	Jvara	Fever	*Ci.* 1: 127, 128, 129–131, 140
17	Kaphaduṣṭa stanyam	Post-partum mother suffering from *kapha* vitiation of breast milk	*Ut.* 2: 16
18	Karṇanāda	Tinnitus	*Ut.* 18: 24
19	Karṇaśūla	Pricking pain in ear	*Ut.* 18: 9
20	Kāsa	Cough	*Ci.* 3: 2, 90
21	Kuṣṭha	Leprosy and other skin diseases	*Ci.* 19: 14, 75, 78–83
22	Lihya	Small pimple with itching, pus and pain appearing on ear lobule	*Ut.* 18: 49
23	Madātyaya	Alcoholism	*Ci.* 7: 17, 52
24	Manda nidra	Insomnia	*Sū.* 7: 66
25	Mūḍhagarbha pātanam	Expulsion of stillbirth	*Śa.* 2: 46
26	Mūtrarōdha	Diseases arising from restrainment of urge to void urine	*Sū.* 4: 5
27	Rājayakṣma	Kingly consumption	*Ci.* 5: 64, 69, 71
28	Raktapitta	Hemorrhages of obscure origin	*Ci.* 2: 48
29	Sadyōvraṇa	Traumatic wound	*Ut.* 26: 54
30	Sarvajatrūrdhva rōga	All diseases above collar bone	*Ut.* 24: 48–49
31	Śiśiracarya	Regimen for winter season	*Sū.* 3: 17
32	Śvayathu	Edema	*Ci.* 17: 24
33	Śvitra	Vitiligo	*Ci.* 20: 2–3
34	Udara	Enlargement of abdomen	*Ci.* 15: 113
35	Unmantha	Edema with itching in ear lobule	*Ut.* 18: 45
36	Vātaśōṇita	*Vāta*-induced blood disorders	*Ci.* 22: 22
37	Vātavyādhi	*Vāta* diseases	*Ci.* 21: 4
38	Yōnirōga	Diseases of vagina	*Ut.* 34: 26

TABLE 5.9
Diseases and Conditions in Which *sēka* Is Recommended

Sl. No.	Sanskrit Term	English Equivalent	Reference
	Name of Disease or Condition		
1	Abhiṣyanda	Conjunctivitis	*Ut.* 16: 8
2	Ardhāvabhēdaka	Hemicrania	*Ut.* 24: 9
3	Arśas	Hemorrhoids	*Ci.* 8: 12–16
4	Bhagandara	Anal fistula	*Ut.* 28: 23
5	Bhaṅga	Fracture	*Ut.* 27: 18–19
6	Dāhajvara	Fever with burning sensation all over body	*Ci.* 1: 133
7	Garbhakāla upadravaḥ	Discomfort during pregnancy	*Śa.* 1: 61
8	Garbhiṇī paricarya	Preparing pregnant mother for parturition	*Śa.* 1: 77
9	Grahabādha	Spiritual afflictions	*Ut.* 5: 7
10	Gudakuṭṭaka	Napkin rash	*Ut.* 2: 72
11	Jīrṇajvara	Chronic fever	*Ci.* 1: 129
12	Kīṭaviṣa	Insect bite	*Ut.* 37: 21
13	Kṛmi	Worm infestation	*Ci.* 20: 23
14	Kukūṇaka	Neonatal conjunctivitis	*Ut.* 9: 27–28
15	Kumbikā vartma	Meibomian cyst	*Ut.* 9: 2
16	Liṅganāśa	Blurred vision (near blindness)	*Ut.* 14: 20–21
17	Nābhināḷa chēdanam	Wound on the navel of the newborn caused by cutting of umbilical cord	*Ut.* 1: 6
18	Niruddhamaṇi	Phimosis	*Ut.* 34: 18
19	Pōthaki	Pimples on eyelid	*Ut.* 9: 21–22
20	Rājayakṣma	Kingly consumption	*Ci.* 5: 72
21	Raktārśas	Bleeding piles	*Ci.* 8: 125
22	Raktātisāra	Dysentery with bleeding	*Ci.* 9: 82–84
23	Śaṅkhaka	Pain in the forehead with heat and puffiness of the temples	*Ut.* 24: 13
24	Sarpaviṣa	Snake bite	*Ut.* 36: 47–48
25	Saviṣānna sparśanam	Discomforts arising from touching poisonous food	*Sū.* 7: 19
26	Śiraḥkampa	Shivering of head	*Ut.* 24: 18
27	Śiraḥstāpa	Headache	*Ut.* 24: 3, 12, 13, 14
28	Śītajvara	Fever with shivering fits	*Ci.* 1: 142
29	Śukḷarōga	Diseases of sclera	*Ut.* 11: 29
30	Sūryāvarta	Migraine	*Ut.* 24: 11
31	Śvayathu	Edema	*Ci.* 17: 24–25
32	Timira	Cataract	*Ut.* 13: 37
33	Udara	Enlargement of abdomen	*Ci.* 15: 49
34	Vāta kāsa	Cough with predominance of *vāta*	*Ci.* 3: 2
35	Vātaja mūtrakṛchra	Dysuria with predominance of *vāta*	*Ci.* 11: 1
36	Vātaśōṇita	*Vāta*-induced blood disorders	*Ci.* 22: 23–26
37	Vātavyādhi	*Vāta* diseases	*Ci.* 21: 51
38	Visarpa	Cellulitis	*Ci.* 18: 24
39	Vraṇa	Wound, sore, ulcer	*Ut.* 25: 24–25, *Ut.* 26: 12
40	Yōnirōga	Diseases of vagina	*Ut.* 34: 27

TABLE 5.10
Diseases and Conditions in Which *avagāha* Is Indicated

	Name of Disease or Condition		
Sl. No.	**Sanskrit Term**	**English Equivalent**	**Reference**
1	Arśas	Hemorrhoids	*Ci.* 8: 12–13
2	Jvara	Fever	*Ci.* 1: 133, 142
3	Kāsa	Cough	*Ci.* 3: 2
4	Mūtrakṛchra	Dysuria	*Ci.* 11: 1
5	Mūtrarōdha	Discomforts arising from restrainment of urge to void urine	*Sū.* 4: 5
6	Rājayakṣma	Kingly consumption	*Ci.* 5: 77
7	Vātavyādhi	*Vāta* diseases	*Ci.* 21: 29
8	Vātaśōṇita	*Vāta*-induced blood disorders	*Ci.* 22: 38

TABLE 5.11
Diseases and Conditions in Which *lēpa* Is Indicated

	Name of Disease or Condition		
Sl. No.	**Sanskrit Term**	**English Equivalent**	**Reference**
1	Ākhuviṣa	Rat bite	*Ut.* 38: 18–19
2	Alaji	Red eruptions with intolerable pain appearing on penis	*Ut.* 34: 10–12
3	Alasa	Dhobi's itch	*Ut.* 32: 12–13
4	Apaci	Cervical glandular swelling	*Ut.* 30: 14–16
5	Arśas	Hemorrhoids	*Ci.* 8: 11, 21–26
6	Arūmṣika	Pityriasis	*Ut.* 24: 22
7	Aṣṭhīlika	Hard, round boil with uneven surface, appearing at base of penis	*Ut.* 34: 14
8	Bhagandara	Anal fistula	*Ut.* 28: 33
9	Dantavidradhi	Dental abscess	*Ut.* 22: 33
10	Dāruṇaka	Dandruff	*Ut.* 24: 26
11	Dhūmara	Smoky vision	*Ut.* 13: 91
12	Duṣṭakṣīra	Post-partum mother suffering from vitiation of breast milk	*Ut.* 2: 16
13	Gaḷagaṇḍa	Swelling in neck	*Ut.* 22: 69–70
14	Garbhakāla upadravaḥ	Discomforts during pregnancy	*Śa.* 1: 60
15	Garbha vyāpat	Mishaps in pregnancy	*Śa.* 2: 24
16	Grahabādha	Spiritual afflictions	*Ut.* 5: 7
17	Granthi	Glandular swelling	*Ut.* 30: 2–3
18	Gulma	Phantom tumor. Abdominal mass characterized by tumor-like, hard, round mass, unstable in size and consistency, moving or immobile, situated in the abdomen	*Ci.* 14: 49
19	Indralupta	Alopecia	*Ut.* 24: 28–32
20	Jvara	Fever	*Ci.* 1: 129, 132–133, 141, 143, 150

(Continued)

TABLE 5.11 (*Continued*)
Diseases and Conditions in Which *lēpa* Is Indicated

Sl. No.	Name of Disease or Condition: Sanskrit Term	English Equivalent	Reference
21	Kaphaja pratiśyāya	Common cold due to enragement of *kapha*	*Ut*. 20: 13
22	Karṇaśōpha	Suppuration in ear	*Ut*. 18: 34
23	Karṇaśūla	Pricking pain in ear	*Ut*. 18: 10
24	Kīṭaviṣa	Insect stings	*Ut*. 37: 21
25	Kṛmidanta	Dental caries	*Ut*. 22: 19
26	Kumbhika	Large boil appearing on penis	*Ut*. 34: 10–12
27	Kuṣṭha	Leprosy and other skin diseases	*Ci*. 19: 62–74
28	Lāñchana	Congenital hyperpigmentation	*Ut*. 32: 15
29	Lihya	Small pimples with itching, pus and pain appearing on ear lobule	*Ut*. 18: 48–49
30	Madyapāna vidhi	Topical application of pastes of fragrant substances is recommended as a preparation before drinking wine	*Ut*. 7: 95
31	Mūḍhagarbha pātanam	Expulsion of stillbirth	*Śa*. 2: 27
32	Mukhadūṣika	Pimples	*Ut*. 32: 3
33	Nāḍīvraṇa	Sinus	*Ut*. 30: 33–35
34	Nīlika	Naevi	*Ut*. 32: 15
35	Padmakaṇṭaka	Papilloma of skin	*Ut*. 32: 4–5
36	Raktapitta	Hemorrhages of obscure origin	*Ci*. 2: 49
37	Sankhaka	Pain in the forehead with heat and puffiness of the temples	*Ut*. 24: 13
38	Sarpaviṣa	Snake bite	*Ut*. 36: 47
39	Śatapōnaka	Venereal disease with several perforations in the base of penis	*Ut*. 34: 20
40	Saviṣānna sparśanam	Discomforts arising from touching poisoned food	*Sū*. 7: 20
41	Śiraḥstāpa	Headache	*Ut*. 24: 12–14
42	Ślīpada	Elephantiasis	*Ut*. 30: 12
43	Śvayathu	Edema	*Ci*. 17: 27, 29–30, 35, 36, 41
44	Śvitra	Vitiligo	*Ci*. 20: 9–17
45	Timira	Cataract	*Ut*. 13: 47
46	Udara	Enlargement of abdomen	*Ci*. 15: 48
47	Upadamśa	Venereal ulcer	*Ut*. 34: 2–6
48	Uttama	Small pimple of the size of black gram or green gram seeds, appearing on the penis	*Ut*. 34: 11–12
49	Valmīka	Actinomycosis	*Ut*. 32: 9
50	Vātaśōṇita	*Vāta*-induced blood disorders	*Ci*. 22: 28–29, 33–34, 36–40
51	Vātavyādhi	*Vāta* diseases	*Ci*. 21: 51–53
52	Visarpa	Cellulitis	*Ci*. 18: 10–20, 25
53	Vṛddhi	Inguinoscrotal swelling	*Ci*. 13: 31
54	Vraṇa	Wound, sore, ulcer	*Ut*. 25: 24 *Ut*. 26: 12
55	Vyaṅga	Black pigmentation	*Ut*. 32: 15
56	Yavaprākhya	Hard mole	*Ut*. 32: 1
57	Yōnirōga	Diseases of vagina	*Ut*. 34: 27

TABLE 5.12
Fumigants Used in Treatment of Diseases

Sl. No.	Fumigant	Indication/Application	Reference
1	*Acorus calamus* and *Brassica juncea*	Fumigation of bed, bed sheet and blanket of the newborn infant	*Ut*. 1: 25
2	Aparājita dhūpa	Fumigation of body of patients suffering from all types of fever	*Ci*. 1: 164
3	*Aquilaria agallocha*	Fumigation of body of healthy persons in pre-winter season	*Sū*. 3: 11
		Fumigation of body of *vātaja madātyaya* patient	*Ci*. 7: 18
4	*Brassica juncea* and clarified butter	Fumigation of the mother of the baby suffering from neonatal conjunctivitis	*Ut*. 9: 25–27
5	Brahmyādi varti	Insanity	*Ut*. 6: 38–40
6	*Calotropis gigantea*, *Prosopsis cineraria*, human hair, snakeskin, hide of cat and clarified butter	Fumigation of anal area in hemorrhoids	*Ci*. 8: 18
7	Cat droppings	Fumigation of body of patients suffering from all types of fever	*Ci*. 1: 163
8	*Crocus sativus*	Fumigation of young women in pre-winter season as a cosmetic measure	*Sū*. 3: 15
9	Dead crow, clarified butter, sesame oil and bone marrow	Fumigation of bed, bed sheet and blanket of the newborn infant	*Ut*. 1: 26
10	Flesh of dog, cow and fish	Insanity	*Ut*. 6: 44
11	Fragrant substances (*vāsadhūpa*)	Prelude to drinking of wine	*Ci*. 7: 95
12	Gōśṛṅgādi dhūpa	Spiritual afflictions of children, irregular fever	*Ut*. 3: 55–57
13	Hides of panther, tiger, snake lion, bear; clarified butter	Spiritual afflictions of children	*Ut*. 3: 47–47
14	Hiṅguvyōṣādi dhūpa	Spiritual afflictions coupled with epilepsy and insanity	*Ut*. 5: 2–8
15	Kārpāsāsthyādi dhūpa	Fever due to insanity and spiritual afflictions	*Ut*. 5: 18
16	Mixture of *Hordeum vulgare*, *Betula edulis*, *Cedrus deodara*, *Pinus longifolia*, clarified butter and beeswax	Fumigation of sores and ulcers	*Ut*. 25: 45
17	Mixture of powders of *Acorus calamus*, *Azadirachta indica*, *Brassica juncea*, *Hordeum vulgare*, *Saussurea lappa*, *Terminalia chebula*, *Tribulus terrestris* and clarified butter	Fumigation of body of patients suffering from all types of fever	*Ci*. 1: 162–163
18	Mixture of *Withania somnifera*, *Ocimum sanctum*, *Piper longum*, *Solanum indicum* and clarified butter	Fumigation of anal area in hemorrhoids	*Ci*. 8: 18
19	Nakulōlūkādi dhūpa	Epilepsy	*Ut*. 7: 33
20	Pūtidaśāṅgādi dhūpa	Spiritual afflictions of children	*Ut*. 3: 47–48
21	Sarṣapanimbādi dhūpa	Spiritual afflictions of children	*Ut*. 3: 48–49
22	Sṛgālaśalyakādi dhūpa	Insanity	*Ut*. 6: 42–43

TABLE 5.13
Diseases and Conditions in Which *udvartana* Is Indicated

Sl. No.	Name of Disease or Condition: Sanskrit Term	English Equivalent	Reference
1	Bhēṣajakṣaya	Mishaps in treatment	*Sū.* 4: 29
2	Grahabādha	Spiritual afflictions	*Ut.* 5: 13–14
3	Kuṣṭha	Leprosy and other skin diseases	*Ci.* 19: 66, 68, 72
4	Madātyaya	Alcoholism	*Ci.* 7: 17, 42, 52
5	Madyapāna vidhi	Preparation before drinking wine	*Ci.* 7: 95
6	Nētra samrakṣaṇam	Conservation of eye health	*Ut.* 16: 66
7	Pāḷīśōṣam	Thinning of ear lobule	*Ut.* 18: 38
8	Pramēha	Polyuria	*Ci.* 12: 33
9	Rājayakṣma	Kingly consumption	*Ci.* 5: 78–81
10	Svayathu	Edema	*Ci.* 17: 35
11	Unmāda	Insanity	*Ut.* 5: 13–14. *Ut.* 6: 21–22
12	Vasanta carya	Part of regimen for spring season	*Sū.* 3: 19
13	Vātavyādhi	*Vāta* diseases	*Ci.* 21: 51

is administered as medicated food. Frequent administration of medicines is advised in treatment of poisons, vomiting, hiccup, excessive thirst, respiratory diseases and cough. Distaste for food is to be treated with medicines incorporated into different types of food. Different modes are advised for normalizing the destabilization of the five subtypes of *vāta*. Medicine should be administered at the end of breakfast to stabilize *vyāna* and at the end of dinner in the case of derangement of *udāna*. Derangement of *apāna* and *samāna* is pacified by administration of medicines at the beginning and middle of meals, respectively. Medicine administered along with each morsel of food stabilizes *prāṇa*. Consuming medicines at night, before retiring to bed, cures diseases of throat and those above the collar bone (*Sū.* 13: 37–41). *Vātaja karṇaśūla* patient should consume *vāta*-pacifying *ghṛta* at night, after having dinner with meat soup (*Ut.* 18: 1). Similarly, *vātaja śiraḥstāpa* patient is advised to consume *ghṛta* at night (*Ut.* 24: 1). Collyria is also to be applied at night (*Sū.* 23: 19).

5.9 SIGNS OF IMMINENT DEATH

Just as flowers appear before fruiting, smoke precedes fire and clouds foretell the arrival of rain, *riṣṭaṃ* (signs of imminent death) give warning of imminent death. This knowledge is detailed in *Vikṛtivijñānīyādhyāyaḥ*, the fifth chapter of *Śārīrasthānam*. The patient does not survive after the appearance of *riṣṭaṃ*. *Riṣṭaṃ* is defined as the change in shape of body, efficiency of sense organs, voice, lustre of body and behavior happening without any particular cause. Imminent death can be inferred from the abnormal color and shape of eyeballs, nose, tongue, teeth and lips; altered odor of body and mouth, appearance of veins and lines on forehead, visions of ghosts, devils and such bizarre beings while staying in wakeful state; anosmia and disturbed sensory perception; extrasensory perception without any formal training in yoga; weak, plaintive or unclear voice and several other signs. A physician should always be aware of *riṣṭaṃ*, while treating patients (*Śā.* 5: 1–132).

5.10 MANAGEMENT OF MISHAPS

Elimination therapies like emesis, purgation and medicinal enema may not sometimes work in the desired way. The main causes of these are stale or ineffective medicines, overlooking of precautions and noncompliant patients. Instead of eliminating *dōṣa*s, such improperly applied measures

TABLE 5.14
Excellent Remedies for Diseases and Conditions (*Ut.* 40: 48–58)

Sl. No.	Disease/Condition	Remedy
1	Fever (*jvara*)	Kvātha of Mustā (*Cyperus rotundus*) and Parpaṭakaṃ (*Mollugo cerviana*)
2	Vomiting (*chardi*)	Water boiled with powder of parched rice (*lāja*)
3	Urinary diseases (*vastirōga*), all diseases (*sarvarōga*)	Mineral bitumen (*śilājit, śilāhvayaṃ*)
4	Polyuric diseases (*mēha*)	*Emblica officinalis* (Dhātri) and *Curcuma longa* (Niśa)
5	Morbid pallor (*pāṇḍu*)	Calcined iron
6	Vāta-kapha enragement (*anilakaphē*)	*Terminalia chebula* (Abhaya)
7	Enragement of spleen (*plīhāmaya*)	*Piper longum* (Pippali)
8	Fractures (*sandhāna*)	Lac (Kṛmijaṃ)
9	Poisons (*viṣa*)	*Albizia lebbeck* (Śukataru)
10	*Vāta* residing in fat (*mēdōnila*)	Gum resin of *Commiphora mukul* (Gugguluḥ)
11	Hemorrhages of obscure origin (*asrapitta*)	*Adhatoda vasica* (Vṛṣa)
12	Dysentery (*atisāra*)	*Holarrhena antidysenterica* (Kuṭaja)
13	Hemorrhoids (*arśas*)	*Semearpus anacardium* (Bhallātaka)
14	Poisoning with criminal intent (*gara*)	Calcined gold (*hēma*)
15	Obesity (*sthoulya*)	Powder of emerald (*tārkṣyaṃ*)
16	Worm infestation (*kṛmi*)	*Embelia ribes* (Kṛmighnaṃ)
17	Emaciation of body (*śōṣa*)	*Sura* wine, milk and meat of goat
18	Eye diseases (*akṣyāmaya, nētrarujaḥ*)	Nasyaṃ, añjanaṃ, tarpaṇaṃ, Triphala
19	Vātaśōṇita (*vātāsrarōga*)	*Tinospora cordifolia* (Guḍūci)
20	Disease characterized by voiding of loose and hard stool (*grahaṇi*)	Buttermilk prepared without adding water to curd (*mathitaṃ*)
21	Skin diseases (*kuṣṭha*)	Heartwood of *Acacia catechu* (Khadirasya sāra)
22	Insanity (*unmāda*)	Aged clarified butter (*ghṛtam anavaṃ*)
23	Melancholy (*śōkaṃ*)	Wine (*madyaṃ*)
24	Loss of memory (*vismṛti*)	*Bacopa monnieri* (Brahmi)
25	Sleeplessness (*nidrānāśaṃ*)	Milk
26	Rhinitis (*pratiśyāya*)	The beverage *Rasāḷa* mentioned in *Sūtrasthānam*
27	Leaning of body (*kārśyaṃ*)	Meat (*māṃsaṃ*)
28	Inability to move body and limbs (*stabdhata*)	Sudation (*svēdaḥ*)
29	Pain in shoulder, arm and hand	Nasyaṃ with gum exudate of *Bombax malabaricum* (Khuḍamañjari)
30	Facial palsy (*arditam*)	Butter mixed with candied sugar (*navanītakhaṇḍaṃ*)
31	Enlargement of abdomen (*udaram*)	Urine and milk of camel (*auṣṭraṃ mūtraṃ payaḥ*)
32	Acute abscess (*vidradhim acirōtthitam*)	Bloodletting (*Asra visravaḥ*)
33	Diseases of mouth (*mukharōga*)	Nasyaṃ and kabaḷaṃ
34	Old age (*vārddhakya*)	Milk and clarified butter
35	Fainting (*mūrcha*)	Cold water, wind and shade
36	Dull appetite (*manda vahni*)	Equally dry and moist food
37	Tiredness (*śrama*)	Wine
38	Grief (*duḥkha*)	Bathing (*snānam*)
39	Strength of body (*sthairya*)	Exercise (*vyāyāma*)
40	Dysuria (*kṛcchram*)	*Tribulus terrestris* (Gōkṣura)
41	Cough	*Solanum xanthocarpum* (Nidigdhikā)
42	Pain in flanks (*pārśvaśūla*)	*Inula racemosa* (Puṣkaramūla)
43	Maintenance of youthfulness	*Emblica officinalis* (Dhātri)
44	Wounds and ulcers (*vraṇa*)	Triphala, Gugguluḥ
45	*Vāta* diseases	*Vasti*, sesame oil, garlic (*Laśūnaḥ*)
46	*Pitta* diseases	*Virēcana*, clarified butter
47	*Kapha* diseases	*Vamana*, honey

produce untoward effects and even death. *Vamanavirēcanavyāpatsiddhir adhyāyaḥ* (*Ka.* 3: 1–39) and *Bastivyāpatsiddhir adhyāyaḥ* (*Ka.* 5: 1–54) advise how to prevent and treat such mishaps. A competent physician should begin elimination therapies only after ensuring that all medicines, equipment and requirements are in stock. Damage to body arising from treatments and consumption of medicines was known in Ayurveda. *Abhyaṅga*, *udvartana*, bathing, *nirūhavasti* and *snēhavasti* are recommended to overcome these problems. Thereafter, health is to be regained with consumption of food prepared with *Śāli*, *Ṣaṣṭika* varieties of rice, wheat, green gram and clarified butter (*Sū.* 4: 28–30). Administration of clarified butter is also advised in case of weakness caused by thermal and chemical cautery (*Sū.* 5: 38).

5.11 TREATMENT OF NEW DISEASES

Many new diseases appear over the course of time. Sometimes it is not possible to assign a name to a new disease based on ayurvedic jargon. AH advises that in such cases the physician should not feel inferior at all. As all diseases spring from the destabilization of *tridōṣa*, knowledge of the etiopathogenesis of the disease in relation to the *dhātu*, *mala* and the symptoms produced will enable a physician to treat the disease successfully, even if he may not be able to assign a name to it (*Sū.* 12: 64). This point is amply illustrated by the example of Covid-19 pandemic. Corona virus disease 2019 (Covid-19) was not recorded in ancient ayurvedic literature. The outbreak of Covid-19 viral fever became a global health emergency on a pandemic scale in 2019. The most common symptoms of Covid-19 are fever, dry cough, fatigue and dyspnea. Severe symptoms are accompanied by systemic infection and pneumonia (Jiang et al., 2020).

Based on the natural history of the disease (*vikāra prakṛti*), site of the pathological process (*adhiṣṭhānam*) and etiological features (*samutthāna viśēṣam*), Nechiyil et al. (2020) observed that Covid-19 infection progresses in five stages. In the first stage, the disease can be considered as *kapha-vāta samsargaja jvara*, with *pitta* association. Patient in this stage can be treated by pacifying fever and *kapha*, inducing *vāta* to move in the normal direction (*vātānulōmana*) and conserving stamina of patient. If untreated, the disease progresses to the second stage. Line of treatment comprises pacification of fever and *kapha*, *vātānulōmana*, protection of stamina and bringing *pitta* to normal state. This is especially important in the case of elderly patients with co-morbidities. Fatal pathological changes begin at this stage. Stage three is characterized by *dhātupāka* (tissue necrosis), due to aggravation of *pitta*. This stage can be controlled by measures that pacify fever, prevent the progress of *dhātupāka* and conserve the stamina of patient. Administration of medicines for bronchodilation and expectoration is also required. In stage four, *dhātupāka* progresses uncontrolled, generating respiratory distress. This stage can be managed by the same treatments recommended for stage three, but in more formidable manner. Measures for pacifying *sannipāta jvara* such as sudation with boluses of rice bran and special decoctions are recommended. If the disease is not controlled at the fourth stage, it will progress into the nearly fatal fifth stage. As patient cured of the infection will be weak due to medications and the disease itself, electuaries such as *Agastya rasāyana*, *Cyavanaprāśa*, *Kūśmāṇḍa rasāyana* and *Indukānta ghṛta* are to be administered to convalescing patients (Nechiyil et al., 2020). Thus, based on the principles of Ayurveda, Covid-19 can be classed as a form of *kapha-vāta jvara* with *pitta* association, which progresses into full-blown *sannipāta jvara* if left untreated.

5.12 CONCLUSION

Comprehensive knowledge of a disease is gained using five parameters such as *nidāna* (cause of the disease), *pūrvarūpa* (prodromes), *rūpa* (signs and symptoms), *upaśaya* (confirmation by trial and error) and *samprāpti* (etiopathogenesis). AH teaches that symptoms and signs are the only tools for gauging the underlying pathology of a disease. Therefore, a physician should collect as many symptoms and signs as possible, using the principles of observation, palpation and interrogation. He should then study and evaluate these symptoms at the levels of increase or decrease of *tridōṣa*, tissue elements, waste products, *pūrvarūpa*, *rūpa*, *āvaraṇa* (envelopment) and *saṃprāpti*. This exercise helps the physician to pinpoint the disease entity. Once the disease is identified, the physician follows the line of treatment recommended in its treatment. The prognosis of diseases can be of four types: curable, difficult to cure, manageable with medicines and incurable. In addition to elimination measures collectively called *pañcakarma*, several other measures such as fasting, exercise, fumigation and so on are also adopted in treatment. After conditioning the patient with these measures, compatible medicines from various dosage forms are administered. Suitable post-prandial drinks, desirable diets and behavioral restrictions are also recommended.

REFERENCES

Jiang, F., L. Deng, L. Zhang, Y. Cai, C.W. Cheung and Z. Xia. 2020. Review of the clinical characteristics of coronavirus disease 2019 (COVID-19). *Journal of General Internal Medicine* 35:1545–1549.

Nechiyil, S., L. Mahadevan and S. Saji. 2020. *Ayurvedic Diagnostic & Management Protocol*. Koottanad: Working Group of Ayurvedic Physicians.

Sharma, P.V. 1983. Chapter 5 *Gulma cikitsa*. In *Caraka Samhita*, Volume 2, 97–116. Varanasi: Chaukhamba Orientalia.

6 Medicinal Substances

6.1 INTRODUCTION

AH recommends many substances of vegetable, animal and mineral origin to be used in the prevention and treatment of diseases. They are intended to serve as food, beverages and medicinal formulations. These ingredients are processed as different dosage forms such as paste (*kalka*), juice (*svarasa*), decoctions (*kvātha*), powder (*cūrṇa*), medicated oil (*taila*), medicated clarified butter (*ghṛta*), fermented liquids (*ariṣṭa, āsava*), electuary (*lēhya*), pills (*guḷika*) and collyrium (*añjana*).

Of all the ingredients used in the preparation of medicines, herbs form the major segment. They are procured from different agroclimatic regions of India. Whole herbs, leaf, flower, stamens, stem, fruit, seed, heartwood, bark, root, root stock, rhizome, tuber, resin, gum-exudate, gall and oil go into the production of medicines. *Kalpasthānam* gives clear instructions for collecting herbs. One should not collect herbs from burial grounds, temple premises, sacred groves and ant hills. They should not have been eaten by animals or burnt by forest fire. These herbs are to be collected by a clean, skilled and virtuous person. The collected fresh herbs are used for preparing juices, pastes and decoctions. Dry herbs can be used for this purpose, if fresh herbs are unavailable. But they should not be more than one year old. Nevertheless, jaggery, clarified butter, honey, coriander, grains like paddy, *Piper longum* fruits and seeds of *Embelia ribes* are potent when they are more than one year old (*Ka.* 6: 1–6).

6.2 THE *GAṆAS* OF VĀGBHAṬA

In *Śōdhanādigaṇasaṅgrahādhyāyaḥ* (Chapter 15) of *Sūtrasthānam*, Vāgbhaṭa describes 33 groups of medicinal substances. The *Śōdhanādigaṇa*s are composed of 230 specific herbs. However, animal and mineral products such as cow milk, cow urine, honey, shells of pearl oyster, alkaline earth, rock salt, antimony sulphide, blue vitriol, green vitriol, melanterite and mineral bitumen (*Śilājit*) occur in some *gaṇas*. The constituent herbs are arranged according to their therapeutic properties. For example, *Guḍūcyādi gaṇa* cures fever, *Paṭōlādi* pacifies jaundice, *Varaṇādi* heals internal abscesses and so on. Their curative properties are not reduced, even if a few ingredients are left out and used as medicine. All the *gaṇa*s advised by Vāgbhaṭa are intended to be used both internally and topically. Nevertheless, traditional physicians of Kerala use some *gaṇa*s for internal and external use, while some others are used for external or internal use alone (Mooss, 1980).

Varaṇādi gaṇa is an example of a *gaṇa* commonly used for internal administration. A decoction is prepared for internal administration in various kinds of headache. A *ghṛta* is also prescribed for the same indication. Clarified butter prepared directly from curdled milk medicated with powder of these herbs is used for *nasya*. *Āragvadhādi gaṇa* is traditionally used both internally and externally. A decoction is generally prepared for internal administration. An *ariṣṭa* or *āsava* is also recommended. In the treatment of cutaneous eruptions and ulcers, a decoction is prescribed for washing and cleaning the affected parts. *Ēlādi gaṇa* is a group traditionally used for external application alone. *Taila* prepared using sesame oil or a mixture of sesame oil, castor oil and clarified butter is recommended for topical application in skin diseases including leprous conditions. A powder of herbs of *Ēlādi gaṇa* is used for *udvartana* or dry massage all over the body. The *gaṇa*s of Vāgbhaṭa are used as *kalka*, *kvātha*, *cūrṇa*, *snēha* (*taila* and *ghṛta*), *lēhya* and so on compatible with the *dōṣa*s and other factors (Mooss, 1980). The 33 *gaṇa*s of Vāgbhaṭa are presented in Tables 6.1–6.33. Information on herbs is available in all the six *sthāna*s of AH, excluding *Nidānasthānam*, which deals with etiology and classification of diseases. Such herbs are listed in Table 6.34. Herbs described in AH were identified based on Vaidya (1936), Anonymous (1978), Warrier et al. (2007a–d, 2008)

DOI: 10.1201/9781003148296-6

TABLE 6.1
Madanādi gaṇa

Sl. No.	Sanskrit Name	Latin Binomial/English Equivalent
1	*Madanaḥ*	*Randia dumetorum* Lamk.
2	*Madhukaṃ*	*Glycyrrhiza glabra* L.
3	*Lambā*	*Lagenaria siceraria* Standl. Bitter variety
4	*Nimbaḥ*	*Azadirachta indica* A. Juss.
5	*Bimbī*	*Coccinia grandis* (L.) Voigt.
6	*Viśālā*	*Citrullus colocynthis* Schrad.
7	*Trapusaṃ*	*Cucumis sativus* L.
8	*Kuṭajaḥ*	*Holarrhena antidysenterica* (Heyne ex Roth.) A. DC
9	*Mūrvā*	*Marsdenia volubilis* T. Cooke
10	*Dēvadāḷī*	*Luffa echinata* Roxb.
11	*Kṛmighnaṃ*	*Embelia ribes* Burm. f.
12	*Vidulaḥ*	*Acacia sinuata* (Lour.) Merr.
13	*Dahanaḥ*	*Plumbago indica* L.
14	*Citrā*	*Ricinus communis* L.
15	*Dhamārgavaḥ*	*Luffa cylindrica* (L.) M. Roem.
16	*Kṣvēḍaḥ*	*Luffa acutangula* (L.) Roxb. var. *amara* Cl.
17	*Karañjaḥ*	*Pongamia glabra* Vent.
18	*Kaṇā*	*Piper longum* L.
19	*Lavaṇaṃ*	Rock salt
20	*Vacā*	*Acorus calamus L.*
21	*Ēlā*	*Elettaria cardamomum* Maton.
22	*Sarṣapaḥ*	*Brassica juncea* L.

Indications: These 22 ingredients are used as emetics.

TABLE 6.2
Nikumbhādi gaṇa

Sl. No.	Sanskrit Name	Latin Binomial/English Equivalent
1	*Nikumbhaḥ*	*Baliospermum montanum* Muell.
2	*Kumbhaḥ*	*Ipomoea turpethum* R. Br.
3	*Harītaki*	*Terminalia chebula* Retz.
4	*Āmalaki*	*Emblica officinalis* Gaertn.
5	*Vibhītaki*	*Terminalia belerica* Roxb.
6	*Gavākṣi*	*Citrullus colocynthis* Schrad.
		Cucumis trigonus Roxb.
		Trichosanthes palmata Roxb.
7	*Snuk*	*Euphorbia nivullia* Buch.-Ham.
8	*Śaṅkhini*	*Clitoria ternatea* L.
9	*Nīlini*	*Ipomoea nil* (L.) Roth.
10	*Tilvakaḥ*	*Excoecaria agallocha* L.
		Symplocos racemosa Roxb.
11	*Śamyākaḥ*	*Cassia fistula* L.
12	*Kampilyakaḥ*	*Mallotus philippensis* Muell.
13	*Hēmadugdha*	*Euphorbia thomsoniana* Boiss.
		Garcinia morella Desr.
14	*Dugdhaṃ*	Milk
15	*Mūtraṃ*	Cow urine

Indications: The 15 ingredients are used as purgatives.

TABLE 6.3
Madanakuṭajādi gaṇa

Sl. No.	Sanskrit Name	Latin Binomial/English Equivalent
1	*Madanaḥ*	*Randia dumetorum* Lamk.
2	*Kuṭajaḥ*	*Holarrhena antidysenterica* (Heyne ex Roth.) A. DC.
3	*Kuṣṭha*	*Saussurea lappa* C.B. Cl.
4	*Dēvadāḷi*	*Luffa echinata* Roxb.
5	*Madhukaṃ*	*Glycyrrhiza glabra* L.
6	*Vaca*	*Acorus calamus* L.
7	*Vilvaḥ*	*Aegle marmelos* Corr.
8	*Kāśmaryaḥ*	*Gmelina arborea* Roxb.
9	*Takkāri*	*Premna serratifolia* L.
10	*Pāṭala*	*Stereospermum tetragonum* DC.
11	*Duṇḍukaḥ*	*Oroxylum indicum* Vent.
12	*Bṛhati*	*Solanum indicum* L.
13	*Śvētabṛhati*	*Solanum xanthocarpum* Sch. & Wendl.
14	*Śālaparṇi*	*Desmodium gangeticum* DC.
15	*Pṛśniparṇi*	*Pseudarthria viscida* W. & A.
16	*Gōkṣuraḥ*	*Tribulus terrestris* L.
17	*Dāru*	*Cedrus deodara* Roxb.
18	*Rāsnā*	*Alpinia galanga* Willd.
19	*Yavaḥ*	*Hordeum vulgare* L.
20	*Misiḥ*	*Peucedanum graveolens* L.
21	*Kṛtavēdhana*	*Luffa acutangula* (L.) Roxb. var. *amara* Cl.
22	*Kulatthaḥ*	*Dolichos biflorus* L.
23	*Madhu*	Honey
24	*Lavaṇaṃ*	Rock-salt
25	*Tṛvṛtā*	*Ipomoea turpethum* R. Br.

Indications: The 25 ingredients are useful for *nirūhavasti.*

TABLE 6.4
Vēllādi gaṇa

Sl. No.	Sanskrit Name	Latin Binomial/English Equivalent
1	*Vēllaḥ*	*Embelia-tsjeriam-cottam* A. DC.
2	*Apāmārgaḥ*	*Achyranthes aspera* L.
3	*Maricam*	*Piper nigrum* L.
4	*Pippali*	*Piper longum* L.
5	*Nāgaram*	*Zingiber officinale* Roscoe
6	*Dārvī*	*Coscinium fenestratum* (Gaertn.) Colebr.
7	*Surālā*	*Aconitum heterophyllum* Wall.
8	*Śairīṣam bījam*	*Albizia lebbeck* (L.) Benth. seeds
9	*Bārhatam bījam*	*Solanum indicum* L. seeds
10	*Śaigravam bījam*	*Moringa oleifera* Lamk. seeds
11	*Mādhūkaḥ sāraḥ*	*Madhuca longifolia* (Koenig) Mcbr. heartwood
12	*Saindhavam*	Rock salt
13	*Tārkṣya śailam*	*Rasāñjana* (concentrated water extract of *Berberis aristata*)
14	*Ēlā*	*Elettaria cardamomum* Maton
15	*Bṛhad ēlā*	*Amomum subulatum* Roxb.
16	*Pṛdhvīkā*	*Gardenia gummifera* L.

Indications: These 16 ingredients are useful for performing *nasya*.

TABLE 6.5
Bhadradārvādi gaṇa

Sl. No.	Sanskrit Name	Latin Binomial
1	*Bhadradāru*	*Cedrus deodara* Roxb.
2	*Natam*	*Valeriana wallichii* D C.
3	*Kuṣṭham*	*Saussurea lappa* C.B. Cl.
4	*Vilvah*	*Aegle marmelos* Corr.
5	*Kāśmaryaḥ*	*Gmelina arborea* Roxb.
6	*Takkāri*	*Premna serratifolia* L.
7	*Pāṭalā*	*Stereospermum tetragonum* D C.
8	*Duṇḍukaḥ*	*Oroxylum indicum* Vent.
9	*Bṛhati*	*Solanum indicum* L.
10	*Śvētabṛhati*	*Solanum xanthocarpum* Sch. & Wendl.
11	*Śālaparṇi*	*Desmodium gangeticum* DC.
12	*Pṛśniparṇi*	*Pseudarthria viscida* W. & A.
13	*Gōkṣuraḥ*	*Tribulus terrestris* L.
14	*Balā*	*Sida rhombifolia* L. ssp. *retusa* (L.) Boiss.
15	*Atibalā*	*Abutilon indicum* G. Don.

Indications: These 15 herbs are useful in the alleviation of *vāta*.

TABLE 6.6
Dūrvādi gaṇa

Sl. No.	Sanskrit Name	Latin Binomial
1	*Dūrvā*	*Cynodon dactylon* Pers.
2	*Anantaḥ*	*Tragia involucrata* L. var. *angustifolia* Hook.f.
3	*Nimbaḥ*	*Azadirachta indica* A. Juss.
4	*Vāśā*	*Adhatoda vasica* Nees
5	*Ātmaguptā*	*Mucuna pruriens* DC.
6	*Gundrā*	*Typha angustata* Bory & Chaub.
7	*Abhīruḥ*	*Asparagus racemosus* Willd.
8	*Śītapāki*	*Calycopteris floribunda* Lamk.
9	*Priyaṅguḥ*	*Aglaia roxburghiana* Miq.
		Callicarpa macrophylla Vahl.
		Prunus mahaleb L.

Indications: This group alleviates *pitta*.

TABLE 6.7
Sthirādi gaṇa

Sl. No.	Sanskrit Name	Latin Binomial
1	*Śālaparṇi*	*Desmodium gangeticum* DC.
2	*Pṛśniparṇi*	*Pseudarthria viscida* W. & A.
3	*Padmam*	*Nelumbo nucifera* Gaertn.
4	*Vanyam*	*Nymphaea stellata* Willd.

Indications: These four herbs alleviate *pitta*.

TABLE 6.8
Vidāryādi gaṇa

Sl. No.	Sanskrit Name	Latin Binomial
1	*Vidāri*	*Ipomoea digitata* L.
2	*Pañcāṅgulaḥ*	*Ricinus communis* L.
3	*Vṛścikāḷi*	*Heliotropium indicum* L.
4	*Vṛścivaḥ*	*Boerhaavia diffusa* L.
5	*Dēvāhvayaṃ*	*Cedrus deodara* Roxb.
6	*Mudgaparṇi*	*Phaseolus adenanthus* G.F. Meyer
7	*Māṣaparṇi*	*Phaseolus sublobatus* Roxb.
8	*Kaṇḍukāri*	*Mucuna pruriens* DC.
9	*Abhīruḥ*	*Asparagus racemosus* Willd.
10	*Vīrā*	*Coccinia grandis* (L.) Voigt.
11	*Jīvanti*	*Holostemma annulare* (Roxb.) K. Schum.
12	*Jīvakaḥ*	*Malaxis acuminata* D. Don.
13	*Ṛṣabhakaḥ*	*Malaxis muscifera* (Lindley) O. Kuntze
14	*Bṛhati*	*Solanum indicum* L.
15	*Śvētabṛhati*	*Solanum xanthocarpum* Sch. & Wendl.
16	*Śālaparṇi*	*Desmodium gangeticum* DC.
17	*Pṛśniparṇi*	*Pseudarthria viscida* W. & A.
18	*Gōkṣuraḥ*	*Tribulus terrestris* L.
19	*Gōpasūtā*	*Hemidesmus indicus* R. Br.
20	*Tripādi*	*Desmodium triflorum* DC.

Indications: These herbs alleviate *vāta*, *pitta*, abnormal upward movement of *vāta*, dyspnoea and cough.

TABLE 6.9
Jīvanīya gaṇa

Sl. No.	Sanskrit Name	Latin Binomial
1	*Jīvanti*	*Holostemma annulare* (Roxb.) K. Schum.
2	*Kākōḷi*	*Fritillaria roylei* Hook.
3	*Kṣīrakākōḷi*	*Lilium polyphyllum* D. Don.
4	*Mēda*	*Polygonatum cirrhifolium* Royle
5	*Mahāmēda*	*Polygonatum verticillatum* (L.) Allioni
6	*Mudgaparṇi*	*Phaseolus adenanthus* G.F. Meyer
7	*Māṣaparṇi*	*Phaseolus sublobatus* Roxb.
8	*Ṛṣabhakaḥ*	*Malaxis muscifera* (Lindley) O. Kuntze
9	*Jīvakaḥ*	*Malaxis acuminata* D. Don.
10	*Madhukaṃ*	*Glycyrrhiza glabra* L.

Indications: These herbs are vitalizers.

TABLE 6.10
Śāribādi gaṇa

Sl. No.	Sanskrit Name	Latin Binomial
1	*Śāribā*	*Hemidesmus indicus* R. Br.
2	*Uśīram*	*Vetiveria zizanioides* Nash.
3	*Kāśmaryaḥ*	*Gmelina arborea* L.
4	*Madhūkaḥ*	*Madhuca longifolia* (Koenig) Macb.
5	*Candanam*	*Santalum album* L.
6	*Rakta candanam*	*Pterocarpus santalinus* L.
7	*Yaṣṭi*	*Glycyrrhiza glabra* L.
8	*Parūṣakam*	*Grewia asiatica* L.

Indications: This group of herbs cures burning sensation, hemorrhages of obscure origin, excessive thirst and fever.

TABLE 6.11
Padmakadi gaṇa

Sl. No.	Sanskrit Name	Latin Binomial
1	*Padmakam*	*Prunus puddum* Roxb.
2	*Puṇḍraḥ*	*Saccharum officinarum* L.
3	*Vṛddhiḥ*	*Habenaria intermedia* D. Don.
4	*Tukā*	Starch from rhizomes of *Curcuma angustifolia* Roxb.
5	*Ṛddhiḥ*	*Habenaria edgeworthii* Hook f. ex Collett
6	*Śṛṅgi*	Galls on *Pistacia integerrima* Stew. ex Brandis
7	*Amṛtā*	*Tinospora cordifolia* (Willd.) Miers. ex Hook. f. & Th.
8	*Jīvanti*	*Holostemma annulare* (Roxb.) K. Schum.
9	*Kākōḷi*	*Fritillaria roylei* Hook.
10	*Kṣīrakākōḷi*	*Lilium polyphyllum* D. Don.
11	*Mēda*	*Polygonatum cirrhifolium* Royle
12	*Mahāmēda*	*Polygonatum verticillatum* (L.) Allioni
13	*Mudgaparṇi*	*Phaseolus adenanthus* G.F. Meyer
14	*Māṣaparṇi*	*Phaseolus sublobatus* Roxb.
15	*Ṛṣabhakaḥ*	*Malaxis muscifera* (Lindley) O. Kuntze
16	*Jīvakaḥ*	*Malaxis acuminata* D. Don.
17	*Madhukaṃ*	*Glycyrrhiza glabra* L.

Indications: This *gaṇa* is galactagogue, nourishing, roborant and virilific. It pacifies *vāta* and *pitta*.

TABLE 6.12
Parūṣakādi gaṇa

Sl. No.	Sanskrit Name	Latin Binomial
1	*Parūṣakam*	*Grewia asiatica* L.
2	*Harītaki*	*Terminalia chebula* Retz.
3	*Āmalaki*	*Emblica officinalis* Gaertn.
4	*Vibhītaki*	*Terminalia belerica* Roxb.
5	*Drākṣā*	*Vitis vinifera* L.
6	*Kaṭphalam*	*Gmelina arborea* L.
7	*Katakāt phalam*	*Strychnos potatorum* L.
8	*Rājāhvam*	*Mimusops hexandra* Roxb.
9	*Dāḍimam*	*Punica granatum* L.
10	*Śākam*	*Tectona grandis* L.

Indications: Fruits of these ten herbs alleviate excessive thirst, urinary disorders and *vāta* enragement.

TABLE 6.13
Añjanādi gaṇa

Sl. No.	Sanskrit Name	Latin Binomial/English Equivalent
1	*Añjana*	Antimony sulphide
2	*Phalinī*	*Aglaia roxburghiana* Miq.
		Callicarpa macrophylla Vahl.
		Prunus mahaleb L.
3	*Māmsī*	*Nardostachys jatamansi* DC.
4	*Padmam*	*Nelumbo nucifera* Gaertn.
5	*Utpalam*	*Kaempferia rotunda* L.
6	*Rasāñjanam*	Solidified extract of *Berberis aristata*
7	*Ēlā*	*Elettaria cardamomum* Maton.
8	*Madhukaṃ*	*Glycyrrhiza glabra* L.
9	*Nāgāhvam*	*Mesua ferrea* L.

Indications: This group dispels internal burning sensation, enragement of *pitta* and effects of poisoning.

TABLE 6.14
Paṭōlādi gaṇa

Sl. No.	Sanskrit Name	Latin Binomial
1	*Paṭōlaḥ*	*Trichosanthes cucumerina* L.
2	*Kaṭurōhiṇi*	*Picrorhiza kurroa* Royle.
3	*Candanam*	*Santalum album* L.
4	*Madhusravaḥ*	*Moringa oleifera* Lamk.
5	*Guḍūci*	*Tinospora cordifolia* (Willd.) Miers. ex Hook. f. & Th.
6	*Pāṭhā*	*Cyclea peltata* Diels.

Indications: This group pacifies *kapha*, *pitta*, skin diseases, fever, poisoning, vomiting, anorexia and jaundice.

TABLE 6.15
Guḍūcyādi gaṇa

Sl. No.	Sanskrit Name	Latin Binomial
1	*Guḍūci*	*Tinospora cordifolia* (Willd.) Miers. ex Hook. f. & Th.
2	*Padmakam*	*Prunus puddum* Roxb.
3	*Ariṣṭaḥ*	*Azadirachta indica* A. Juss.
4	*Dhānakā*	*Coriandrum sativum* L.
5	*Rakta candanam*	*Pterocarpus santalinus* L.

Indications: This group pacifies *pitta*, *kapha*, fever, vomiting, burning sensation, excessive thirst and stimulates abdominal fire.

TABLE 6.16
Āragvadhādi gaṇa

Sl. No.	Sanskrit Name	Latin Binomial
1	*Āragvadha*	*Cassia fistula* L.
2	*Indrayavaḥ*	*Holarrhena antidysenterica* (Heyne ex Roth) A. DC.
3	*Pāṭalīh*	*Stereospermum tetragonum* DC.
4	*Kākatiktā*	*Hydnocarpus laurifolia* (Dennst.) Sleumer
5	*Nimbaḥ*	*Azadirachta indica* A. Juss.
6	*Amṛtā*	*Tinospora cordifolia* (Willd.) Miers. ex. Hook. f. & Th.
7	*Madhurasaḥ*	*Marsdenia volubilis* T. Cooke
8	*Sruvavṛkṣa*	*Flacourtia ramontchi* L'Herit
9	*Pāṭhā*	*Cyclea peltata* Diels.
10	*Bhūnimbaḥ*	*Andrographis paniculata* (Burm.f.) Nees.
11	*Sairyakaḥ*	*Ecbolium linneanum* Kurz.
12	*Paṭōlaḥ*	*Trichosanthes cucumerina* L.
13	*Karañjaḥ*	*Pongamia glabra* Vent.
14	*Pūtikarañjaḥ*	*Holoptelia integrifolia* Planch.
15	*Saptacchadaḥ*	*Alstonia scholaris* R. Br.
16	*Agniḥ*	*Plumbago indica* L.
17	*Suṣavi*	*Calycopteris floribunda* Lamk.
18	*Phalam*	*Randia dumetorum* Lamk.
19	*Bāṇaḥ*	*Tephrosia purpurea* Pers.
20	*Ghōṇṭu*	*Bridelia stipularis* Blume.

Indications: This group cures vomiting, skin diseases, poisoning, fever, polyuria and cleanses indolent ulcers.

TABLE 6.17
Asanādi gaṇa

Sl. No.	Sanskrit Name	Latin Binomial
1	*Asanaḥ*	*Pterocarpus marsuplum* Roxb.
2	*Tiniśaḥ*	*Ougeinia oojeinensis* (Roxb.) Hochr.
3	*Bhūrjaḥ*	*Betula utilis* D. Don.
4	*Śvētavāhaḥ*	*Terminalia arjuna* W. & A.
5	*Prakīryā*	*Pongamia glabra* Vent.
6	*Khadiraḥ*	*Acacia catechu* Willd.
7	*Kadarah*	*Flagellaria indica* L.
8	*Bhāṇḍī*	*Albizia chinensis* (Osb.) Merrill.
9	*Śiṃśapā*	*Dalbergia sissoo* Roxb.
10	*Mēṣaśṛṅgi*	*Gymnema sylvestre* R. Br.
11	*Candanam*	*Santalum album* L.
12	*Raktacandanam*	*Pterocarpus santalinus* L.
13	*Pītacandanam*	*Coscinium fenestratum* (Gaertn.) Colebr.
14	*Natam*	*Valeriana wallichii* DC.
15	*Palāśāḥ*	*Butea monosperma* Kuntze.
16	*Jōṅgakaḥ*	*Aquilaria agallocha* Roxb.
17	*Śākaḥ*	*Tectona grandis* L.
18	*Sālaḥ*	*Vateria indica* L.
19	*Dhavaḥ*	*Anogeissus latifolia* Wall.
20	*Kabukaḥ*	*Areca catechu* L.
21	*Kaliṅgaḥ*	*Holarrhena antidysenterica* (Heyne ex Roth.) A. DC.
22	*Chāgakarṇaḥ*	*Terminalia crenulata* (Heyne) Roxb.
23	*Aśvakarṇaḥ*	*Terminalia paniculata* Roth.

Indications: This group cures vitiligo, skin diseases, *kapha* diseases, anemia, polyuria and disorders due to deranged fat.

TABLE 6.18
Varaṇādi gaṇa

Sl. No.	Sanskrit Name	Latin Binomial
1	*Varaṇaḥ*	*Crataeva religiosa* Forst.
2	*Nīlasahacaraḥ*	*Ecbolium linneanum* Kurz.
3	*Śvētasahacaraḥ*	*Justicia betonica* L.
4	*Śatāvari*	*Asparagus racemosus* Willd.
5	*Dahanaḥ*	*Plumbago indica* L.
6	*Mōraṭaḥ*	*Marsdenia volubilis* T. Cooke
7	*Vilvaḥ*	*Aegle marmelos* Corr.
8	*Viṣāṇika*	*Gymnema sylvestre* R. Br.
9	*Bṛhati*	*Solanum indicum* L.
10	*Śvētabṛhati*	*Solanum xanthocarpum* Sch. & Wendl.
11	*Karañjaḥ*	*Pongamia glabra* Vent.
12	*Pūtikarañjaḥ*	*Holoptelia integrifolia* Pl.
13	*Agnimanthaḥ*	*Premna serratifolia* L.
14	*Harītaki*	*Terminalia chebula* Retz.
15	*Bahaḷa pallavaḥ*	*Moringa oleifera* Lamk.
16	*Darbhaḥ*	*Saccharum arundinaceum* Retz.
17	*Rujākaraḥ*	*Semecarpus anacardium* L.

Indications: *Varaṇādi gaṇa* cures *āḍhyavāta*, *gulma*, *antarvidradhi* and improves abdominal fire.

TABLE 6.19
Ūṣakādi gaṇa

Sl. No.	Sanskrit Name	Latin Binomial/English Equivalent
1	*Ūṣakaḥ*	Alkaline earth
2	*Tutthakaḥ*	Blue vitriol
3	*Hiṅgu*	*Ferula foetida* Regel
4	*Kāsīsam*	Ferrous sulphate (green vitriol)
5	*Puṣpakāsisam*	Melanterite
6	*Saindhavam*	Rock salt
7	*Śilājatu*	Mineral bitumen

Indications: This group cures dysuria, urinary stone, *gulma* and disorder of fat.

TABLE 6.20
Vīratarādi gaṇa

Sl. No.	Sanskrit Name	Latin Binomial
1	*Vēllantaraḥ*	*Dichrostachys cinerea* W. & A.
2	*Araṇikā*	*Premna serratifolia* L.
3	*Ṭūkaḥ*	*Oldenlandia auricularia* K. Schum.
4	*Vṛṣaḥ*	*Adhatoda vasica* Nees.
5	*Aśmabhēdaḥ*	*Rotula aquatica* Lour.
6	*Gōkaṇṭakaḥ*	*Tribulus terrestris* L.
7	*Itkaṭaḥ*	*Commiphora caudata* (W. & A.) Engl.
8	*Sahacaraḥ*	*Ecbolium linneanum* Kurz.
9	*Bāṇaḥ*	*Tephrosia purpurea* Pers.
10	*Kāśaḥ*	*Saccharum spontaneum* L.
11	*Vṛkṣādani*	*Loranthus elasticus* Desr.
12	*Nalaḥ*	*Phragmites karka* Trin.
13	*Darbhaḥ*	*Saccharum arundinaceum* Retz.
14	*Kuśaḥ*	*Desmostachya bipinnata* (L.) Stapf.
15	*Guccahaḥ*	*Typha elephantina* Roxb.
16	*Gundrā*	*Typha angustata* Bory & Chaub
17	*Bhallūkaḥ*	*Oroxylum indicum* Vent.
18	*Mōraṭaḥ*	*Marsdenia volubilis* T. Cooke
19	*Kurūṭakaḥ*	*Pergularia extensa* N.E. Br.
20	*Rambhā*	*Musa paradisiaca* L.
21	*Pārthaḥ*	*Terminalia arjuna* W. & A.

Indications: This combination cures *vāta* diseases as well as urinary stones, gravels, dysuria and anuria.

TABLE 6.21
Lōdhrādi gaṇa

Sl. No.	Sanskrit Name	Latin Binomial
1	*Lōdhraḥ*	*Symplocos spicata* Roxb.
2	*Śābara lōdhraḥ*	*Symplocos laurina* Wall.
3	*Palāśāḥ*	*Butea monosperma* Kuntze
4	*Jhiñjhinī*	*Odina woodier* Roxb. *Crotolaria juncea* L.
5	*Saraḷaḥ*	*Pinus longifolia* Roxb.
6	*Kaṭphalaḥ*	*Gmelina arborea* Roxb.
7	*Yuktā*	*Alpinia galanga* (L.) Willd.
8	*Kutsitāṅgaḥ*	*Anthocephalus indicus* A. Rich.
9	*Kadaḷī*	*Musa paradisiaca* L.
10	*Gataśōkaḥ*	*Saraca asoca* (Roxb.) de Wilde
11	*Ēlāvālu*	*Prunus cerasus* (Mill.) A. Gray
12	*Parivelavam*	*Cyperus esculentus* L.
13	*Mōcaḥ*	*Salmalia malabarica* Schott & Endl.

Indications: This *gaṇa* cures *kapha* as well as vaginal diseases. It cures poisoning and improves complexion.

TABLE 6.22
Arkkādi gaṇa

Sl. No.	Sanskrit Name	Latin Binomial
1	*Arkkaḥ*	*Calotropis gigantea* (L.) R. Br. ex Ait.
2	*Alarkaḥ*	*Calotropis gigantea* (L.) R. Br. ex Ait.var. *alba*
3	*Nāgadanti*	*Baliospermum montanum* Muell. Arg.
4	*Viśalyā*	*Gloriosa superba* L.
5	*Bhārṅgī*	*Clerodendrum serratum* (L.) Moon.
6	*Rāsnā*	*Alpinia galanga* Willd.
7	*Vṛścikāḷi*	*Heliotropium indicum* L.
8	*Prakīryā*	*Pongamia glabra* Vent.
9	*Pratyakpuṣpi*	*Achyranthes aspera* L.
10	*Pītataila*	*Celastrus paniculata* Willd.
11	*Udakīryā*	*Holoptelia integrifolia* Planch.
12	*Kṣudraśvēta*	*Cassia angustifolia* Wall.
13	*Mahāśvēta*	*Cassia obtusa* Roxb.
14	*Tāpasavṛkṣaḥ*	*Butea monosperma* (Lam.) Kuntze.

Indications: Arkkādi gaṇa pacifies *kapha*, fat and effects of poisoning. It kills parasites and helps in healing of wounds.

TABLE 6.23
Surasādi gaṇa

Sl. No.	Sanskrit Name	Latin Binomial
1	*Kṛṣṇasurasaḥ*	*Ocimum sanctum* L.
2	*Śvētasurasaḥ*	*Ocimum sanctum* L. with greenish stems
3	*Phaṇijaḥ*	*Origanum marjorana* L.
4	*Kālamālaḥ*	*Ocimum kilimandscharicum* Guerke
5	*Viḷaṅgaḥ*	*Embelia tsjeriam-cottam* A. DC.
6	*Kharabusaḥ*	*Achyranthes aspera* L.
		Leucas cephalotes Spr.
7	*Vṛṣakarṇī*	*Merremia emarginata* (Burm.f.) Hallier f.
8	*Kaṭphalaḥ*	*Gmelina arborea* L.
9	*Kāsamardaḥ*	*Cassia sophera* L.
10	*Kṣavakaḥ*	*Leucas aspera* Spr.
11	*Jharasī*	*Valeriana wallichii* DC.
12	*Bhārṅgī*	*Clerodendrum serratum* Spr.
13	*Kāmukaḥ*	*Jasminum pubescens* Willd.
14	*Kākamācī*	*Geophilia repens* (L.) I. M. Johnston
15	*Kulahalaḥ*	*Sphaeranthus indicus* L.
16	*Viṣamuṣṭi*	*Ageratum conyzoides* L.
17	*Bhūstṛṇaḥ*	*Hyptis suaveolens* (L.) Poit.
18	*Bhūtakēśi*	*Nardostachys jatamansi* DC.
		Corydalis govaniana Wall.
		Corydalis ramosa Wall.

Indications: This group cures diseases arising from *kapha* and fat, intestinal worms, rhinitis, anorexia, dyspnoea and cough. It hastens healing of ulcers.

TABLE 6.24
Muṣkakādi gaṇa

Sl. No.	Sanskrit Name	Latin Binomial
1	*Muṣkakaḥ*	*Schrebera swietenioides* Roxb.
2	*Snuk*	*Euphorbia nivulia* Buch.-Ham.
3	*Harītaki*	*Terminalia chebula* Retz.
4	*Āmalaki*	*Emblica officinalis* Gaertn.
5	*Vibhītaki*	*Terminalia belerica* Roxb.
6	*Dvīpi*	*Plumbago indica* L.
7	*Palāśāḥ*	*Butea monosperma* (Lam.) Taub.
8	*Dhavaḥ*	*Anogeissus latifolia* Wall.
9	*Śiṃśapā*	*Xylia xylocarpa* Taub.

Indications: Cures *gulma*, polyuria, urinary stone and hemorrhoids.

TABLE 6.25
Vatsakādi gaṇa

Sl. No.	Sanskrit Name	Latin Binomial
1	*Vatsakaḥ*	*Holarrhena antidysenterica* (Heyne ex Roth.) A. DC.
2	*Mūrvā*	*Marsdenia volubilis* T. Cooke
3	*Bhārṅgī*	*Clerodendrum serratum* Spr.
4	*Kaṭukā*	*Picrorhiza kurroa* Royle.
5	*Maricam*	*Piper nigrum* L.
6	*Ghuṇapriyā*	*Aconitum heterophyllum* Wall.
7	*Gaṇḍīram*	*Euphorbia nivulia* Buch.-Ham. *Naravelia zeylanica* DC. *Scindapsus officinalis* Schott.
8	*Ēlā*	*Elettaria cardamomum* Maton.
9	*Pāṭhā*	*Cyclea peltata* Diels.
10	*Ajājī*	*Nigella sativa* L.
11	*Kaṭvaṅga phalam*	*Ailanthus malabarica* DC.
12	*Ajamōjaḥ*	*Ptychotis ajowan* DC.
13	*Siddhārthaḥ*	*Brassica alba* Boiss.
14	*Vacā*	*Acorus calamus* L.
15	*Jīrakam*	*Cuminum cyminum* L.
16	*Hiṅgu*	*Ferula foetida* Regel.
17	*Viḷaṅgam*	*Embelia tsjeriam-cottam* A. DC.
18	*Paśugandhā*	*Cleome viscosa* L.
19	*Pippali*	*Piper longum* L. fruits
20	*Pippalīmūlam*	*Piper longum* L. roots
21	*Cavyam*	*Piper chaba* Hunter
22	*Citrakaḥ*	*Plumbago indica* L.
23	*Nāgaram*	*Zingiber officinale* Rosc.

Indications: Cures derangement of *vāta*, *kapha*, fat, rhinitis, *gulma*, fever, colic and hemorrhoids.

TABLE 6.26
Vacādi gaṇa

Sl. No.	Sanskrit Name	Latin Binomial
1	*Vacā*	*Acorus calamus* L.
2	*Jaladaḥ*	*Cyperus rotundus* L.
3	*Dēvāhvaṃ*	*Cedrus deodara* Roxb.
4	*Nāgaram*	*Zingiber officinale* Rosc.
5	*Ativiṣa*	*Aconitum heterophyllum* Wall.
6	*Abhayā*	*Terminalia chebula* Retz.

Indications: This *gaṇa* cures *āmātisāra*, *āḍhyavāta*, diseases of *mēdas* and *kapha* and vitiation of breast milk.

TABLE 6.27
Haridrādi gaṇa

Sl. No.	Sanskrit Name	Latin Binomial
1	*Haridra*	*Curcuma longa* L.
2	*Dāruharidra*	*Coscinium fenestratum* (Gaertn.) Colebr.
3	*Yaṣṭyāhvam*	*Glycyrrhiza glabra* L.
4	*Kalaśi*	*Pseudarthria viscida* W. & A.
5	*Kuṭajōdbhavaḥ*	*Holarhena antidysenterica* (Heyne ex Roth) A. DC.

Indications: This *gaṇa* cures *āmātisāra*, *āḍhyavāta*, diseases of *mēdas* and *kapha* and vitiation of breast milk.

TABLE 6.28
Priyaṅgvādi gaṇa

Sl. No.	Sanskrit Name	Latin Binomial/English Equivalent
1	*Priyaṅgupuṣpam*	*Aglaia roxburghiana* Miq. *Callicarpa macrophylla* Vahl. *Prunus mahleb* L.
2	*Rasāñjanam*	Solidified extract of *Berberis aristata* DC.
3	*Srōtōñjanam*	Antimony sulphide
4	*Padmā*	*Habenaria grandiflora* Lindl.
5	*Padmād rajaḥ*	Stamens of *Nelumbo nucifera* Gaertn.
6	*Yōjanavallī*	*Rubia cordifolia* L.
7	*Anantā*	*Tragia involucrata* L. var. *angustifolia* Hook. f.
8	*Śāladrumaḥ*	*Salmalia malabarica* Schott & Endl.
9	*Mōcarasaḥ*	*Salmalia malabarica* Schott & Endl. gum exudate
10	*Samaṅga*	*Biophytum sensitivum* (L.) DC.
11	*Punnāgam*	*Calophyllum inophyllum* L.
12	*Śītam*	*Santalum album* L.
13	*Madanīyahētuḥ*	*Woodfordia fruticosa* (L.) Kurz.

Indications: This *gaṇa* cures *pakvātisāra*, heals wounds and pacifies *pitta*.

TABLE 6.29
Ambaṣṭhādi gaṇa

Sl. No.	Sanskrit Name	Latin Binomial
1	*Ambaṣṭhā*	*Spondias mangifera* Willd.
2	*Madhukaṃ*	*Glycyrrhiza* glabra L.
3	*Namaskarī*	*Biophytum sensitivum* (L.) DC.
4	*Nandīvṛkṣaḥ*	*Cedrela toona* Roxb.
5	*Palāśāḥ*	*Butea monosperma* (Lam.) Taub.
6	*Kacchūraḥ*	*Tragia involucrata* L. var. *angustifolia* Hook. f.
7	*Lōdhraḥ*	*Symplocos spicata* Roxb.
8	*Dhātaki*	*Woodfordia fruticosa* (L.) Kurz.
9	*Vilvapēśika*	Young fruits of *Aegle marmelos* Corr.
10	*Kaṭvaṅgaḥ*	*Oroxylum indicum* Vent.
11	*Kamalōdbhavam rajaḥ*	Stamens of *Nelumbo nucifera* Gaertn.

Indications: This *gaṇa* cures diarrhea and hastens healing of broken bones and wounds.

TABLE 6.30
Mustādi gaṇa

Sl. No.	Sanskrit Name	Latin Binomial
1	*Mustā*	*Cyperus rotundus* L.
2	*Vacā*	*Acorus calamus* L.
3	*Agniḥ*	*Plumbago indica* L.
4	*Haridra*	*Curcuma longa* L.
5	*Dāruharidra*	*Coscinium fenestratum* (Gaertn.) Colebr.
6	*Kaṭurōhiṇi*	*Picrorhiza kurroa* Royle.
7	*Śārṅgaṣṭhā*	*Clerodendrum serratum* (L.) Moon.
8	*Bhallātakaḥ*	*Semecarpus anacardium* Linn. f.
9	*Pāṭhā*	*Cyclea peltata* Diels.
10	*Harītaki*	*Terminalia chebula* Retz.
11	*Āmalaki*	*Emblica officinalis* Gaertn.
12	*Vibhītaki*	*Terminalia belerica* Roxb.
13	*Viṣākhyā*	*Aconitum heterophyllum* Wall.
14	*Kuṣṭham*	*Saussurea lappa* C.B. Cl.
15	*Tutī*	*Elettaria cardamomum* Maton
16	*Haimavati*	*Iris germanica* L.

Indications: Cures uterine diseases and vitiation of breast milk.

TABLE 6.31
Nyagrōdhādi gaṇa

Sl. No.	Sanskrit Name	Latin Binomial
1	*Nyagrōdhaḥ*	*Ficus benghalensis* L.
2	*Pippalaḥ*	*Ficus religiosa* L.
3	*Sadāphalaḥ*	*Ficus racemosa* L.
4	*Lōdhraḥ*	*Symplocos spicata* Roxb.
5	*Śvētalōdhraḥ*	*Symplocos crataegoides* Buch.-Ham.
6	*Rājajambūḥ*	*Syzygium cumini* (L.) Skeels.
7	*Hrasvajambūḥ*	*Syzygium caryophyllatum* (L.) Alston.
8	*Arjunaḥ*	*Terminalia arjuna* W. & A.
9	*Kapītanaḥ*	*Ficus tjakela* Burm.
10	*Sōmavalkaḥ*	*Acacia suma* (Roxb.) Voigt.
11	*Pḷakṣaḥ*	*Ficus retusa* L.
12	*Āmraḥ*	*Mangifera indica* L.
13	*Vañculaḥ*	*Salix tetrasperma* Roxb.
14	*Priyāḷaḥ*	*Buchanania lanzan* Spreng.
15	*Palāśāḥ*	*Butea monosperma* (Lam.) Taub.
16	*Nandī*	*Cedrela toona* Roxb.
17	*Kōlī*	*Zizyphus jujuba* Lam.
18	*Kadambaḥ*	*Anthocephalus indicus* A. Rich.
19	*Viraḷā*	*Diospyros embryopteris* Pers.
20	*Madhukaṃ*	*Glycyrrhiza glabra* L.
21	*Madhūkaṃ*	*Madhuca longifolia* (Koenig) Macb.

Indications: This group is especially beneficial in healing fractured bones and ulcers.

and Shajahan and Nikita (2019). A unique feature of AH is its ingenious use of herbs as single herb remedies. Four hundred and fifty-four recipes involving 185 herbs are described in the text (Table 6.35). For furthering ease of application in therapeutics, AH also makes use of several short groupings of herbs such as *Mahat pañcamūla*, *Hrasva pañcamūla*, *Jīvana pañcamūla* and so on. Fourteen of such reputed groupings are listed in Table 6.36.

6.3 HERBS AND NUTRITION

In the *Annasavarūpavijñānīyādhyāyaḥ* of AH, Vāgbhaṭa classified herbs based on their dietetic value into *śūkadhānya varga* (awned grain), *śimbidhānya varga* (grains with pods), *śāka varga* (pot herbs/vegetables) and *phala varga* (fruits). *Śūkadhānya varga* is further grouped into *śāli*, *vrīhi* and *tṛṇadhānya*.

6.3.1 Varieties of Red Rice

These are 16 varieties of red rice (*Śāli*), the grains of which are red on account of their anthocyanin content (Priya et al., 2019; Bhuvaneswari et al., 2020). They bear names like *Raktaśāli*, *Mahān*, *Kaḷamam*, *Tūrṇakam*, *Śakunāhṛtam*, *Sāramukham*, *Dīrghaśūkam*, *Lōdhraśūkam*, *Sugandhikam*, *Pataṅgam*, *Tapanīyam*, *Yavakā*, *Hāyanam*, *Pāmsu*, *Bāṣpa* and *Naiṣadhakam*. Of all the varieties of *Śāli*, *Raktaśāli* is the best one. It pacifies thirst and derangement of *tridōṣa* (*Sū*. 6: 1–7).

TABLE 6.32
Ēlādi gaṇa

Sl. No.	Sanskrit Name	Latin Binomial/English Equivalent
1	*Sūkṣmaila*	*Elettaria cardamomum* Maton.
2	*Sthūlaila*	*Amomum subulatum* Roxb.
3	*Turuṣkaḥ*	Storax balsam of *Liquidambar orientalis* Mill.
4	*Kuṣṭham*	*Saussurea lappa* C.B. Clarke
5	*Phalinī*	*Callicarpa macrophylla* Vahl.
6	*Māṃsī*	*Nardostachys jatamansi* DC.
7	*Jalam*	*Coleus vettiveroides* K.C. Jacob
8	*Dhyāmakam*	*Cymbopogon martini* (Roxb.) Wats.
9	*Spṛkkaḥ*	*Cymbopogon citratus* (DC.) Stapf.
10	*Cōrakaḥ*	*Kaempferia galanga* L.
11	*Cōcam*	*Cinnamomum zeylanicum* Blume.
12	*Patram*	*Cinnamomum tamala* Nees & Eborm.
13	*Tagaram*	*Valeriana wallichii* DC.
14	*Sthauṇēyam*	*Taxus baccata* L.
15	*Jāti*	*Myristica fragrans* Houtt.
16	*Rasam*	Gum resin of *Commiphora myrrha* (Nees) Engl.
17	*Śuktiḥ*	Shells of pearl oyster
18	*Vyākhranakhī*	*Ipomoea pes-tigridis* L.
19	*Surāhvaṃ*	*Cedrus deodara* Roxb.
20	*Agaru*	*Aquilaria agallocha* Roxb.
21	*Śrīvāsakaḥ*	Resinous exudate of *Pinus longifolia* Roxb.
22	*Kuṅkumam*	*Crocus sativus* L.
23	*Caṇḍā*	*Costus speciosus* Sm.
24	*Gulguluḥ*	Gum resin of *Commiphora mukul* Engl.
25	*Dēvadhūpaḥ*	Resin of *Shorea robusta* Gaertn.
26	*Khapuraḥ*	Gum resin of *Boswellia serrata* Roxb.
27	*Punnāgam*	*Calophyllum inophyllum* L.
28	*Nāgāhvayam*	*Mesua ferrea* L.

Indications: Cures *vāta*, *kapha*, effects of poison and promotes complexion.

TABLE 6.33
Śyāmādi gaṇa

Sl. No.	Sanskrit Name	Latin Binomial
1	*Śyāma*	*Operculina turpethum* (L.) Silva Manso
2	*Danti*	*Baliospermum montanum* Muell. Arg.
3	*Dravanti*	*Croton tiglium* L.
4	*Kramukam*	*Areca catechu* L.
5	*Kuṭaraṇa*	*Ipomoea turpethum* R. Br.
6	*Śaṅkhini*	*Clitorea ternatea* L.
7	*Carmasāhva*	*Acacia sinuata* (Lour.) Merr.
8	*Svarṇakṣīri*	*Euphorbia thomsoniana* Boiss. *Garcinia morella* (Gaertn.) Desr.
9	*Gavākṣi*	*Citrullus colocynthis* (L.) Schrad.
10	*Śikhari*	*Achyranthes aspera* L.
11	*Rajanikaḥ*	*Mallotus philippinensis* Muell. Arg.
12	*Chinnarōha*	*Tinospora cordifolia* (Willd.) Miers. ex Hook. f. & Th.
13	*Karañjaḥ*	*Pongamia glabra* Vent.
14	*Bastāntri*	*Argyreia nervosa* (Burm. f.) Boj.
15	*Vyādhighātaḥ*	*Cassia fistula* L.
16	*Bahalaḥ*	*Moringa oleifera* L.
17	*Bahurasaḥ*	*Saccharum officinarum* L.
18	*Tīkṣṇavṛkṣāt phalāni*	Fruit of *Salvadora persica* L.

Indications: This *gaṇa* cures *gulma*, effects of poison, distaste for food, disorders of *kapha*, pain in precordial area and dysuria.

TABLE 6.34
Other Medicinal Plants Mentioned in AH

Sl. No.	Sanskrit Name	Latin Binomial	Reference
1	*Abhiṣuka*	*Aporosa acuminata* Thwaites	*Sū.* 6: 120
2	*Āḍhakī*	*Cajanus cajan* (L.) Millsp.	*Sū.* 6: 17
3	*Adrikarṇi*	*Haldina cordifolia* (Roxb.) Ridsdale	*Ut.* 5: 20
4	*Airāvatam*	*Citrus reticulata* Blanco	*Sū.* 6: 138
5	*Akṣōṭam*	*Juglans regia* L.	*Sū.* 6: 120
6	*Ālukam*	*Colocasia esculenta* (L.) Schott	*Sū.* 6: 94
7	*Amḷavētasa*	*Solena amplexicaulis* (Lam.) Gandhi	*Sū.* 10: 25
8	*Amḷika*	*Tamarindus indica* L.	*Sū.* 6: 139
9	*Āmrātā*	*Spondias pinnata* (L.f.) Kurz	*Sū.* 6: 119
10	*Aṅkōla*	*Alangium salvifolium* (L.f.) Wangorin	*Sū.* 6: 120
11	*Araḷuka*	*Ailanthus excelsa* Roxb.	*Ut.* 3: 60
12	*Āraṇyakulattham*	*Cajanus scarabaeoides* (L.) Thouars	*Ut.* 16: 13
13	*Arimēda*	*Acacia leucophloea* (Roxb.) Willd.	*Ut.* 22: 105
14	*Ārjakam*	*Ocimum basilicum* L.	*Sū.* 6: 106
15	*Ārtagaḷam*	*Barleria strigosa* Willd.	*Ci.* 5: 36
16	*Ārukam*	*Prunus domestica* L.	*Sū.* 6: 135
17	*Āśāḷika*	*Lepidium sativum* L.	*Ut.* 34: 30
18	*Aśmāntakam*	*Ficus rumphii* Bl.	*Ci.* 11: 18
19	*Aśōkam*	*Saraca asoca* (Roxb.) Willd.	*Ci.* 3: 10
20	*Asphōta*	*Vallaris solanacea* (Roth) Kuntze	*Sū.* 30: 9
21	*Aśvagandha*	*Withania somnifera* (L.) Dunal	*Ci.* 14: 14
22	*Aśvamārakam*	*Nerium indicum* Mill.	*Sū.* 30: 9
23	*Avalguja*	*Psoralea corylifolia* L.	*Sū.* 6: 75
24	*Āvartaki*	*Grewia tiliifolia* Vahl.	*Ci.* 19: 22
25	*Barhiśikha*	*Actinopteris dichotoma* (Forssk.) Kuhn	*Ci.* 11: 34
26	*Bhavyam*	*Averrhoa carambola* L.	*Sū.* 10: 26
27	*Bhūkadambam*	*Sphaeranthus amaranthoides* L.	*Ci.* 12: 20
28	*Brahmacāriṇi*	*Tricholepis glaberrima* DC.	*Sū.* 29: 31
29	*Caṇaka*	*Cicer arietinum* L.	*Ci.* 1: 71
30	*Cañcu*	*Pistia stratiotes* L.	*Ut.* 30: 37
31	*Cāpya*	*Corchorus capsularis* L.	*Sū.* 6: 84
32	*Cāṅgēri*	*Oxalis corniculata* L.	*Sū.* 6: 74
33	*Chatrā*	*Foeniculum vulgare* Mill.	*Sū.* 29: 31
34	*Cōca*	*Artocarpus heterophyllus* L.	*Sū.* 3: 31
35	*Cōrakam*	*Angelica glauca* Edgew.	*Ci.* 4: 44
36	*Darbha*	*Imperata cylindrica* (L.) P. Beauv.	*Ci.* 3: 36
37	*Dīpyakam*	*Trachyspermum roxburghianum* (DC) Craib	*Ci.* 3: 54
38	*Ērvāru*	*Cucumis sativus* L.	*Sū.* 6: 87
39	*Gajapippali*	*Piper retrofractutm Vahl.*	*Sū.* 6: 165
40	*Gavēthukam*	*Coix lachryma-jobi* L.	*Sū.* 6: 93
41	*Gōdhūmah*	*Triticum aestivum* L.	*Sū.* 6: 16
42	*Gōjihva*	*Elephantopus scaber* L.	*Sū.* 6: 77
43	*Gourasarṣapa*	*Brassica alba* (L.) Rabenh.	*Sū.* 22: 19
44	*Granthika*	*Piper brachystachyum* Wall.	*Ci.* 8: 46
45	*Gṛdhranakhi*	*Hugonia mystax* L.	*Ut.* 35: 46

(Continued)

TABLE 6.34 (*Continued*)
Other Medicinal Plants Mentioned in AH

Sl. No.	Sanskrit Name	Latin Binomial	Reference
46	*Gṛñjanakaḥ*	*Allium cepa* var. *aggregatum* G. Don.	*Sū.* 6: 113
47	*Guñja*	*Abrus precatorius* L.	*Sū.* 15: 78
48	*Harēṇu*	*Corchorus trilocularis* L.	*Ut.* 26: 55
49	*Haricandanam*	*Jateorhiza palmata* Lam. Miers.	*Sū.* 3: 40
50	*Hutāśa*	*Plumbago zeylanica* L.	*Ci.* 19: 46
51	*Ikṣuram*	*Hygrophila auriculata* Schumach.	*Ci.* 3: 136
52	*Indravatsam*	*Aconitum variegatum* L.	*Ut.* 35: 4
53	*Indravṛkṣa*	*Toona ciliata* M. Roem.	*Sū.* 30: 9
54	*Ingudi*	*Sarcostigma kleinii* Wight & Arn.	*Ut.* 18: 48
55	*Itkaṭa*	*Commiphora caudata* (Wight & Arn.) Engl.	*Ci.* 11: 22
56	*Jambīram*	*Citrus limon* (L.) Osbeck.	*Sū.* 6: 106
57	*Jīvantakam*	*Amaranthus viridis* L.	*Sū.* 6: 94
58	*Jūrṇam*	*Zea mays* L.	*Sū.* 14: 21
59	*Jyōtiṣmati*	*Celastrus paniculatus* Willd.	*Sū.* 21: 17
60	*Kadambam*	*Neolamarckia cadamba* (Roxb.) Bosser	*Sū.* 6: 93
61	*Kākajaṅgha*	*Cucumis trigonus* Roxb.	*Sū.* 30: 10
62	*Kākamāci*	*Solanum nigrum* L.	*Sū.* 6: 74
63	*Kākāṇḍakī*	*Trichosanthes tricuspidata* Lour.	*Sū.* 6: 22
64	*Kāḷakūṭam*	*Strychnos nux-vomica* L.	*Ut.* 35: 4
65	*Kāḷānusāri*	*Ichnocarpus frutescens* (L.) W.T. Aiton	*Śā.* 2: 49
66	*Kāḷaśākam*	*Murraya koenigii* (L.) Sprengel	*Sū.* 6: 97
67	*Kalayah*	*Pisum sativum* L.	*Sū.* 6: 18
68	*Kalhāram*	*Nymphaea rubra* Roxb. ex Salisb.	*Sū.* 22: 20
69	*Kāliṅgam*	*Citrullus lanatus* (Thunb.) Matsum. & Nakai	*Sū.* 6: 87
70	*Kaṅgu*	*Setaria italica* (L.) P. Beauvois	*Sū.* 6: 11
71	*Kaṅkōla*	*Piper cubeba* L. f.	*Ci.* 21: 78
72	*Kapittham*	*Feronia elephantum* Correa	*Sū.* 6: 126
73	*Karamardakam*	*Carissa carandas* L.	*Sū.* 6: 137
74	*Kāravēlam*	*Momordica charantia* L.	*Sū.* 6: 76
75	*Kāravi*	*Nigella sativa* L.	*Ci.* 8: 46
76	*Karīram*	*Capparis zeylanica* L.	*Sū.* 6: 77
77	*Karkāru*	*Cucurbita pepo* L.	*Sū.* 6: 87
78	*Karkaśam*	*Senna occidentalis* (L.) Link	*Sū.* 6: 78
79	*Karkōṭa*	*Momodica dioica* Roxb. ex Willd.	*Sū.* 6: 76
80	*Kārpāsi*	*Gossypium herbaceum* L.	*Ci.* 9: 24
81	*Karpūra*	*Cinnamomum camphora* (L.) J. Presl.	*Sū.* 3: 40
82	*Kāsamardam*	*Cassia occidentalis* L. var. *aristata* Collard	*Sū.* 6: 100
83	*Kēbukam*	*Costus speciosus* (J. Koenig) Sm.	*Sū.* 14: 26
84	*Kētaki*	*Pandanus odoratissimus* L. f.	*Ut.* 30: 39
85	*Kharjūra*	*Phoenix dactylifera* L.	*Sū.* 6: 119
86	*Kiṇihi*	*Careya arborea* Roxb.	*Ci.* 20: 26
87	*Kōdravam*	*Paspalum scrobiculatum* L.	*Sū.* 6: 11
88	*Kōvidāram*	*Bauhinia variegata* (L.) Benth.	*Ci.* 9: 96
89	*Krauñcādanam*	*Nymphaea pubescens* Willd.	*Sū.* 6: 92
90	*Kṣīraśukḷa*	*Pueraria tuberosa* (Willd.) DC.	*Ci.* 3: 155

(*Continued*)

TABLE 6.34 (*Continued*)
Other Medicinal Plants Mentioned in AH

Sl. No.	Sanskrit Name	Latin Binomial	Reference
91	*Kṣīriṇi*	*Euphorbia thymifolia* L.	*Ci.* 22: 6
92	*Kṣuraka*	*Vitex altissima* L. f.	*Ci.* 15: 95
93	*Kukkuṭi*	*Celosia argentea* L.	*Ut.* 5: 3
94	*Kulakam*	*Trichosanthes cucumerina* L.	*Sū.* 6: 77
95	*Kumbhayōna*	*Sesbania grandiflora* (L.) Poireti	*Ut.* 13: 90
96	*Kumuda*	*Nymphaea nouchali* Burm. f.	*Śa.* 2: 3
97	*Kuravakam*	*Barleria cristata* L.	*Sū.* 26: 13
98	*Kuruḍakam*	*Daemia extensa* (Jacq.) R. Br. ex Schult.	*Sū.* 6: 93
99	*Kūśmāṇḍam*	*Benincasa hispida* (Thunb.) Cogn.	*Sū.* 6: 87
100	*Kusumbham*	*Carthamus tinctorius* L.	*Sū.* 6: 24
101	*Kuṭhēram*	*Orthosiphon aristatus* (Blume) Miq.	*Sū.* 6: 106
102	*Kuṭiñjaram*	*Plectranthus rotundifolius* (Poir.) Spreng.	*Sū.* 6: 93
103	*Kuṭilī*	*Sesamum radiatum* Schumach. and Thonn.	*Sū.* 6: 76
104	*Lakṣmaṇa*	*Ipomoea sepiaria* Roxb.	*Sū.* 6: 94
105	*Lakṣmi*	*Nervilia aragoana* Gaudich.	*Sū.* 29: 31
106	*Laśūnam*	*Allium sativum* L.	*Sū.* 6: 109
107	*Lavaṅga*	*Syzygium aromaticum* (L.) Merr. & L.M. Perry	*Ci.* 8: 149
108	*Likucam*	*Artocarpus hirsutus* Lam.	*Sū.* 6: 137
109	*Lōṇikā*	*Portulacca oleracea* L.	*Sū.* 6: 93
110	*Mādhavi*	*Jasminum multiflorum* (Burm. f.) Andrews	*Sū.* 3: 34
111	*Madhūli*	*Chonemorpha fragrans* (Moon) Alston	*Ci.* 4: 34
112	*Madhuśigru*	*Moringa concanensis* Nimmo	*Ci.* 13: 10
113	*Mahādrōṇam*	*Anisomeles malabarica* (L.) R. Br. ex Sims.	*Ut.* 36: 92
114	*Mahāśrāvaṇi*	*Sphaeranthus africanus* L.	*Sū.* 10: 24
115	*Mālati*	*Jasminum grandiflorum* L.	*Sū.* 17: 8
116	*Mallikā*	*Jasminum sambac* (L.) Aiton	*Sū.* 3: 40
117	*Maṇḍūkaparṇi*	*Centella asiatica* (L.) Urban	*Sū.* 6: 76
118	*Mārkavam*	*Eclipta alba* Hassk.	*Ci.* 5: 36
119	*Mārṣam*	*Amaranthus blitum* L.	*Sū.* 6: 93
120	*Māṣam*	*Phaseolus mungo* L.	*Sū.* 6: 21
121	*Masūram*	*Lens culinaris* Medik.	*Sū.* 6: 17
122	*Mātuḷuṅgam*	*Citrus medica* L.	*Sū.* 6: 131
123	*Mēthika*	*Trigonella foenum-graecum* L.	*Ut.* 13: 87
124	*Mṛgaliṇḍikam*	*Baccaurea courtallensis* (Wight) Muell. Arg.	*Sū.* 6: 138
125	*Mudgam*	*Vigna radiata* (L.) R. Wilczek	*Sū.* 6: 17
126	*Mukūlakam*	*Schleichera oleosa* (Lour.) Oken	*Sū.* 6: 120
127	*Mūlakaṃ*	*Raphanus sativus* L.	*Sū.* 6: 102
128	*Muñjātam*	*Dioscorea esculenta* (Lour.) Burkill	*Sū.* 6: 83
129	*Muram*	*Selinum tenuifolium* Franch	*Ci.* 1: 138
130	*Nāgabala*	*Sida vernonicaefolia* Lam.	*Ci.* 3: 120
131	*Nāgapurīṣachatram*	*Psilocybe subcubensis* Guzman	*Ut.* 37: 42
132	*Nakham*	*Capparis sepiaria* L.	*Ci.* 1: 138
133	*Naḷikā*	*Ochlandra travancorica* (Bedd.) Benth. ex Gamble	*Ci.* 7: 27
134	*Nāḷikā*	*Ipomoea aquatica* Forssk.	*Sū.* 6: 93
135	*Nāḷīkaḷāyam*	*Alternanthera sessilis* (L.) R. Br. ex DC.	*Sū.* 6: 77

(*Continued*)

TABLE 6.34 (*Continued*)
Other Medicinal Plants Mentioned in AH

Sl. No.	Sanskrit Name	Latin Binomial	Reference
136	*Nāḷikēram*	*Cocos nucifera* L.	*Sū*. 5: 19
137	*Namaskāri*	*Mimosa pudica* L.	*Ut*. 27: 24
138	*Nandī*	*Tabernamontana divaricata* R. Br. ex Roem & Schult.	*Sū*. 6: 77
139	*Nikōcakam*	*Pistacia vera* L.	*Sū*. 6: 120
140	*Nīlini*	*Indigofera tinctoria* L.	*Ka*. 2: 21
141	*Niṣpāvam*	*Dolichos lablab* L.	*Sū*. 6: 20
142	*Nīvārā*	*Hygrorhiza aristata* Nees	*Sū*. 6: 11
143	*Padmakam*	*Prunus cerasoides* D. Don.	*Sū*. 21: 17
144	*Palāṇḍu*	*Allium cepa* L.	*Sū*. 6: 112
145	*Pālāṅkya*	*Amaranthus tricolor* L.	*Sū*. 6: 84
146	*Pālēvatam*	*Diospyros malabarica* (Desr.) Kostel	*Sū*. 6: 135
147	*Pāribhadram*	*Erythrina variegata* L.	*Ut*. 3: 45
148	*Parpaṭaḥ*	*Oldenlandia corymbosa* L.	*Sū*. 6: 76
149	*Pattaṅga*	*Caesalpinia sappan* L.	*Sū*. 27: 48
150	*Paṭōla*	*Trichosanthes anguina* L.	*Sū*. 3: 51
151	*Phalgu*	*Ficus hispida* L.f.	*Sū*. 6: 120
152	*Pītāṅgi*	*Coptis teeta* Wall.	*Ut*. 22: 98
153	*Plavaṅga*	*Ficus arnottiana* (Miq.) Miq.	*Ci*. 12: 18
154	*Prasāriṇi*	*Merremia tridentata* (L.) Hallier f.	*Ci*. 21: 65
155	*Puṣkaramūlam*	*Inula racemosa* Hook. f.	Ci. 3: 163
156	*Pūtimatsyakaḥ*	*Sterculia foetida* L.	*Ut*. 20: 16
157	*Rājādana*	*Manilkara hexandra* (Roxb.) Dubard	*Sū*. 6: 119
158	*Rājakṣavam*	*Brassica nigra* L.	*Sū*. 6: 72–73
159	*Rājamāṣaḥ*	*Vigna unguiculata* (L.) Walp.	*Sū*. 6: 19
160	*Ravivalli*	*Ventilago maderasaptana* Gaertn.	*Ut*. 1: 44
161	*Rōhītakam*	*Amoora rohitaka* (Roxb.) Wight & Arn.	*Ci*. 6: 53
162	*Śābaraka*	*Aporosa cardiosperma* (Gaertn. Merr.)	*Ut*. 16: 5
163	*Sahacaram*	*Nilgirianthus ciliatus* (Nees) Bremek	*Ka*. 4: 55
164	*Sahadēva*	*Vernonia cinerea* (L.) Less.	*Ut*. 1: 44
165	*Śailēyam*	*Parmelia perlata* (Huds.) Ach.	*Sū*. 21: 13
166	*Sairyakam*	*Barleria prionitis* L.	*Sū*. 1: 39
167	*Śaivalā*	*Ceratophyllum demersum* L.	*Ci*. 6: 48
168	*Śāli*	*Oryza sativa* L.	*Sū*. 5: 3
169	*Sallaki*	*Phoenix pusilla* Roxb.	*Ut*. 34: 3
170	*Śami*	*Prosopsis juliflora* (Sw.) DC.	*Ut*. 3: 59
171	*Śālūkam*	*Alocasia macrorrhizos* (L.) G. Don.	*Sū*. 6: 91
172	*Śaṇapuṣpi*	*Crotolaria retusa* L.	*Ka*. 1: 8
173	*Śaṅkhini*	*Euphorbia tirucalli* L.	*Ci*. 15: 13
174	*Sarpagandha*	*Aristolochia indica* L.	*Ut*. 5: 3
175	*Śatāhva*	*Anethum graveolens* L.	*Sū*. 17: 2
176	*Śēphālika*	*Nyctanthes arbor-tristis* L.	*Ut*. 9: 5
177	*Sinduvārita*	*Vitex negundo* L.	*Sū*. 7: 25
178	*Śḷēṣmātaka*	*Cordia dichotoma* G. Forst.	*Sū*. 6: 120
179	*Śṛṅgāṭakam*	*Trapa bispinosa* Roxb.	*Sū*. 6: 92
180	*Śṛṅgi*	*Aconitum chasmanthum* Stapf. ex Holmes	*Ut*. 35: 4

(*Continued*)

TABLE 6.34 (*Continued*)
Other Medicinal Plants Mentioned in AH

Sl. No.	Sanskrit Name	Latin Binomial	Reference
181	*Sthauṇēyaka*	*Clerodendrum infortunatum* L.	*Sū*. 21: 13
182	*Sudha*	*Euphorbia neriifolia* L.	*Ka*. 2: 42
183	*Sumukham*	*Ocimum canum* Sims.	*Sū*. 6: 106
184	*Suniṣaṇṇakam*	*Marsilea quadrifolia* L.	*Sū*. 6: 72
185	*Sūraṇam*	*Amorphophallus paeoniifolius* (Dennst.) Nicolson	*Sū*. 6: 113
186	*Sūṣā*	*Cassia tora* L.	*Sū*. 6: 72
187	*Suvarṇam*	*Datura metel* L.	*Sū*. 21: 17
188	*Śvētapatram*	*Mussaenda frondosa* L.	*Ut*. 5: 35
189	*Śyāmākam*	*Panicum miliare* Auct. non Lam.	*Sū*. 6: 11
190	*Tālam*	*Borassus flabellifer* L.	*Sū*. 3: 33
191	*Tālapatri*	*Curculigo orchioides* Gaertn.	*Sū*. 30: 22
192	*Tālīsapatram*	*Abies webbiana* (Wall. ex D. Don.) Lindl.	*Ci*. 4: 54
193	*Tāmalaki*	*Phyllanthus amarus* Schumm. & Thonn.	*Ci*. 1: 90
194	*Tāmbūlam*	*Piper betle* L.	*Ut*. 40: 45
195	*Taṇḍulīyaḥ*	*Amaranthus spinosus* L.	*Ci*. 2: 27
196	*Tiktōttama*	*Trichosanthes dioica* Roxb.	*Ci*. 1: 24
197	*Tilaḥ*	*Sesamum indicum* L.	*Sū*. 6: 23
198	*Tilaparṇika*	*Ziziphus oenoplia* (L.) Mill.	*Sū*. 6: 76
199	*Trāyanti*	*Bacopa monnieri* (L.) Pennell	*Sū*. 10: 28
200	*Tugā*	*Maranta arundinacea* L.	*Ci*. 4: 34
201	*Tumbam*	*Lagenaria siceraria* (Molina) Standl.	*Sū*. 6: 87
202	*Tumburūṇi*	*Zanthoxylum armatum* DC.	*Ci*. 8: 50
203	*Udumbaram*	*Ficus glomerata* Roxb.	*Ci*. 6: 69
204	*Uma/Kṣuma*	*Linum usitatissimum* L.	*Sū*. 6: 24
205	*Upakuñjika*	*Physalis angulata* L.	*Ut*. 20: 5
206	*Upōdakā*	*Basella alba* L.	*Sū*. 6: 84
207	*Ūrumāṇam*	*Musa sapientum* L.	*Sū*. 6: 121
208	*Utpala*	*Monochoria vaginalis* (Burm. f.) C. Presl. ex Kunth	*Ut*. 39: 49
209	*Vamśam*	*Bambusa bambos* (L.) Voss	*Sū*. 6: 99
210	*Vanasūraṇam*	*Amorphophallus commutatus* L.	*Ci*. 8: 156
211	*Vārāhi*	*Dioscorea bulbifera* L.	*Ci*. 18: 11
212	*Varṣā*	*Trianthema portulacastrum* L.	*Sū*. 6: 97
213	*Vārtākam*	*Solanum melongena* L.	*Sū*. 6: 77
214	*Vārtākini*	*Solanum indicum* L.	*Ci*. 1: 75
215	*Vasukaḥ*	*Borreria hispida* Spruce ex. K. Schum.	*Ci*. 11: 18
216	*Vāstukam*	*Chenopodium murale* L.	*Sū*. 6: 72
217	*Vātāmam*	*Prunus dulcis* (Mill.)	*Sū*. 6: 120
218	*Vētāgram*	*Calamus rotang* L.	*Sū*. 6: 76
219	*Vīra*	*Lasia spinosa* (L.) Thw.	*Sū*. 10: 23
220	*Viṣa*	*Aconitum ferox* Wall. ex. Ser.	*Ci*. 19: 64
221	*Viṣamūṣika*	*Melia azedarach* (L.)	*Ut*. 36: 91
222	*Vṛkṣādani*	*Dolichandrone falcata* Seem.	*Śa*. 2: 55
223	*Vṛkṣāmḷam*	*Garcinia gummi-gutta* (L.) N. Robson	*Sū*. 6: 129
224	*Yavam*	*Hordeum vulgare* L.	*Sū*. 6: 15
225	*Yavāni*	*Trachyspermum ammi* (L.) Sprague ex Turrill	*Sū*. 14: 25
226	*Yūthika*	*Jasminum coarctatum* Roxb.	*Ci*. 9: 24

TABLE 6.35
Single Herb Remedies Recommended in AH

Sl. No.	Name of Herb/Drug	Mode of Administration and Indications	Reference
1	*Abrus precatorius*	Application of paste of root and seed on scalp, after multiple puncturing, induces hair growth in alopecia.	*Ut*. 24: 29
2	*Abutilon indicum*	Paste of roots mixed with honey and clarified butter is to be consumed for one month according to *rasāyana vidhi*. The treatment improves wisdom, intellect, strength, stability and longevity.	*Ut*. 39: 60–61, 104–105
3	*Acacia catechu*	*Ghṛta* prepared with heartwood and milk cures leprosy and other skin diseases with predominance of *vāta*.	*Ci*. 19: 39
		Keeping inside mouth a reasonably large volume of decoction of heartwood pacifies inflammation of mouth and sinus.	*Ut*. 22: 106
4	*Acacia leucophloea*	Keeping inside mouth a reasonably large volume of decoction of bark pacifies inflammation of mouth and sinus.	*Ut*. 22: 106
5	*Acacia sinuata*	Consumption of *ghrta* prepared with decoction and paste of bark cures *vāta* diseases.	*Ci*. 21: 10
6	*Achyranthus aspera*	Paste of the root of herb with white stem, suspended in water and administered orally on *Puṣya* star in the first month of pregnancy, causes the birth of a boy child.	*Śa*. 1: 39–40
		Paste of seeds suspended in rice washing cures bleeding piles.	*Ci*. 8: 104
		Powder of this herb belonging to *Surasādi gaṇa* is to be mixed with honey and consumed for pacifying worm infestation in head.	*Ci*. 20: 27
7	*Acorus calamus*	Consumption of root powder with honey pacifies epilepsy.	*Ut*. 7: 34
		Powder of the root mixed with milk, sesame oil or clarified butter is to be consumed for one month according to *rasāyana vidhi*. He who undergoes this treatment will live long with strength and oratory skill.	*Ut*. 39: 163
8	*Actinopteris dichotoma*	Consumption of paste of roots suspended in rice-washing breaks urinary stones.	*Ci*. 11: 34
9	*Adhatoda vasica*	*Ghṛta* prepared with juice of leaves and paste of flowers cures fever with predominance of *pitta*, irregular fever and *gulma* due to enragement of *pitta*.	*Ci*. 1: 92, 94, 156 *Ci*. 14: 62
		Ghṛta prepared with decoction of stem, root and paste of flowers, administered with honey, cures hemorrhages of obscure origin, fever, cough, cataract and several other diseases.	*Ci*. 2: 42–44
		Juice of leaves mixed with sugar and honey, the juice alone or decoction of root, cures hemorrhages of obscure origin.	*Ci*. 2: 25–26
		Keeping inside mouth a reasonably large volume of decoction of roots pacifies inflammation of mouth and sinus.	*Ut*. 22: 106
10	*Aegle marmelos*	Milk boiled with immature fruits cures chronic fever.	*Ci*. 1: 112–113
		Consumption of unripe fruit cures dysentery.	*Ci*. 9: 108–109
		Snēhavasti (medicinal enema using oleaginous substances) using *taila* prepared with root or fruit cures dysentery with *kapha–vāta* enragement.	*Ci*. 9: 119–120
		Consumption of *taila* prepared with ash derived from aerial parts and root pacifies various kinds of pain experienced in enlargement of abdomen.	*Ci*. 15: 45
		Consumption of fish cooked with immature fruits pacifies *vāta* enraged in the navel area.	*Ci*. 21: 15

(*Continued*)

TABLE 6.35 (*Continued*)
Single Herb Remedies Recommended in AH

Sl. No.	Name of Herb/Drug	Mode of Administration and Indications	Reference
		Juice obtained from leaves subjected to *puṭapāka* should be warmed and poured into auditory canal to pacify earache due to *vāta* enragement.	*Ut*. 18: 1–2
11	*Ageratum conyzoides*	Powder of this herb belonging to *Surasādi gaṇa* is to be mixed with honey and consumed for pacifying worm infestation in head.	*Ci*. 20: 27
12	*Alangium salvifolium*	Decoction of root is administered to induce emesis in treatment of rat bite.	*Ut*. 38: 21
		Topical application of paste of root or its consumption as a suspension in goat urine pacifies effects of stings and bites of all venomous creatures.	*Ut*. 38: 29
		Decoction of root mixed with clarified butter is an antidote for bite of rabid dog.	*Ut*. 38: 36
13	*Albizia lebbeck*	Topical application of paste of bark cures leprosy and other skin diseases.	*Ci*. 19: 63
		Consumption of juice of leaves, mixed with honey cures worm infestation in head.	*Ci*. 20: 26–27
		Decoction of root is administered to induce emesis in treatment of rat bite.	*Ut*. 38: 21
		Nasya with decoction of heartwood and fruits pacifies effects of rat bite.	*Ut*. 38: 24
14	*Allium cepa*	*Nasya* using juice of bulbs, mixed with breast milk, cures respiratory distress and hiccup.	*Ci*. 4: 47–48
		Regular consumption of the bulbs along with thin gruel, vegetable soup and meat soup pacifies bleeding piles.	*Ci*. 8: 121
15	*Allium cepa* var. *aggregatum*	*Nasya* using juice of bulbs, mixed with breast milk, cures respiratory distress and hiccup.	*Ci*. 4: 47–48
		Consumption of soup of bulbs acidified with lemon juice and seasoned with *yamaka snēha* (combination of two oleaginous substances) cures dysentery.	*Ci*. 9: 32
16	*Allium sativum*	Consumption of paste of cloves, mixed with sesame oil in the early morning, cures irregular fever.	*Ci*. 1: 155
		Nasya using juice of garlic cloves, mixed with breast milk, cures respiratory distress and hiccup.	*Ci*. 4: 47–48
		Oral administration of juice of bulbs pacifies coma (*sannyāsa*).	*Ci*. 7: 112
		Consumption of milk boiled with cloves of garlic cures *gulma* due to enragement of *vāta*, *udāvarta*, sciatica, irregular fever, heart disease and abscess.	*Ci*. 14: 45–47
		Consumption of cloves according to *rasāyana vidhi* cures envelopment of *vāta*.	*Ci*. 22: 70–71
		Consumption of paste of garlic cloves mixed with sesame oil, pacifies epilepsy.	*Ut*. 7: 34
		Filling auditory canal with warmed juice pacifies pricking pain in ear due to enragement of *kapha*.	*Ut*. 18: 12
		Cloves of garlic are to be consumed as a *rasāyana*, according to *rasāyana vidhi*.	*Ut*. 39: 110–129
17	*Alpinia galanga*	Paste of rhizomes mixed with honey and clarified butter is to be consumed for one month according to *rasāyana vidhi*. The treatment improves wisdom, intellect, strength, stability and longevity.	*Ut*. 39: 60–61

(*Continued*)

TABLE 6.35 (*Continued*)
Single Herb Remedies Recommended in AH

Sl. No.	Name of Herb/Drug	Mode of Administration and Indications	Reference
18	*Alternanthera sessilis*	Consumption of juice of whole herb, mixed with honey, cures worm infestation in head.	*Ci*. 20: 26–27
19	*Amaranthus spinosa*	Consumption of *ghṛta* prepared with juice of the herb pacifies effects of rat bite.	*Ut*. 38: 25
20	*Amorphophallus commutatus*	The yam is to be covered with fine mud and dried. Thereafter, it is processed according to the directions given in *puṭapāka vidhi*. Consumption of the cooked yam with sesame oil and rock salt cures hemorrhoids.	*Ci*. 8: 156
21	*Anethum graveolens*	Topical application of seeds ground to paste with milk, pacifies pain associated with *vāta*-induced blood disorder having predominance of *vāta*.	*Ci*. 22: 34
22	*Anogeissus latifolia*	Paste of roots is to be smeared inside an iron pot and left overnight. It is washed in the next morning with water and the suspension consumed along with food or beverages. He who follows this instruction lives for hundred years free from diseases.	*Ut*. 39: 104–105
23	*Aquilaria agallocha*	Powder of heartwood administered with honey cures cough with predominance of *kapha*.	*Ci*. 3: 48
		Paste of heartwood is to be smeared inside an iron pot and left overnight. It is washed in the next morning with water, and the suspension is consumed along with food or beverages. He who follows this instruction lives for hundred years free from diseases.	*Ut*. 39: 104–105
24	*Asparagus racemosus*	Consumption of paste of root suspended in milk or *ghṛta* prepared with the root, followed by milk porridge cures dysentery with bleeding.	*Ci*. 9: 89, 99
		Consumption of powder or paste of roots suspended in milk pacifies epilepsy.	*Ut*. 7: 34
		Consumption of tender leaves sautéed in clarified butter cures night blindness (*dōṣāndha*).	*Ut*. 13: 89–90
		Paste of roots mixed with honey and clarified butter is to be consumed for one month according to *rasāyana vidhi*. The treatment improves wisdom, intellect, strength, stability and longevity.	*Ut*. 39: 60–61, 156
25	*Azadirachta indica*	Decoction or juice of bark mixed with honey cures jaundice.	*Ci*. 16: 43–44
		Ghṛta prepared with bark and milk cures leprosy and other skin diseases with predominance of *vāta*.	*Ci*. 18: 39
		Consumption of juice of leaves, mixed with honey, cures worm infestation in head.	*Ci*. 20: 26–27
		Keeping inside mouth a reasonably large volume of decoction of bark pacifies inflammation of mouth and sinus.	*Ut*. 22: 106
		Nasya with neem oil for one month induces the growth of hair in alopecia. The patient should consume only rice cooked in milk, remaining celibate during the period.	*Ut*. 24: 34
		Paste of heartwood is to be smeared inside an iron pot and left overnight. It is washed in the next morning with water and the suspension consumed along with food or beverages. He who follows this instruction lives for hundred years free from diseases.	*Ut*. 39: 104–105
26	*Bacopa monnieri*	*Ghṛta* prepared with juice and paste of the herb cures fever with predominance of *pitta*.	*Ci*. 1: 92
		Ghṛta prepared with decoction and paste of herb and administered with honey cures hemorrhages of obscure origin.	*Ci*. 2: 44–45

(*Continued*)

TABLE 6.35 (*Continued*)
Single Herb Remedies Recommended in AH

Sl. No.	Name of Herb/Drug	Mode of Administration and Indications	Reference
		Decoction of the herb is to be consumed, followed by drinking of lukewarm milk cures dysentery with bleeding.	*Ci*. 9: 70
		Consumption of juice of herb pacifies epilepsy.	*Ut*. 7: 34
		Ghṛta prepared with juice and paste of herb is to be administered to infants, for eliminating *dōṣa*s associated with *kukūṇaka*.	*Ut*. 9: 31
27	*Baliospermum montanum*	Paste of root suspended in cold water and mixed with jaggery cures morbid pallor.	*Ci*. 16: 42
28	*Bambusa bambos*	*Taila* prepared with outer skin of stem, goat urine and sheep urine is poured into auditory canal to pacify pricking pain in ear due to enragement of *kapha*.	*Ut*. 18: 15
29	*Barleria prionitis*	Paste of the root, suspended in water and administered orally on *Puṣya* star in the first month of pregnancy, causes the birth of a boy child.	*Śa*. 1: 39–40
		Consumption of paste of root suspended in rice-washing and mixed with honey pacifies effects of rat bite.	*Ut*. 38: 31
30	*Barleria strigosa*	Consuming post-prandially the *ghṛta* prepared with roots cures cough and lassitude of voice experienced by patients of kingly consumption.	*Ci*. 5: 36
31	*Bauhinia variegata*	Root bark powder suspended in buttermilk cures hemorrhoids.	*Ci*. 8: 31–32
32	*Berberis aristata*	Consumption of cow urine mixed with paste of bark cures hydrocele due to enragement of *kapha*.	*Ci*. 13: 33
		Decoction of bark mixed with honey cures jaundice.	*Ci*. 16: 43–44
		Application of concentrated aqueous extract of the herb (*rasāñjana*) mixed with honey pacifies the eye disease *sirāharṣa*.	*Ut*. 11: 11
		Pouring in a thin stream concentrated decoction into the eyes, mixed with honey, pacifies conjunctivitis due to *tridōṣa*.	*Ut*. 16: 8
		Concentrated decoction of bark, mixed with red ochre and honey, is to be held in mouth to pacify inflammation of mouth and sinus.	*Ut*. 22: 105
33	*Boerhaavia diffusa*	Milk boiled with the root breaks urinary stones.	*Ci*. 11: 33
		Paste of fresh root suspended in milk is to be consumed for fourteen days, two months, six months or one year according to *rasāyana vidhi*.	*Ut*. 39: 154
34	*Borassus flabellifer*	Consumption of *ghṛta* prepared with the fruit kernel pacifies voiding of discolored urine or obstructed urination experienced by patients having cough.	*Ci*. 3: 155
35	*Brassica alba*	Paste of seeds is to be applied on body during bath to improve skin health of patients of kingly consumption.	*Ci*. 5: 81
		Topical application of paste of seeds on scalp pacifies common cold due to enragement of *kapha*.	*Ut*. 20: 13
36	*Brassica juncea*	Consumption of mustard oil pacifies all types of skin diseases.	*Ci*. 19: 12
		Powder of mustard seeds is to be blown into the nasal passage to pacify insanity.	*Ut*. 6: 41
		Applying mustard oil all over body pacifies insanity.	*Ut*. 6: 41
		Filling auditory canal with mustard oil pacifies tinnitus, deafness, *pūtikarṇa* and *kṛmikarṇa*.	*Ut*. 18: 26, 35
37	*Butea monosperma*	Decoction of bark mixed with sugar cures hemorrhages of obscure origin.	*Ci*. 2: 29
		Ghṛta prepared with juice and paste of pedunculus of flowers, administered with honey, cures hemorrhages of obscure origin.	*Ci*. 2: 44–45

(*Continued*)

TABLE 6.35 (*Continued*)
Single Herb Remedies Recommended in AH

Sl. No.	Name of Herb/Drug	Mode of Administration and Indications	Reference
		Milk is to be filled in an earthen pot, the inside of which is smeared with paste of the root. Consumption of buttermilk prepared from this curd cures hemorrhoids.	*Ci.* 8: 49
		Consumption of decoction of fruits mixed with or without milk, followed by drinking of lukewarm milk, cures dysentery with bleeding.	*Ci.* 9: 68–69
		Consumption of *ghṛta* prepared with aerial parts and root cures hemorrhages of obscure origin.	*Ci.* 14: 122–123
		Ash obtained from aerial parts and roots is to be suspended in water and mixed with thickened sugarcane juice. Consumption of this liquid cures vitiligo.	*Ci.* 20: 5
		Consumption of juice of seeds, mixed with honey, cures worm infestation in head.	*Ci.* 20: 26–27
38	*Cajanus scarabaeoides*	The seeds are packed in a small cloth bag and steamed with juice of cow dung. Thereafter, the seed coats are removed and the seeds powdered. Dusting with this powder in the eye once in the middle of the night is recommended to cure conjunctivitis.	*Ut.* 16: 6
39	*Callicarpa macrophylla*	Paste of flowers mixed with honey and rice-washing is to be administered to cure dysentery with bleeding.	*Ci.* 9: 92
40	*Calotropis gigantea*	Juice obtained from leaves subjected to *puṭapāka* should be warmed and poured into auditory canal to pacify earache due to *vāta* enragement.	*Ut.* 18: 1–2
41	*Careya arborea*	Consumption of juice of leaves, mixed with honey, cures worm infestation in head.	*Ci.* 20: 26–27
		Ghṛta prepared with decoction and paste of roots pacifies effects of rat bite.	*Ut.* 38: 26
42	*Cassia fistula*	Consumption of *ghṛta* prepared by repeating 100 times the addition of root paste and milk cures leprosy and other skin diseases fast.	*Ci.* 19: 13–14
		Topical application of paste of leaves cures leprosy and other skin diseases.	*Ci.* 19: 63
43	*Cassia sophera*	Powder of this herb belonging to *Surasādi gaṇa* is to be mixed with honey and consumed for pacifying worm infestation in head.	*Ci.* 20: 27
44	*Cassia tora*	Seeds are to be ground to a paste with lemon juice and applied over forehead, to pacify hemicrania.	*Ut.* 24: 10
45	*Cedrus deodara*	Decoction of heartwood cures respiratory distress and hiccup.	*Ci.* 4: 28–29
		Ghṛta prepared with heartwood and milk cures leprosy and other skin diseases with predominance of *vāta*.	*Ci.* 19: 39
		Powder of heartwood is to be subjected many times to *bhāvana* with goat urine. Application of the paste in clarified butter, as a collyrium cures *pillarōga*.	*Ut.* 16: 54
		Filling auditory canal with oil extracted from heartwood pacifies earache due to *vāta* enragement.	*Ut.* 18: 5
		Paste of heartwood is to be smeared inside an iron pot and left overnight. It is washed in the next morning with water and the suspension consumed along with food or beverages. He who follows this instruction lives for hundred years free from diseases.	*Ut.* 39: 104–105
46	*Celosia argentea*	Consumption of fine powder of seeds, suspended in buttermilk, cures dysuria due to enragement of *kapha*.	*Ci.* 11: 11

(*Continued*)

TABLE 6.35 (*Continued*)
Single Herb Remedies Recommended in AH

Sl. No.	Name of Herb/Drug	Mode of Administration and Indications	Reference
47	*Centella asiatica*	6 g of herb powder are to be boiled with milk and consumed on the first day. 12 g of the powder are to be consumed on the next day. Thus everyday 6 g more of powder are to be consumed till the quantity of powder reaches 48 g in eight days. Thereafter, 48 g of powder are consumed everyday for one month. During the treatment period, the patient should consume only milk and no solid food. Treatment cures injury to thorax and improves stamina, complexion and longevity.	*Ci.* 3: 118–120
		Paste of root suspended in wine or milk boiled with roots breaks urinary stones.	*Ci.* 11: 32–33
		After pacifying enlargement of abdomen caused by *tridōṣa* using vegetable and inanimate poisons, the patient should eat for one month leaves cooked in the leaf juice, without adding salt, sour ingredients and oil. No other food should be served during this period. Leaf juice can be consumed for quenching thirst.	*Ci.* 15: 82–83
		The juice is to be consumed as *rasāyana*, according to *rasāyana vidhi*.	*Ut.* 39: 44–45, 60–61, 164
48	*Chenopodium murale*	After pacifying enlargement of abdomen caused by *tridōṣa* using vegetable and inanimate poisons, the patient should eat for one month leaves cooked in the leaf juice, without adding salt, sour ingredients and oil. No other food should be served during this period. Leaf juice can be consumed for quenching thirst.	*Ci.* 15: 82 83
49	*Citrus medica*	Carpels of the fruit ground to a paste with rock salt, clarified butter and kept in mouth, cures distaste for food, associated with fever.	*Ci.* 1: 128
		Consumption of *ghṛta* prepared with fruit cures *grahaṇi*.	*Ci.* 10: 30
		Filling auditory canal with fruit juice pacifies edema, suppuration and pricking pain in ear due to enragement of *kapha*.	*Ut.* 18: 12, 14
50	*Cleome viscosa*	Paste of whole herb suspended in water cures irregular fever.	*Ci.* 1: 160
51	*Clerodendrum serratum*	Milk is to be filled in an earthen pot, the inside of which is smeared with paste of the root. Consumption of buttermilk prepared from this curd cures hemorrhoids.	*Ci.* 8: 49
		Powder of this herb belonging to *Surasādi gaṇa* is to be mixed with honey and consumed for pacifying worm infestation in head.	*Ci.* 20: 27
52	*Clitoria ternatea*	Paste of the root and flower is to be consumed as *rasāyana* according to *rasāyana vidhi*, to improve, health, intelligence and longevity.	*Ut.* 39: 44–45, 60–61
53	*Coleus vettiveroides*	Cooled decoction of roots is to be administered in alcoholism due to enragement of *kapha*.	*Ci.* 7: 34
		Topical application of ash of whole herb mixed with oil of *Terminaliia belerica* pacifies vitiligo.	*Ci.* 20: 12
54	*Commiphora mukul*	Gum resin is to be burnt, and the smoke is inhaled through nostril for the cure of respiratory distress and hiccup.	*Ci.* 4: 13
		Consumption of gum resin suspended in cow urine cures *gulma* due to enragement of *kapha* and stiffness of thighs.	*Ci.* 14: 99 *Ci.* 21: 49
		Oral administration of purified gum resin is recommended to cure envelopment of *vāta* by all *dhātus*.	*Ci.* 22: 65
55	*Corchorus capsularis*	Topical application of seeds ground with water heals sinuses.	*Ut.* 30: 37
56	*Cordyalis govaniana*	Powder of this herb belonging to *Surasādi gaṇa* is to be mixed with honey and consumed for pacifying worm infestation in head.	*Ci.* 20: 27

(*Continued*)

TABLE 6.35 (*Continued*)
Single Herb Remedies Recommended in AH

Sl. No.	Name of Herb/Drug	Mode of Administration and Indications	Reference
57	*Coriandrum sativum*	Drinking of water boiled with the seeds, along with sugar and honey, cures polydipsia that is secondary to a major disease (*upasargaja tṛṣṇa*).	*Ci.* 6: 82
58	*Costus speciosus*	Consumption of juice of leaves, mixed with honey, cures worm infestation in head.	*Ci.* 20: 26–27
59	*Curculigo orchioides*	Root to be ground in goat milk, mixed with honey and applied topically to cure black discoloration on face. Alternatively, powder of root and bone of cow may be mixed with honey and clarified butter and applied.	*Ut.* 32: 21
		Paste of root is to be suspended in milk and consumed for one month according to *rasāyana vidhi*. The treatment retards aging.	*Ut.* 39: 58–59
60	*Curcuma longa*	Decoction of rhizome is to be administered with honey and sugar to pacify polydipsia due to enragement of *kapha*.	*Ci.* 6: 73
		Ghṛta prepared with rhizomes and milk cures leprosy and other skin diseases with predominance of *vāta*.	*Ci.* 19: 39
		Paste of rhizome is to be smeared inside an iron pot and left overnight. It is washed in the next morning with water, and the suspension is consumed along with food or beverages. He who follows this instruction lives for hundred years free from diseases.	*Ut.* 39: 104–105
61	*Cyclea peltata*	Consumption of powder of dried roots mixed with jaggery or paste of roots suspended in buttermilk cures hemorrhoids.	*Ci.* 8: 54–55, 61
62	*Cynodon dactylon*	*Nasya* with juice of herb, mixed with or without milk, cures hemorrhages of obscure origin.	*Ci.* 2: 48–49
		Oral and topical administration of *ghṛta* prepared with juice cures *kardama visarpa* with predominance of *kapha*.	*Ci.* 18: 36
63	*Cyperus rotundus*	Powders of tubers mixed with honey is to be administered in stupor (*mada*).	*Ci.* 7: 105–106
		20 tubers are to be boiled in milk diluted with water. Consumption of this milk cures dysentery with pain.	*Ci.* 9: 39–40
		Consumption of decoction of tubers, mixed with honey, cures dysentery with predominance of *pitta*.	*Ci.* 9: 61
		Pouring of cooled decoction of tubers in a thin stream pacifies *agnivisarpa*.	*Ci.* 18: 22
		Paste of tubers is to be smeared inside an iron pot and left overnight. It is washed in the next morning with water, and the suspension is consumed along with food or beverages. He who follows this instruction lives for hundred years free from diseases.	*Ut.* 39: 104–105
64	*Dalbergia sissoo*	Milk boiled with heartwood cures fever.	*Ci.* 1: 115
65	*Datura metel*	Application of leaf juice on scalp, after multiple puncturing, induces hair growth in alopecia.	*Ut.* 24: 30
66	*Desmodium gangeticum*	Consumption of milk boiled with roots pacifies *vāta* enraged in heart.	*Ci.* 21: 17
		Milk boiled with the root is to be sweetened and fed to the newborn, if the mother does not have sufficient breast milk.	*Ut.* 1: 20
		Paste of roots mixed with honey and clarified butter is to be consumed for one month according to *rasāyana vidhi*. The treatment improves wisdom, intellect, strength, stability and longevity.	*Ut.* 39: 60–61
67	*Desmodium triflorum*	*Nasya* using *taila* prepared with roots cures lassitude of voice experienced in kingly consumption	*Ci.* 5: 38

(*Continued*)

TABLE 6.35 (*Continued*)
Single Herb Remedies Recommended in AH

Sl. No.	Name of Herb/Drug	Mode of Administration and Indications	Reference
68	*Diospyros malabarica*	Paste of fruits ground with leaf juice removes discoloration of skin.	*Ut*. 32: 22
69	*Eclipta alba*	Juice of the herb mixed with honey cures cough with predominance of *kapha.*	*Ci*. 3: 48–49
		Consumption of cakes prepared with tender leaves and paste of soaked rice grains pacifies worm infestation.	*Ci*. 20: 30
		Juice of the herb is to be consumed for one month according to *rasāyana vidhi*, with milk alone as food. He who undergoes this treatment will live for hundred years with strength and valor.	*Ut*. 39: 162
70	*Elettaria cardamomum*	Powder of seeds suspended in wine cures dysuria due to enragement of *kapha*.	*Ci*. 11: 10
71	*Embelia ribes*	Performing *snēhavasti* using *taila* prepared with seeds pacifies worm infestation.	*Ci*. 20: 23
		Consumption of cakes prepared with seeds and paste of soaked rice grains pacifies worm infestation.	*Ci*. 20: 31
		Paste of seeds is to be smeared inside an iron pot and left overnight. It is washed in the next morning with water, and the suspension is consumed along with food or beverages. He who follows this instruction lives for hundred years free from diseases.	*Ut*. 39: 104–105
72	*Embelia tsjeriam-cottam*	Powder of this herb belonging to *Surasādi gaṇa* is to be mixed with honey and consumed for pacifying worm infestation in head.	*Ci*. 20: 27
73	*Emblica officinalis*	Powder of pericarp is to be boiled in milk, cooled, mixed with honey and consumed for curing *kṣata kāsa* (cough due to trauma to thorax).	*Ci*. 3: 79
		Ghṛta prepared with juice of berries is to be administered in *mada.*	*Ci*. 7: 107
		Juice of berries mixed with honey cures polyuria.	*Ci*. 12: 7
		Consumption of juice of berries pacifies morbid pallor and jaundice.	*Ci*. 16: 32
		Ash of pericarps is to be suspended in water and filtered. Consumption of the concentrated filtrate followed by eating of cooked millet (*Paspalum scrobiculatum*) cures swelling in neck due to enragement of *kapha*.	*Ut*. 22: 70
		Consumption of mixture of decoction and paste of pericarp cures lymphangitis.	*Ut*. 32: 6
		The berries are to be consumed as *rasāyana*, according to *rasāyana vidhi* for improving health and longevity.	*Ut*. 39: 28–32, 60–61, 148
74	*Euphorbia neriifolia*	*Ghṛta* prepared with milk curdled using latex of the herb cures enlargement of abdomen.	*Ci*. 15: 31–32
75	*Feronia elephantum*	Filling auditory canal with fruit juice pacifies pricking pain in ear due to enragement of *kapha.*	*Ut*. 18: 14
		Consumption of juice of fruit with honey and juice of cow dung pacifies effects of rat bite.	*Ut*. 38: 25
		Ghṛta prepared with root, bark, leaf, flower and fruit pacifies effects of rat bite.	*Ut*. 38: 26
		Juice of fruit mixed with honey pacifies respiratory distress and hiccup.	*Ci*. 4: 40
76	*Ferula foetida*	Consumption of mixture of paste and decoction of the herb cures respiratory distress and hiccup.	*Ci*. 4: 32

(*Continued*)

TABLE 6.35 (*Continued*)
Single Herb Remedies Recommended in AH

Sl. No.	Name of Herb/Drug	Mode of Administration and Indications	Reference
77	*Ficus benghalensis*	Paste of eight apical buds of the tree suspended in milk and administered through mouth or nostrils in the first month of pregnancy causes the birth of a boy child.	*Śā*. 1: 41–42
		Milk boiled with apical buds or powder of adventitious roots cures hemorrhages of obscure origin.	*Ci*. 2: 40
78	*Ficus glomerata*	Juice of the fruit or its decoction is to be administered along with sugar to pacify polydipsia due to enragement of *pitta.*	*Ci*. 6: 69
79	*Ficus religiosa*	Juice obtained from leaves subjected to *puṭapāka* should be warmed and poured into auditory canal to pacify earache due to *vāta* enragement.	*Ut*. 18: 1–2
80	*Fritillaria roylei*	Paste of roots mixed with honey and clarified butter is to be consumed for one month according to *rasāyana vidhi*. The treatment improves wisdom, intellect, strength, stability and longevity.	*Ut*. 39: 60–61
81	*Geophila repens*	Powder of this herb belonging to *Surasādi gaṇa* is to be mixed with honey and consumed for pacifying worm infestation in head.	*Ci*. 20: 27
82	*Gloriosa superba*	Application of paste of rhizome on scalp, after multiple puncturing, induces hair growth in alopecia.	*Ut*. 24: 29
83	*Glycyrrhiza glabra*	6 g of root powder are to be boiled with milk and consumed on the first day. 12 g of the powder are to be consumed on the next day. Thus everyday 6 g more of powder are to be consumed till the quantity of powder reaches 48 g in eight days. Thereafter, 48 g of powder are consumed everyday for one month. During the treatment period, the patient should consume only milk and no solid food. Treatment cures injury to thorax and improves stamina, complexion and longevity.	*Ci*. 3: 118–120
		Taila prepared with sesame oil, milk and paste of roots can be consumed or administered as medicinal enema, to cure heart disease with *vāta* predominance.	*Ci*. 6: 38–39, 49
		Snēhavasti using *taila* prepared with decoction and root of paste cures hydrocele due to enragement of *vāta.*	*Ci*. 13: 31
		Pouring of cooled decoction of roots in a thin stream pacifies *agnivisarpa.*	*Ci*. 18: 21
		Ghṛta prepared with the roots is to be administered orally three times to the newborn infant on second and third day of birth.	*Ut*. 1: 13
		Applying in eyes powder of root, mixed with visceral fat of vulture, snake or rooster, cures cataract.	*Ut*. 13: 56
		Topical application of root powder mixed with sesame oil pacifies inflammation of lips due to enragement of *vāta*.	*Ut*. 22: 4
		Keeping inside mouth a reasonably large volume of paste of root powder mixed with sesame oil or honey is recommended after extraction of tooth.	*Ut*. 22: 24
		Keeping inside mouth a reasonably large volume of decoction of roots pacifies inflammation of mouth and sinus.	*Ut*. 22: 106
		Ghṛta prepared with roots may be poured in a thin stream to pacify pain in traumatic wounds.	*Ut*. 26: 6
		Uttaravasti with mixture of milk, decoction and paste of roots cures *pittaduṣṭayōni*.	*Ut*. 34: 60
		The powder of root mixed with milk is to be consumed as *rasāyana*, according to *rasāyana vidhi* for improving health, intelligence and longevity. It is an aphrodisiac as well.	*Ut*. 39: 44–45, 60–61, 104–105 *Ut*. 40: 28–29

(*Continued*)

TABLE 6.35 (*Continued*)
Single Herb Remedies Recommended in AH

Sl. No.	Name of Herb/Drug	Mode of Administration and Indications	Reference
84	*Gmelina arborea*	Consumption of paste of fruits mixed with honey cures dysentery with predominance of *kapha*.	*Ci*. 9: 108
		Powder of this herb belonging to *Surasādi gaṇa* is to be mixed with honey and consumed for pacifying worm infestation in head.	*Ci*. 20: 27
85	*Gossypium herbaceum*	Topical application of paste of flowers cures leprosy and other skin diseases.	*Ci*. 19: 63
86	*Grewia tiliifolia*	*Ghṛta* prepared with decoction of bark and paste of roots is to be consumed on alternate days. Recommended food during the treatment period is cooked millet (*Paspalum scrobiculatum*) mixed with rice-washing. The *ghṛta* cures leprosy and other skin diseases, vitiligo and multiple glandular swelling around neck.	*Ci*. 19: 22–23
87	*Gymnema sylvestre*	Consumption of paste of root suspended in goat urine cures hemorrhoids	*Ci*. 8: 57
88	*Holoptelia integrifolia*	Consumption of juice of bark, mixed with honey, cures worm infestation in head.	*Ci*. 20: 26–27
89	*Hordeum vulgare*	Milk boiled with undried grains and mixed with clarified butter cures fever and burning sensation experienced in injury to thorax.	*Ci*. 3: 77
		After pacifying enlargement of abdomen caused by *tridōsa* using vegetable and inanimate poisons, the patient should eat for one month leaves cooked in the leaf juice, without adding salt, sour ingredients and oil. No other food should be served during this period. Leaf juice can be consumed for quenching thirst.	*Ci*. 15: 82–83
90	*Holarrhena antidysenterica*	Consumption of bark powder suspended in buttermilk cures hemorrhoids.	*Ci*. 8: 162
91	*Holostemma annulare*	Root of the herb is to be included in the food of post-partum mother, to induce milk production.	*Ut*. 1: 18
		He who drinks decoction of seeds, followed by food and meat soup, will be cured of dysentery with predominance of *pitta*.	*Ci*. 9: 60
		Consumption of tender leaves sautéed in clarified butter cures night blindness.	*Ut*. 13: 89–90
		Paste of roots mixed with honey and clarified butter is to be consumed for one month according to *rasāyana vidhi*. The treatment improves wisdom, intellect, strength, stability and longevity.	*Ut*. 39: 60–61
		Roots are cooked in milk and then ground to a paste. Consumption of this paste mixed with clarified butter and honey improves sexual vigor and vitality.	*Ut*. 40: 30–31
92	*Hordeum vulgare*	Topical application of paste of grains with honey heals non-healing ulcers.	*Ut*. 25: 56
93	*Hydnocarpus laurifolia*	Consumption of oil extracted from seeds pacifies all types of skin diseases.	*Ci*. 19: 12
		Consumption of paste of seeds according to the procedure for rejuvenation therapy (*rasāyana vidhi*) cures leprosy and other skin diseases.	*Ci*. 19: 53
		The seed oil is to be consumed according to *rasāyana vidhi* for curing leprosy and all skin diseases.	*Ut*. 39: 84–95
94	*Hygrophila auriculata*	Regular consumption of decoction of herb and its leaves as food cures *vāta*-induced blood disorder, just as compassion overcomes anger.	*Ci*. 22: 18–19
95	*Hyptis suaveolens*	Powder of this herb belonging to *Surasādi gaṇa* is to be mixed with honey and consumed for pacifying worm infestation in head.	*Ci*. 20: 27

(*Continued*)

TABLE 6.35 (*Continued*)
Single Herb Remedies Recommended in AH

Sl. No.	Name of Herb/Drug	Mode of Administration and Indications	Reference
96	*Indigofera tinctoria*	Paste of leaves suspended in water cures irregular fever.	*Ci.* 1: 160
		Decoction of whole herb is administered to induce emesis in treatment of rat bite.	*Ut.* 38: 21
97	*Ipomoea digitata*	Consumption of *ghrta* prepared with the tuber pacifies voiding of discolored urine or obstructed urination experienced by patients having cough.	*Ci.* 3: 155
		Paste of tuber mixed with honey and clarified butter is to be consumed for one month according to *rasāyana vidhi*. The treatment improves wisdom, intellect, strength, stability and longevity.	*Ut.* 39: 60–61
		Powder of the tuber is to be subjected many times to *bhāvana* with its own juice. Consumption of this fortified powder mixed with sugar, honey and clarified butter improves sexual vigor and vitality.	*Ut.* 40: 26
98	*Ipomoea sepiaria*	Paste of root suspended in milk and administered through mouth or instilled into nostrils in the first month of pregnancy causes the birth of a boy child.	*Śā*.1: 41–42
99	*Jasminum grandiflorum*	Keeping inside mouth a reasonably large volume of decoction of leaves pacifies inflammation of mouth and sinus.	*Ut.* 22: 106
100	*Jasminum multiflorum*	Consumption of tender leaves sautéed in clarified butter cures night blindness.	*Ut.* 13: 89–90
101	*Jasminum pubescens*	Powder of this herb belonging to *Surasādi gaṇa* is to be mixed with honey and consumed for pacifying worm infestation in head.	*Ci.* 20: 27
102	*Lagenaria siceraria*	Consumption of water stored in the hollowed-out fruit overnight pacifies effects of rat bite.	*Ut.* 38: 31
103	*Lens culinaris*	Topical application of roasted seeds ground in milk and mixed with honey and clarified butter cures black discoloration on face.	*Ut.* 32: 19
104	*Leucas aspera*	Powder of this herb belonging to *Surasādi gaṇa* is to be mixed with honey and consumed for pacifying worm infestation in head.	*Ci.* 20: 27
105	*Leucas cephalotus*	Powder of this herb belonging to *Surasādi gaṇa* is to be mixed with honey and consumed for pacifying worm infestation in head.	*Ci.* 20: 27
106	*Linum usitatissimum*	Topical application of seeds ground to paste with milk pacifies pain associated with *vāta*-induced blood disorder having predominance of *vāta*.	*Ci.* 22: 34
107	*Luffa echinata*	Pills are rolled from a mass of seeds ground to a fine paste with fermented rice-water and dried under shade. These pills are soaked in water, turned into a paste and applied in hemorrhoids.	*Ci.* 8: 19–20
108	*Madhuca longifolia*	Consumption of concentrated juice of flowers, fermented with honey cures *grahaṇi*.	*Ci.* 10: 51–52
		Application of powder of heartwood mixed with honey in eyes cures *śukḷarōga* (diseases of sclera).	*Ut.* 11: 47
109	*Malaxis acuminata*	Paste of the root, suspended in water and administered orally on *Puṣya* star in the first month of pregnancy, causes the birth of a boy child.	*Śā*.1: 39–40
110	*Malaxis muscifera*	Paste of the root, suspended in water and administered orally on *Puṣya* star in the first month of pregnancy, causes the birth of a boy child.	*Śā*.1: 39–40
111	*Mangifera indica*	*Nasya* using juice of seed kernel, mixed with or without milk, cures hemorrhages of obscure origin.	*Ci.* 2: 48–49

(*Continued*)

TABLE 6.35 (*Continued*)
Single Herb Remedies Recommended in AH

Sl. No.	Name of Herb/Drug	Mode of Administration and Indications	Reference
112	*Merremia emarginata*	Consumption of cakes prepared with tender leaves and paste of soaked rice grains pacifies worm infestation.	*Ci*. 20: 29
		Powder of this herb belonging to *Surasādi gaṇa* is to be mixed with honey and consumed for pacifying worm infestation in head.	*Ci*. 20: 27
113	*Monochoria vaginalis*	*Ghṛta* prepared with stem, rootstock, leaf, stamens and gold leaf improves intelligence when consumed according to *rasāyana vidhi*.	*Ut*. 39: 49, 104–105
114	*Moringa concanensis*	Bark of the stem can be administered internally as decoction, powders and topically as *taila*, *ghṛta* and *lēpa* for the cure of abscess.	*Ci*. 13: 10, 23
		Topical application of paste of seeds ground with fermented rice-water pacifies pain associated with *vāta*-induced blood disorder having predominance of *vāta* and *kapha*.	*Ci*. 22: 37
		Filling auditory canal with warmed juice of bark pacifies pricking pain in ear due to enragement of *kapha*.	*Ut*. 18: 12
115	*Moringa oleifera*	Consumption of warm decoction of root breaks urinary stones.	*Ci*. 11: 31
		Slightly warmed paste of bark is to be applied topically for curing *granthi visarpa*.	*Ci*. 18: 25
		Pouring into eyes the leaf juice mixed with honey pacifies various eye diseases caused by derangement of *tridōṣa*.	*Ut*. 16: 9 *Ut*. 16: 37–38
		Filling auditory canal with warmed juice of bark pacifies pricking pain in ear due to enragement of *kapha*.	*Ut*. 18: 12
116	*Murraya koenigii*	After pacifying enlargement of abdomen caused by *tridōṣa* using vegetable and inanimate poisons, the patient should eat for one month leaves cooked in the leaf juice, without adding salt, sour ingredients and oil. No other food should be served during this period. Leaf juice can be consumed for quenching thirst.	*Ci*. 15: 82–83
117	*Musa paradisiaca*	Filling auditory canal with warmed juice of inflorescence pacifies pricking pain in ear due to enragement of *kapha*.	*Ut*. 18: 12
118	*Nardostachys jatamansi*	Powder of this herb belonging to *Surasādi gaṇa* is to be mixed with honey and consumed for pacifying worm infestation in head.	*Ci*. 20: 27
119	*Nelumbo nucifera*	Consuming stamens of flowers, mixed with sugar and butter, cures bleeding piles	*Ci*. 8: 118
		Ghṛta prepared with rhizome, leaf stalks, leaves, seeds, stamens and gold leaf improves valor, strength and vivid imagination.	*Ut*. 39: 48, 104–105
120	*Neolamarckia cadamba*	Consumption of *ghṛta* prepared with the heartwood pacifies voiding of discolored urine or obstructed urination experienced by patients having cough.	*Ci*. 3: 155
		Consumption of cakes prepared with tender leaves and paste of soaked rice grains pacifies worm infestation.	*Ci*. 20: 30–31
121	*Nerium indicum*	Application of juice of root on scalp, after multiple puncturing, induces hair growth in alopecia.	*Ut*. 24: 29
122	*Nymphaea rubra*	Paste of flower is to be smeared inside an iron pot and left overnight. It is washed in the next morning with water, and the suspension is consumed along with food or beverages. He who follows this instruction lives for hundred years free from diseases.	*Ut*. 39: 104–105
123	*Ocimum kilimandsharicum*	Powder of this herb belonging to *Surasādi gaṇa* is to be mixed with honey and consumed for pacifying worm infestation in head.	*Ci*. 20: 27

(*Continued*)

TABLE 6.35 (*Continued*)
Single Herb Remedies Recommended in AH

Sl. No.	Name of Herb/Drug	Mode of Administration and Indications	Reference
124	*Ocimum sanctum*	Juice of the herb mixed with honey cures cough with predominance of *kapha*.	*Ci*. 3: 48–49
		Powder of this herb belonging to *Surasādi gaṇa* is to be mixed with honey and consumed for pacifying worm infestation in head.	*Ci*. 20: 26–27
125	*Ocimum sanctum* with green stems	Powder of this herb belonging to *Surasādi gaṇa* is to be mixed with honey and consumed for pacifying worm infestation in head.	*Ci*. 20: 27
126	*Operculina turpethum*	Paste of leaves suspended in water cures irregular fever.	*Ci*. 1: 160
		Electuary prepared with decoction, powder and sugar cures hemorrhages of obscure origin.	*Ci*. 2: 9–10
		After pacifying enlargement of abdomen caused by *tridōṣa* using vegetable and inanimate poisons, the patient should eat for one month leaves cooked in the leaf juice, without adding salt, sour ingredients and oil. No other food should be served during this period. Leaf juice can be consumed for quenching thirst.	*Ci*. 15: 82–83
		Powder of stem mixed with milk cures *vāta*-induced blood disorder with predominance of *pitta*.	*Ci*. 22: 12
		Ghṛta prepared with decoction and paste of stems cures *kṣataśukḷa*.	*Ut*. 1: 30
127	*Origanum marjorana*	Powder of this herb belonging to *Surasādi gaṇa* is to be mixed with honey and consumed for pacifying worm infestation in head.	*Ci*. 20: 27
128	*Oroxylum indicum*	Paste of bark pieces is to be rolled in leaves of *Gmelina arborea* and processed according to *puṭapāka vidhi*. Consumption of juice mixed with honey or sugar cures dysentery.	*Ci*. 9: 79–80
		Paste of the bark mixed with clarified butter is to be cooked by steaming. Consumption of the cooked paste, mixed with honey, cures dysentery.	*Ci*. 9: 81
		Filling auditory canal with juice of leaves pacifies earache due to *vāta* enragement.	*Ut*. 18: 2–3
129	*Oryza sativa*	Performing sudation over eyes using warmed balls of rice bran pacifies quickly pain due to conjunctivitis with *vāta* predominance.	*Ut*. 16: 10
		Inhalation of smoke from a mixture of parched rice and clarified butter pacifies catarrh.	*Ut*. 20: 8
130	*Paspalum scrobiculatum*	Washing of head with ash of hay suspended in water cures pityriasis.	*Ut*. 24: 27
131	*Phoenix dactylifera*	Consumption of juice of fruits, fermented with honey, cures *grahaṇi*.	*Ci*. 10: 52
132	*Phyllanthus amarus*	Consumption of water boiled with the herb is beneficial to patients of kingly consumption	*Ci*. 5: 13
133	*Picrorhiza kurroa*	Consumption of juice obtained by heating in a closed container, mixed with clarified butter, cures fever with predominance of *vāta* and *pitta*.	*Ci*. 1: 59, 160
134	*Pinus longifolia*	Filling auditory canal with oil extracted from heartwood pacifies earache due to *vāta* enragement.	*Ut*. 18: 5
135	*Piper chaba*	Paste of roots mixed with honey and clarified butter is to be consumed for one month according to *rasāyana vidhi*. The treatment improves wisdom, intellect, strength, stability and longevity.	*Ut*. 39: 60–61
136	*Piper longum*	Milk boiled with powder of fruits cures *āmajvara* (fever arising from impaired digestion).	*Ci*. 1: 111

(*Continued*)

TABLE 6.35 (*Continued*)
Single Herb Remedies Recommended in AH

Sl. No.	Name of Herb/Drug	Mode of Administration and Indications	Reference
		Powder of fruits cures irregular fever.	*Ci.* 1: 154
		Powder of fruits suspended in warm water and mixed with rock salt cures cough.	*Ci.* 3: 16
		Powder of fruits mixed with curd cures cough,	*Ci.* 3: 16
		Paste of fruits mixed with rock salt, sautéed in clarified butter and administered in wine, curd or whey cures cough.	*Ci.* 3: 17
		Consumption of fruit powder suspended in warm water or milk cures chronic and painful dysentery.	*Ci.* 9: 40–41
		Powder of fruits mixed with honey cures dysentery.	*Ci.* 9: 108–109
		Consumption of buttermilk mixed with powder of fruits and rock salt pacifies enlargement of abdomen caused by enragement of *vāta*.	*Ci.* 15: 127
		Consumption of buttermilk mixed with honey and powder of fruits pacifies *chidrōdara*.	*Ci.* 15: 129
		Powder of fruits, rock salt and honey mixed together are to be applied in eyes after surgery of the eye disease *upanāha*.	*Ut.* 11: 1–2
		Ghṛta prepared with paste and decoction of fruits is administered orally to patient suffering from pricking pain in ear due to enragement of *kapha*, before subjecting to emesis.	*Ut.* 18: 11
		Consumption of fruit powder according to *rasāyana vidhi* cures cough, respiratory distress, kingly consumption, polyuria, *grahaṇi* and several other diseases.	*Ut.* 39: 96–103 *Ci.* 3: 79
137	*Piper nigrum*	Powder of fruits administered with honey cures cough with predominance of *kapha*.	*Ci.* 3: 48
		Jaggery water is to be heated and brought to thick consistency. On cooling, honey and powder of the fruits are to be added to it and consumed. This treatment cures *kṣata kāsa*.	*Ci.* 3: 78
		Powder of fruits mixed with honey, clarified butter and sugar pacifies all forms of respiratory distress and cough.	*Ci.* 3: 173
		Consumption of fruit powder suspended in warm water and mixed with *yavakṣāra* (alkali from barley straw) cures respiratory distress and hiccup.	*Ci.* 4: 32
		Consumption of fruit powder suspended in warm water or milk cures chronic and painful dysentery.	*Ci.* 9: 40–41
		Consumption of buttermilk mixed with powder of fruits and sugar pacifies enlargement of abdomen caused by *pitta*.	*Ci.* 15: 127
		Applying in eyes paste of fruits, ground with curd, cures *dōṣāndha*.	*Ut.* 13: 85
138	*Pistacia integerrima*	Consumption of paste of leaf galls suspended in milk improves sexual vigor and vitality. Food should include sugar, clarified butter and milk.	*Ut.* 40: 29–30
139	*Plumbago zeylanica*	Paste of roots suspended in milk is to be administered in alcoholism.	*Ci.* 7: 103
		Regular consumption of buttermilk, prepared from milk boiled with pieces of the roots, cures hemorrhoids.	*Ci.* 8: 30–31
		Milk is to be filled in an earthen pot, the inside of which is smeared with paste of the root. Consumption of buttermilk prepared from this curd cures hemorrhoids.	*Ci.* 8: 48–49
		Consumption of buttermilk mixed with root powder cures dysentery.	*Ci.* 9: 108–109

(*Continued*)

TABLE 6.35 (*Continued*)
Single Herb Remedies Recommended in AH

Sl. No.	Name of Herb/Drug	Mode of Administration and Indications	Reference
		Consumption *ghṛta* prepared with cow urine and paste of roots, with *yavakṣāra* as adjuvant, cures enlargement of abdomen.	*Ci.* 15: 7
		Milk boiled with roots is to be curdled and the butter isolated. The resultant clarified butter, the remaining buttermilk and the paste of roots are processed into a *ghṛta*, the consumption of which cures edema.	*Ci.* 17: 13
		Consumption of paste of root according to the procedure for rejuvenation therapy (*rasāyana vidhi*) cures leprosy and other skin diseases.	*Ci.* 19: 53
		Root powder of the blue variety is to be consumed for one month in clarified butter, sesame oil, cow urine, buttermilk, water, milk or honey according to *rasāyana vidhi*. The treatment cures all chronic *vāta* diseases, morbid pallor, hemorrhoids, leprosy and other skin diseases.	*Ut.* 39: 62–65, 104–105
140	*Pongamia glabra*	Tender leaves are to be sautéed in *yamaka snēha* and sprinkled with powder of parched rice. Consumption of this preparation before food cures hemorrhoids.	*Ci.* 8: 53–54
		Slightly warmed paste of bark is to be applied topically for curing *granthi visarpa*.	*Ci.* 18: 25
		Milk boiled with seeds is to be poured in a thin stream over eyes on the third, fifth and seventh days after surgical removal of pterygium.	*Ut.* 11: 21
141	*Pseudarthria viscida*	Cooled decoction of roots is to be administered in alcoholism due to enragement of *kapha*.	*Ci.* 7: 34
		Taila prepared with decoction and paste of root is to be poured in a thin stream to pacify pain and edema caused by scorpion sting.	*Ut.* 37: 29
142	*Psoralea corylifolia*	Consumption of buttermilk mixed with seed powder cures leprosy and other skin diseases.	*Ci.* 19: 26
		Consumption of paste of seeds according to the procedure for rejuvenation therapy (*rasāyana vidhi*) cures leprosy and other skin diseases.	*Ci.* 19: 53 *Ut.* 39: 107–110
143	*Pterocarpus marsupium*	Paste of heartwood is to be smeared inside an iron pot and left overnight. It is washed in the next morning with water, and the suspension is consumed along with food or beverages. He who follows this instruction lives for hundred years free from diseases.	*Ut.* 39: 104–105, 153
144	*Punica granatum*	*Nasya* using juice of flowers mixed with or without milk cures hemorrhages of obscure origin.	*Ci.* 2: 48–49
145	*Raphanus sativus*	Slightly warmed paste of root is to be applied topically for curing *granthi visarpa*.	*Ci.* 18: 25
		Filling auditory canal with juice of radish root pacifies earache due to *vāta* and *kapha* enragement.	*Ut.* 18: 2–3, 12
146	*Ricinus communis*	Milk boiled with root cures chronic fever.	*Ci.* 1: 112–113
		Topical application of seeds ground to paste with milk pacifies pain associated with *vāta*-induced blood disorder having predominance of *vāta*.	*Ci.* 22: 34
		Consumption of tender leaves of white variety sautéed in clarified butter cures night blindness.	*Ut.* 13: 89–90

(*Continued*)

TABLE 6.35 (*Continued*)
Single Herb Remedies Recommended in AH

Sl. No.	Name of Herb/Drug	Mode of Administration and Indications	Reference
		Performing sudation over eyes using tender leaves or roots of the white variety, cooked in goat milk pacifies pain due to conjunctivitis with *vāta* predominance.	*Ut.* 16: 10
		Juice obtained from leaves of white variety subjected to *puṭapāka* should be warmed and poured into auditory canal to pacify earache due to *vāta* enragement.	*Ut.* 18: 1–2
		Applying over lips steam emerging from tender leaves boiled in milk pacifies inflammation of lips due to enragement of *vāta.*	*Ut.* 22: 4
147	*Rubia cordifolia*	Topical application of paste of roots mixed with honey cures black pigmentation on face.	*Ut.* 32: 16
148	*Saccharum officinarum*	Sugarcane juice may be administered to patients suffering from vomiting.	*Ci.* 6: 14
		Dysentery with bleeding is cured by performing *snēhavasti*, using *ghṛta* prepared with roots.	*Ci.* 9: 98
		Consumption of sugarcane juice, fermented with honey, cures *grahaṇi.*	*Ci.* 10: 52
		Pouring of sugarcane juice in a thin stream pacifies *agnivisarpa.*	*Ci.*18: 22
149	*Salmalia malabarica*	Milk boiled with gum exudate cures hemorrhages of obscure origin.	*Ci.* 2: 39
		Petioles of leaves are kept immersed in water overnight. They are crushed in the morning and the fluid consumed with honey, for curing dysentery with predominance of *pitta.*	*Ci.* 9: 61
		Consumption of goat milk boiled with gum exudate cures dysentery with bleeding.	*Ci.* 9: 83
		Paste of conical spines on trunk is to be mixed with milk and applied topically to cure black discoloration on face.	*Ut.* 32: 19
150	*Salvadora persica*	Hemorrhoids will be cured by eating the fruit, followed by drinking of buttermilk	*Ci.* 8: 36
		Ghṛta prepared with fruits cures bloating of abdomen experienced by patients suffering from enlargement of abdomen.	*Ci.* 15: 38
151	*Santalum album*	*Nasya* using juice filtered from aqueous paste, mixed with breast milk, cures respiratory distress and hiccup.	*Ci.* 4: 48
		Paste of heartwood mixed with sugar and honey is to be suspended in rice-washing. Consumption of this mixture cures dysentery with bleeding, burning sensation, polydipsia and polyuria.	*Ci.* 9: 93–94
152	*Saussurea lappa*	*Taila* prepared with the root is poured in a thin stream to heal the wound caused by cutting of the umbilical cord of newborn infant.	*Ut.* 1: 6
		Consumption of decoction of root pacifies epilepsy.	*Ut.* 7: 34
		Filling auditory canal with oil extracted from root pacifies earache due to *vāta* enragement.	*Ut.* 18: 5
		Roasted and powdered root mixed with sesame oil is to be applied on head to cure pityriasis.	*Ut.* 24: 23
		A piece of fruit is kept embedded in fruit of *Citrus medica* for seven days. Thereafter, it is ground with honey and the paste applied to cure black discoloration on face.	*Ut.* 32: 20
153	*Semecarpus anacardium*	Administration of the seed in different dosage forms, mixed with jaggery, cures irregular fever.	*Ci.* 1: 154

(*Continued*)

TABLE 6.35 (*Continued*)
Single Herb Remedies Recommended in AH

Sl. No.	Name of Herb/Drug	Mode of Administration and Indications	Reference
		Consumption of buttermilk prepared with powder of seeds cures non-bleeding hemorrhoids.	*Ci.* 8: 35, 162
		Consumption of oil extracted from seeds pacifies all types of skin diseases.	*Ci.* 19: 12
		Consumption of paste of seeds according to the procedure for rejuvenation therapy (*rasāyana vidhi*) cures leprosy and other skin diseases.	*Ci.* 19: 53
		Application of juice of fruit on scalp, after multiple puncturing, induces hair growth in alopecia.	*Ut.* 24: 30
		Decoction of the seeds, the seed oil or *ghṛta* and *taila* prepared with the seeds are to be consumed according to *rasāyana vidhi*. The treatment cures polyuria, hemorrhoids, leprosy and other skin diseases.	*Ut.* 39: 66–83
154	*Senna occidentalis*	Juice of the herb mixed with honey cures cough with predominance of *kapha*.	*Ci.* 3: 48–49
155	*Sesamum indicum*	Consuming paste of sesame seeds along with butter cures bleeding piles.	*Ci.* 8: 118
		Consumption of paste of seeds mixed with five parts of sugar and suspended in goat milk cures at once dysentery with bleeding.	*Ci.* 9: 92–93
		Topical application of seed paste and honey heals non-healing ulcers in buttocks and thighs.	*Ut.* 25: 54
		Topical application of seed paste mixed with honey and clarified butter heals sinuses caused by sharp objects.	*Ut.* 30: 35
		Topical application of paste of seeds mixed with honey and clarified butter hastens healing of operated *upadaṃśa*.	*Ut.* 34: 2
		He who consumes one *pala* (48 g) of sesame seeds, followed by drinking of cold water, will have strong teeth and well-built body till the end of his life.	*Ut.* 39: 158
156	*Sesbania grandiflora*	*Ghṛta* prepared with juice and paste of leaves can be administered orally or as *nasya* to pacify night blindness.	*Ut.* 13: 90
157	*Sida rhombifolia* ssp. *retusa*	*Ghṛta* prepared with decoction and paste of roots cures chronic fever.	*Ci.* 1: 94
		Milk and paste of the roots are mixed with sesame oil and heated over slow fire, with frequent stirring, so as to remove most of the water and bring the sediment to mud-like consistency. Same quantities of milk and paste of roots are added again and the process continued. The oil obtained after repeating this process 100 or 1,000 times is administered internally or applied topically to cure *vāta*-induced blood disorders.	*Ci.* 22: 45–46
		Taila prepared with root paste, milk and sesame oil may be poured in a thin stream to pacify pain in traumatic wounds.	*Ut.* 26: 6
		Paste of roots mixed with honey and clarified butter is to be consumed for one month according to *rasāyana vidhi*. The treatment improves wisdom, intellect, strength, stability and longevity.	*Ut.* 39: 60–61, 104–15

(*Continued*)

TABLE 6.35 (*Continued*)
Single Herb Remedies Recommended in AH

Sl. No.	Name of Herb/Drug	Mode of Administration and Indications	Reference
158	*Sida vernonicaefolia*	6 g of root powder are to be boiled with milk and consumed on the first day. 12 g of the powder are to be consumed on the next day. Thus everyday 6 g more of powder are to be consumed till the quantity of powder reaches 48 g (in eight days). Thereafter, 48 g of powder are consumed everyday for one month. During the treatment period, the patient should consume only milk and no solid food. Treatment cures injury to thorax and improves stamina, complexion and longevity.	*Ci*. 3: 118–120
		Root powder consumed for a year, mixed with milk or honey-clarified butter mixture, improves health and longevity.	*Ut*. 39: 54–55
159	*Śilājatu*	Consumption of mineral bitumen mixed with cow urine for one month cures *kumbhakāmala* (advanced stage of jaundice characterized by an enlarged abdomen).	*Ci*. 16: 52–53
		Consumption of mineral bitumen according to *rasāyana vidhi* cures leprosy and other skin diseases.	*Ci*. 19: 53
		Regular consumption of mineral bitumen suspended in cow urine cures stiffness in thigh muscles (*ūrustambha*).	*Ci*. 21: 49
		Oral administration of mineral bitumen along with milk is recommended to cure envelopment of *vāta* by all *dhātus*.	*Ci*. 22: 65
		Consumption of mineral bitumen according to *rasāyana vidhi* cures all diseases.	*Ut*. 39: 130–142
160	*Solanum indicum*	Paste of root suspended in milk and administered through right nostril by the pregnant mother herself in the first month of pregnancy causes the birth of a boy child. Administration through left nostril causes the birth of a girl child.	*Śa*. 1: 40–41
		Juice of the herb mixed with honey cures cough with predominance of *kapha*.	*Ci*. 3: 48–49
161	*Solanum nigrum*	Topical application of paste of whole herb cures leprosy and other skin diseases.	*Ci*. 19: 63
162	*Solanum xanthocarpum*	Juice of the herb mixed with honey cures cough with predominance of *kapha*.	*Ci*. 3: 48–49
		Buttermilk is to be filled in an earthen pot, the inside of which is smeared with paste of the fruits. Consumption of this buttermilk after one night cures hemorrhoids.	*Ci*. 8: 44–45
		Consumption of the juice of fruits, mixed with honey, cures dysuria due to enragement of *kapha*.	*Ci*. 11: 11
		Application of juice of fruit with honey on scalp, after multiple puncturing, induces hair growth in alopecia.	*Ut*. 24: 30
163	*Sphaeranthus indicus*	Powder of this herb belonging to *Surasādi gaṇa* is to be mixed with honey and consumed for pacifying worm infestation in head.	*Ci*. 20: 27
164	*Stereospermum tetragonum*	Ash of the herb consumed with sesame oil cures dysuria due to enragement of *kapha*.	*Ci*. 11: 13
165	*Symplocos crataegoides*	Bark is sautéed in clarified butter, powdered and packed in small cloth bag. After dipping the bag in warm water, it is squeezed and the fluid poured into eyes for pacifying pain due to *saśōpha* and *alpaśōpha akṣipāka*.	*Ut*. 16: 32–33
166	*Symplocos spicata*	*Ghṛta* prepared with decoction and paste of bark cures *vāta* diseases.	*Ci*. 21: 10

(*Continued*)

TABLE 6.35 (*Continued*)
Single Herb Remedies Recommended in AH

Sl. No.	Name of Herb/Drug	Mode of Administration and Indications	Reference
167	*Tephrosia purpurea*	*Nasya* or topical application of roots ground in rice-washing pacifies obstinate ulcers, cervical glandular swellings and poisoning.	*Ut*. 30: 26
		Consumption of powder of seeds suspended in buttermilk pacifies effects of rat bite.	*Ut*. 38: 28
168	*Terminalia arjuna*	Consumption of decoction of bark cures all types of dysuria.	*Ci*. 11: 37
		Topical application of paste of bark mixed with honey cures black pigmentation on face.	*Ut*. 32: 16
169	*Terminalia belerica*	Slightly warmed paste of bark is to be applied topically for curing *granthi visarpa*.	*Ci*. 18: 25
		Application of powder of seed kernels, mixed with honey, in eyes cures *śukḷarōga*.	*Ut*. 11: 47
		Keeping pericarps in mouth and slowly sucking the juice is desirable in all forms of respiratory distress and hiccup.	*Ci*. 3: 17
170	*Terminalia chebula*	Powder or decoction of pericarp cures irregular fever.	*Ci*. 1: 154
		Powder of pericarp, mixed with honey, pacifies vomiting with predominance of *pitta*.	*Ci*. 6: 17
		Ghṛta prepared with pericarps and sonchal salt cures heart disease with *vāta* predominance, respiratory distress and *gulma*.	*Ci*. 6: 29–30
		Ghṛta prepared with decoction of pericarps is to be administered in *mada*.	*Ci*. 7: 107
		Fruits are to be soaked overnight in cow urine. Eating the fruits with jaggery cures hemorrhoids.	*Ci*. 8: 54–55
		200 fruits are to be cooked over slow fire in 13 l of cow urine, till all the liquid evaporates. Consuming daily two of such fruits, along with honey, cures hemorrhoids and several other diseases.	*Ci*. 8: 55–57
		Consumption of powder of pericarps suspended in buttermilk cures hemorrhoids.	*Ci*. 8: 58
		Consumption of powder of pericarps, suspended in warm water, cures *grahaṇi*.	*Ci*. 10: 8
		Milk boiled with seed kernels breaks urinary stones.	*Ci*. 11: 33
		Ghṛta prepared with powder and decoction of pericarp and curd cures enlargement of abdomen, poisoning, *aṣṭhīla* (hard and round swelling below the navel), *gulma*, abscess, skin diseases, insanity and epilepsy.	*Ci*. 15: 28–31
		Consumption of paste of pericarp suspended in cow urine pacifies morbid pallor and stiffness of thighs.	*Ci*. 16: 7 *Ci*. 21: 49
		Powder of pericarp mixed with honey and clarified butter pacifies morbid pallor.	*Ci*. 16: 10
		Ghṛta prepared with decoction of 100 fruits and paste of 50 fruits cures *gulma*, jaundice and morbid pallor.	*Ci*. 16: 40–41
		Consumption of powder of pericarp mixed with jaggery and followed by drinking of buttermilk cures edema.	*Ci*. 17: 5
		Consumption of decoction of pericarps mixed with clarified butter cures *vāta*-induced blood disorders with predominance of *pitta*.	*Ci*. 22: 12
		Consuming daily in the early morning or before food, powder of pericarps, along with sugar, grapes or honey, cures cataract.	*Ut*. 13: 19

(*Continued*)

TABLE 6.35 (*Continued*)
Single Herb Remedies Recommended in AH

Sl. No.	Name of Herb/Drug	Mode of Administration and Indications	Reference
		Consumption of decoction of pericarps, mixed with honey, cures all diseases of throat.	*Ut*. 22: 55
		Paste or decoction of pericarps mixed with honey and clarified butter is to be consumed for one month according to *rasāyana vidhi*. The treatment improves wisdom, intellect, strength, stability and longevity.	*Ut*. 39: 60–61, 106–107
171	*Terminalia paniculata*	Paste of bark is to be smeared inside an iron pot and left overnight. It is washed in the next morning with water, and the suspension is consumed along with food or beverages. He who follows this instruction lives for hundred years free from diseases.	*Ut*. 39: 104–105
172	*Tinospora cordifolia*	*Ghṛta* prepared with milk, juice and paste of stem cures chronic fever, skin diseases with predominance of *vāta* and *vāta*-induced blood disorders with predominance of *vāta*.	*Ci*. 1: 94 *Ci*. 19: 39 *Ci*. 22: 7
		Juice or decoction of stem cures irregular fever.	*Ci*. 1: 154
		Water boiled with stems may be administered to patients suffering from vomiting with predominance of *pitta*.	*Ci*. 6: 14
		Milk is to be filled in an earthen pot, the inside of which is smeared with paste of the root. Consumption of buttermilk prepared from this curd cures hemorrhoids.	*Ci*. 8: 49
		Juice of stem mixed with honey cures polyuria.	*Ci*. 12: 7
		Decoction or juice of stem, mixed with honey and administered in the morning, cures jaundice.	*Ci*. 16: 43–44
		Consumption of *ghṛta* prepared with juice of stem, cow milk and clarified butter from buffalo milk pacifies *halīmaka*.	*Ci*. 16: 53–54
		Paste, juice, powder or decoction of the stem is to be administered for pacifying *vāta*-induced blood disorder with predominance of *kapha*.	*Ci*. 22: 15
		The juice of stem is to be consumed as *rasāyana* according to *rasāyana vidhi*, to improve health, intelligence and longevity.	*Ut*. 39: 44–45, 60–61, 104–105
173	*Trachyspermum ammi*	Consumption of buttermilk prepared with powder of seeds cures non-bleeding hemorrhoids.	*Ci*. 8: 35
		Upanāha with seeds roasted and ground in clarified butter and mixed with rock salt pacifies effects of scorpion sting.	*Ut*. 37: 31
174	*Tragia involucrata*	Cooled decoction of roots is to be administered in alcoholism due to enragement of *kapha*.	*Ci*. 7: 34
		Consumption of juice of herb cures all types of dysuria.	*Ci*. 11: 37
175	*Tribulus terrestris*	Consumption of fruit powder mixed with honey and suspended in sheep milk, for seven days, breaks urinary stones.	*Ci*. 11: 30
		Consumption of powder of whole herb, subjected to *bhāvana* in its own juice, improves health, vigor, vitality and longevity.	*Ut*. 39: 56–57
176	*Trichosanthes cucumerina*	Keeping inside mouth a reasonably large volume of decoction of roots pacifies inflammation of mouth and sinus.	*Ut*. 22: 106
177	*Triticum aestivum*	Wheat powder mixed with copious quantity of clarified butter is to be suspended in milk. Consumption of this mixture cures *atyagni* (excessive appetite).	*Ci*. 10: 88
		Root powder mixed with honey is administered in *mada*.	*Ci*. 7: 105–106

(*Continued*)

TABLE 6.35 (*Continued*)
Single Herb Remedies Recommended in AH

Sl. No.	Name of Herb/Drug	Mode of Administration and Indications	Reference
178	*Valeriana wallichii*	Powder of this herb belonging to *Surasādi gaṇa* is to be mixed with honey and consumed for pacifying worm infestation in head.	*Ci.* 20: 27
179	*Vateria indica*	Gum resin is to be burnt, and the smoke is inhaled through nostril for the cure of respiratory distress and hiccup.	*Ci.* 4: 13
		A *taila* is prepared using sesame oil, fermented wheat-water mixture, and the gum resin. A large volume of water is added to this *taila* and agitated continuously with a churner. Topical application of the resultant cream pacifies heat, pain and burning sensation associated with *vāta* -induced blood disorders.	*Ci.* 22: 21
180	*Vitex altissima*	Consumption of paste of inflorescence suspended in milk acts as antidote for rat poisoning.	*Ut.* 38: 30
181	*Vitex negundo*	*Ghṛta* prepared with juice and paste of leaves cures cough.	*Ci.* 3: 57
		Consumption of cakes prepared with tender leaves and paste of soaked rice grains pacifies worm infestation.	*Ci.* 20: 30
		Consumption of tender leaves sautéed in clarified butter cures night blindness.	*Ut.* 13: 89–90
182	*Vitis vinifera*	*Ghṛta* prepared with juice and paste of fruits cures chronic fever.	*Ci.* 1: 94
		Fruits are to be eaten with honey to pacify vomiting with predominance of *pitta*.	*Ci.* 6: 14, 17
		Consumption of fruit juice pacifies polydipsia experienced in alcoholism with enragement of *vāta* and *pitta*.	*Ci.* 7: 27
		Consumption of juice of fruits, fermented with honey, cures *grahaṇi*.	*Ci.* 10: 52
		Paste of the fruits is to be suspended in water that has been stored overnight. Consumption of this suspension in the morning cures dysuria due to enragement of *pitta*.	*Ci.* 11: 8
		Consumption of fruit juice pacifies morbid pallor and jaundice.	*Ci.* 16: 32
		Ghṛta prepared with aged clarified butter (more than five years old), juice and paste of fruits is to be administered with sugar, for pacifying *kṛchrōnmīla*.	*Ut.* 9: 1
183	*Withania somnifera*	Butter isolated from milk boiled with the roots is to be consumed with sweetened milk for the cure of kingly consumption.	*Ci.* 5: 25
		Paste of roots mixed with honey and clarified butter is to be consumed for one month according to *rasāyana vidhi*. The treatment improves wisdom, intellect, strength, stability and longevity.	*Ut.* 39: 60–61, 157
184	*Zingiber officinale*	Powder of dried rhizomes mixed with sugar and whey cures cough.	*Ci.* 3: 16
		6 g of powder of rhizomes are to be boiled with milk and consumed on the first day. 12 g of the powder are to be consumed on the next day. Thus everyday 6 g more of powder are to be consumed till the quantity of powder reaches 48 g in eight days. Thereafter, 48 g of powder are consumed every day for one month. During the treatment period, the patient should consume only milk and no solid food. Treatment cures injury to thorax and improves stamina, complexion and longevity.	*Ci.* 3: 118–120
		Consumption of powder of dry rhizomes, mixed with equal quantity of jaggery, cures respiratory distress and hiccup.	*Ci.* 4: 47–48
		Cooled decoction of rhizomes is to be administered in alcoholism due to enragement of *kapha*.	*Ci.* 7: 34

(*Continued*)

TABLE 6.35 (*Continued*)
Single Herb Remedies Recommended in AH

Sl. No.	Name of Herb/Drug	Mode of Administration and Indications	Reference
		Consumption of powder of dried rhizomes mixed with jaggery cures hemorrhoids and edema.	*Ci.* 8: 54–55 *Ci.* 17: 5
		Powder of dried rhizomes is to be mixed with jaggery and administered in curd, sesame oil, clarified butter or milk. This treatment cures dysentery in small volume with froth, sliminess and pain.	*Ci.* 9: 18
		Mixture of powder of rhizomes, jaggery and whey is seasoned with *yamaka snēha*. Consumption of this beverage cures dysentery.	*Ci.* 9: 31
		Consumption of powder of dried rhizomes suspended in warm water cures *grahaṇi*.	*Ci.* 10: 8
		Ghṛta prepared with powder of rhizomes, clarified butter, sesame oil and whey cures all types of enlargement of abdomen and *gulma* due to enragement of *vāta–kapha*.	*Ci.* 15: 5–6
		Paste of 24 g (half *pala*) of green ginger and 24 g of jaggery are administered on the first day. The doses of ginger and jaggery are thereafter increased by 24 g every day till the dose reaches 240 g of ginger and 240 g of jaggery or 5 *pala* each. This dose is maintained for one month. Vegetable soup, meat soup and milk are to be included in food. This treatment cures *gulma*, enlargement of abdomen, hemorrhoids, edema, polyuria and several other diseases.	*Ci.* 17: 6–7
		Consumption of *ghṛta* prepared with ginger juice, dry ginger paste and milk cures edema, enlargement of abdomen and lowering of abdominal fire.	*Ci.* 17: 8
		Applying in eyes paste of rhizome ground in breast milk pacifies *śuṣkākṣipāka*.	*Ut.* 16: 29
		Applying in eyes paste of rhizome and rock salt ground with visceral fat of animals that live in marshy land pacifies *śuṣkākṣipāka*.	*Ut.* 16: 30
		Filling auditory canal with warmed juice of rhizome pacifies pricking pain in ear due to enragement of *kapha*.	*Ut.* 18: 12
		Paste of rhizome is to be smeared inside an iron pot and left overnight. It is washed in the next morning with water, and the suspension is consumed along with food or beverages. He who follows this instruction lives for hundred years free from diseases.	*Ut.* 39: 104–105
185	*Ziziphus mauritiana*	Paste of kernel administered with wine, curd or whey cures cough.	*Ci.* 3: 17
		Paste of leaves sautéed in clarified butter is to be mixed with rock salt and administered to pacify lassitude of voice experienced in kingly consumption.	*Ci.* 5: 37
		Consumption of paste of seed kernel mixed with honey pacifies vomiting with predominance of *pitta*.	*Ci.* 6: 17
		Eating fruits boiled in water, along with jaggery and sesame oil, cures dysentery in small volume, with froth, sliminess and pain.	*Ci.* 9: 18

TABLE 6.36
Reputed Groupings of Herbs

Sl. No.	Name of Group	Sanskrit Name of Constituent Herb	Latin Binomial	Reference
1	*Balātrayam*	*Bala*	*Sida rhombifolia* (L.) ssp. *retusa* (L.) Boiss.	*Sū.* 10: 23 Upadhyaya (1975)
		Atibala	*Abutilon indicum* D. Don	
		Nāgabala	*Sida vernonicaefolia* Lam.	
2	*Cāturjātam*	*Tvak*	*Cinnamomum zeylanicum* Blume	*Sū.* 6: 160
		Patra	*Cinnamomum tamala* Nees. And Eberm.	
		Ēlā	*Elettaria cardamomum* Maton.	
		Kēsaram	*Mesua ferrea* L.	
3	*Catuṣparṇinī*	*Māṣaparṇi*	*Phaseolus sublobatus* Roxb.	*Ci.* 5: 13
		Mudgaparṇi	*Phaseolus adenanthus* G.F. Meyer	
		Śālaparṇi	*Desmodium gangeticum* DC	
		Pṛśniparṇi	*Pseudarthria viscida* W. & A.	
4	*Hrasva pañcamūla*	*Bṛhati*	*Solanum indicum* L.	*Sū.* 6: 168–169
		Bṛhati	*Solanum xanthocarpum* Sch. & Wendl.	
		Amśumati	*Desmodium gangeticum* DC	
		Amśumati	*Pseudarthria viscida* W. & A.	
		Gōkṣuraḥ	*Tribulus terrestris* L.	
5	*Jīvana pañcamūla*	*Abhīruḥ*	*Asparagus racemosus* Willd.	*Sū.* 6: 170–171
		Vīrā	*Coccinia grandis* (L.) Voigt.	
		Jīvanti	*Holostemma annulare* (Roxb.) K. Schum.	
		Jīvakaḥ	*Malaxis acuminata* D. Don	
		Ṛṣabhakaḥ	*Malaxis muscifera* (Lindley) O. Kuntze	
6	*Madhyama pañcamūla*	*Bala*	*Sida rhombofolia* L. ssp. *retusa* (L.) Boiss.	*Sū.* 6: 169–170
		Punarnava	*Boerhaavia diffusa* L.	
		Ēraṇḍa	*Ricinus communis* L.	
		Sūpyaparṇi	*Phaseolus adenanthus* G.F. Meyer	
		Sūpyaparṇi	*Phaseolus sublobatus* Roxb.	
7	*Mahat pañcamūla*	*Vilva*	*Aegle marmelos* Corr.	*Sū.* 6: 167–168
		Kāśmari	*Gmelina arborea* L.	
		Tarkāri	*Premna serratifolia* L.	
		Pāṭalā	*Stereospermum tetragonum* DC	
		Duṇḍukaḥ	*Oroxylum indicum* Vent.	
8	*Pañcakōla*	*Cavika*	*Piper chaba* Hunter	*Sū.* 6: 166
		Pippali	*Piper longum* L.	
		Pippalīmūlam	*Piper longum* L. root	
		Citraka	*Plumbago zeylanica* L.	
		Nāgara	*Zingiber officinale* Rosc.	
9	*Pañcāmḷaka*	*Kōla*	*Ziziphus mauritiana* Lam.	*Ci.* 7: 31
		Dāḍimam	*Punica granatum* L.	
		Vṛkṣāmḷa	*Garcinia-gummi-gutta* (L.) N. Robson	
		Cukrīkā	*Tamarindus indica* L.	
		Cukrikā	*Solena amplexicaulis* (Lam.) Gandhi	

(*Continued*)

TABLE 6.36 (*Continued*)
Reputed Groupings of Herbs

Sl. No.	Name of Group	Sanskrit Name of Constituent Herb	Latin Binomial	Reference
10	*Pañcatiktaka*	*Nimba*	*Azadirachta indica* A. Juss.	*Ka.* 4: 23–24
		Guḍūcī	*Tinospora cordifolia* (Willd.) Miers. Ex Hook. F. & Th.	
		Vṛṣaḥ	*Adhatoda vasica* Nees.	
		Paṭōlaḥ	*Trichosanthes cucumerina* L.	
		Nidigdhika	*Solanum xanthocarpum* Sch. & Wendl.	
11	*Trijātakam*	*Tvak*	*Cinnamomum zeylanicum* Blume	*Sū.* 6: 160
		Patra	*Cinnamomum tamala* Nees. and Eberm.	
		Ēlā	*Elettaria cardamomum* Maton.	
12	*Trikaṭukam*	*Pippali*	*Piper longum* L.	*Sū.* 6: 164
		Maricam	*Piper nigrum* L.	
		Nāgaram	*Zingiber officinale* Rosc.	
13	*Triphala*	*Harītaki*	*Terminaia chebula* Retz.	*Sū.* 6: 159
		Āmalaka	*Emblica officinalis* Gaertn.	
		Akṣaḥ	*Terminalia belerica* Roxb.	
14	*Tṛṇapañcamūla*	*Darbhaḥ*	*Imperata cylindrica* (L.) P. Beauv.	*Sū.* 6: 171
		Kāśā	*Saccharum spontaneum* L.	
		Ikṣu	*Saccharum officinarum* L.	
		Śāra	*Saccharum arundinaceum* Retz.	
		Nīvārā	*Hygroryza aristata* Nees.	

6.3.2 Varieties of Rice

The best variety of *vrīhi* rice is *Ṣaṣṭika*. It is unctuous, cold, sweet in taste and causes constipation. It pacifies *tridōṣa*. There are two types of *Ṣaṣṭika*: white (*gauram*) and black (*asitagauram*). They are called so on the color of the grain. The other 15 varieties of *vrīhi* rice are *Mahāvrīhi*, *Kṛṣṇavrīhi*, *Jatumukham*, *Kukkuṭāṇḍakam*, *Lāvākhya*, *Pārāvatakam*, *Sukaram*, *Varakam*, *Uddālakam*, *Ujjvālam*, *Cīnam*, *Śāradam*, *Darduram*, *Gandhanāḥ* and *Kuruvindāḥ* (*Sū.* 6: 7–10).

6.3.3 Grassy Grains

This group of grassy grains (*tṛṇadhānya*) consists of *Kaṅgu* (*Setaria italica*), *Kōdravam* (*Paspalum scrobiculatum*), *Nīvāra* (*Hygroryza aristata*), *Śyāmāka* (*Panicum miliare*), *Yavam* (*Hordeum vulgare*), *Vaṃśajaḥ* (seeds of *Bambusa bambos*), *Gōdhūmaḥ* (*Triticum aestivum*) and *Nandīmukhi* (a variety of wheat). *Tṛṇadhānya* are light, cold in potency, increases *vāta*, scrubs off fat and pacifies *kapha–pitta* (*Sū.* 6: 11–15).

6.3.4 Legumes

The 13 legumes included in this group of *śimbidhānya varga* are *Mudgaḥ* (*Vigna radiata*), *Āḍhaki* (*Cajanus cajan*), *Masūram* (*Lens culinaris*), *Kaḷāyaḥ* (*Pisum sativum*), *Rājamāṣaḥ* (*Vigna unguiculata*), *Kulatthāḥ* (*Dolichos biflorus*), *Niṣpāvam* (*Dolichos lablab*), *Māṣam* (*Phaseolus mungo*), *Kākāṇḍaki* (*Trichosanthes tricuspidata*), *Ātmagupta* (*Mucuna pruriens*), *Tilaḥ* (*Sesamum indicum*), *Uma* (*Linum usitatissimum*) and *Kusumbham* (*Carthamus tinctorius*) (*Sū.* 6: 17–26).

6.3.5 Pot-Herbs

Bulbs, tubers, stem, leaves, flowers and fruits of these plants are used as pot herbs or vegetables (*śāka varga*). Some others are used for seasoning foods also. This group includes *Pāṭha* (*Cyclea peltata*), *Śaṭhi* (*Kaempferia galanga*), *Sūṣā* (*Cassia tora*), *Suniṣaṇṇakam* (*Marsilea quadrifolia*), *Satīnajam* (*Pisum sativum*), *Rājakṣavam* (*Brassica nigra*), *Vāstukam* (*Chenopodium murale*), *Kākamāci* (*Solanum nigrum*), *Cāṅgēri* (*Oxalis corniculata*), *Paṭōla* (*Trichosanthes cucumerina*), *Saptala* (*Acacia sinuata*), *Ariṣṭam* (*Azadirachta indica*), *Śārṅgaṣṭha* (*Clerodendrum serratum*), *Avalguja* (*Psoralea corylifolia*), *Amṛtāḥ* (*Tinospora cordifolia*), *Vētāgram* (*Calamus rotang*), *Bṛhatīdvayam* (*Solanum indicum, Solanum xanthocarpum*), *Vāśā* (*Adhatoda vasica*), *Kuṭilī* (*Sesamum radiatum*), *Tilaparṇika* (*Ziziphus oenoplia*), *Maṇḍūkaparṇi* (*Centella asiatica*), *Karkōṭa* (*Momordica dioica*), *Kāravēlam* (*Momordica charantia*), *Nāḷīkalāyam* (*Alternanthera sessilis*), *Gōjihva* (*Elephantopus scaber*), *Vārtākaṃ* (*Solanum melongena*), *Vanatiktakam* (*Andrographis paniculata*), *Karīram* (*Capparis zeylanica*), *Kulakam* (*Trichosanthes cucumerina*), *Nandi* (*Tabernamontana divaricata*), *Śakulādanī* (*Picrorhiza kurroa*), *Kaṭhillaṃ* (*Boerhaavia diffusa*), *Kēmbukaṃ* (*Costus speciosus*), *Kōśātakam* (*Luffa acutangula*), *Karkaśam* (*Senna occidentalis*), *Taṇḍulīyaḥ* (*Amaranthus spinosus*), *Muñjātam* (*Dioscorea esculenta*), *Pālāṅkyā* (*Amaranthus tricolor*), *Upōdakā* (*Basella alba*), *Vidāri* (*Ipomoea digitata*), *Jīvanti* (*Holostemma annulare*), *Kūśmāṇḍam* (*Benincasa hispida*), *Tumbam* (*Lagenaria siceraria*), *Kāliṅga* (*Citrullus lanatus*), *Karkāru* (*Cucurbita pepo*), *Ērvāru* (*Cucumis sativus*), *Tindiśa* (*Coccinia grandis*), *Mṛṇāḷa* (*Nelumbo nucifera*), *Śālukam* (*Alocasia macrorrhizos*), *Kumudam* (*Nymphaea stellata*), *Utpala* (*Monochoria vaginalis*), *Nandīmāṣaka* (*Saccharum arundinaceum*), *Śṛṅgāṭakam* (*Trapa bispinosa*), *Kuṭiñjaram* (*Plectranthus rotundifolius*), *Laṭvākam* (*Carthamus tinctorius*), *Lōṇikā* (*Portulacca oleracea*), *Kuṭumbakam* (*Daemia extensa*), *Gavēthukam* (*Coix lacryma-jobi*), *Jīvantakam* (*Amaranthus viridis*), *Yavaśākam* (*Hordeum vulgare*), *Suvarcala* (*Linum usitatissimum*), *Lakṣmaṇa* (*Ipomoea sepiaria*), *Tarkāri* (*Premna serratifolia*), *Varuṇa* (*Crataeva religiosa*), *Kāḷaśākam* (*Murraya koenigii*), *Cirivilvāṅkura* (apical buds of *Holoptelia integrifolia*), *Śatāvaryaṅkura* (apical buds of *Asparagus racemosus*), *Vaṃśakarīraḥ* (tender leaves of *Bambusa bambos*), *Sarṣapaśākam* (tender leaves of *Brassica juncea*), *Mūlakam* (*Raphanus sativus*), *Kāsamarda* (*Cassia occidentalis* var. *aristata*), *Laśūna* (*Allium sativum*), *Palāṇḍu* (*Allium cepa*), *Gṛñjanaka* (*Allium cepa* var. *aggregatum*), *Sūraṇaḥ* (*Amorphophallus paeoniifolius*), *Kuṭhēram* (*Orhthosiphon aristatus*), *Śigru* (*Moringa oleifera*), *Surasa* (*Ocimum sanctum*), *Sumukham* (*Ocimum canum*), *Bhūstṛṇam* (*Hyptis suaveolens*), *Phaṇijjam* (*Origanum marjorana*), *Ārjakam* (*Ocimum basilicum*) and *Jambīram* (*Citrus limon*) (*Sū*. 6: 72–115).

6.3.6 Fruits

In addition to their nutritive value, the various fruits included in this group of *phala varga* have the ability to cure diseases. For example, *Drākṣā* (*Vitis vinifera*) cures hemorrhages of obscure origin, alcoholism, excessive thirst, cough, fever, respiratory distress, lassitude of voice, trauma to thorax and kingly consumption (*Sū*. 6: 115–117). The group includes fruits such as *Drākṣā* (*Vitis vinifera*), *Dāḍimam* (*Punica granatum*), *Panasa* (*Artocarpus heterophyllus*), *Kharjūra* (*Phoenix dactylifera*), *Mōcaḥ* (*Musa paradisiaca*), *Nārīkēlam* (*Cocos nucifera*), *Parūṣakam* (*Phoenix pusilla*), *Āmrāta* (*Spondias pinnata*), *Tāla* (*Borassus flabellifer*), *Kāśmaryaḥ* (*Gmelina arborea*), *Rājādana* (*Manilkara hexandra*), *Madhūka* (*Madhuca longifolia*), *Sauvīram* (*Ziziphus mauritiana*), *Aṅkōla* (*Alangium salvifolium*), *Phalgu* (*Ficus hispida*), *Śḷēṣmātaka* (*Cordia dichotoma*), *Vātāmam* (*Prunus dulcis*), *Abhiṣuka* (*Aporosa acuminata*), *Akṣōṭam* (*Juglans regia*), *Mukūlakam* (*Schleichera oleosa*), *Nikōcakam* (*Pistacia vera*), *Ūrumāṇaṃ* (*Musa sapientum*), *Priyāḷam* (*Buchanania lanzan*), *Vilvam* (*Aegle marmelos*), *Kapittha* (*Feronia elephantum*), *Jambu* (*Syzygium cumini*), *Āmram* (*Mangifera indica*), *Vṛkṣāmḷam* (*Garcinia gummi-gutta*), *Pīlu* (*Salvadora persica*), *Mātuḷuṅgam* (*Citrus medica*), *Bhallātaka* (*Semecarpus anacardium*), *Pālēvatam* (*Diospyros malabarica*), *Ārukam* (*Prunus domestica*), *Karamardakam* (*Carissa carindas*), *Karkandhu* (*Ziziphus oenoplia*), *Likuca*

(*Artocarpus hirsutus*), *Airāvatam* (*Citrus reticulata*), *Dantaśaṭham* (*Citrus limon*), *Mṛgaliṇḍikam* (*Baccaurea courtallensis*) and *Amḷīkā* (*Tamarindus indica*). Fresh and dried fruits of *Drākṣā* (*Vitis vinifera*) are to be used. Use of unripe *Aegle marmelos* is recommended, as ripe ones enrage the *dōṣas*. All other fruits are to be used while they are fresh (*Sū.* 6: 117–143).

6.4 ANIMALS AND BIRDS

Many animals and birds are recommended to be used as food, medicine or therapeutic tool. They are classified into eight groups such as grazing animals (*mṛga*), birds that scatter their food and peck (*viṣkiravarga*), birds that peck and gobble their food (*pratūdavarga*), animals that live in burrows (*vilēśayavarga*), creatures that grab and tear off their food (*prasahavarga*), large animals (*mahāmṛga*), birds that live around or on the surface of water (*apcaravarga*) and animals that live under water (*matsyavarga*). They are listed in Table 6.37.

TABLE 6.37
Animals Which Are Recommended as Food, Medicine or Therapeutic Tool

Sl. No.	Sanskrit Name	Common Name	Latin Binomial	Reference
I			***Mṛgva* (Grazing Animals)**	
1	*Aja*	Goat	*Capra hircus* L.	*Sū.* 6: 54
2	*Āvi*	Sheep	*Ovis aries* L.	*Sū.* 6: 54
3	*Caruṣka*	Indian gazelle	*Procapra picticaudata* Hodgson	*Sū.* 6: 43
4	*Ēṇa*	Black deer	*Antilope cervicapra* L.	*Sū.* 6: 43
5	*Gōkarṇa*	Cow-eared deer	*Antilope picta* Pallas	*Sū.* 6: 43
6	*Hariṇa*	Chinkara	*Gazella bennetti* Sykes	*Sū.* 6: 43
7	*Kuraṅga*	Roe deer	*Capreolus capreolus* L.	*Sū.* 6: 43
8	*Mṛgamātṛka*	Red deer	*Cervus elaphus* L.	*Sū.* 6: 43
9	*Ṛśyam*	Nilgai	*Boselaphus tragocamelus* Pallas	*Sū.* 6: 43
10	*Śambara*	Sambhar	*Rusa unicolor* Kerr.	*Sū.* 6: 43
11	*Śarabha*	Wapiti	*Cervus canadensis* Erxleben	*Sū.* 6: 43
12	*Śaśa*	Rabbit	*Oryctolagus cuniculus domesticus* L.	*Sū.* 6: 43
II			***Prasaha varga* (Creatures That Grab and Tear off Their Food)**	
1	*Aśva*	Horse	*Equus cabellus* L.	*Sū.* 6: 48
2	*Aśvatara*	Mule	Hybrid of male ass and female horse	*Sū.* 6: 48
3	*Babhru*	Indian grey mongoose	*Urva edwardsii* E. Geffroy Saint-Hilaire	*Sū.* 6: 48
4	*Chuchundari*	Asian house shrew	*Suncus murinus* L.	*Ut.* 18: 57
5	*Dvīpi*	Panther	*Panthera pardus* L.	*Sū.* 6: 48
6	*Gō*	Cow	*Bos taurus* L.	*Sū.* 6: 48
7	*Jambūka*	Jackal	*Canis aureus* L.	*Sū.* 6: 49
8	*Khara*	Donkey	*Equus asinus* L.	*Sū.* 6: 48
9	*Lōpāka*	Fox	*Vulpus bengalensis* Shaw	*Sū.* 6: 49
10	*Mārjāra*	Cat	*Felis catus* L.	*Sū.* 6: 48
11	*Mūṣika*	Mouse	*Mus musculus* L.	*Sū.* 6: 48
12	*Ṛkṣa*	Bear	*Melursus labiatus* de Blainville	*Sū.* 6: 48
13	*Siṃha*	Lion	*Panthera leo* L.	*Sū.* 6: 48
14	*Tarakṣu*	Hyena	*Hyaena hyaena* L.	*Sū.* 6: 48
15	*Unduru*	Rat	*Rattus rattus* L.	*Ci.* 9: 53–54

(*Continued*)

TABLE 6.37 (*Continued*)
Animals Which Are Recommended as Food, Medicine or Therapeutic Tool

Sl. No.	Sanskrit Name	Common Name	Latin Binomial	Reference
16	*Uṣṭra*	Camel	*Camelus dromedarius* L.	*Sū.* 6: 48
17	*Vānara*	Indian monkey	*Semnopithecus entellus* Dufresne	*Sū.* 6: 48
18	*Vāntāda*	Dog	*Canis familiaris* L.	*Sū.* 6: 49
19	*Vṛka*	Wolf	*Canis lupus* L.	*Sū.* 6: 48
20	*Vyāghra*	Tiger	*Panthera tigris* L.	*Sū.* 6: 48
III		***Mahāmṛga* (Large Animals)**		
1	*Camara*	Yak	*Bos grunniens* L.	*Sū.* 6: 50–51
2	*Gavaya*	Gayal	*Bos frontalis* Lambert	*Sū.* 6: 51
3	*Khaḍga*	Rhinoceros	*Rhinoceros unicornis* L.	*Sū.* 6: 51
4	*Mahiṣa*	Buffalo	*Bubalus bubalis* L.	*Sū.* 6: 50–51
5	*Nyaṅku*	Hog deer	*Axis porcinus* Zimmermann	*Sū.* 6: 50–51
6	*Rōhita*	Screw-horned goat	*Capra falconeri* Wagner	*Sū.* 6: 50–51
7	*Rūru*	Swamp deer	*Rucervus duvaucelli* G. Cuvier	*Sū.* 6: 50–51
8	*Śṛmara*	Indian wild boar	*Sus scrofa cristatus* Wagner	*Sū.* 6: 50–51
9	*Varāha*	Pig	*Sus scrofa* L.	*Sū.* 6: 50–51
10	*Vāraṇa*	Indian elephant	*Elephas maximus indicus* Cuvier	*Sū.* 6: 50–51
IV		***Viṣkiravarga* (Birds That Scatter Their Food and Peck)**		
1	*Bakara*	Sarus crane	*Antigone antigone* L.	*Sū.* 6: 44–46
2	*Cakōram*	Crow pheasant	*Centropus sinensis* Stephens	*Sū.* 6: 44–46
3	*Girivartika*	Painted bush quail	*Perdicula erythrorhyncha* Sykes	*Sū.* 6: 44–46
4	*Gōnarda*	Great egret	*Ardea alba* L.	*Sū.* 6: 44–46
5	*Indrāhva*	Hedge sparrow	*Leucocerca aureola* Less.	*Sū.* 6: 44–46
6	*Kapiñjala*	Black partridge	*Francolinus francolinus* L.	*Sū.* 6: 44–46
7	*Krakara*	Snipe	*Lymnocryptes minimus* Brunnich	*Sū.* 6: 44–46
8	*Kukkubha*	Tibetan eared pheasant	*Crossoptilon harmani* Elwes	*Sū.* 6: 44–46
9	*Kuruvāhaka*	Grey jungle fowl	*Gallus sonneratii* Temminck	*Sū.* 6: 44–46
10	*Lāva*	Bustard quail	*Turnix suscicator* Gmelin	*Sū.* 6: 44–46
11	*Raktavartmaka*	Red jungle fowl	*Gallus gallus* L.	*Sū.* 6: 44–46
12	*Śarapada*	Stork	*Ciconia boyciana* R. Swinhoe	*Sū.* 6: 44–46
13	*Śikhi*	Peacock	*Pavo cristatus* L.	*Sū.* 6: 44–46
14	*Tāmracūḍam*	Rooster	*Gallus gallus domesticus* L.	*Sū.* 6: 44–46
15	*Tittiri*	Hill partridge	*Arborophila torqueola* Valenciennes	*Sū.* 6: 44–46
16	*Upacakra*	Chukar partridge	*Alectoris chukar* J.E. Gray	*Sū.* 6: 44–46
17	*Vāraṭam*	Bar-headed goose	*Anser indicus* Latham	*Sū.* 6: 44–46
18	*Vartāka*	Indian button quail	*Turnix sylvaticus* Desfontaines	*Sū.* 6: 44–46
19	*Vārtīka*	Jungle bush quail	*Perdicula asiatica* Latham	*Sū.* 6: 44–46
20	*Vartikā*	Bush quail	*Coturnix coturnix* L.	*Sū.* 6: 44–46
21	*Vārtīraka*	Rain quail	*Coturnix coromandelica* Gmelin	*Sū.* 6: 44–46
V		***Pratūdavarga* (Birds That Peck and Gobble Their Food)**		
1	*Bhṛṅgāhvam*	Common racket-tailed drongo	*Dicrurus paradiseus* L.	*Sū.* 6: 46–47
2	*Caṭaka*	Tree sparrow	*Passer montanus* L.	*Sū.* 6: 46–47
3	*Dātyūham*	Common moorhen	*Gallinula chloropus* L.	*Sū.* 6: 46–47
4	*Hārīta*	Yellow-footed green pigeon	*Treron phoenicoptera* Latham	*Sū.* 6: 46–47
5	*Jīvanjīvaka*	Grey peacock- pheasant	*Polyplectron bicalcaratum* L.	*Sū.* 6: 46–47

(*Continued*)

TABLE 6.37 (*Continued*)
Animals Which Are Recommended as Food, Medicine or Therapeutic Tool

Sl. No.	Sanskrit Name	Common Name	Latin Binomial	Reference
6	*Kapōta*	Common pigeon	*Columba livia* Gmelin	*Sū.* 6: 46–47
7	*Kōkila*	Indian koel	*Eudynamys scolopaceus* L.	*Sū.* 6: 46–47
8	*Laṭvā*	Little pied flycatcher	*Ficedula westermanni* Sharpe	*Sū.* 6: 46–47
9	*Śārika*	Mynah	*Acridotherus tristis* L.	*Sū.* 6: 46–47
10	*Śuka*	Parrot	*Psittacula krameri* Scopoli	*Sū.* 6: 46–47
VI		***Prasahavarga*** **(Birds That Grab and Tear Off Their Food)**		
1	*Bhāsa*	Bearded vulture	*Gypaetus barbatus* L.	*Sū.* 6: 49–50
2	*Cāṣa*	Blue jay	*Coracias benghalensis* L.	*Sū.* 6: 49
3	*Dhūmika*	Spotted owlet	*Athena brama* Temminck	*Sū.* 6: 49–50
4	*Gṛdhra*	Indian vulture	*Gyps indicus* Scopoli	*Sū.* 6: 49–50
		Red-headed vulture	*Sarcogyps calvus* Scopoli	*Ut.* 13: 37
5	*Kuliṅgaka*	Sparrow hawk	*Accipiter nisus* L.	*Sū.* 6: 49–50
6	*Kuraram*	Osprey	*Pandion haliaetus* L.	*Sū.* 6: 49–50
7	*Madhuha*	Honey buzzard	*Pernis apivorus* L.	*Sū.* 6: 49–50
8	*Śaśaghni*	Golden eagle	*Aquila chrysaetos* L.	*Sū.* 6: 49–50
9	*Śyēna*	Black kite	*Milvus migrans* Boddaert	*Sū.* 6: 49 *Śa.* 2: 18–19
10	*Ulūka*	Indian owl	*Bubo bengalensis* Franklin	*Sū.* 6: 49–50
11	*Vāyasa*	Crow	*Corvus splendens* Vieillot	*Sū.* 6: 49
VII		***Apcaravarga*** **(Birds That Live around or on the Surface of Water)**		
1	*Baka*	Common crane	*Grus grus* L.	*Sū.* 6: 51–52
2	*Cakrāhvā*	Ruddy shelduck	*Tadorna ferruginea* Pallas	*Sū.* 6: 51–52
3	*Hamsa*	Swan	*Cygnus olor* Gmelin	*Sū.* 6: 51–52
4	*Kādambam*	Grey -leg goose	*Anser anser* L.	*Sū.* 6: 51–52
5	*Kāraṇḍava*	White-fronted goose	*Anser albifrons* Scopoli	*Sū.* 6: 51–51
6	*Krouñcam*	Demoiselle crane	*Grus virgo* L.	*Sū.* 6: 51–52
7	*Madgu*	Pygmy cormorant	*Microcarbo pygmaeus* Pallas	*Sū.* 6: 51–52
8	*Pḷava*	Pelican	*Pelecanus onocrotalus* L.	*Sū.* 6: 51–52
9	*Sārasa*	Indian crane	*Megalornis grus* L.	*Sū.* 6: 51–52
10	*Utkrōśa*	Sea eagle	*Haliaeetus albicilla* L.	*Sū.* 6: 51–52 *Śa.* 2: 18–19
11	*Valāka*	Crane	*Ardea nivea* Cuvier	*Sū.* 6: 51–52
VIII		***Matsya*** **(Animals That Live under Water)**		
1	*Candrika*	Common ponyfish	*Leiognathus equulus* Forsskål	*Sū.* 6: 52–54
2	*Ciḷicima*	Prawn	*Fenneropenaeus indicus* H. Milne-Edwards	*Sū.* 6: 52–54
3	*Culūki*	Porpoise	*Neomeris phocaenoides* Cuvier	*Sū.* 6: 52–54
4	*Karkaṭaka*	Crab	*Scylla serrata* Forsskål	*Sū.* 6: 52–54
5	*Kumbhīra*	Saltwater crocodile	*Crocodylus porosus* Schneider	*Sū.* 6: 52–54
6	*Kūrma*	Crowned river turtle	*Hardella thurjii* Gray	*Sū.* 6: 52–54
7	*Makara*	Mugger crocodile	*Crocodylus palustris* Lesson	*Sū.* 6: 52–54
8	*Nakra*	Ghariyal	*Gavialis gangeticus* Gmelin	*Sū.* 6: 52–54
9	*Pāṭhina*	Boal	*Wallago atu* Bloch and Schneider	*Sū.* 6: 52–54
10	*Rāji*	Mullet	*Mugil cephalus* L.	*Sū.* 6: 52–54
11	*Rōhīta*	Red carp	*Labeo rohita* F. Hamilton	*Sū.* 6: 52–54
12	*Śambūka*	Common snail	*Cornu aspersum* O.F. Müller	*Sū.* 6: 52–54

(*Continued*)

TABLE 6.37 (*Continued*)
Animals Which Are Recommended as Food, Medicine or Therapeutic Tool

Sl. No.	Sanskrit Name	Common Name	Latin Binomial	Reference
13	*Śaṅkha*	Conch shell	*Turbinella pyrum* L.	*Sū.* 6: 52–54
14	*Śaphari*	Schilbid catfish	*Eutropiichthys vacha* Hamilton	*Sū.* 6: 52–54
15	*Śiśumāra*	Ganges river dolphin	*Platanista gangetica* Lebeck	*Sū.* 6: 52–54
16	*Śukti*	Pearl oyster	*Margaritifera margarifiera* L.	*Sū.* 6: 52–54
17	*Timiṅgala*	Whale	*Balaenoptera musculus* L.	*Sū.* 6: 52–54
18	*Udra*	Otter	*Lutra lutra* L.	*Sū.* 6: 52–54
19	*Varmi*	Catfish	*Rita rita* F. Hamilton Forsskål	*Sū.* 6: 52–54
IX			***Vilēśayavarga* (Animals That Live in Burrows)**	
1	*Ahi*	Snake	–	*Sū.* 6: 47
2	*Bhēka*	Frog	*Rana hexadactyla* Lesson	*Sū.* 6: 47
3	*Gōdha*	Monitor lizard	*Varanus bengalensis* Daudin	*Sū.* 6: 47
4	*Kṛṣṇasarpa*	King cobra	*Ophiophagus hannah* Cantor	*Ut.* 13: 41 *Ci.* 20: 12
5	*Śvāviṭ*	Indian pangolin	*Manis crassicaudata* E. Geoffroy	*Sū.* 6: 47
X			**Unclassified**	
1	*Daśana*	Wasp	*Amata huebneri* Boisduval	*Ci.* 7: 111
2	*Jalūka*	Indian leech	*Hirudo manillensis* Lesson	*Sū.* 26: 35–46
3	*Jatuka*	Indian pipistrelle	*Pipistrellus coromandra* Gray	*Ut.* 6: 42 *Ut.* 18: 58
4	*Lākṣā*	Lac insect	*Kerria lacca* Kerr	*Ci.* 3: 73
5	*Mākṣika*	Honeybee	*Apis cerana indica* Fabricius	*Ci.* 4: 48
6	*Phēna*	Cuttlefish bone	*Sepia officinalis* L.	*Ut.* 11: 12
7	*Pravāḷa*	Coral	*Heliopora coerulea* Pallas	*Ci.* 18: 14
8	*Śallyaka*	Indian crested porcupine	*Hystrix indica* Kerr	*Ut.* 2: 60 *Ka.* 4: 49
9	*Vṛścika*	Scorpion	*Scorpio hottentotta* Fabricius	*Ci.* 7: 111 *Ut.* 7: 33

6.5 ANIMAL PRODUCTS

Several animal products such as whole bodies of animals, animal heads, legs of birds, beeswax, liver, meat, milk, blood, gall bladder, gall stone, visceral fat, bone, claws, teeth, snakeskin, dung, droppings, feather, quill, hair, hide, hoof, horn, sinews, ivory, urine, conch shell, coral, pearl, shell of pearl oyster and cuttlefish bone are used as medicinal ingredients. Urine is a waste product that contains compounds that are not beneficial for the animal which voids it. Nevertheless, it will be of use to another creature. That is the reason why urines of animals such as cow, horse, donkey, elephant, camel, goat and so on are used in the treatment of diseases of human beings. This view is also strengthened by the observation that human urine is not mentioned as a medicinal substance anywhere in AH, while human hair is recommended for external use, as fumigant or collyrium. Medicinal uses of these products are presented in Table 6.38.

6.6 LIQUID AND SEMI-SOLID INGREDIENTS

Liquid ingredients like various kinds of water, milk, milk products, vegetable oils, wines, fermented liquids and urine are used in treatment of diseases. Similarly, many solid and semi-solid ingredients such as jaggery, candied sugar, thickened sugarcane juice, honey, butter, cream, fresh and aged clarified butter are also used (Table 6.39).

TABLE 6.38
Animal Products Used as Medicine

Sl. No.		Particulars	Reference
I	**Animal Head**		
1	Vulture	Ash of head mixed with antimony sulphide and applied as collyrium cures cataract.	*Ut*. 13: 37
2	Black lance-hooded cobra	Calcined powder of head of *Kṛṣṇasarpa* is ingredient of a collyrium that cures cataract.	*Ut*. 13: 38, 39–40
II	**Beeswax**		
1		Smoke emanating from a mixture containing beeswax is to be inhaled to pacify respiratory distress and hiccup.	*Ci*. 4: 12
2		Consumption of milk or clarified butter along with molten beeswax cures polydipsia of patients suffering from *grahaṇi*.	*Ci*. 10: 87
3		Beeswax is one of the five ingredients of medicated mustard oil that quickly heals itching and dry and weeping eczema.	*Ci*. 19: 84
4		Beeswax is one of the ingredients of *Piṇḍa taila* that cures *vāta*-induced blood disorders.	*Ci*. 22: 22
5		Beeswax is ingredient of a *taila* that cures *vātōṣṭhakōpa.*	*Ut*. 22: 3
6		Beeswax is ingredient of *Mañjiṣṭhādi tṛvṛta* that cures black pigmentation on face and naevi and improves complexion.	*Ut*. 32: 31–32
III	**Blood**		
1	Buffalo	Freshly drawn blood is to be fed in voiding of blood due to excessive purgation.	*Ka*. 3: 36
2	Cow	Freshly drawn blood is to be fed in voiding of blood due to excessive purgation.	*Ka*. 3: 36
3	Deer	Blood is to be sautéed in clarified butter and consumed with milk, as post-prandial drink in dysentery with bleeding.	*Ci*. 9: 87–88
4	Elephant	Freshly drawn blood is to be fed in voiding of blood due to excessive purgation.	*Ka*. 3: 36
5	Goat	Freshly drawn blood is to be fed in voiding of blood due to excessive purgation.	*Ka*. 3: 36
6		Blood is to be sautéed in clarified butter and consumed with milk, as post-prandial drink in dysentery with bleeding.	*Ci*. 9: 87–88
7	Porcupine	Blood is ingredient of *Utkārika* (balls of rice paste cooked in boiling water) to be consumed for curing respiratory distress and hiccup associated with *vāta* enragement.	*Ci*. 4: 35
8	Rabbit	Topical application of kernel of *Ziziphus mauritiana* ground to a fine paste with blood of rabbit cures black pigmentation on face.	*Ut*. 32: 20
IV	**Bone**		
1	Camel	Bone is ingredient of a powder that cures urinary gravel.	*Ci*. 11: 28
2		Ingredient of a collyrium that cures discoloration of sclera.	*Ut*. 11: 53
3		Frontal bone of camel is ingredient of *Śaśadantādi varti* that cures cataract.	*Ut*. 13: 72
4	Cat	Topical application of cat bone ground to a fine paste, with decoction of *Triphala*, heals ulcers caused by anal fistula.	*Ut*. 28: 33
5		*Nasya* with a paste containing bone of cat cures effects of rat bite.	*Ut*. 38: 32
6	Cow	Powders of *Curculigo orchioides* and bone of cow are to be mixed with honey and clarified butter. Topical application of this paste cures black pigmentation on face.	*Ut*. 32: 21

(*Continued*)

TABLE 6.38 (*Continued*)
Animal Products Used as Medicine

Sl. No.		Particulars	Reference
7		Topical application of ash mixed with sesame oil induces the growth of hair on healed wounds.	*Ut*. 25: 63
8	Cuttlefish bone	Ingredient of collyria that cures eye diseases.	*Ut*. 11: 34, 35, 44–47 *Ut*. 13: 23–25. 72, 82, 87 *Ut*. 14: 32 *Ut*. 16: 52
9	Demoiselle crane	Bone of *Krauñca* bird is one of the ingredients of a powder to be consumed for curing urinary gravel.	*Ci*. 11: 28
10	Donkey	Bone is ingredient of a powder that cures urinary gravel.	*Ci*. 11: 28
11		Bone is ingredient of *Tāmalapatrādi* pill indicated in diseases of sclera.	*Ut*. 11: 35
12		Frontal bone of donkey is ingredient of *Śaśadantādi varti* that cures cataract.	*Ut*. 13: 72
13	Indian crane	Bone of *Sārasa* bird (*Megalomis grus*) is one of the ingredients of the powder to be consumed for curing dysuria.	*Ci*. 11: 10
14	Lion	Frontal bone of lion is ingredient of *Śaśadantādi varti* that cures cataract.	*Ut*. 13: 72
15	Rabbit	Frontal bone of rabbit is ingredient of *Śaśadantādi varti* that cures cataract.	*Ut*. 13: 72
V	**Claws**		
1	Bat	Fumigation, inhalation of smoke, topical application as paste and as *taila* are recommended in treatment of insanity.	*Ut*. 6: 42
2	Bear	Claws are ingredients of *taila* indicated in spiritual afflictions and epilepsy.	*Ut*. 5: 4–5
3	Cat	Claws are ingredients of *taila* indicated in spiritual afflictions and epilepsy.	*Ut*. 5: 4–5
4	Fox	Fumigation, inhalation of smoke, topical application as paste and as *taila* are recommended in treatment of insanity.	*Ut*. 6: 42
5	Lion	Claws are ingredients of *taila* indicated in spiritual afflictions and epilepsy.	*Ut*. 5: 4–5
6	Mongoose	Claws are ingredients of *taila* indicated in spiritual afflictions and epilepsy.	*Ut*. 5: 4–5
7	Monitor lizard	Claws are ingredients of *taila* indicated in spiritual afflictions and epilepsy.	*Ut*. 5: 4–5
8	Owl	Talons are used for fumigation, inhalation of smoke, topical application as paste and as ingredient of *taila* in treatment of insanity.	*Ut*. 6: 42
9	Panther	Claws are ingredients of *taila* indicated in spiritual afflictions and epilepsy.	*Ut*. 5: 4–5
10	Porcupine	Claws are ingredients of *taila* indicated in spiritual afflictions and epilepsy.	*Ut*. 5: 4–5
11	Porcupine	Fumigation, inhalation of smoke, topical application as paste and as *taila* are recommended in treatment of insanity.	*Ut*. 6: 42
12	Tiger	Claws are ingredients of *taila* indicated in spiritual afflictions and epilepsy.	*Ut*. 5: 4–5
13	Wolf	Fumigation, inhalation of smoke, topical application as paste and as *taila* are recommended in treatment of insanity.	*Ut*. 6: 42

(*Continued*)

TABLE 6.38 (*Continued*)
Animal Products Used as Medicine

Sl. No.		Particulars	Reference
VI	**Conch Shell**		
1		Topical application of paste of conch shell, mixed with clarified butter, is recommended in the treatment of cellulitis.	*Ci.* 18: 13–14
2		Powder is ingredient of collyria that cures eye diseases.	*Ut.* 11: 12, 34, 35, 44, 47–47 *Ut.* 13: 23–24, 41, 70, 72, 74 *Ut.* 16: 23–24, 35, 54
VII	**Coral**		
1		Topical application of paste of coral, mixed with clarified butter, is recommended in the treatment of cellulitis.	*Ci.* 18: 13–14
2		Ingredient of *Srōtōja piṇḍāñjana*, a collyrium that improves vision	*Ut.* 14: 32
VIII	**Dung**		
1	Bat	Fumigation of body, inhalation of smoke and topical application as paste and as *taila* with droppings are recommended in treatment of insanity.	*Ut.* 6: 42–43
2	Bull calf	Juice of dung mixed with honey to be administered to patient of cough with *pitta* enragement.	*Ci.* 3: 30
3	Camel	Juice of dung mixed with honey cures respiratory distress and hiccup associated with *kapha*.	*Ci.* 4: 38
4	Cat	Fumigation of body with cat stool pacifies irregular fever and spiritual afflictions.	*Ci.* 1: 163 *Ut.* 3: 55 *Ut.* 5: 18
5		Fumigation of body with droppings pacifies epilepsy.	*Ut.* 7: 33
6	Cow	Juice of dung is ingredient of *Pañcagavya ghṛta* that cures epilepsy, fever and jaundice.	*Ut.* 7: 18–19
7		Internal administration of cow dung juice mixed with sugar, clarified butter and honey pacifies discomforts arising from spider sting.	*Ut.* 37: 72
8		Consumption of juice of dung mixed with honey and clarified butter pacifies hemorrhages of obscure origin.	*Ci.* 2: 29
9		Consumption of juice of dung mixed with honey pacifies *kapha*-dominant respiratory distress and hiccup.	*Ci.* 4: 38
10	Crow	Fumigation of body with droppings pacifies epilepsy.	*Ut.* 7: 33
11	Donkey	Juice of dung mixed with honey cures respiratory distress and hiccup associated with *kapha*.	*Ci.* 4: 38
12	Elephant	Juice of dung mixed with honey cures respiratory distress and hiccup associated with *kapha*.	*Ci.* 4: 38
13		Dung is ingredient of a paste that is to be applied topically for the cure of vitiligo.	*Ci.* 20: 14–15
14	Fox	Fumigation of body, inhalation of smoke and topical application as paste and as *taila* using dung are recommended in treatment of insanity.	*Ut.* 6: 42–43
15	Goat	Juice of dung mixed with honey cures respiratory distress and hiccup associated with *kapha*.	*Ci.* 4: 38

(*Continued*)

TABLE 6.38 (*Continued*) Animal Products Used as Medicine

Sl. No.		Particulars	Reference
16		Fumigation of body, inhalation of smoke and topical application as paste and as *taila* with droppings are recommended in treatment of insanity.	*Ut.* 6: 42–43
17	Honey bee	*Nasya* of honey bees excrement, mixed with breast milk, cures respiratory distress and hiccup.	*Ci.* 4: 48
18	Horse	Consumption of juice of dung mixed with honey and clarified butter cures hemorrhages of obscure origin.	*Ci.* 2: 29
19		Juice of dung mixed with honey cures respiratory distress and hiccup associated with *kapha.*	*Ci.* 4: 38
20		Dung is ingredient of a medicinal wick that is to be used for inhalation of smoke in treatment of chronic catarrh.	*Ut.* 20: 17
21		Topical application of juice of dung, mixed with rock salt, cures pityriasis.	*Ut.* 24: 21–22
22	Mongoose	Fumigation of body with droppings pacifies epilepsy.	*Ut.* 7: 33
23	Owl	Fumigation of body, inhalation of smoke and topical application as paste and as *taila* with droppings are recommended in treatment of insanity.	*Ut.* 6: 42–43
24		Fumigation of body with droppings pacifies epilepsy.	*Ut.* 7: 33
25	Pig	Juice of dung mixed with honey cures respiratory distress and hiccup associated with *kapha.*	*Ci.* 4: 38
26	Pigeon	Consumption of droppings of pigeon, mixed with honey, cures hemorrhages of obscure origin.	*Ci.* 21: 30
27		Pigeon dropping is ingredient of a paste for topical application in scorpion sting.	*Ut.* 37: 34, 40
28	Porcupine	Fumigation of body, inhalation of smoke and topical application as paste and as *taila* with dung are recommended in treatment of insanity.	*Ut.* 6: 42–43
29	Rooster	Dropping is ingredient of a paste to be applied topically on hemorrhoids.	*Ci.* 8: 22
30	Sheep	Juice of dung mixed with honey cures respiratory distress and hiccup associated with *kapha.*	*Ci.* 4: 38
31	Vulture	Fumigation of body with droppings pacifies epilepsy.	*Ut.* 7: 33
IX	**Feather**		
1	Crow	Fumigation with feathers pacifies epilepsy.	*Ut.* 7: 33
2	Owl	Fumigation with feathers pacifies epilepsy.	*Ut.*7: 33
3	Peacock	Ingredient of a mixture to be used for fumigating the whole body of person suffering from spiritual affliction and irregular fever	*Ut.* 3: 51 *Ut.* 5: 18
4		Consumption of ash mixed with honey cures cough with *kapha* enragement.	*Ci.* 3: 46
5	Rooster	Consumption of ash mixed with honey cures cough with *kapha* enragement.	*Ci.* 3: 46
6	Vulture	Fumigation with feathers pacifies epilepsy.	*Ut.* 7: 33
X	**Gall bladder**		
1	Bat	Fumigation of body, inhalation of smoke and topical application as paste and as *taila* with gall bladder are recommended in treatment of insanity.	*Ut.* 6: 42–43
2	Bear	Ingredient of a *taila* indicated in spiritual afflictions and epilepsy.	*Ut.* 5: 4–6

(*Continued*)

TABLE 6.38 (*Continued*)
Animal Products Used as Medicine

Sl. No.		Particulars	Reference
3		*Nasya* or *abhyaṅga* using *taila* prepared with bile cures epilepsy.	*Ut.* 7: 30–31
4	Camel	Ingredient of a *taila* indicated in spiritual afflictions and epilepsy.	*Ut.* 5: 4–6
5	Cat	Ingredient of a *taila* indicated in spiritual afflictions and epilepsy.	*Ut.* 5: 4–6
6		*Nasya* with bile cures epilepsy.	*Ut.* 7: 30
7		Gall bladder is ingredient of a paste to be used for *nasya*, *abhyaṅga* and *sēka* on persons suffering from spiritual afflictions.	*Ut.* 5: 8–9
8	Cow	Ingredient of *Hiṅguvyōṣādi ghṛta* or *taila* that is to be administered as internal medicine, *nasya* or *abhyaṅga*, to cure epilepsy, spiritual afflictions and insanity.	*Ut.* 5: 2–8 *Ut.* 5: 16 *Ut.* 7: 30–31
9		*Nasya* with bile from tawny-colored cow cures epilepsy.	*Ut.* 7: 29
10	Dog	*Nasya* with bile cures epilepsy.	*Ut.* 7: 30
11	Donkey	Ingredient of a *taila* indicated in spiritual afflictions and epilepsy.	*Ut.* 5: 4–6
12	Elephant	*Nasya* or *abhyaṅga* using *taila* prepared with bile cures epilepsy.	*Ut.* 7: 30–31
13	Fox	*Nasya* with bile cures epilepsy.	*Ut.* 7: 30
14		Fumigation of body, inhalation of smoke and topical application as paste and as *taila* with gall bladder are recommended in treatment of insanity.	*Ut.* 6: 42
15	Goat	Fumigation of body, inhalation of smoke and topical application as paste and as *taila* with gall bladder are recommended in treatment of insanity.	*Ut.* 6: 42–43
16	Horse	Ingredient of a *taila* indicated in spiritual afflictions and epilepsy.	*Ut.* 5: 4–6
17	Lion	Ingredient of a *taila* indicated in spiritual afflictions and epilepsy.	*Ut.* 5: 4–6
18		*Nasya* with bile cures epilepsy.	*Ut.* 7: 30
19	Mongoose	Ingredient of a *taila* indicated in spiritual afflictions and epilepsy.	*Ut.* 5: 4–6
20		Gall bladder is ingredient of a paste to be used for *nasya*, *abhyaṅga* and *sēka* on persons suffering from spiritual afflictions.	*Ut.* 5: 8–9
21		*Nasya* or *abhyaṅga* using *taila* prepared with bile cures epilepsy.	*Ut.* 7: 30–31
22	Monitor lizard	Ingredient of a *taila* indicated in spiritual afflictions and epilepsy.	*Ut.* 5: 4–6
23		Gall bladder is ingredient of a paste to be used for *nasya*, *abhyaṅga* and *sēka* on persons suffering from spiritual afflictions.	*Ut.* 5: 8–9
24		*Nasya* or *abhyaṅga* using *taila* prepared with bile cures epilepsy.	*Ut.* 7: 30–31
25	Owl	Fumigation of body, inhalation of smoke and topical application as paste and as *taila* with gall bladder are recommended in treatment of insanity.	*Ut.* 6: 42–43
26	Panther	Ingredient of a *taila* indicated in spiritual afflictions and epilepsy.	*Ut.* 5: 4–6
27	Peacock	Burnt bile mixed with oil of *Terminalia belerica* seeds is to be applied on skin to cure vitiligo.	*Ci.* 20: 12

(*Continued*)

TABLE 6.38 (*Continued*)
Animal Products Used as Medicine

Sl. No.		Particulars	Reference
28	Pig	Suppository prepared with bile of pig is to be applied in *raktagulma*.	*Ut*. 14: 124–125
29	Porcupine	Ingredient of a *taila* indicated in spiritual afflictions and epilepsy.	*Ut*. 5: 4–6
30		Fumigation of body, inhalation of smoke and topical application as paste and as *taila* with gall bladder are recommended in treatment of insanity.	*Ut*. 6: 42–43
31	Rabbit	Gall bladder is ingredient of a paste to be used for *nasya*, *abhyaṅga* and *sēka* on persons suffering from spiritual afflictions.	*Ut*. 5: 8–9
32	Spotted deer	*Nasya* or *abhyaṅga* using *taila* prepared with bile cures epilepsy.	*Ut*. 7: 30–31
33	Tiger	Ingredient of a *taila* indicated in spiritual afflictions and epilepsy.	*Ut*. 5: 4–6
34	Wolf	Fumigation of body, inhalation of smoke and topical application as paste and as *taila* with gall bladder are recommended in treatment of insanity.	*Ut*. 6: 42–43
XI	**Gall Stone**		
1		Gall stone of cow (*gōrōcana*) is one of the ingredients of a paste that is to be applied on various parts of the newborn infant, during naming ceremony.	*Ut*. 1: 22–23
2		*Gōrōcana* is ingredient of *Mañjiṣṭhādi tṛvṛta* that cures black pigmentation on face, naevi and improves complexion.	*Ut*. 32: 31–32
XII	**Hair**		
1	Cow	Hair from tail switch is ingredient of *Gōśṛṅgādi dhūpa* to be used for fumigating patients suffering from spiritual afflictions and irregular fever.	*Ut*. 3: 55
2		Topical application of ash of hair on tail switch mixed with sesame oil induces the growth of hair on healed wounds.	*Ut*. 25: 63
3		Inhalation of smoke from hair pacifies respiratory distress.	*Ci*. 4: 12
4	Human hair	Ingredient of mixture to be used for fumigating hemorrhoids.	*Ci*.8: 18
5		Ingredient of a mixture to be used for fumigating body of person suffering from spiritual affliction and irregular fever.	*Ut*. 3: 56 *Ut*. 5: 18
6		Ash of human hair, mixed with clarified butter, is to be applied as collyrium to cure *Śuṣkākṣipāka*.	*Ut*. 16: 30–31
7		Ingredient of a mixture to be used for fumigating sites of insect stings.	*Ut*. 37: 23–24
XIII	**Hide**		
1	Bat	Hide is ingredient of a mixture to be used for fumigation of body, inhalation of smoke, topical application as paste and as *taila*, in the treatment of insanity.	*Ut*. 6: 42–43
2	Bear	Inhalation of smoke cures respiratory distress.	*Ci*. 4: 13
3		Hide is ingredient of a *taila* indicated in spiritual afflictions and epilepsy.	*Ut*. 5: 4–5
4		Hide is an ingredient of the mixture that is to be used along with clarified butter to fumigate the body of a child afflicted with evil spirit.	*Ut*. 3: 47
5	Camel	Hide is ingredient of a *taila* indicated in spiritual afflictions and epilepsy.	*Ut*. 5: 4–5

(*Continued*)

TABLE 6.38 (*Continued*)
Animal Products Used as Medicine

Sl. No.		Particulars	Reference
6	Cat	Hide is ingredient of a mixture to be used for fumigating hemorrhoids.	*Ci*. 8: 18
7		Hide is ingredient of a *taila* indicated in spiritual afflictions and epilepsy.	*Ut*. 5: 4–5
8	Cow	Ingredient of *Gōśṛṅgādi dhūpa* to be used for fumigating patients suffering from spiritual afflictions and irregular fever.	*Ut*. 3: 55
9		Topical application of ash mixed with sesame oil induces the growth of hair on healed wounds.	*Ut*. 25: 63
10	Deer	Inhalation of smoke cures respiratory distress.	*Ci*. 4: 13
11	Donkey	Hide is ingredient of a *taila* indicated in spiritual afflictions and epilepsy.	*Ut*. 5: 4–5
12	Elephant	Ash of hide mixed with sesame oil forms a paste that is very efficacious in treatment of vitiligo.	*Ci*. 20: 10
13	Fox	Hide is ingredient of a mixture to be used for fumigation of body, inhalation of smoke and topical application as paste and as *taila* in the treatment of insanity.	*Ut*. 6: 42–43
14	Goat	Hide is ingredient of a mixture to be used for fumigation of body, inhalation of smoke and topical application as paste and as *taila* in the treatment of insanity.	*Ut*. 6: 42–43
15	Horse	Hide is ingredient of a *taila* indicated in spiritual afflictions and epilepsy.	*Ut*. 5: 4–5
16	Lion	Hide is an ingredient of the mixture that is to be used along with clarified butter to fumigate the body of a child afflicted with evil spirit.	*Ut*. 3: 47
17		Hide is ingredient of a *taila* indicated in spiritual afflictions and epilepsy.	*Ut*. 5: 4–5
18	Mongoose	Hide is ingredient of a *taila* indicated in spiritual afflictions and epilepsy.	*Ut*. 5: 4–5
19	Monitor lizard	Inhalation of smoke cures respiratory distress.	*Ci*. 4: 13
20		Ash of hide mixed with honey and clarified butter is to be administered to infants to pacify vomiting.	*Ut*. 2: 60
21		Hide is ingredient of a *taila* indicated in spiritual afflictions and epilepsy.	*Ut*. 5: 4–5
22	Nilgai	Ash of hide mixed with honey and clarified butter is to be administered to infants to pacify vomiting.	*Ut*. 2: 60
23	Panther	Hide is an ingredient of the mixture that is to be used along with clarified butter to fumigate the body of a child afflicted with evil spirit.	*Ut*. 3: 47
24		Hide is ingredient of a *taila* indicated in spiritual afflictions and epilepsy.	*Ut*. 5: 4–5
25	Porcupine	Ash of hide mixed with honey and clarified butter is to be administered to infants to pacify vomiting.	*Ut*. 2: 60
26		Hide is ingredient of a *taila* indicated in spiritual afflictions and epilepsy.	*Ut*. 5: 4–5
27		Hide is ingredient of a mixture to be used for fumigation of body, inhalation of smoke and topical application as paste and as *taila*, in the treatment of insanity.	*Ut*. 6: 42–43
28	Snakeskin	Snakeskin is an ingredient of the mixture that is to be used along with clarified butter to fumigate the body of a child afflicted with evil spirit.	*Ut*. 3: 47

(*Continued*)

TABLE 6.38 (*Continued*)
Animal Products Used as Medicine

Sl. No.		Particulars	Reference
29	Tiger	Hide is an ingredient of the mixture that is to be used along with clarified butter to fumigate the body of a child afflicted with evil spirit.	*Ut*. 3: 47
30		Hide is ingredient of a *taila* indicated in spiritual afflictions and epilepsy.	*Ut*. 5: 4–5
31		Ash of hide mixed with sesame oil forms a paste that is very efficacious in treatment of vitiligo.	*Ci*. 20: 10
32	Wolf	Hide is ingredient of a mixture to be used for fumigation of body, inhalation of smoke and topical application as paste and as *taila* in the treatment of insanity.	*Ut*. 6: 42–43
XIV	**Hoof**		
1	Camel	Ingredient of *taila* is indicated in spiritual afflictions and epilepsy.	*Ut*. 5: 4–5
2	Cow	Ingredient of *Hiṅguvyōṣādi ghṛta* or *taila* that is to be administered as internal medicine, *nasya* or *abhyaṅga*, to cure epilepsy, spiritual afflictions and insanity.	*Ut*. 5: 2–8
3		Topical application of ash mixed with sesame oil induces the growth of hair on healed wounds.	*Ut*. 25: 63
4		Topical application of ash of hoof mixed with mustard oil cures cervical glandular swelling.	*Ut*. 30: 28–29
5	Deer	Smoke from a mixture containing the hoof is to be inhaled to cure respiratory distress and hiccup.	*Ci*. 4: 13
6	Donkey	Ingredient of *taila* indicated in spiritual afflictions and epilepsy.	*Ut*. 5: 4–5
7	Goat	Fumigation of body, inhalation of smoke and topical application as paste and as medicated oil are recommended in treatment of insanity.	*Ut*. 6: 42
8	Goat	Ash of hoof mixed with mustard oil is to be applied on cervical glandular swellings.	*Ut*. 30: 28
9	Horse	Ingredient of *taila* indicated in spiritual afflictions and epilepsy.	*Ut*. 5: 4–5
10	Horse	Ash of hoof mixed with mustard oil is to be applied on cervical glandular swellings.	*Ut*. 30: 28
11	Sheep	Ash of hoof mixed with mustard oil is to be applied on cervical glandular swellings.	*Ut*. 30: 28
XV	**Horn**		
1	Cow	Inhalation of smoke from horn pacifies respiratory distress.	*Ci*. 4: 12
2		Ingredient of *Gōśṛṅgādi dhūpa* to be used for fumigating patients suffering from spiritual afflictions and irregular fever.	*Ut*. 3: 55
3		Topical application of ash mixed with sesame oil induces the growth of hair on healed wounds.	*Ut*. 25: 63
4	Deer	Smoke emanating from a mixture containing the horn is to be inhaled to cure respiratory distress and hiccup.	*Ci*. 4: 13
5	Goat	Topical application of burnt horn mixed with sesame oil blackens gray hairs that appear on skin denuded of hair.	*Ut*. 24: 32
XVI	**Ivory**		
1	Elephant	Topical application of calcined ivory mixed with sesame oil cures alopecia.	*Ut*. 24: 31

(*Continued*)

TABLE 6.38 (*Continued*)
Animal Products Used as Medicine

Sl. No.		Particulars	Reference
XVII	**Lac**		
1		Internal administration of lac with honey and milk is recommended to cure injury to thorax.	*Ci.* 3: 73
2		Lac is ingredient of a paste that cures tinea corporis.	*Ci.* 19: 85
3		Ingredient of *Mañjiṣṭhādi sēka* that pacifies conjunctivitis	*Ut.* 16: 13, 57
4		Ingredient of *Lākṣādi nasya* that cures catarrh	*Ut.* 20: 21–22
5		Ingredient of *Mālatyādi taila* that cures pityriasis	*Ut.* 24: 27
6		Ingredient of a paste that heals wounds quickly	*Ut.* 25: 60–61
XVIII	**Leg of Bird**		
1	Peacock	Consumption of ash of legs, mixed with honey and clarified butter, pacifies cough.	*Ci.* 3: 170
XIX	**Liver**		
1	Buffalo	Consumption of liver sautéed with sesame oil and clarified butter is recommended for the cure of night blindness.	*Ci.* 13: 89
2	Goat	Consumption of uncooked liver along with gall bladder cures hemorrhages of obscure origin.	*Ci.* 2: 31
XX	**Meat**		
1	Asian house shrew	Flesh of *Suncus murinus* is ingredient of a *taila* that enhances the growth of lobule of ear.	*Ut.* 18: 57–59
2	Bat	Flesh of bat is ingredient of a *taila* that enhances the growth of lobule of ear.	*Ut.* 18: 57–59
3	Bearded vulture	Patient suffering from kingly consumption is to be fed with cooked flesh of bearded vulture (*Gypaetus barbatus*).	*Ci.* 5: 6–7
4	Black partridge	Soup of meat of black partridge is recommended in alcoholism with enragement of *pitta*.	*Ci.* 7: 21–22
5	Bustard quail	Soup of meat is a recommended food for patients suffering from hemorrhoids.	*Ci.* 8: 79–80
6		Dried meat is to be powdered, mixed with honey and applied on gums to hasten the appearance of teeth.	*Ut.* 2: 36–37
7	Camel	Patient suffering from kingly consumption is to be fed with cooked meat of camel.	*Ci.* 5: 6–7
8		Cooked camel meat is a recommended food for patients suffering from hemorrhoids.	*Ci.* 8: 85
9	Cat	Flesh is ingredient of fluid for performing *Vājīkaraṇa vasti* to improve sexual vigor.	*Ka.* 4: 49–52
10	Crow	Patient suffering from kingly consumption is to be fed with cooked flesh of crow.	*Ci.* 5: 6–7
11	Demoiselle crane	Meat soup mixed with clarified butter is to be consumed by patients suffering from heart disease.	*Ci.* 6: 35–38
12		Meat is to be consumed for pacifying *vāta gulma* (phantom tumor caused by aggravation of *vāta*).	*Ci.* 14: 59
13	Dog	Fumigation of body with flesh is indicated in insanity.	*Ut.* 6: 44
14	Donkey	Patient suffering from kingly consumption is to be fed with cooked flesh of donkey.	*Ci.* 5: 6–7
15	Fish	Fumigation of body with fish is indicated in insanity.	*Ut.* 6: 44
16	Fox	Soup of flesh is a recommended food for patients suffering from dysentery.	*Ci.* 9: 22

(*Continued*)

TABLE 6.38 (*Continued*)
Animal Products Used as Medicine

Sl. No.		Particulars	Reference
17	Goat	Soup of goat meat is recommended in alcoholism with enragement of *pitta*.	*Ci.* 7: 21–22
18		Cooked goat meat is a recommended food for patients suffering from hemorrhoids.	*Ci.* 8: 85
19		Meat soup is to be administered to lean and weak persons suffering from polydipsia.	*Ci.* 6: 81
20		Soup mixed with pomegranate juice, clarified butter and coriander leaves is to be consumed along with cooked *Śāli* rice in dysentery.	*Ci.* 9: 33–35
21	Hill partridge	Meat cooked in decoction of *Ficus glomerata* bark is to be consumed when *vāta* is enraged in hemorrhages of obscure origin.	*Ci.* 2: 23
22		Soup of meat is recommended to alcoholics suffering from cough and spitting of blood.	*Ci.* 7: 26
23		Soup of meat is a recommended food for patients suffering from hemorrhoids.	*Ci.* 8: 79–80
24		Soup of meat is a recommended food for patients suffering from dysentery.	*Ci.* 9: 22
25		Dried meat is to be powdered, mixed with honey and applied on gums to hasten the appearance of teeth.	*Ut.* 2: 37–37
26		Meat is to be fed to patient suffering from irregular fever, prior to performing emesis.	*Ci.* 1: 158–159
27		Meat soup mixed with clarified butter is to be consumed by patients suffering from heart disease.	*Ci.* 6: 35–36
28		Meat is to be consumed for pacifying *vāta gulma* (phantom tumor caused by aggravation of *vāta*).	*Ci.* 14: 59
29		Soup recommended in treatment of jaundice	*Ci.* 16: 49
30	Horse	Patient suffering from kingly consumption is to be fed with cooked flesh of horse.	*Ci.* 5: 6–7
31	Indian button quail	Soup of meat is a recommended food for patients suffering from hemorrhoids.	*Ci.* 8: 79–80
32		Soup of meat is a recommended food for patients suffering from dysentery.	*Ci.* 9: 22
33		Meat soup mixed with clarified butter is to be consumed by patients suffering from heart disease.	*Ci.* 6: 35–38
34		Meat is to be consumed for pacifying *vāta gulma*.	*Ci.* 14: 59
35	Kite	Soup of flesh is indicated in *līna*.	*Śa.* 2: 18–19
37	Mongoose	Patient suffering from kingly consumption is to be fed with cooked flesh of mongoose.	*Ci.* 5: 6–7
38		Flesh is ingredient of fluid for performing *Vājīkaraṇa vasti* to improve sexual vigor.	*Ka.* 4: 49–52
39	Monitor lizard	Cooked flesh is a recommended food for patients suffering from hemorrhoids.	*Ci.* 8: 85
40		Flesh is ingredient of fluid for performing *Vājīkaraṇa vasti* to improve sexual vigor.	*Ka.* 4: 49–52
41		Patient suffering from edema is to consume food along with soup of flesh.	*Ci.* 17: 19

(*Continued*)

TABLE 6.38 (*Continued*)
Animal Products Used as Medicine

Sl. No.		Particulars	Reference
42		Fruit of *Piper longum* is placed inside the liver of monitor lizard and cooked. Thereafter, it is dried, ground, mixed with honey and applied as collyrium, to cure night blindness.	*Ut.* 13: 88
43	Owl	Patient suffering from kingly consumption is to be fed with cooked flesh of owl.	*Ci.* 5: 6–7
44	Peacock	Soup of flesh is indicated in *līna*.	*Śā.* 2: 18–19
45		Meat cooked in decoction of *Ficus retusa* bark is to be consumed, when *vāta* is enraged in hemorrhages of obscure origin.	
46		Soup of meat is a recommended food for patients suffering from hemorrhoids.	*Ci.* 8: 79–80
47		Soup of meat is a recommended food for patients suffering from dysentery.	*Ci.* 9: 22
48		Meat is ingredient of *Māyūra ghṛta* and *Mahāmāyūra ghṛta* indicated in diseases above the collar.	*Ut.* 24: 47–56
49		Ingredient of medicinal fluid for performing *Māyūramāmsa vasti*, to improve stamina and sexual vigor.	*Ka.* 4: 45–47
50		Meat is to be fed to patient suffering from irregular fever, prior to performing emesis.	*Ci.* 1: 158–159
51		Meat soup mixed with clarified butter is to be consumed by patients suffering from heart disease.	*Ci.* 6: 35–38
52		Meat is to be consumed for pacifying *vāta gulma* (phantom tumor caused by aggravation of *vāta*).	*Ci.* 14: 59
53		Soup recommended in treatment of jaundice	*Ci.* 16: 49
54	Pig	Consumption of gruel cooked in pork soup, mixed with rock salt and clarified butter cures cough with enragement of *vāta*.	*Ci.* 3: 23
55	Pigeon	Consumption of decoction of pigeon meat, *Kaempferia galanga* and *Inula racemosa* pacifies effects of poisoning.	*Ut.* 35: 59
56	Porcupine	Patient suffering from edema is to consume food along with soup of flesh.	*Ci.* 17: 19
57		Flesh is ingredient of fluid for performing *Vājīkaraṇa vasti* to improve sexual vigor.	*Ka.* 4: 49–52
58		Flesh of porcupine is ingredient of *Utkārika* to be consumed for the cure of respiratory distress and hiccup associated with *vāta* enragement.	*Ci.* 4: 34–35
59	Rabbit	Rabbit meat cooked with *Chenopodium murale* is to be consumed to pacify constipation experienced in hemorrhages of obscure origin.	*Ci.* 2: 23
60		Rabbit meat is ingredient of *Utkārika* to be consumed for the cure of respiratory distress and hiccup associated with *vāta* enragement.	*Ci.* 4: 34–35
61		Soup of rabbit meat is recommended in alcoholism with enragement of *pitta*.	*Ci.* 7: 21–22
62		A medicated clarified butter similar to *Mahāmāyūra ghṛta* can be prepared using rabbit meat.	*Ut.* 24: 56
63	Rat	Flesh is an ingredient of a *taila* for internal administration and topical application in prolapse of rectum.	*Ci.* 9: 53–54

(*Continued*)

TABLE 6.38 (*Continued*)
Animal Products Used as Medicine

Sl. No.		Particulars	Reference
64		Flesh is ingredient of fluid for performing *Vājīkaraṇa vasti* to improve sexual vigor.	*Ka.* 4: 49–52
65		A medicated clarified butter similar to *Mahāmāyūra ghṛta* can be prepared using the flesh of rat.	*Ut.* 24: 56
66	Rooster	Meat cooked in decoction of *Ficus benghalensis* bark is to be consumed, when *vāta* is enraged in hemorrhages of obscure origin.	*Ci.* 2: 24
67		Consumption of gruel cooked in rooster meat soup, mixed with rock salt and clarified butter cures cough with enragement of *vāta*.	*Ci.* 3: 23
68		Soup of meat is a recommended food for patients suffering from hemorrhoids.	*Ci.* 8: 79–80
69		Soup of flesh is a recommended food for patients suffering from dysentery.	*Ci.* 9: 22
70		Meat is to be fed to patient suffering from irregular fever, prior to performing emesis.	*Ci.* 1: 158–159
71		Meat soup mixed with clarified butter is to be consumed by patients suffering from heart disease.	*Ci.* 6: 35–38
72		Consumption of gruel prepared in decoction of *Vēllantarādi gaṇa* and soup of rooster meat, mixed with thickened juice of sugarcane, cures dysuria.	*Ci.* 11: 38
73		Consumption of rooster meat is recommended in *Śukḷāśmari*.	*Ci.* 11: 42
74		Meat is to be consumed for pacifying *vāta gulma* (phantom tumor caused by aggravation of *vāta*).	*Ci.* 14: 58
75		Soup recommended in treatment of jaundice	*Ci.* 16: 49
76		A medicated clarified butter similar to *Mahāmāyūra ghṛta* can be prepared using rooster meat.	*Ut.* 24: 56
77	Sea eagle	Soup of flesh of sea eagle (*Haliaeetus albicilla*) is indicated in *līna*.	*Śā.* 2: 18–19
78	Sheep	Soup mixed with pomegranate juice, clarified butter and coriander leaves is to be consumed along with cooked *Śāli* rice in dysentery.	*Ci.* 9: 33–35
79		Consumption of sheep meat pacifies excessive appetite (*atyagni*).	*Ci.* 10: 87
80	Snake	Patient suffering from kingly consumption is to be fed with cooked flesh of snake.	*Ci.* 5: 6–7
81	Spotted deer	Soup of spotted deer meat is recommended in alcoholism with enragement of *pitta*.	*Ci.* 7: 21–22
82	Swan	A medicated clarified butter similar to *Mahāmāyūra ghṛta* can be prepared using the meat of swan.	*Ut.* 24: 56
83	Tiger	Patient suffering from kingly consumption is to be fed with cooked flesh of tiger.	*Ci.* 5: 6–7
84	Tortoise	Soup of flesh is a recommended food for patients suffering from dysentery.	*Ci.* 9: 22
85		Patient suffering from edema is to consume food along with soup of flesh.	*Ci.* 17: 19
86	Vulture	Patient suffering from kingly consumption is to be fed with cooked flesh of vulture.	*Ci.* 5: 6–7

(*Continued*)

TABLE 6.38 (*Continued*)
Animal Products Used as Medicine

Sl. No.		Particulars	Reference
87	Wolf	Patient suffering from kingly consumption is to be fed with cooked flesh of wolf.	*Ci.* 5: 6–7
XXI	**Milk**		
1	Breast milk	Good for applying in eyes and nasal instillation; pacifies eye diseases	*Sū.* 5: 26
2	Buffalo	Pacifies excessive appetite and insomnia	*Sū.* 5: 23
3	Camel	Consumption of camel milk is recommended to cure enlargement of abdomen, hemorrhoids, edema and worm infestation.	*Sū.* 5: 25 *Ci.* 15: 3, 27, 84
4		Patient is to survive solely on camel milk for seven days or a month to cure edema.	*Ci.* 17: 10
5	Cow	Pacifies tiredness, giddiness, respiratory distress, cough, excessive thirst, chronic fever, dysuria and hemorrhages of obscure origin	*Sū.* 5: 20–23
6	Donkey	Pacifies enragement of *vāta* in limbs.	*Sū.* 5: 27–28
7	Elephant	Strengthens body	*Sū.* 5: 27
8	Goat	Pacifies emaciation, fever, respiratory distress, hemorrhages of obscure origin and dysentery	*Sū.* 5: 24
9	Horse	Pacifies enragement of *vāta* in limbs.	*Sū.* 5: 27–28
10	Sheep	Cures *vāta* diseases, hiccup and respiratory distress.	*Sū.* 5: 26–27
XXII	**Pearl**		
1		Ash prepared from pearl oyster shell mixed with cow milk is to be administered in enlargement of spleen.	*Ci.* 15: 86
2		Topical application of paste of pearl, mixed with clarified butter, is recommended in the treatment of cellulitis.	*Ci.* 18: 13–14
3		Topical application of paste of pearl oyster shell, mixed with clarified butter, is recommended in the treatment of cellulitis.	*Ci.* 18: 13–14
4		Pearl is ingredient of collyria that cure eye diseases.	*Ut.* 11: 34 *Ut.* 14: 31
5		Application of paste of pearl on the site of insect sting pacifies edema, pricking pain and wounds.	*Ut.* 37: 27
XXIII	**Quill**		
1	Porcupine	Consumption of ash of quill mixed with honey, clarified butter and sugar cures cough and respiratory distress.	*Ci.* 3: 169–170
XXIV	**Sinews**		
1	Cow	Inhalation of smoke from sinews pacifies respiratory distress.	*Ci.* 4: 12
XXV	**Snakeskin**		
1		Ingredient of mixture to be used for fumigating hemorrhoids	*Ci.* 8: 18
2		Ingredient of mixture to be used for fumigating the whole body of person suffering from spiritual afflictions and irregular fever	*Ut.* 3: 50 *Ut.* 5: 18
3		Smoke from molted skin of *Kṛṣṇasarpa* (black lance-hooded cobra) is to be introduced into the vagina to facilitate the expulsion of feus during childbirth.	*Śa.* 1: 83
XXVI	**Tooth**		
1	Bear	Ingredient of *taila* indicated in spiritual afflictions and epilepsy	*Ut.* 5: 4–5
2	Boar	Tooth is ingredient of *Dantavarti* that cures *Kṣataśukḷa.*	*Ut.* 11: 33–34

(*Continued*)

TABLE 6.38 (*Continued*)
Animal Products Used as Medicine

Sl. No.		Particulars	Reference
3	Camel	Ingredient of *taila* indicated in spiritual afflictions and epilepsy	*Ut*. 5: 4–5
4		Tooth is ingredient of *Dantavarti* that cures *Kṣataśukḷa.*	*Ut*. 11: 33–34, 47–48
5		Ingredient of *Śaśadantādi varti* that cures cataract	*Ut*. 13: 72–73
6		Topical application of paste of camel tooth and *Ceratophyllum demersum* (*Śaivala*) pacifies scorpion sting.	*Ut*. 37: 35
7	Cat	Ingredient of *taila* indicated in spiritual afflictions and epilepsy	*Ut*. 5: 4–5
8	Cow	Ingredient of *Hiṅguvyōṣādi ghṛta* or *taila* to be administered as internal medicine, *nasya* or *abhyaṅga* to cure epilepsy, spiritual afflictions and insanity	*Ut*. 5: 2–8
9		Tooth is ingredient of *Dantavarti* that cures *Kṣataśukḷa.*	*Ut*. 11: 33, 35, 47 *Ut*. 13: 72–73
10	Donkey	Ingredient of *taila* indicated in spiritual afflictions and epilepsy	*Ut*. 5: 4–5
11		Tooth is ingredient of *Dantavarti* that cures *Kṣataśukḷa.*	*Ut*. 11: 33–34, 47–48
12		Ingredient of *Śaśadantādi varti* that cures cataract	*Ut*. 13: 72–73
13	Elephant	Tooth is ingredient of a mixture to be used for fumigating body of person suffering from spiritual affliction and insanity.	*Ut*. 5: 18
14		Tooth is ingredient of *Dantavarti* that cures *Kṣataśukḷa.*	*Ut*. 11: 33–34, 44–45
15	Goat	Tooth is ingredient of *Dantavarti* that cures *Kṣataśukḷa.*	*Ut*. 11: 33–34
16	Horse	Ingredient of *taila* indicated in spiritual afflictions and epilepsy	*Ut*. 5: 4–5
17		Tooth is ingredient of *Dantavarti* that cures *Kṣataśukḷa.*	*Ut*. 11: 33–34, 47–48
18	Lion	Ingredient of *taila* indicated in spiritual afflictions and epilepsy	*Ut*. 5: 4–5
19		Ingredient of *Śaśadantādi varti* that cures cataract	*Ut*. 13: 72–73
20	Mongoose	Ingredient of *taila* indicated in spiritual afflictions and epilepsy	*Ut*. 5: 4–5
21	Monitor lizard	Ingredient of *taila* indicated in spiritual afflictions and epilepsy	*Ut. 5: 4–5*
22	Panther	Ingredient of *taila* indicated in spiritual afflictions and epilepsy	*Ut*. 5: 4–5
23	Porcupine	Ingredient of *taila* indicated in spiritual afflictions and epilepsy	*Ut*. 5: 4–5
24	Rabbit	Ingredient of *Śaśadantādi varti* that cures cataract	*Ut*. 13: 72–73
25	Tiger	Ingredient of *taila* indicated in spiritual afflictions and epilepsy	*Ut*. 5: 4–5
XXVII	**Urine**		
1	Bat	Fumigation of body, inhalation of smoke and topical application as paste and *taila* with urine are recommended in treatment of insanity.	*Ut*. 6: 42–43
2	Buffalo	Ingredient of an alkali which improves abdominal fire	*Ci*. 10: 56–57
3		Urine is to be consumed for one to two months for curing enlargement of abdomen.	*Ci*. 15: 2
4		Buffalo urine mixed with cow milk is to be consumed along with food to pacify edema.	*Ci*. 17: 10
5	Camel	Ingredient of *Bhūtarāva* and *Mahābhūtarāva ghṛta* indicated in affliction with evil spirits.	*Ut*. 5: 19–20
6	Cow	Cures worm infestation, edema, enlargement of abdomen, bloating of abdomen, pricking pain, morbid pallor and several other diseases.	*Sū*. 5: 82–83
7		Paste of *Cedrus deodara* suspended in cow urine is to be consumed to cure inguinoscrotal swelling.	*Ci*. 13: 33
8		Herbs of *Surasādi gaṇa* ground to a paste with cow urine is to be used for performing sudation, before incising inguinoscrotal swelling caused by increase of adipose tissue.	*Ci*. 13: 35

(*Continued*)

TABLE 6.38 (*Continued*)
Animal Products Used as Medicine

Sl. No.		Particulars	Reference
9		Cow urine is used as a purgative in treatment of *gulma*.	*Ci*. 14: 99
10		*Dāśamūlika vasti* performed using milk, cow urine and alkali cures abdominal tumors appearing in women.	*Ci*. 14: 127
11		Cow urine is to be consumed for one to two months for curing enlargement of abdomen.	*Ci*. 15: 2, 74, 102
12		Medicinal enema with cow urine, decoction of herbs of *Muṣkakādi gaṇa*, *Trikaṭu* and sesame oil is recommended in treatment of enlargement of abdomen.	*Ci*. 15: 67–68, 99
13		Gruel prepared with milk and *Ṣaṣṭika* rice subjected to *bhāvana* with cow urine is to be fed to stabilize *tridōṣa* in patients suffering from enlargement of abdomen.	*Ci*. 15: 123
14		Paste of five herbs ground and suspended in cow urine is to be consumed to pacify morbid pallor.	*Ci*. 16: 7–8
15		Powder of calcined iron kept immersed in cow urine for seven days is to be mixed with cow milk and consumed for pacifying morbid pallor.	*Ci*. 16: 9
16		Mineral bitumen mixed with cow urine is to be administered for one month to cure *kumbhakāmala*.	*Ci*. 16: 52–53
17		Cow urine mixed with cow milk is to be consumed along with food to pacify edema.	*Ci*. 17: 10, 34
18		Consumption of cow urine fermented with powders of *Plumbago zeylanica*, *trikaṭu* and honey cures vitiligo.	*Ci*. 20: 7
19		Cow urine is ingredient of several *ghṛta* that cure spiritual afflictions.	*Ut*. 5: 19, 29–30, 38–40
20		Cow urine is ingredient of *ghṛta* that cures insanity, epilepsy, jaundice, enlargement of abdomen and hemorrhoids.	*Ut*. 6: 22–23 *Ut*. 7: 18–24
21	Elephant	Ingredient of *Bhūtarāva* and *Mahābhūtarāva ghṛta* indicated in affliction with evil spirits.	*Ut*. 5: 19–20
22		Ingredient of *Brahmyādi varti* that cures insanity	*Ut*. 6: 38–40
23		Ingredient of a *taila* that enhances growth of ear lobule	*Ut*. 18: 59
24		Ingredient of a medicated wick that is to be used for inhalation of smoke in treatment of chronic catarrh.	*Ut*. 20: 17
25	Fox	Fumigation of body, inhalation of smoke and topical application as paste and *taila* with urine are recommended in treatment of insanity.	*Ut*. 6: 42–43
26	Goat	Fumigation of body, inhalation of smoke and topical application as paste and *taila* with urine are recommended in treatment of insanity.	*Ut*. 6: 42–43
27		Consumption of paste of *Gymnema sylvestre* suspended in goat urine cures hemorrhoids.	*Ci*. 8: 57–58
28		Ingredient of three formulations that can be applied as *nasya* and as collyria.	*Ut*. 5: 31, 36–37, 41–43
29		Ingredient of *Triphalādi taila* that cures epilepsy	*Ut*. 7: 31–32
30		Thirteen herbs are ground in goat urine and pills rolled. This *Vilvādi guḷika* cures all kinds of poisoning, stings of insects, bites of poisonous snakes, fever and spiritual afflictions.	*Ut*. 36: 84–85
31	Horse	Ingredient of *Bhūtarāva* and *Mahābhūtarāva ghṛta* indicated in affliction with evil spirits.	*Ut*. 5: 19–20

(*Continued*)

TABLE 6.38 (*Continued*)
Animal Products Used as Medicine

Sl. No.		Particulars	Reference
32		Ingredient of a medicated wick that is to be used for inhalation of smoke in treatment of chronic catarrh.	*Ut.* 20: 17
33		Ingredient of a *taila* that enhances growth of ear lobule.	*Ut.* 18: 59
34	Porcupine	Fumigation of body, inhalation of smoke, topical application as paste and *taila* with urine are recommended in treatment of insanity.	*Ut.* 6: 42–43
35	Sheep	Alkali prepared from five herbs suspended in sheep urine is to be consumed to break urinary stone.	*Ci.* 11: 31–32
XXVIII	**Visceral Fat**		
1	Bear	Internal administration of a *taila* containing visceral fat of bear cures dysuria.	*Ci.* 11: 2–3
2	Black lance-hooded cobra	Visceral fat is ingredient of a collyrium that improves vision in persons suffering from cataract.	*Ut.* 13: 41
3	Camel	*Sēka* with visceral fat is recommended in treatment of hemorrhoids.	*Ci.* 8: 16
4	Cat	*Sēka* with visceral fat is recommended in treatment of hemorrhoids.	*Ci.* 8: 16
5	Crab	Ingredient of a *taila* that cures the ear disease *unmantha*	*Ut.* 18: 44–46
6	Crocodile	Visceral fat is to be applied topically in *vāta* diseases.	*Ci.* 21: 63
7	Hill partridge	*Tarpaṇa* with visceral fat is recommended in cataract.	*Ut.* 13: 61
8	Lion	*Nasya* with visceral fat pacifies irregular fever.	*Ci.* 1: 161–162
9	Monitor lizard	*Tarpaṇa* with visceral fat is recommended in cataract.	*Ut.* 13: 61
10		Ingredient of a *taila* that cures the ear disease *unmantha*	*Ut.* 18: 44–46
11	Peacock	*Tarpaṇa* with visceral fat is recommended in cataract.	*Ut.* 13: 61
12	Pig	*Sēka* with visceral fat is recommended in treatment of hemorrhoids.	*Ci.* 8: 16
13		Internal administration of a *taila* containing visceral fat of pig cures dysuria.	*Ci.* 11: 2–3
14		Visceral fat is to be used for *nasya* and as collyrium to cure cataract.	*Ut.* 13: 56
15	Porcupine	*Tarpaṇa* with visceral fat is recommended in cataract.	*Ut.* 13: 61
16	Rooster	Visceral fat is to be used as collyrium to cure cataract.	*Ut.* 13: 56
17		*Tarpaṇa* with visceral fat is recommended in cataract.	*Ut.* 13: 61
18	Snake	*Sēka* with visceral fat is recommended in treatment of hemorrhoids.	*Ci.* 8: 16
19		Visceral fat is to be used as collyrium to cure cataract.	*Ut.* 13: 56
20	Spotted deer	Visceral fat is ingredient of a *taila* to be used for *tarpaṇa* in treatment of cataract.	*Ut.* 13: 60
21	Tiger	*Nasya* with visceral fat pacifies irregular fever.	*Ci.* 1: 161–162
22		Visceral fat is to be used for *nasya* and as collyrium to cure cataract.	*Ut.* 13: 56
23	Tortoise	Visceral fat is to be applied topically in *vāta* diseases.	*Ci.* 21: 63
24	Vulture	Visceral fat is to be used as collyrium to cure cataract.	*Ut.* 13: 56
XXIX	**Whole Animal**		
1	Black lance-hooded cobra	Ash of *Kṛṣṇasarpa* (black lance-hooded cobra) mixed with oil of *Terminalia belerica* seeds cures vitiligo.	*Ci.* 20: 12

(*Continued*)

TABLE 6.38 (*Continued*)
Animal Products Used as Medicine

Sl. No.		Particulars	Reference
2		Ash obtained by burning a black lance-hooded cobra that died naturally is to be mixed with oil of *Sarcostigma kleinii* and applied topically to cure cervical glandular swelling.	*Ut.* 30: 28–29
3	Crow	Carcass of a naturally dead crow is to be used for fumigating the bed and blankets of a newborn baby.	*Ut.* 1: 26
4		Ash obtained by burning a crow that died naturally is to be mixed with oil of *Sarcostigma kleinii* and applied topically to cure cervical glandular swelling.	*Ut.* 30: 28–29
5	Scorpion	Scorpions are ingredients of *Kukkuṭa purīṣāñjanam,* a collyrium that cures cataract.	*Ut.* 13: 39

TABLE 6.39
Liquid and Semi-solid Ingredients

Sl.	Name of Ingredient			
No.	Sanskrti Term	English Equivalent	Attributes	Reference
I		***Ikṣuvargam* (Sweet Liquids and Semi-solids)**		
1	*Aparaḥ guḍaḥ*	Unpurified jaggery	Increases *kṛmi*, bone marrow, blood, adipose tissue and muscle	*Sū.* 5: 48
2	*Dhoutaḥ guḍaḥ*	Purified jaggery	Does not increase *kapha* much, eliminates urine and stool	*Sū.* 5: 47
3	*Ikṣōhrasaḥ*	Sugarcane juice	Heavy, cold, unctuous and roborant in nature, increases *kapha* and urine, aphrodisiac, laxative and pacifies hemorrhages of obscure origin	*Sū.* 5: 42–43
4	*Khaṇḍa*	Candied sugar	Aphrodisiac, good for those convalescing from injuries, pacifies *vāta*, *rakta* and *pitta*	*Sū.* 5: 49
5	*Madhu*	Honey	Good for eyes, pacifies thirst, *kapha*, poisoning and hemorrhages of obscure origin; cures skin diseases, polyuria, worm infestation, vomiting, respiratory distress, cough and dysentery; cleanses and heals wounds, increases *vāta*, having dry quality, sweet and astringent tastes	*Sū.* 5: 51–53
6	*Matsyakhaṇḍika*	Sugarcane juice boiled down to solid consistency, but still exuding a little fluid on drawing	Aphrodisiac, good for those convalescing from injuries, pacifies *vāta*, *rakta* and *pitta*	*Sū.* 5: 49
7	*Navaḥ guḍaḥ*	Fresh jaggery	Increases *kapha* and lowers abdominal fire	*Sū.* 5: 48
8	*Phāṇitam*	Thickened sugarcane juice	Heavy in nature, causes exudation of mucus, eliminates water	*Sū.* 5: 47
9	*Purāṇaḥ guḍaḥ*	Aged jaggery	Good for heart, a desirable food ingredient	*Sū.* 5: 48
10	*Sitāḥ*	Cane sugar	Aphrodisiac, good for those convalescing from injuries, pacifies *vāta*, *rakta* and *pitta*	*Sū.* 5: 49
11	*Yāṣaśarkara*	A variety of sugar	Has qualities of cane sugar, bitter, sweet and astringent in taste	*Sū.* 5: 50

(*Continued*)

TABLE 6.39 (*Continued*)
Liquid and Semi-solid Ingredients

Sl. No.	Name of Ingredient: Sanskrti Term	Name of Ingredient: English Equivalent	Attributes	Reference
II		***Jalavargam* (Water)**		
1	*Gaṅgāmbu*	Rain water	Maintains life, pleasing, good for heart, induces happiness, enlivens intelligence, light in nature, cold and comparable to ambrosia.	*Sū*. 5: 1–2
2	*Kvathita śītaḷam tōyam*	Boiled and cooled water	Light in nature, induces flow of *kapha*, useful in *pitta*-related conditions	*Sū*. 5: 18
3	*Nārīkēlōdakaṃ*	Coconut water	Unctuous, sweet, cold, light, aphrodisiac, carminative, cleanses urinary bladder, pacifies *pitta- vāta* and thirst	*Sū*. 5: 19
4	*Śītāmbu*	Cold water	Pacifies discomforts of alcoholism, tiredness, fainting, vomiting, giddiness, thirst, body heat, burning sensation, hemorrhages of obscure origin and poisoning	*Sū*. 5: 15–16
5	*Uṣṇāmbu*	Warm water	Improves abdominal fire, digests *dōṣas*, good for throat, light, cleanses urinary bladder; useful in hiccup, bloating of abdomen, *vāta –kapha* conditions, following elimination measures, acute fever, cough, discomforts caused by improper digestion of food, rhinitis, respiratory discomforts and pain in flanks	*Sū*. 5: 16–17
III		***Kṣīravargam* (Milk and Milk Products)**		
1	*Aikaśapham payaḥ*	Donkey milk	Highly hot and light, cures *vāta* associated with limbs and tissue elements, slightly sour and saltish in taste	*Sū*. 5: 27–28
2		Horse milk	Highly hot and light, cures *vāta* associated with limbs and tissue elements, slightly sour and saltish in taste	*Sū*. 5: 27–28
3	*Ājam payaḥ*	Goat milk	Light, cures emaciation, fever, respiratory distress, dysentery and hemorrhages of obscure origin	*Sū*. 5: 24
4	*Auṣṭrikam*	Camel milk	Stimulates digestion, light, slightly dry and hot in nature with salty taste, ideal in *vāta–kapha* diseases, bloating of abdomen, edema, enlargement of abdomen and hemorrhoids	*Sū*. 5: 25
5	*Āvikam payaḥ*	Sheep milk	Hot in nature, cures *vāta* diseases, causes hiccup, respiratory distress and *pitta–kapha* diseases	*Sū*. 5: 26–27
6	*Dadhi*	Curd	Sour in taste, constipating, heavy and hot, pacifies *vāta*, promotes formation of adipose tissue, seminal plasma, stamina, blood, abdominal fire and edema; preferred in distaste for food, *śītajvara*, irregular fever, rhinitis and dysuria; fat-free curd is useful in *grahaṇi*	*Sū*. 5: 29–31

(*Continued*)

TABLE 6.39 (*Continued*)
Liquid and Semi-solid Ingredients

Sl. No.	Name of Ingredient: Sanskrti Term	Name of Ingredient: English Equivalent	Attributes	Reference
7	*Ghṛta*	Clarified butter	Improves intelligence, memory, comprehension, abdominal fire, stamina, longevity, sexual vigor and vision; pacifies *vāta–pitta* aggravation, poisoning, insanity edema and fever; maintains youthfulness	*Sū.* 5: 37–39
8	*Hastinyā payaḥ*	Elephant milk	Bestows stability to body	*Sū.* 5: 27
9	*Kīlāṭam*	Prepared by mixing small quantity of milk with more buttermilk	Causes *viṣṭambha*, improves stamina, virility and sleep	*Sū.* 5: 41
10	*Kṣīrōdbhavam*	Cream from milk	Constipating, cures hemorrhages of obscure origin and eye diseases	*Sū.* 5: 36
11	*Kūcika*	Prepared by mixing warm milk with buttermilk	Causes *viṣṭambha*, improves stamina, virility and sleep	*Sū.* 5: 41
12	*Māhiṣam payaḥ*	Buffalo milk	Very heavy, cold, suitable for those with high appetite and insomnia	*Sū.* 5: 23
13	*Mānuṣam*	Breast milk	Good for instilling in eye and nose, cures eye diseases arising from enragement of *vāta*, *pitta*, *rakta* and physical trauma	*Sū.* 5: 26
14	*Mastu*	Whey	Possesses qualities similar to those of buttermilk, cleanses body channels, light in nature, cures *viṣṭambha*	*Sū.* 5: 35
15	*Mōraṇa*	Prepared by mixing equal parts of milk and buttermilk	Causes *viṣṭambha*, improves stamina, virility and sleep	*Sū.* 5: 41
16	*Navam navanītam*	Fresh butter	Cold in nature, aphrodisiac, improves complexion, stamina and appetite; constipating, cures *vāta* diseases, hemorrhages of obscure origin, emaciation, hemorrhoids, facial palsy and cough	*Sū.* 5: 35–36
17	*Payaḥ*	Cow milk	Unctuous, sweet, heavy, cold, pacifies *vāta–pitta*, nourishes tissue elements, aphrodisiac, increases *kapha*, maintains life, improves intelligence and stamina, galactagogue, laxative, cures tiredness, giddiness, respiratory distress, cough, excessive thirst, chronic fever, dysuria and hemorrhages of obscure origin	*Sū.* 5: 20–23
18	*Pīyūṣam*	Prepared by mixing milk of past-partum cow with buttermilk	Causes *viṣṭambha*, improves stamina, virility and sleep	*Sū.* 5: 41
19	*Purāṇa ghṛta*	Aged *ghṛta* (more than five years old)	Hastens healing of wounds, cures epilepsy, fainting and stupor	*Sū.* 5: 40
20	*Takra*	Buttermilk	Light in nature, astringent and sour in taste, improves digestion, pacifies *kapha–vāta* diseases, useful in edema, enlargement of abdomen, *grahaṇi*, dysuria, distaste for food, *gulma*, splenomegaly, poisoning, morbid pallor and diseases arising from overuse of clarified butter	*Sū.* 5: 33–34

(*Continued*)

TABLE 6.39 (*Continued*)
Liquid and Semi-solid Ingredients

Sl. No.	Name of Ingredient: Sanskrti Term	English Equivalent	Attributes	Reference
IV		*Madyavargam* (Wines and Fermented Liquids)		
1	*Ariṣṭaḥ*	Medicated wine	Best among all wines, astringent and pungent in taste, cures fever, skin diseases, *grahaṇi*, morbid pallor, hemorrhoids, edema, emaciation and enlargement of abdomen	*Sū*. 5: 70–71
2	*Dhānyāmḷa*	Grains, ginger, lemon slices and ajowan mixed with large quantity of water and heated mildly for seven days	Disrupts *dōṣas*, *tīkṣṇa*, light, hot; increases *pitta*, appetizer, improves digestion, pacifies *vāta* and *kapha*, cures pain in urinary bladder, good for use in *nirūhavasti*	*Sū*. 5: 79–80
3	*Gauḍaḥ*	Brewed from jaggery	Improves digestion, eliminates *vāta*, urine and stool	*Sū*. 5: 74
4	*Khārjūram*	Brewed from dates	Inferior to *Mārdvīkam*, heavy, increases *vāta*	*Sū*. 5: 73
5	*Madhvāsavaḥ*	Brewed from honey	Penetrating in nature, cures *pramēha*, rhinitis and cough	*Sū*. 5: 75
6	*Mārdvīkam*	Brewed from grapes	Sweet in taste, not very hot, removes fat, laxative, slightly raises *pitta* and *vāta*, cures morbid pallor, *pramēha*, hemorrhoids and worm infestation	*Sū*. 5: 72
7	*Śaṇḍāki*	Coconut oil cake, curry leaves, onion, cumin and mustard ground and mixed with lemon juice	Light in nature and improves appetite	*Sū*. 5: 78
8	*Śārkaraḥ*	Brewed from sugar	Fragrant and tasty, light, not very stupefying	*Sū*. 5: 73
9	*Sīdhuḥ*	Brewed from sugarcane juice	Increases *vāta* and *pitta*, cures *kapha* diseases, adipose tissue, edema, enlargement of abdomen and hemorrhoids	*Sū*. 5: 74
10	*Śuktam*	Vinegar	Induces *vāta* to move in the normal direction, hot, penetrating, dry and sour in nature, stimulates digestion, cures morbid pallor and worm infestation, harms eyes	*Sū*. 5: 76–77
11	*Sura*	Brewed from rice powder	Cures *gulma*, hemorrhoids, enlargement of abdomen, *grahaṇi* and emaciation; heavy and unctuous; pacifies *vāta*; increases adipose tissue, blood, breast milk, urine and *kapha*	*Sū*. 5: 67
12	*Vaibhītaki*	Brewed from bark of *Terminalia belerica*	Not very stupefying, light in nature, compatible in wounds, sores, morbid pallor and skin diseases	*Sū*. 5: 69
13	*Vāruṇi*	Brewed from diluted honey	Similar to *Sura* in qualities, light and penetrating in nature, cures pricking pain, cough, respiratory distress, vomiting, obstruction of body channels, bloating of abdomen and rhinitis	*Sū*. 5: 68

(*Continued*)

TABLE 6.39 (*Continued*)
Liquid and Semi-solid Ingredients

Sl. No.	Name of Ingredient: Sanskrti Term	Name of Ingredient: English Equivalent	Attributes	Reference
V		***Mūtravargam* (Urine of Various Animals)**		
1	*Ajamūtram*	Goat urine	All these urines are light, dry, penetrating, hot and pungent in nature, having mildly saltish taste; cures worm infestation, edema, enlargement of abdomen, bloating of abdomen, pricking pain, morbid pallor, *kapha–vāta* diseases, *gulma*, distaste for food, poisoning, vitiligo, hemorrhoids, leprosy and other skin diseases.	*Sū.* 5: 82–83
2	*Aśvamūtram*	Horse urine		
3	*Āvimūtram*	Sheep urine		
4	*Gajamūtram*	Elephant urine		
5	*Gōmūtram*	Cow urine		
6	*Kharamūtram*	Donkey urine		
7	*Mahiṣīmūtram*	Buffalo urine		
8	*Uṣṭramūtram*	Camel urine		
VI		***Tailavargam* (Oils and Fats)**		
1	*Airaṇḍa tailam*	Castor oil	Bitter, hot, sweet and heavy; cures inguinoscrotal swelling, *gulma*, *vāta–kapha* diseases, enlargement of abdomen, irregular fever and diseases of hip, pelvic area, back and alimentary canal; pacifies pain and edema	*Sū.* 5: 57–58
2			*Snēhavasti* with medicated castor oil to be performed in treatment of enlargement of abdomen	*Ci.* 15: 57
3			Castor oil to be mixed with cow urine or cow milk and consumed to cure enlargement of abdomen	*Ci.* 15: 2
4	*Akṣatailam*	Oil from *Terminalia belerica* seeds	Sweet, cold and heavy, good for hair, pacifies *pitta* and *vāta*	*Sū.* 5: 60
5			To be used for achieving oleation prior to emesis and purgation in treatment of *pramēha*	*Ci.* 12: 1
6	*Aruṣkarajam snēham*	Oil from *Semecarpus anacardium* seeds	Internal administration is recommended in leprosy and all skin diseases.	*Ci.* 19: 12–13
7	*Dravantītailam*	Oil from *Croton tiglium* seeds	Internal administration of the oil is recommended in treatment of enlargement of abdomen.	*Ci.* 15: 76–77
8	*Iṅgudītailam*	Oil from *Sarcostigma kleinii* seeds	Ingredient of *Viḷaṅgādi taila* that cures headache	*Ut.* 24: 16–17
9	*Jyōtiṣmatītailam*	Oil from *Celastrus paniculatus* seeds	Topical application of *taila* prepared with oil of *Celastrus paniculatus* and ash of *Achyranthes aspera* cures *Sidhma kuṣṭha*.	*Ci.* 19: 75
10	*Karañjatailam*	Oil from *Pongamia glabra* seeds	To be used for achieving oleation prior to emesis and purgation in treatment of *pramēha*	*Ci.* 12: 1
11	*Kōśāmra Tailam*	Oil from *Schleichera oleosa* seeds	To cause oleation in treatment of inguinoscrotal swelling	*Ci.* 13: 29–30
12	*Kusumbhatailam*	Oil from *Carthamus tinctorius* seeds	Hot in nature, causes skin diseases and *kapha–pitta* diseases	*Sū.* 5: 61
13	*Majja*	Bone marrow	Pacifies *vāta*, increases stamina, *pitta* and *kapha*	*Sū.* 5: 61–62

(*Continued*)

TABLE 6.39 (*Continued*)
Liquid and Semi-solid Ingredients

Sl. No.	Name of Ingredient: Sanskrti Term	Name of Ingredient: English Equivalent	Attributes	Reference
14	*Nikumbhatailam*	Oil from *Baliospermum montanum* seeds	To be used for achieving oleation prior to emesis and purgation in treatment of *pramēha*	*Ci.* 12: 1
15			Internal administration of the oil is recommended in treatment of enlargement of abdomen.	*Ci.* 15: 76–77
16	*Nimbatailam*	Neem oil	Bitter, cures worm infestation, *kapha* diseases and skin diseases	*Sū.* 5: 60
17			To be used for achieving oleation prior to emesis and purgation in treatment of *pramēha*	*Ci.* 12: 1
18			Ingredient of *Viḷaṅgādi taila* that cures headache	*Ut.* 24: 16–17
19	*Pīlūtailam*	Oil from *Salvadora persica* seeds	Ingredient of *Viḷaṅgādi taila* that cures headache	*Ut.* 24: 16–17
20	*Sārṣapam*	Mustard oil	Pungent, hot, penetrating and light in nature; destroys *kapha*, *vāta* and seminal plasma; cures *kōṭha*, leprosy and other skin diseases, hemorrhoids sores, ulcers and worm infestation	*Sū.* 5: 59
21			To be used for achieving oleation prior to emesis and purgation in treatment of *pramēha*	*Ci.* 12: 1
22			Internal administration is recommended in leprosy and all skin diseases.	*Ci.* 19: 12–13
23			Ingredient of *Viḷaṅgādi taila* that cures headache	*Ut.* 24: 16–17
24	*Surakāṣṭhasnēham*	Oil from heartwood of *Cedrus deodara*	Consumption of oil mixed with *Trikaṭu* and *yavakṣāra* is recommended to cure cough with enragement of *kapha*.	*Ci.* 3: 41
25	*Tailam*	Sesame oil	Penetrating, subtle, hot and spreading in nature; dispels skin diseases, does not increase *kapha*, good for weight control, when processed with herbs cures all diseases	*Sū.* 5: 55–56
26			*Snēhavasti* with medicated sesame oil to be performed in treatment of enlargement of abdomen	*Ci.* 15: 57
27			*Nirūhavasti* with cow urine, decoction of *Muṣkakādi gaṇa*, *Trikaṭu* and sesame oil is recommended in treatment of enlargement of abdomen.	*Ci.* 15: 67–68
28	*Tauvaram snēham*	Oil from *Hydnocarpus laurifolia* seeds	Internal administration is recommended in leprosy and all skin diseases.	*Ci.* 19: 12–13
29	*Tilvakatailam*	Oil from *Symplocos racemosa* seeds	To cause oleation in treatment of inguinoscrotal swelling	*Ci.* 13: 29–30
30	*Umātailam*	Oil from *Linum usitatissimum* seeds	Hot in nature, causes skin diseases and *kapha–pitta* diseases	*Sū.* 5: 61
31	*Vasa*	Visceral fat	Pacifies *vāta*, increases stamina, *pitta* and *kapha*	*Sū.* 5: 61–62

6.7 SALTS AND ALKALIS

Salts form an integral part of formulary of AH. They are either collected from nature or prepared by special methods. The eight salts described in AH are *saindhava lavaṇa* (rock-salt), *souvarcala lavaṇa* (sonchal salt), *viḍ lavaṇa* (salt prepared by specific method), *sāmudra lavaṇa* (sea salt), *audbhīda lavaṇa* (salt prepared from vegetable alkali), *kṛṣṇalavaṇa* (black salt prepared by specific method), *rōmaka lavaṇa* (sambhar salt) and *pāmsuja lavaṇa* (earth salt) (*Sū*. 6: 143–149).

Caustic alkali (*kṣāra*) is one of the dosage forms recommended in AH. It is prepared by a special method called *kṣāra kalpana*. *Sarjikṣāra* is an example of a *kṣāra*. It is prepared from the creeper, *Dhanvayāsa* (*Fagonia cretica*) or *Tragia involucrata* (Shajahan and Nikita, 2019). Dried roots, leaves, stems, flowers and fruits are cut into small pieces and burnt in a big iron cauldron. The white ashes are collected and mixed with eight times of fresh water. The suspension is agitated with hand and left undisturbed for three hours. Thereafter, the clear supernatant fluid is decanted and filtered seven times through three-layered muslin cloth. The residual ash is dried, mixed and agitated in eight volumes of water, left undisturbed for three hours and filtered as earlier. This process is repeated three times. The three filtrates are pooled together and dried. When completely dried, the *kṣāra* is scraped out and stored in a clean container (Gupta et al., 2020). All *kṣāra* are caustic in nature and are used in chemical cautery (*kṣāra karma*). Salts and *kṣāra* described in AH are listed in Table 6.40.

TABLE 6.40
Salts and Alkalis Used as Medicine

Sl. No.	Herb/Animal Part/Medicinal Ingredient/Compound Formulation	Therapeutic Value	Reference
I	***Lavaṇa* (Salt)**		
1	*Audbhīda lavaṇa* (Salt prepared from vegetable alkali)	Alkaline in nature, possesses bitter and pungent tastes, penetrating in nature, causes exudation of fluid	*Sū*. 6: 148
2	*Kṛṣṇalavaṇa* (Black salt prepared by specific method)	Very similar to sonchal salt; however, it does not have any smell.	*Sū*. 6: 148
3	*Pāmsuja lavaṇa* (Earth salt)	Alkaline in nature, heavy and increases *kapha*	*Sū*. 6: 149
4	*Rōmaka lavaṇa* (Sambhar salt)	Light in nature	*Sū*. 6: 149
5	*Saindhava lavaṇa* (Rock salt)	Slightly sweet in taste, virilific and pleasing, pacifies *tridōṣa*, light and mildly hot in nature, stimulator of digestion, good for eyes	*Sū*. 6: 144–145
6	*Sāmudra lavaṇa* (Sea salt)	Turns sweet in *vipāka*, heavy and increases *kapha*	*Sū*. 6: 147
7	*Souvarcala lavaṇa* (Sonchal salt)	Light, pleasing and possessing good aroma; removes obstructions in body channels, stimulator of digestion, improves taste	*Sū*. 6: 145–146
8	*Viḍ lavaṇa* (A salt prepared by specific method)	Stimulates the natural movement of *kapha* upward and *vāta* downward, stimulator of digestion, removes obstructions in body channels, digestive disturbances, pain and heaviness of body	*Sū*. 6: 146–147
II	***Kṣāra* (Caustic Alkali)**		
1	*Achyranthes aspera*	Topical application of *taila* prepared with *kṣāra* of the herb and oil of *Celastrus paniculatus* cures *sidhma kuṣtha*	*Ci*. 19: 75
2	*Aegle marmelos*	Consumption of *taila* prepared with *kṣāra* of leaves cures enlargement of abdomen.	*Ci*. 15: 45

(*Continued*)

TABLE 6.40 (*Continued*)
Salts and Alkalis Used as Medicine

Sl. No.	Herb/Animal Part/Medicinal Ingredient/Compound Formulation	Therapeutic Value	Reference
3	*Butea monosperma*	*Kṣāra* of bark is an ingredient of a *ghṛta* that cures hemorrhoids.	*Ci.* 8: 74–75
4		*Taila-ghṛta* combination prepared using the *kṣāra* of bark is useful in treatment of *raktagulma*.	*Ci.* 14: 122–123
5		Consumption of *kṣāra* of bark mixed with *phāṇitam* cures vitiligo	*Ci.* 20: 5
6		Powders of seven herbs are to be ground to a fine paste with aqueous suspension of *kṣāra* of *Butea monosperma*. Topical application of this paste cures *gaḷagaṇḍa*.	*Ut.* 2: 69–70
7	*Coleus vettiveroides*	*Kṣāra* of the herb mixed with oil of *Terminalia belerica* cures vitiligo.	*Ci.* 20: 12
8	Elephant	Topical application of *kṣāra* of hide mixed with sesame oil cures vitiligo.	*Ci.* 20: 10
9		Topical application of paste prepared with *Psoralea corylifolia* seeds, *kṣāra* of dung and urine cures vitiligo.	*Ci.* 20: 14–15
10	*Emblica officinalis*	*Kṣāra* of pericarps is suspended in cow urine, filtered and the filtrate concentrated by slow heating. On acquiring thick consistency, it is cooled and consumed to cure *gaḷagaṇḍa*.	*Ut.* 22: 70
11	*Erythrina variegata*	*Kṣāra* of leaves is to be consumed in wine, along with the powder of three herbs, to cure dysuria.	*Ci.* 11: 14–15
12	*Gloriosa superba*	*Kṣāra* prepared from the rhizome is to be soaked for a day in juices of *Emblica officinalis* and *Ocimum sanctum*. The dried powder known as *Lāṅgalaki kṣārāñjana* is to be applied in eyes as a collyrium to cure diseases of sclera.	*Ut.* 11: 45–46
13	*Hordeum vulgare*	*Kṣāra* of barley leaves (*yakakṣāra*) is ingredient of a medicine that cures hemorrhoids.	*Ci.* 8: 20–21
14		*Kṣāra* of leaves is to be consumed in wine, along with the powder of three herbs, to cure dysuria.	*Ci.* 11: 14–15
15	*Kalyāṇaka kṣāra*	This compound formulation involving *kṣāra* of 13 ingredients cures many diseases including hemorrhoids.	*Ci.* 8: 140–143
16	King cobra	Ash of *Kṛṣṇasarpa* mixed with oil of *Terminalia belerica* cures vitiligo.	*Ci.* 20: 12
17	*Kṣārāgadam*	This compound formulation involving *kṣāra* of 16 herbs, *sarjikṣāra*, *yavakṣāra*, 4 oleaginous substances and 5 salts cures *gulma*, *udāvarta*, hemorrhoids, enlargement of abdomen and *grahaṇi*.	*Ci.* 14: 103–107
18	*Kṣāra guḷika*	Pill prepared using three salts and alkali from four herbs cures all diseases of digestive tract, including cholera.	*Ci.* 10: 58–60
19	*Ocimum sanctum*	Consumption of *kṣāra* of leaves mixed with *Trikaṭu*, sesame oil and jaggery cures cough.	*Ci.* 3: 171

(*Continued*)

TABLE 6.40 (*Continued*)
Salts and Alkalis Used as Medicine

Sl. No.	Herb/Animal Part/Medicinal Ingredient/Compound Formulation	Therapeutic Value	Reference
20	*Paspalum scrobiculatum*	Ash of the straw is suspended in water and filtered. Washing the head with this filtrate cures pityriasis.	*Ut*. 24: 27
21	Peacock	Consumption of *kṣāra* of leg of peacock mixed with honey and clarified butter cures cough.	*Ci*. 3: 170
22		*Kṣāra* of gall bladder of peacock mixed with oil of *Terminalia belerica* cures vitiligo.	*Ci*. 20: 12
23	*Pongamia pinnata*	*Kṣāra* of bark suspended in acidic liquids and mixed with *viḍ lavaṇa* and powder of *Piper longum* cures enlargement of abdomen.	*Ci*. 15: 87
24	Porcupine	Consumption of *kṣāra* of quill of porcupine (*Hystrix indica* Kerr) mixed with clarified butter, honey and sugar cures respiratory disorders including cough.	*Ci*. 3: 169–170
25	*Pūtika kṣāram*	Consumption of *kṣāra* of seven herbs cures *gulma*, enlargement of abdomen, edema, morbid pallor and hemorrhoids.	*Ci*. 14: 38
26	*Ricinus communis*	Consumption of *kṣāra* of leaves mixed with *Trikaṭu*, sesame oil and jaggery cures cough.	*Ci*. 3: 170–171
27	*Sarcogyps calvus* Scopoli	Calcined powder of the head of red-headed vulture is to be mixed with antimony sulphide and applied as a collyrium (*Gṛdhrāñjanam*) to cure cataract and to improve acuity of vision.	*Ut*. 13: 37
28	*Sesamum indicum*	*Kṣāra* of leaves is to be consumed in wine, along with the powder of three herbs, to cure dysuria.	*Ci*. 11: 14–15
29	Shell of pearl oyster	*Kṣāra* suspended in milk is to be administered to patient suffering from enlargement of abdomen.	*Ci*. 15: 86
30	*Stereospermum tetragonum*	*Kṣāra* of leaves is to be consumed in wine, along with the powder of three herbs, to cure dysuria.	*Ci*. 11: 14–15
31	Tiger	Topical application of *kṣāra* of hide mixed with sesame oil cures vitiligo.	*Ci*. 20: 10
32	*Triphala* powder	Topical application of *kṣāra* of *Triphala* mixed with clarified butter cures venereal ulcer (*upadaṃśa*).	*Ut*. 34: 6
33	Unnamed collyrium	*Kṣāra* of coconut shell, seeds of *Semecarpus anacardium*, apical tender leaves of *Bambusa arundinacea* and *Borassus flabellifer* are mixed with water and filtered. Powder of camel bone is to be added to the filtrate and dried. Application of this powder as a collyrium cures all diseases of the sclera.	*Ut*. 11: 53–54
34	Unnamed electuary	This electuary prepared with alkali of 16 herbs, powders of 6 herbs and *yavakṣāra cures gulma*, splenomegaly, hemorrhoids, skin diseases and *pramēha* and lowering of abdominal fire.	*Ci*. 8: 151–152
35	Unnamed *ghṛta*	*Ghṛta* prepared using three salts, alkali of three herbs, *sarjikṣāra* and *yavakṣāra* cures *grahaṇi*.	*Ci*. 10: 63–65

(*Continued*)

TABLE 6.40 (*Continued*)
Salts and Alkalis Used as Medicine

Sl. No.	Herb/Animal Part/Medicinal Ingredient/Compound Formulation	Therapeutic Value	Reference
36	Unnamed *kṣāra*	*Andrographis paniculata*, *Azadirachta indica*, *Cyclea peltata*, *Picrorhiza kurroa*, *Trichosanthes cucumerina* and *Oldenlandia corymbosa* are to be powdered and ground into a thick paste using buffalo urine. The *kṣāra* obtained by calcining this mixture improves digestive efficiency and cures *grahaṇi*.	*Ci.* 10: 56–57
37		*Acorus calamus*, *Berberis aristata*, *Curcuma longa*, *Cyperus rotundus*, *Picrorhiza kurroa*, *Plumbago zeylanica* and *Saussurea lappa* are powdered and ground to a paste with goat urine. The *kṣāra* obtained by calcining this mixture is to be consumed to improve digestive efficiency and to cure *grahaṇi*.	*Ci.* 10: 57–58
38		Consumption of *kṣāra* prepared from 16 herbs, 5 salts, curd and 4 oleaginous substances cures enlargement of abdomen, edema, heart disease and hemorrhoids.	*Ci.* 15: 70–73
39		*Kṣāra* of *Acorus calamus*, *Embelia ribes*, *Plumbago zeylanica*, parched rice mixed with clarified butter and rock salt, administered in milk cures *gulma* and splenomegaly.	*Ci.* 15: 89–90
40	Unnamed pill	Consumption of pill prepared from *kṣāra* of dried droppings of goat, cow urine, 12 herbs, 5 salts and *sarjikṣāra* cures advanced ascites and edema.	*Ci.* 15: 103–106
41	Unnamed *taila*	Consumption of *taila* prepared with *kṣāra* of *Achyranthes aspera*, *Butea monosperma*, *Musa paradisiaca*, *Oroxylum indicum*, *Premna serratifolia*, *Sesamum indicum* and *Sida rhombifolia* cures enlargement of abdomen.	*Ci.* 15: 46
42		*Taila* prepared with *kṣāra* of *Musa paradisiaca*, *Sesamum indicum* and seeds of *Vitex altissima* is reputed to cure splenomegaly with *kapha–vata* aggravation.	*Ci.* 15: 95–96
43	*Withania somnifera*	Consumption of *kṣāra* of root mixed with honey cures respiratory diseases.	*Ci.* 4: 39

6.8 INORGANIC SUBSTANCES

Various metals like iron, bronze, gold, silver; minerals such as chalcopyrite, alum; and precious stones like ruby, emerald and diamond are considered as medicinal ingredients in AH. Interestingly, mercury is recommended as an ingredient of three formulations such as *Rasēndryādyañjanam* (*Ut.* 13: 36), *Savarṇakaraṇa pralēpa* (*Ut.* 25: 61) and *Mañjiṣṭhādi tṛvṛta* (*Ut.* 32: 31) intended for topical application. In the chapter on rejuvenation (*Rasāyanādhyāyaḥ*), mercury is indicated as one of the ingredients of a formula for rejuvenation. A person with weak body and tissue elements should consume for fifteen days, powder of mineral bitumen, chalcopyrite, calcined iron, mercury, seeds of *Embelia ribes* and pericarps of *Terminalia chebula*, mixed with honey and clarified butter. By undergoing such a treatment, he will acquire a perfectly healthy body (*Ut.* 39: 161). Table 6.41 lists the inorganic substances mentioned in AH.

TABLE 6.41
Inorganic Substances Mentioned in AH

Sl. No.	Sanskrit Name	English Equivalent	Reference
1	*Āgāradhūma*	Chimney soot	*Ci.* 21: 23
2	*Ālaṃ*	Orpiment	*Sū.* 21: 18
3	*Añjana*	Antimony sulphide	*Sū.* 10: 32
4	*Ayaḥ*	Iron	*Sū.* 10: 29
5	*Ayaskāntam*	Magnetite	*Sū.* 25: 39
6	*Bhujaga*	Tin	*Ut.* 13: 36
7	*Gairikam*	Red ochre	*Sū.* 10: 32
8	*Gandhōpalaḥ*	Sulphur	*Ci.* 19: 67
9	*Hēmam*	Gold	*Sū.* 10: 22
10	*Kāmsya*	Bronze	*Sū.* 10: 29
11	*Karkētanam*	Ruby	*Ut.* 36: 90
12	*Kāsīsam*	Green vitriol (melanterite)	*Ci.* 18: 26
13	*Lōham*	Steel	*Ci.* 12: 32
14	*Maṇḍūram*	Rust of iron	*Ci.* 16: 15
15	*Manōhva*	Realgar	*Sū.* 21: 18
16	*Maratakam*	Emerald	*Ut.* 36: 90
17	*Mṛt*	Black soil	*Ci.* 2: 27
18	*Muktā*	Pearl	*Sū.* 10: 32
19	*Padminī kardamaḥ*	Mud from ponds where lotus grows	*Ci.* 18: 13
20	*Phēna*	Cuttlefish bone	*Sū.* 24: 16
21	*Pravāḷam*	Coral	*Sū.* 10: 32
22	*Rajatam*	Silver	*Sū.* 10: 26
23	*Rasēndra*	Mercury	*Ut.* 13: 36 *Ut.* 25: 61 *Ut.* 32: 31 *Ut.* 39: 161
24	*Śaṅkham*	Conch shell	*Sū.* 24: 16
25	*Saurāṣṭrika*	Double sulphate of potassium and aluminium (alunite)	*Ut.* 34: 52
26	*Śilājatu*	Mineral bitumen	*Ci.* 12: 43
27	*Souvīrāñjanam*	Galena	*Ut.* 11: 12
28	*Sphaṭikaḥ*	Potash alum	*Ut.* 11: 12
29	*Srōtōjāñjana*	Stibnite	*Sū.* 24: 16
30	*Śukti*	Pearl oyster shell	*Ci.* 18: 14
31	*Tāmrarajam*	Copper filings	*Ut.* 9: 32
32	*Tāpyam*	Chalcopyrite	*Ci.* 16: 16
33	*Trapu*	Lead	*Sū.* 29: 58
34	*Tuttham*	Copper sulphate	*Ci.* 20: 16
35	*Vaiḍūryam*	Cat's eye	*Ut.* 13: 22
36	*Vajram*	Diamond	*Ut.* 36: 90

6.9 WEIGHTS AND MEASURES

Weights and measures are standardized in modern world to promote uniformity in national and international legal metrology laws, standards and test procedures, so as to ensure consumer protection and facilitate transparency in trade. However, international trade did not exist in ancient days, and a physician's activity was restricted only to his village or town. Therefore, authors of

Ayurveda texts like AH devised their own measurement systems applicable to both solids and liquids. AH advocates the use of measures such as *śāṇam*, *palam*, *kuḍubam*, *prastham*, *āḍhakam*, *drōṇam*, *tulām*, *bhāram* and so on (*Ka.* 6: 22–29). In contemporary Ayurveda, the commonly used measure is *palam*, and it is considered to be equivalent to 48 g. In the case of liquids, 1 kg will be equal to 1 l and 1 g to 1 ml. Weights and measures approved by *The Ayurvedic Formulary of India* (Anonymous, 1978) are presented in Box 6.1.

6.10 POTENTIATION OF MEDICINES

BOX 6.1: WEIGHTS AND MEASURES APPROVED BY *THE AYURVEDIC FORMULARY OF INDIA* (ANONYMOUS, 1978)

1 *Ratti* or *Guñja*		= 125 mg
8 *Ratti*	= 1 *Māṣa*	= 1 g
12 *Māṣa*	= 1 *Karṣa* (*Tōla*)	= 12 g
2 *Karṣa*	= 1 *Śukti*	= 24 g
2 *Śukti*	= 1 *Palam*	= 48 g
2 *Palam*	= 1 *Prasṛti*	= 96 g
2 *Prasṛti*	= 1 *Kuḍava*	= 192 g
2 *Kuḍava*	= 1 *Mānika*	= 384 g
2 *Mānika*	= 1 *Prastha*	= 768 g
4 *Prastha*	= 1 *Āḍhaka*	= 3.072 kg
4 *Āḍhaka*	= 1 *Drōṇa*	= 12.288 kg
2 *Drōṇa*	= 1 *Śūrpa*	= 24.576 kg
2 *Śūrpa*	= 1 *Drōṇi*	= 49.152 kg
4 *Drōṇi*	= 1 *Khari*	= 196.608 kg
1 *Palam*		= 48 g
100 *Palam*	= 1 *Tula*	= 4.800 kg
20 *Tula*	= 1 *Bhāra*	= 96 kg

6.10.1 Bhāvana

AH describes two processing methods for enhancing the potency of medicinal powders (*cūrṇa*), powders of minerals and medicinal oils (*taila*). The method recommended for powders or dried pericarps is known as *bhāvana*. It is carried out by mixing homogeneously the powdered drug with specific liquids, followed by drying. This process is repeated seven or several times, as specified in the formula. The liquids recommended are expressed juices of herbs or fruits, decoctions, latex, bile, cow urine, goat urine, alkaline water and milk. References to *bhāvana* are listed in Table 6.42.

6.10.2 Repetition of Process

The method adopted for potentiating *taila* and *ghṛta* is based on repetition of a step in the process of their preparation. For example, if *Balā taila* is to be potentiated by repetition of the process, first of all sesame oil, milk and decoction or paste of *Balā* (*Sida rhombifolia* ssp. *retusa*) roots are poured into a cauldron. The mixture is, thereafter, heated over slow fire, with frequent stirring. Heating is stopped when the sediment in the mixture attains the consistency of mud (Gopinath, 2021). The mixture is then left in the cauldron. On the next day, fresh quantities of milk, decoction

TABLE 6.42
Potentiation of Medicines with *bhāvana*

Sl. No.	Absorbent	Description of Potentiation	Reference
1	Broken grains	Broken grains of barley are subjected to *bhāvana* in latex of *Calotropis gigantea*. Administration of the treated grains mixed with honey cures respiratory distress and hiccup.	*Ci.* 4: 26–27
2	Camel bone	Ash of ingredients such as coconut shells, seeds of *Semecarpus anacardium*, dried tender leaves of *Borassus flabellifer* and *Bambusa bambos* are suspended in water and filtered. Powder of camel bone is subjected to *bhāvana* in the filtrate and applied in eyes as collyrium to pacify obstinate diseases of sclera.	*Ut.* 11: 53–54
3	Cotton wad	Cotton wad is subjected to *bhāvana* in juice or decoctions of lac, *Vitex negundo*, *Eclipta alba* and *Berberis aristata*. The treated wad of cotton is dried and used as the wick of an oil lamp, in which clarified butter serves as fuel source. Soot from the lit lamp is collected by placing an earthen plate over the flame. Application of collyrium prepared from the soot cures *pillarōga*.	*Ut.* 16: 57
4	Dried herbs	Pericarps of *Terminalia chebula* and fruits of *Piper longum* are subjected to *bhāvana* in latex of *Euphorbia neriifolia*. Powder of the treated herbs is incorporated into *Utkārika* and consumed to cure enlargement of abdomen.	*Ci.* 15: 44
5	Fruits	Fruits of *Piper longum* subjected to *bhāvana* in latex of *Euphorbia neriifolia* are to be consumed in gradually increasing dose, according to *rasāyana vidhi* to cure enlargement of abdomen.	*Ci.* 15: 40
6		Ash of *Erythrina variegata* is suspended in water and the mixture filtered. *Piper longum* fruits subjected to *bhāvana* in the filtrate are to be consumed as *rasāyana*.	*Ut.* 39: 97–98
7	Green vitriol	Powdered green vitriol is put in a copper dish, subjected to *bhāvana* for ten days in juice of *Ocimum sanctum* and dried. Application of the paste of this powder in eyes cures madarosis and several eye diseases collectively known as *pillarōga*.	*Ut.* 16: 55
8	Meat	Piece of meat subjected to *bhāvana* in an appropriate alkali or latex of *Euphorbia neriifolia* is to be introduced into the vagina of patient suffering from *raktagulma*.	*Ci.* 14: 123–124
9	Mineral bitumen	*Śilājit* (mineral bitumen) subjected to *bhāvana* in decoctions of appropriate herbs is to be administered as *rasāyana* in treatment of several diseases.	*Ut.* 39: 130–142
10	Piece of cloth	A piece of cloth subjected to *bhāvana* many times with bile of pig and fish is to be introduced into the vagina of patient suffering from *raktagulma*.	*Ci.* 14: 124–125
11	Powder	Medicinal powders to be used in treatment of *kapha–vāta gulma* are to be subjected to *bhāvana* in juice of *Citrus medica* and turned into pills.	*Ci.* 14: 30–31

(Continued)

TABLE 6.42 (*Continued*)
Potentiation of Medicines with *bhāvana*

Sl. No.	Absorbent	Description of Potentiation	Reference
12		Powder of *Operculina turpethum* is to be subjected to *bhāvana* with latex of *Euphorbia neriifolia*. 36 g (3 *karṣa*) of this powder mixed with clarified butter and honey are to be administered to cause purgation in *gulma* patients.	*Ci*. 14: 98
13		Consumption of powder of horse dung subjected to *bhāvana* in decoction of *Embelia ribes* or *Triphala*, mixed with honey cures worm infestation.	*Ci*. 20: 27–28
14		Powdered heartwood of *Cedrus deodara* is subjected to *bhāvana* in goat urine and dried. Application of this powder mixed with clarified butter cures several eye diseases collectively known as *pillarōga*.	*Ut*. 16: 54
15		*Guggulu pañcapalam cūrṇa* mixed with ginger powder and subjected to *bhāvana* in decoction of *Daśamūla* herbs is indicated in *vāta* diseases.	*Ut*. 28: 41
16		Powder of *Triphala* and heartwood of *Acacia catechu* is subjected to *bhāvana* in decoction of *Pterocarpus marsupium*. Thereafter, it is mixed with equal quantity of guggul gum and consumed along with honey. This formulation is indicated in skin diseases, *pramēha*, diabetic carbuncles and anal fistula.	*Ut*. 28: 42
17		Powder of entire *Tribulus terrestris* herb subjected to *bhāvana* in its own juice is to be consumed as a *rasāyana* to improve health, vigor, vitality and longevity.	*Ut*. 39: 56–57
18		Powder of *Curculigo orchioides* root subjected to *bhāvana* in its own juice is to be consumed as a *rasāyana* to delay ageing.	*Ut*. 39: 58–59
19		Powder of *Psoralea corylifolia* seeds subjected to *bhāvana* in decoctions of *Pterocarpus marsupium* and *Acacia catechu* heartwood is to be consumed as a *rasāyana*.	*Ut*. 39: 107
20		*Triphala* powder subjected to *bhāvana* in decoctions of *Acacia catechu* and *Pterocarpus marsupium* heartwood is recommended as a *rasāyana*.	*Ut*. 39: 152
21		Powder of *Ipomoea digitata* roots subjected to *bhāvana* in its own juice improves sexual vigor.	*Ut*. 40: 26
22		Powder of *Piper longum* and *Emblica officinalis* fruits subjected to *bhāvana* in juices of the same herbs, when consumed with sugar, honey and clarified butter turns even an eighty-year old person into a youth.	*Ut*. 40: 27–28
23		Powder of *Randia dumetorum* fruits subjected to *bhāvana* in decoction of the same fruit is recommended for inducing emesis in patients suffering from fever, tumors, enlargement of abdomen and distaste for food.	*Ka*. 1: 10–11
24		*Mantha* prepared with powder of roasted barley, subjected to *bhāvana* in juice of *Lagenaria siceraria*, is ideal for inducing emesis in patients of fever, cough, diseases of throat and distaste for food.	*Ka*. 1: 32

(*Continued*)

TABLE 6.42 (*Continued*)
Potentiation of Medicines with *bhāvana*

Sl. No.	Absorbent	Description of Potentiation	Reference
25		Powder of *Citrullus colocynthis* roots subjected to *bhāvana* in decoctions of *Symplocos racemosa* and *Daśamūla* herbs is suitable for inducing purgation.	*Ka.* 2: 38–40
26		Powders of herbs such as *Operculina turpethum*, *Cassia fistula*, *Clitoria ternatea* and six others are to be subjected to *bhāvana* for seven days in latex of *Euphorbia neriifolia*. Administration of the dried powder mixed with meat soup or clarified butter induces purgation.	*Ka.* 2: 47–48
27	Rice grains	Patients suffering from enlargement of abdomen are to be fed with *Ṣaṣṭika* rice subjected to *bhāvana* with cow urine.	*Ci.* 15: 123
28	Seeds	Sesame seeds subjected to *bhāvana* in milk and decoction of *Glycyrrhiza glabra* are used in the preparation of *Gandha taila*, indicated in fractures.	*Ut.* 27: 36–41
29		Sesame seeds, subjected to *bhāvana* in milk boiled with testicles of goat, are to be mixed with sugar and consumed to improve sexual vigor and vitality.	*Ut.* 40: 25
30		Seeds of *Moringa oleifera* are subjected to *bhāvana* in juice of *Albizia lebbeck* flowers and ground to a paste with water. Oral administration of the paste is recommended in bites of serpents. The paste can also be administered as *nasya* or collyrium.	*Ut.* 36: 72

or root paste, but no sesame oil, are added. Heating is continued and stopped when the desired end point is reached. In this way, the process is repeated usually 101 times. The resultant *taila* is called *Śatapāka Balā taila*. *Balā taila* prepared by repeating the process 7, 21, 41 and 101 times are used by Ayurveda practitioners of Kerala. AH recommends *Balā taila* prepared by repeating the process 100 or 1,000 times for oral administration and topical application in *vātaśōṇita* (*Ci.* 22: 45–46). *Taila* of *Yaṣṭimadhu* (*Glycyrrhiza glabra*) prepared by repeating the process 100 times is indicated in heart disease (*Ci.* 6: 38–39). *Ghṛta* prepared with *Cassia fistula* root and repeating the process 100 times is to be consumed by patients suffering from skin diseases (*Ci.* 19: 13–14).

6.11 CONCLUSION

AH recommends numerous substances of vegetable, animal and mineral origin to be used in the preparation of medicines. Nevertheless, herbs form the major segment. The 33 groupings of herbs based on therapeutic properties (*gaṇa*s of Vāgbhaṭa) described in *Śōdhanādigaṇasaṅgrahādhyāyaḥ* (Chapter 15) of *Sūtrasthānam* make use of 230 herbs. In addition to them, 226 herbs are described in the other chapters of AH. Thus, 456 herbs are used in total in AH. However, pharmacognosy of many herbs is in a disorderly state. Ayurvedic lexicons such as *Abhidānamañjari*, *Bhāvaprakāśa Nighaṇṭu*, *Dhanvantari Nighaṇṭu* and *Rāja Nighaṇṭu* provide information on the botanical characteristics and medicinal properties of herbs. This information is usually given in the form of a string of Sanskrit synonyms. Several synonyms are often ascribed to a single herb. Thus, *Citrullus colocynthis* Schrad. has eight synonyms such as *Amāra*, *Indravāruṇi*, *Aindri*, *Gajacirbhaṭa*, *Gavākṣi*, *Viśālā*, *Suravāruṇi* and *Kāliṅgam*. Sometimes same synonym is shared by several herbs. For example, the name *Amḷīka* is shared by *Tamarindus indica*, *Garcinia gummi-gutta* and *Oxalis corniculata* (Shajahan and Nikita, 2019).

Because of this, the identities of many herbs remain controversial. Bapalal Vaidya lists ten plants used in different parts of India, for the Sanskrit entity *Pāṣāṇabhēda*. They include *Aerva lanata* (L.) Juss., *Aerva javanica* (Burm. f.) Schult. (Tamil Nadu, Andhra Pradesh, Rajasthan, Gujarat), *Ammannia baccifera* L. (Kerala), *Rotula aquatica* Lour. (Karnataka), *Bergenia pacumbis* (Buch. -Ham. ex D. Don) C.Y. Wu & J. T. Pan (North India, Gujarat, Kashmir), *Coleus aromaticus* Benth., *Bryophyllum pinnatum* (Lam.) Oken (Bengal), *Bridelia montana* (Roxb.) Willd. (Goa), *Ocimum basilicum* L. and *Homonoia riparia* Lour. (Thomas and Shankar, 2020).

One way of clearing this confusion is by adopting the concept of pharmaco-linguistics proposed by Krishnamurthy (1971). This is the study of linguistic aspects of herbs. Using this method, Krishnamurthy (1971) ascertained the identities of *Haimavati* and *Kuliñjan*. Based on pharmaco-linguistics, he identified *Haimavati* as *Acorus gramineus* Sol. Aiton and *Kuliñjan* as *Alpinia galanga* (L.) Willd. Based on the pharmaco-linguistic approach, Pillai (1976) ascertained the correct identity of the herb *Śaṅkhapuṣpi*, by scoring the Sanskrit synonyms against the exomorphy of 30 contestants.

Another way of solving the problem is analyzing the etymology, synonyms, their clinical application and chronological differences. A survey on living folk traditions and living Ayurveda traditions can also help in correlating vernacular, Sanskrit and botanical names of herbs. Review of regional literature can establish the identity of regionally used plants from those with controversial identity. Non-medical Sanskrit literature is another source of information on herbs. Trade-related studies and review of recent botanical correlates can also be of help (Thomas and Shankar, 2020). Pharmacognostical, phytochemical and pharmacological studies are good tools for identifying the right candidate from a set of controversial herbs. *In vitro* and *in vivo* studies can confirm whether the controversial plants possess the same biological activities. For example, pharmacological studies on the lithotritic properties of controversial plants can identify the correct *Pāṣāṇabhēda*.

REFERENCES

Anonymous. 1978. *The Ayurvedic Formulary of India, Part I*, 1st Edition. New Delhi: Ministry of Health & Family Planning.

Bhuvaneswari, S., S. Gopala Krishnan, H. Bollinedi et al. 2020. Genetic architecture and anthocyanin profiling of aromatic rice from Manipur reveals divergence of Chakhao landraces. *Frontiers in Genetics* 11:570731. doi: 10.3389/fgene.2020.570731.

Gopinath, N. 2021. Industrial manufacture of traditional ayurvedic medicines. In *Ayurveda in the New Millennium: Emerging Roles and Future Challenges*, ed. D.S. Kumar, 41–69. Boca Raton: CRC Press.

Gupta, T., U.U. Zala, B.D. Kalsariya, and B.L. Umrethia. 2020. Preliminary pharmaceutico- analytical study of *sarjika kshara. World Journal of Pharmacy and Pharmaceutical Sciences* 9:2310–2317.

Krishnamurthy, K.H. 1971. Botanical identification of ayurvedic medicinal plants: A new method of pharmaco-linguistics – A preliminary report. *Indian Journal of Medical Research* 59:90–103.

Mooss, N.S. 1980.Ganās *of Vāhaṭa*, 1–215. Kottayam: Vaidya Sarathy Press (P) Ltd.

Pillai, N.G. 1976. On the botanical identity of *Śaṅkhapuṣpi. Journal of Research in Indian Medicine, Yoga and Homoeopathy* 11:67–76.

Priya, T.S.R., A.R.L.E. Nelson, K. Ravichandran and U. Antony. 2019. Nutritional and functional properties of coloured rice varieties of South India: A review. *Journal of Ethnic Foods* 6:11. https://doi.org/10.1186/s42779-019-0017–3.

Shajahan, M.A. and S. Nikita. 2019. *Aṣṭāṅgahṛdayam - Sūcika*, 1–1256. Thiruvananthapuram: Cyberveda Technologies.

Thomas, V. and D. Shankar. 2020. Controversial identities of medicinal plants in classical literature of Ayurveda. *Journal of Ayurveda and Integrative Medicine* 11:565–572.

Upadhyaya, Y. 1975. *Rasabhēdīyam* (Chapter on combination of tastes). In *Aṣṭāṅgahṛdaya*. 82–85. Varanasi: Chowkhamba Sanskrit Sansthan.

Vaidya, K.M. 1936. *Ashtanga Hridaya Kosha with the Hridaya Prakasha (A Critical and Explanatory Commentary)*, 1–654. Trichur: The Mangalodayam Press.

Warrier, P.K., V.P.K. Nambiar, and C. Ramankutty. 2007a. *Indian Medicinal Plants – A Compendium of 500 Species*, Volume 1, 1–420. Hyderabad: Orient Longman Private Ltd.

Warrier, P.K., V.P.K. Nambiar, and C. Ramankutty. 2007b. *Indian Medicinal Plants – A Compendium of 500 Species*, Volume 2, 1–416. Hyderabad: Orient Longman Private Ltd.

Warrier, P.K., V.P.K. Nambiar and C. Ramankutty. 2007c. *Indian Medicinal Plants – A Compendium of 500 Species*, Volume 3, 1–423. Hyderabad: Orient Longman Private Ltd.

Warrier, P.K., V.P.K. Nambiar and C. Ramankutty. 2007d. *Indian Medicinal Plants – A Compendium of 500 Species*, Volume 4, 1–444. Hyderabad: Orient Longman Private Ltd.

Warrier, P.K., V.P.K. Nambiar and C. Ramankutty. 2008. *Indian Medicinal Plants – A Compendium of 500 Species*, Volume 5, 1–592. Hyderabad: Orient Longman Private Ltd.

7 Medicinal Foods in *Aṣṭāṅgahṛdaya*

7.1 INTRODUCTION

With the change in lifestyles due to rise in income, increased leisure time and reduced physical activity, there is a global rise in the incidence of diseases such as obesity, diabetes mellitus, cardiovascular diseases and rheumatoid arthritis. Physicians all over the world, therefore, look for better treatment options. Parallelly, there is a global increase in health awareness and interest in herbal products. Therefore, functional foods have emerged as an effective way of prevention of diseases. These food products are brought with the intention of reducing body weight, lowering cholesterol levels and improving disease resistance through the immune system. Fortified bread, sausages, biscuits, tea, coffee and chocolates are becoming increasingly popular among consumers (Arvanitoyannis and van-Houwelingen-Koukaliaroglou, 2005; Kumar, 2016).

7.2 ANCIENT CULINARY TRADITION

India has a rich tradition in culinary arts. King Naḷa, whose story is narrated in the *Mahabharata*, was an accomplished chef. *Pāka Darpaṇam* (Mirror of Cookery) penned by him is a veritable vault of information on the foods that were popular in ancient India (Madhulika, 2013). AH also follows this tradition. Many varieties of food and beverages having therapeutic value are mentioned by Vāgbhaṭa. Following is a brief description of some such preparations. *Śrīkukkuṭa* is a food that was famous in the Malwa region of Madhya Pradesh. It is prepared by mixing together pastes of millets such as *Kaṅgu* (*Setaria italica*), *Kōdrava* (*Paspalum scrobiculatum*), *Śyāmāka* (*Panicum miliare*), green gram, lentil, *Śāli*, *Ṣaṣṭika* rice grains and sesame and mustard oil cakes. The mixture is made sour by adding lemon juice. *Śrīkukkuṭa* is choice food for *pramēha* patients (*Sū*. 12: 12; Upadhyaya, 1975).

Vēśavāram is a food as well as medicinal ingredient. Boneless meat is cooked and then ground to a paste. Thereafter, it is mixed with powder of *Trikaṭu*, jaggery and clarified butter. Sometimes powders of green gram, black gram and lentil are also added, in which case it is called *Mudgādijaḥ* (Vaidyan, 2017a; *Sū*. 6: 41). *Vēśavāram* is heavy and unctuous, justifying its ability to increase stamina. It is recommended for patients of kingly consumption (*Ci*. 5: 8–9). *Upanāha* carried out with *Vēśavāram* pacifies pricking pain, edema and immobility of limbs associated with *vātaśōṇita* (*Ci*. 22: 31).

In addition to milk, clarified butter, curd, cream and buttermilk, six other dairy products are also mentioned in the chapter dealing with liquid food ingredients. *Kīlāṭa* is prepared by mixing less milk and more buttermilk. Buttermilk mixed with fresh or boiled milk drawn from a cow that has given birth to a calf recently is called *Pīyūṣa*. *Kūcika* is a dairy product prepared by mixing buttermilk with warm milk. When milk and buttermilk are mixed in equal proportions, the preparation is called *Mōraṇa*. These four dairy products increase virility, *kapha* and sleep (*Sū*. 5: 41–42). The fifth product is known as *Śaśāṅkakiraṇam*. Milk is boiled with sugar and brought to thick consistency. Thereafter, it is rolled into small balls. *Śaśāṅkakiraṇam* is recommended to be eaten at night as a part of regimen for summer season (Vaidyan, 2017b; *Sū*. 3:32). *Rasāla*, the sixth dairy product, is prepared by blending undiluted curd with sugar and powders of spices. It is unctuous, heavy, roborant, virilific, diuretic and tasty. It pacifies tiredness, hunger and thirst (Vaidyan, 2017a; *Sū*. 6: 35–36).

DOI: 10.1201/9781003148296-7

Grain-based products are also described in AH. Prominent among them is *Utkārika*, which is prepared by cooking balls of finely ground paste of soaked rice in boiling water. It is also used as a vehicle for delivering medicines in the treatment of respiratory distress (*Ci*. 4: 34–35). *Vāṭya* is a porridge prepared with roasted barley. It is a choice food for *pramēha* patients (*Ci*. 12: 10–11). *Apūpa* and *Pūpalika* are rice cakes. They are served to patients of *pramēha* and worm infestation, respectively (*Ci*. 12: 10, *Ci*. 20: 29–31). *Kulmāṣa* is half-cooked barley. It is served to patients of alcoholism and *gulma* (*Ci*. 7: 15, Ci. 14: 53). *Saktu* (powder of fried barley) is a food that is ideal for those suffering from *pramēha* (*Ci*. 12: 10).

In the *Sūtrasthānam*, there is mention of the different ways in which delicacies from grains can be processed. *Kukūla*, *karpara*, *bhrāṣṭra*, *kaṇḍu* and *aṅgāra* are the five processing modes. Food that is prepared by steaming is called *kukūlapācita*. Those that are cooked on heated plates are called *karparapācita*. Sautéed food is called *bhrāṣṭrapācita*. Food that is baked in an oven, just like common bread is called *kaṇḍupācita*. Food that is cooked over red hot cinders is known as *aṅgārapācita* (*Sū*. 6: 42).

7.3 FORTIFIED FOODS

AH applies the concept of food fortification in the prevention and treatment of diseases. Various kinds of food and beverages are recommended in this regard. The important matrices for delivering the therapeutic agents are gruels (*pēya*, *vilēpi*, *mantha*), milk, curd, buttermilk, butter, sweet pudding, cake, chutney, curry, soup, wine and sweet beverage.

7.3.1 Gruels

They are prepared by cooking grains of rice, barley or wheat in water. The clear fluid that can be decanted from the cooked grains is called *maṇḍa*. It is very light in nature, promotes the normal movement of *vāta*, stimulates abdominal fire, softens body channels, quenches thirst and pacifies tiredness. Gruel with low content of cooked grains and more of *maṇḍa* is called *pēya*. It is heavier than *maṇḍa*. *Pēya* pacifies fever and abdominal diseases, promotes normal bowel movement and stimulates abdominal fire. *Pēya* may be prepared with broken grains as well. Gruel with more of cooked grains and less of *maṇḍa* is called *vilēpi* or *yavāgu*. *Vilēpi* is very heavy in nature. It is ideal for patients being treated for *vraṇa*, eye diseases and those who have been subjected to oleation, emesis or purgation. It improves stamina in weak individuals. *Pēya* and *vilēpi* can also be cooked with milk and meat (*Sū*. 6: 26–31). Medicated gruels can be served as such (*akṛta* or un-processed), with sugar and honey or seasoned in hot clarified butter with mustard seeds and curry leaves (*kṛta* or processed) (*Ci*. 1: 75–77). The various medicated gruels recommended in AH are described below.

7.3.1.1 Medicated Gruels

1. Parched rice (*lāja*) is cooked with *Zingiber officinale*, *Coriandrum sativum*, *Piper longum* and rock salt. Those who appreciate sour taste can add juice of *Punica granatum*. This gruel is served to the patient when signs of optimum fasting are evident in the treatment of fever. If stools are loose or *pitta* is increased, gruel with ginger can be served, but only after cooling it and adding honey (*Ci*. 1: 26–27).
2. Patients suffering from fever associated with pain in lower abdomen, flanks and head are to be served with rice gruel fortified with *Solanum xanthocarpum* and *Tribulus terrestris* (*Ci*. 1: 28).
3. In *jvarātisāra*, the rice gruel is to be cooked along with *Sida rhombifolia* ssp. *retusa*, *Pseudarthria viscida*, *Aegle marmelos*, *Zingiber officinale*, *Monochoria vaginalis* and *Coriandrum sativum*. It is to be served after adding some lemon juice (*Ci*. 1: 28–29).

4. Patient suffering from fever, associated with hiccup, pain and respiratory distress is to consume rice gruel cooked in the decoction of *Hrasva pañcamūla* (see Table 6.36) (*Ci.* 1: 29).
5. In fever with aggravation of *kapha*, the patient is to be served with barley gruel fortified with *Mahat pañcamūla* (see Table 6.36) (*Ci.* 1: 30)
6. Patient suffering from fever associated with constipation should consume gruel prepared with barley, mixed with powders of *Piper longum* and *Emblica officinalis* and seasoned with clarified butter (*Ci.* 1: 30–31).
7. Rice gruel fortified with roots of *Piper chaba* and *Piper longum*, *Vitis vinifera* fruits, *Emblica officinalis* and *Zingiber officinale* is to be served in fever with constipation and pain (*Ci.* 1: 31–32).
8. If patient suffering from fever experiences cutting pain in the alimentary tract, the rice gruel should be fortified with *Ziziphus mauritiana*, *Garcinia gummi-gutta*, *Solanum xanthocarpum* and fruits of *Aegle marmelos* (*Ci.* 1: 32).
9. Patient suffering from fever associated with sleeplessness and lack of sweating should consume rice gruel mixed with sugar, *Emblica officinalis* and *Zingiber officinale* (*Ci.* 1: 33).
10. In fever associated with excessive thirst and vomiting, the patient should consume rice gruel mixed with honey, sugar, kernel of *Ziziphus mauritiana*, *Vitis vinifera*, *Hemidesmus indicus*, *Cyperus rotundus* and *Santalum album* (*Ci.* 1: 33–34).
11. Rice gruel prepared with stamens of *Nelumbo nucifera* and *Monochoria vaginalis*, *Pseudarthria viscida* root and *Callicarpa macrophylla* flowers pacifies hemorrhages of obscure origin manifesting in the lower parts of body (*Ci.* 2: 16).
12. Consumption of rice gruel prepared with roots of *Vettiveria zizanioides*, bark of *Symplocos laurina*, *Zingiber officinale* and heartwood of *Pterocarpus marsupium* pacifies hemorrhages of obscure origin manifesting in the lower parts of body (*Ci.* 2: 16).
13. Rice gruel prepared with root of *Coleus vettiveroides*, flowers of *Woodfordia fruticosa*, immature fruits of *Aegle marmelos* and root of *Tragia involucrata* pacifies hemorrhages of obscure origin manifesting in the lower parts of body (*Ci.* 2: 17).
14. Rice gruel prepared with decoction of *Andrographis paniculata*, *Vettiveria zizanioides* and *Cyperus rotundus* pacifies hemorrhages of obscure origin manifesting in the upper parts of body (*Ci.* 2: 18).
15. Rice gruel prepared with *Pseudarthria viscida* and *Lens culinaris* or *Pseudarthria viscida* and *Vigna radiata* or *Sida rhombifolia* ssp. *retusa*, *Corchorus trilocularis* and clarified butter pacifies hemorrhages of obscure origin manifesting in the upper parts of body (*Ci.* 2: 18).
16. Rice gruel is to be prepared with decoction of *Daśamūla* herbs, powder of *Pañcakōla* herbs, jaggery and equal parts of rice and sesame seeds. Consumption of this medicated gruel cures cough with predominance of *vāta* (*Ci.* 3: 22).
17. Consumption of rice gruel prepared with meat of pig, rooster and fish, mixed with clarified butter and rock salt cures cough with predominance of *vāta* (*Ci.* 3: 22–23).
18. *Fritillaria roylei*, *Lilium polyphyllum*, *Solanum indicum*, *Solanum xanthocarpum*, *Polygonatum cirrhifolium*, *Polygonatum verticillatum*, *Adhatoda vasica* and *Zingiber officinale* are the herbs that are to be added to the medicated gruel which cures cough with predominance of *pitta* (*Ci.* 3: 35).
19. Medicated rice gruel is to be prepared in the decoction of *Vitis vinifera*, *Piper longum* and *Tṛṇapañcamūla* (see Table 6.36). It is to be consumed after cooling and adding honey. This fortified food cures cough with predominance of *pitta* (*Ci.* 3: 36–37).
20. Rice gruel is prepared and it is fortified by adding powders of *Daśamūla* herbs, *Kaempferia galanga*, *Alpinia galanga*, *Clerodendrum serratum*, *Aegle marmelos*, *Habenaria edgeworthii*, *Inula racemosa*, *Pistacia integerrima*, *Piper longum*, *Phyllanthus amarus*, *Tinospora cordifolia* and *Zingiber officinale*. This gruel cures respiratory distress and hiccup (*Ci.* 4: 24–25).

21. Consumption of gruel prepared with barley and fortified with powder of *Piper longum* and *Emblica officinalis*, mixed with sesame oil, clarified butter, visceral fat and bone marrow cures lassitude of voice caused by *kapha* (*Ci.* 5: 45).
22. Rice gruel fortified with juice of *Punica granatum*, powder of *Pseudarthria viscida*, *Sida rhombifolia* ssp. *retusa*, *Aegle marmelos* and *Desmodium gangeticum* cures dysentery with *kapha-pitta* involvement (*Ci.* 9: 13–14).
23. Rice gruel cooked with *Terminalia chebula*, root of *Piper longum* and root of *Aegle marmelos* initiates the normal movement of *vāta* in dysentery patients (*Ci.* 9: 14).
24. Consumption of rice gruel cooked with buttermilk and pastes of tender *Aegle marmelos* fruits, *Cyperus rotundus*, *Symplocos spicata*, *Woodfordia fruticosa* and *Zingiber officinale* cures *pakvātisāra* (*Ci.* 9: 23).
25. Similarly, rice gruel cooked with tender leaves of *Feronia elephantum*, *Tragia involucrata*, *Clerodendrum serratum*, *Jasminum coarctatum*, *Ficus benghalensis*, *Cordia dichotoma*, *Punica granatum*, *Crotolaria juncea*, *Gossypium herbaceum*, *Salmalia malabarica* and *Artocarpus heterophyllus* also cures *pakvātisāra* (*Ci.* 9: 24).
26. Gruel prepared with bark of *Oroxylum indicum*, *Glycyrrhiza glabra*, *Callicarpa macrophylla* flowers, tender leaves of *Punica granatum*, fruits of *Punica granatum* and curd pacifies dysentery (*Ci.* 9: 65–66).
27. Gruel prepared with unripe fruits of *Feronia elephantum*, *Aegle marmelos*, *Syzygium cumini* and *Mangifera indica* also cures dysentery (*Ci.* 9: 66).
28. Those who suffer from *grahaṇi* are to be served with rice gruel fortified with digestive herbs like *Pañcakōla* (*Ci.* 10: 2).
29. *Pañcakōla* herbs, *Terminalia chebula*, *Coriandum sativum*, *Cyclea peltata*, *Kaempferia galanga* and tender leaves of *Citrus medica* are to be boiled in water and turned into a decoction. Gruel prepared by boiling rice grains in this decoction is to be consumed for curing *grahaṇi* (*Ci.* 10: 46).
30. Decoction of *Anogeissus latifolia*, *Alstonia scholaris*, *Holarrhena antidysenterica*, *Tinospora cordfolia*, *Cassia fistula*, *Costus speciosus*, *Elettaria cardamomum* and *Pongamia glabra* is to be prepared. Rice gruel prepared in this decoction is to be administered with honey for curing dysuria (*Ci.* 11: 12–13).
31. Consumption of rice gruel prepared with herbs recommended for treating *vāta*-dominant urinary stone also cures the disease (*Ci.* 11: 18–20).
32. *Pitta*-dominant urinary stone is also cured by consumption of rice gruel prepared with herbs recommended in the treatment of the disease (*Ci.* 11: 22–24).
33. Herbs recommended for disintegration of *kapha*-dominant urinary stone can also be incorporated into rice gruel and consumed for breaking the urinary stone (*Ci.* 11: 25–26).
34. Consumption of rice gruel prepared with bark of *Moringa concanensis* is recommended for the healing of internal abscess (*Ci.* 13: 23).
35. Medicated gruel prepared with *Foeniculum vulgare*, *Kaempferia galanga*, *Holostemma annulare*, *Cuminum cyminum*, *Inula racemosa*, *Plumbago zeylanica*, unripe fruits of *Aegle marmelos*, *yavakṣāra*, fruits of *Garcinia gummi-gutta*, clarified butter and sesame oil is to be consumed for curing edema, dysentery, heart disease, *gulma*, hemorrhoids, lowering of abdominal fire and *pramēha* (*Ci.* 17: 20–21).
36. Consumption of medicated gruel prepared with *Pañcakōla* herbs also cures edema, dysentery, heart disease, *gulma*, hemorrhoids, lowering of abdominal fire, worm infestation and *pramēha* (*Ci.* 17: 22, Ci. 20: 22).
37. Rice gruel prepared with *Embelia ribes*, *Piper longum*, *Piper nigrum*, *Piper chaba*, *Moringa oleifera* and buttermilk is to be consumed, with sonchal salt to cure worm infestation (*Ci.* 20: 25).
38. Consumption of rice gruel fortified with *Piper chaba*, *Cyclea peltata*, *Foeniculum vulgare*, *Pañcakōla* herbs, *Zanthoxylum armatum*, *Cuminum cyminum*, *Coriandrum sativum* and unripe fruit of *Aegle marmelos* is beneficial in hemorrhoids (*Ci.* 8: 50–51).

39. Patients suffering from dysentery should consume rice gruel prepared with unripe fruit of *Aegle marmelos, Kaempferia galanga, Coriandrum sativum, Ferula foetida, Garcinia gummi-gutta* and *Punica granatum* (*Ci.* 9: 11–12).
40. In the same way, rice gruel fortified with *Butea monosperma, Sphaeranthus indicus, Cuminum cyminum, Trachyspermum ammi*, vida salt and rock-salt can be administered to patients suffering from dysentery (*Ci.* 9: 12).
41. Rice gruel for dysentery patients can also be fortified with *Hrasva pañcamūla* herbs, *Pañcakōla* herbs or root of *Cyclea peltata* (*Ci.* 9: 13).
42. Patients suffering from dysentery due to aggravation of *pitta* should be served rice gruel fortified with *Solanum indicum, Solanum xanthocarpum, Desmodium gangeticum, Pseudarthria viscida, Tribulus terrestris, Asparagus racemosus, Sida rhombifolia* ssp. *retusa* and *Vigna pilosa*. It improves digestive efficiency (*Ci.* 9: 55–56).
43. Decoction of *Coleus vettiveroides, Monochoria vaginalis, Zingiber officinale* and *Pseudarthria viscida* is to be mixed with diluted goat milk. Consumption of medicated gruel prepared in this mixture cures dysentery with bleeding (*raktātisāra*) (*Ci.* 9: 86).
44. In the event of abortion of the fetus, the mother should consume wine, followed by oil-free gruel fortified with *Pañcamūla* herbs (*Śā.* 2: 10).
45. If the mother is not habituated to drinking wine, she should consume gruel prepared with *Uddālaka* variety of rice, decoction of *Daśamūla* herbs and pastes of *Pañcakōla* herbs and sesame seeds (*Śā.* 2: 11).
46. A medicated gruel is recommended in the treatment of poisoning. It makes use of ingredients such as *Luffa acutangula, Premna serratifolia, Cyclea peltata, Ventilago maderasapatana, Tinospora cordifolia, Terminalia chebula, Cordia dichotoma, Albizia lebbeck, Crotolaria retusa, Curcuma longa, Berberis aristata*, chalcopyrite, *Boerhaavia diffusa, Trianthema portulacastrum, Zingiber officinale, Piper longum, Piper nigrum, Solanum indicum, Solanum xanthocarpum, Hemidesmus indicus, Ichnocarpus frutescens* and *Acorus calamus*. Gruel is prepared in the decoction of these herbs. Clarified butter and honey are added on cooling. It neutralizes all poisons (*Ut.* 35: 21–23).
47. *Madhūkādi yavāgu* is prepared in the decoction of *Madhuca longifolia, Glycyrrhiza glabra*, stamens of *Nelumbo nucifera* and *Santalum album*. It is an effective remedy for poisoning (*Ut.* 35: 23).

7.3.1.2 Churned Gruels

Gruel is contra-indicated in fever caused by alcoholism, fever manifesting in a regular user of *madya*, fever due to *kapha* enraged at site of *pitta*, fever appearing in summer season, fever due to aggravation of *pitta-kapha*, fever with thirst, vomiting and burning sensation and hemorrhages of obscure origin manifesting in the upper parts of body. *Mantha* (churned gruel) is to be consumed in such conditions with sugar and honey, after mixing it with water or juices of fever-pacifying fruit (*Ci.* 1: 35). *Mantha* is prepared by cooking one part of parched rice (*lāja*) in 14 parts of water, followed by churning. Fortification is carried out by cooking in decoction of herbs instead of water.

1. Mantha prepared with *Vitis vinifera* fruits, *Madhuca longifolia* fruits, *Glycyrrhiza glabra* root, *Symplocos spicata* bark, *Gmelina arborea* fruits, *Hemidesmus indicus* root, *Cyperus rotundus* tubers, *Emblica officinalis* fruits, *Coleus vettiveroides* root, stamens of *Nelumbo nucifera, Santalum album* heartwood, *Vettiveria zizanioides* root, *Monochoria vaginalis* rhizome and *Phoenix pusilla* fruits is to be administered to patients suffering from hemorrhages of obscure origin (*Ci.* 1: 5–58, *Ci.* 2: 14).
2. Honey, *Phoenix dactylifera* fruits, *Phoenix pusilla* fruits, *Vitis vinifera* fruits and sugar syrup are the ingredients of a *mantha* that can be administered in hemorrhages of obscure origin (*Ci.* 2: 14).

3. *Mantha* prepared with juices of fruits such as *Vitis vinifera*, *Madhuca longifolia*, *Phoenix dactylifera*, *Gmelina arborea* and *Grewia asiatica*, collectively called *Pañcasāra*, is useful for the treatment of hemorrhages of obscure origin. The *mantha* may be made slightly sour by adding juices of *Punica granatum* and *Emblica officinalis* (*Ci.* 2: 15).
4. Grains of barley (*Hordeum vulgare*) are to be subjected to *bhāvana* in decoction of anti-emetic herbs. *Mantha* prepared with the processed grains can be administered in vomiting due to aggravation of *kapha* (*Ci.* 6: 19).
5. Consumption of *mantha* prepared with powders of *Hordeum vulgare* and *Ziziphus mauritiana* seed kernels, mixed with sugar pacifies thirst due to sunstroke (*Ci.* 6: 77).
6. Consumption of *mantha* prepared with antimony sulphide, *Santalum album*, *Vettiveria zizanioides*, sugared water and blood of goat pacifies discomforts arising from excessive purgation (*Ka.* 3: 26).
7. *Mantha* prepared with juice of *Ziziphus mauritiana*, clarified butter, honey and sugar is to be consumed for countering effects of excessive emesis (*Ka.* 3: 27).
8. Powders of *Operculina turpethum* and *Cassia fistula* are to be mixed with *mantha* and administered, for eliminating *dōṣa*s through purgation in patients of kingly consumption (*Ci.* 5: 2–4).

7.3.2 Medicated Milk

Cow milk possesses several beneficial properties. It pacifies *vāta-pitta*, increases *kapha*, nourishes tissue elements, maintains life activities and pacifies several diseases (*Sū.* 5: 20–23). Therefore, milk is considered in AH as a suitable vehicle for delivery of bioactive compounds. Generally, one part of herb powder is mixed with eight parts of milk and thirty-two parts of water. The mixture is cooked till the volume is reduced to one-fourth. The medicated milk is filtered and administered.

1. Consumption of milk boiled with *Zingiber officinale*, *Phoenix dactylifera*, sugar and clarified butter pacifies fever, thirst and burning sensation. However, the medicated milk should be administered cooled and mixed with honey (*Ci.* 1: 109–110).
2. Milk fortified with *Vitis vinifera*, *Sida rhombifolia* ssp. *retusa*, *Glycyrrhiza glabra*, *Hemidesmus indicus*, *Piper longum* and *Santalum album* also pacifies fever associated with thirst and burning sensation (*Ci.* 1: 110).
3. Consumption of milk boiled with powder of *Piper longum* also pacifies fever associated with thirst and burning sensation (*Ci.* 1: 111).
4. Consumption of milk boiled with *Hrasva pañcamūla* frees the chronic fever-affected patient from cough, respiratory distress, headache and pain in the flanks (*Ci.* 1: 111–112).
5. Consumption of milk boiled with root of *Ricinus communis* cures fever associated with obstruction of stool, thirst and voiding of blood-stained, slimy stool (*Ci.* 1: 112–113).
6. Consumption of milk boiled with unripe fruit of *Aegle marmelos* also cures fever associated with obstruction of stool, thirst and voiding of blood-stained, slimy stool (*Ci.* 1: 112–113).
7. Milk boiled with *Zingiber officinale*, *Sida rhombifolia* ssp. *retusa*, *Tribulus terrestris* and jaggery pacifies edema, obstruction of urine and stool, fever and cough (*Ci.* 1: 114).
8. Fever and edema are also cured by milk boiled with roots of *Boerhaavia diffusa* and *Aegle marmelos* (*Ci.* 1: 115).
9. Fever can be pacified by milk boiled with heartwood of *Dalbergia sissoo* (*Ci.* 1: 115).
10. Milk boiled with roots of *Hrasva pañcamūla* can be administered with sugar and honey to cure hemorrhages of obscure origin (*Ci.* 2: 37).
11. *Malaxis acuminata*, *Malaxis muscifera*, *Vitis vinifera*, *Sida rhombifolia* ssp. *retusa*, *Tribulus terrestris* and *Zingiber officinale* are to be boiled individually in milk. Each of these medicated milks is to be administered with clarified butter or sugar for pacifying hemorrhages of obscure origin (*Ci.* 2: 37–38).

12. Medicated milk prepared with *Tribulus terrestris* and *Asparagus racemosus* cures painful bleeding through the urinary tract (*Ci.* 2: 38–39).
13. Similarly, medicated milk prepared with *Desmodium gangeticum*, *Pseudarthria viscida*, *Vigna radiata* and *Vigna pilosa* also cures painful bleeding through the urinary tract (*Ci.* 2: 38–39).
14. Bleeding through anus can be pacified by medicated milk prepared with *Zingiber officinale*, *Vettiveria zizanioides* and *Monochoria vaginalis* (*Ci.* 2: 39–40).
15. Medicated milk boiled with gum exudate of *Salmalia malabarica* cures bleeding through anus (*Ci.* 2: 39–40).
16. Consumption of milk boiled with adventitious root or apical buds of *Ficus benghalensis* cures bleeding through anus (*Ci.* 2: 39–40).
17. *Vitis vinifera*, *Piper longum* and roots of *Tṛṇapañcamūla* are to be boiled in water and then filtered. Milk is added to the filtrate, slightly concentrated, cooled and mixed with honey and sugar. Consumption of this medicated milk cures cough with predominance of *pitta* (*Ci.* 3: 36).
18. Lac, clarified butter, beeswax, powder of herbs of *Jīvanīya gaṇa* (see Table 6.36), sugar and bamboo manna are boiled in milk. Consumption of this medicated milk heals injury to thorax (*Ci.* 3: 75).
19. Medicated milk is prepared with pieces of sugarcane stem, stamens and rhizomes of *Nelumbo nucifera* and heartwood of *Santalum album*. Injury to thorax is cured by consuming this milk with honey (*Ci.* 3: 76).
20. Consumption of milk boiled with powder of undried barley, mixed with clarified butter, cures injury to thorax associated with fever and burning sensation (*Ci.* 3: 77).
21. Powder of *Boerhaavia diffusa* root and powder of red rice (*Raktaśāli*) are to be mixed with grape juice, milk and clarified butter. Consumption of this medicated milk cures hemoptysis experienced by those having injury to thorax (*Ci.* 3: 84).
22. *Madhuca longifolia* fruits, *Glycyrrhiza glabra* and *Amaranthus spinosus* are to be boiled in milk. Consumption of this medicated milk cures hemoptysis experienced in injury to thorax (*Ci.* 3: 85).
23. Consumption of milk boiled with juice of *Linum usitatissimum*, *Trikaṭu* and clarified butter cures respiratory distress and hiccup caused by aggravation of *vāta* and *pitta* (*Ci.* 4: 35).
24. Rhizomes of *Nelumbo nucifera*, *Monochoria vaginalis* and *Biophytum sensitivum* are to be boiled in goat milk and consumed mixed with sugar and honey. This medicated milk cures *raktātisāra* (*Ci.* 9: 82–83).
25. Consumption of goat milk boiled with *Hemidesmus indicus*, *Glycyrrhiza glabra* and *Symplocos spicata*, mixed with sugar and honey, cures *raktātisāra* (*Ci.* 9: 82–84).
26. Tender leaves of *Ficus glomerata*, *Ficus retusa*, *Ficus benghalensis* and *Ficus religiosa* are to be boiled in goat milk and administered with sugar and honey. It cures *raktātisāra* (*Ci.* 9: 82–84).
27. Medicated milk prepared with roots of *Hrasva pañcamūla* is recommended for the cure of all diseases of the urinary system. However, quantity of *Tribulus terrestris* needs to be doubled (*Ci.* 11: 35).
28. Paste of *Plumbago zeylanica* and *Cedrus deodara* is to be suspended in milk and consumed for the cure of enlargement of abdomen (*Ci.* 15: 42).
29. *Piper chaba* root and *Zingiber officinale* are to be ground to paste and suspended in milk. Consumption of this medicated milk for one month, with dietary restrictions, cures enlargement of abdomen (*Ci.* 15: 42).
30. Patient suffering from edema associated with thirst, burning sensation and mental confusion should consume milk fortified with herbs of *Nyagrōdhādi gaṇa* (*Ci.* 17: 30–31).
31. Administration of milk boiled with *Gmelina arborea* root and *Glycyrrhiza glabra* root cures emaciation of infants *in utero* (*Ci.* 21: 21–22).

32. *Vāta*-dominant *vātaśōṇita* is cured by milk boiled with *Sida rhombifolia* ssp. *retusa*, *Asparagus racemosus*, *Alpinia galanga*, *Daśamūla* herbs, *Salvadora persica* fruit, *Operculina turpethum*, *Ricinus communis* and *Pseudarthria viscida* (*Ci*. 22: 8–9).
33. Consumption of milk boiled with roots of *Saccharum spontaneum*, *Ipomoea digitata*, *Saccharum officinarum* and *Desmostachya bipinnata* cures epilepsy caused by aggravation of *vāta* and *pitta* (*Ut*. 7: 28).
34. After ablation of *mēdōvarti*, the patient should consume milk boiled with *Glycyrrhiza glabra*, *Ricinus communis*, *Tribulus terrestris*, lac and sugar for pacifying pain and burning sensation (*Ut*. 26: 53–54).
35. Consumption of milk boiled with *Alpinia galanga* rhizome, *Tribulus terrestris* root and *Adhatoda vasica* root pacifies pricking pain in vagina (*Ut*. 34: 33).
36. In the event of mishap in pregnancy named *līnam*, the pregnant mother should consume milk mixed with pastes of unripe *Aegle marmelos* fruit, *Sesamum indicum* seeds, *Vigna mungo* beans and powder of parched rice (*Śā*. 2: 19).
37. If abortion takes place during any of the ten months of pregnancy, medicated milks prepared with the following herbs are to be administered to the pregnant mother in the respective month:

Glycyrrhiza glabra, seeds of *Tectona grandis*, *Lilium polyphyllum* and *Cedrus deodara* (first month); *Schrebera swietenoides*, *Sesamum indicum*, *Rubia cordifolia* and *Asparagus racemosus* (second month); *Dolichandrone falcata*, *Lilium polyphyllum*, *Callicarpa macrophylla* and *Ichnocarpus frutescens* (third month); *Cynodon dactylon*, *Hemidesmus indicus*, *Alpinia galanga*, *Nervilia aragoana* and *Glycyrrhiza glabra* (fourth month); *Solanum indicum*, *Solanum xanthocarpum*, *Gmelina arborea*; apical buds and barks of *Ficus glomerata*, *Ficus retusa*, *Ficus benghalensis*, *Ficus religiosa* and clarified butter (fifth month); *Pseudarthria viscida*, *Sida rhombifolia* ssp. *retusa*, *Moringa oleifera*, *Tribulus terrestris* and *Tinospora cordifolia* (sixth month); *Trapa bispinosa*, *Nelumbo nucifera*, *Vitis vinifera*, *Cyperus rotundus*, *Glycyrrhiza glabra* and sugar (seventh month); *Feronia elephantum*, *Aegle marmelos*, *Solanum indicum*, *Solanum xanthocarpum*, *Trichosanthes cucumerina* and *Saccharum officinarum* (eighth month); *Hemidesmus indicus*, *Cynodon dactylon*, *Lilium polyphyllum* and *Glycyrrhiza glabra* (ninth month); *Lilium polyphyllum* or *Glycyrrhiza glabra*, *Zingiber officinale* and *Cedrus deodara* (tenth month) (*Śā*. 2: 54–60).

7.3.3 Medicated Curd

Curd is sour in taste as well as after transformation (*vipāka*). It is heavy and hot in quality and causes constipation. It increases fat, sexual vigor, stamina, abdominal fire, *kapha*, *pitta* and *rakta*. On account of its ability to improve appetite, it is useful in the treatment of distaste for food. Curd is suitable as a food in *śītajvara*, irregular fever, rhinitis and dysuria. Defatted curd is good for *grahaṇi* patients. Curd should not at all be consumed at night. It should be avoided in spring, summer and autumn seasons. Curd should not be consumed without adding green gram soup, honey, sugar or gooseberry. It is not a food that can be consumed every day. Violation of these pieces of advice can result in the appearance of fever, hemorrhages of obscure origin, cellulitis, skin diseases and morbid pallor (*Sū*. 5: 29–33). Nevertheless, fortified curd is medicinal in nature.

1. Pastes of roots of *Ricinus communis*, *Solanum indicum*, *Solanum xanthocarpum*, *Tribulus terrestris* and *Hygrophila auriculata* are to be mixed with sweet curd. Urinary stone will be disintegrated on consumption of this medicated curd (*Ci*. 11: 21).
2. Curd mixed with jaggery and powder of *Zingiber officinale* is to be administered to patient suffering from dysentery and who voids with pain small volumes of frothy, slimy stool (*Ci*. 9: 18).

3. Powders of *Piper longum*, *Zingiber officinale*, *Cyclea peltata*, *Hemidesmus indicus*, *Solanum indicum*, *Solanum xanthocarpum*, *Plumbago zeylanica*, *Holarrhena antidysenterica*, *yavakṣāra*, rock salt, sonchal salt, vida salt, sea salt and salt prepared from vegetable alkali are to be mixed with curd and consumed. Digestive efficiency will be improved in *grahaṇi* patients (*Ci.* 10: 12–13).

7.3.4 Fortified Buttermilk

Buttermilk is light, astringent and sour in tastes. It pacifies *kapha* and *vata*. Edema, enlargement of abdomen, hemorrhoids, *grahaṇi*, *gulma*, splenomegaly, morbid pallor and distaste for food are cured by its consumption. Whey has properties similar to those of buttermilk. It is light and laxative in nature (*Sū.* 5: 33–35). Buttermilk fortified with herbs is effective in hemorrhoids and enlargement of abdomen.

1. Inside of an earthen pot is first of all smeared with clarified butter. Thereafter, paste of *Piper longum* (fruit, root), *Terminalia chebula*, *Embelia ribes* and *Plumbago zeylanica* is smeared inside the pot, which is then filled with buttermilk and its mouth closed tightly. Consumption of the fortified buttermilk after one month pacifies respiratory distress and hiccup (*Ci.* 4: 29–30).
2. Milk should be boiled with root of *Plumbago zeylanica*, cooled and then curdled. Consumption of buttermilk prepared from this curd cures hemorrhoids (*Ci.* 8: 30–31).
3. Consumption of buttermilk mixed with powders of *Terminalia chebula*, *Embelia ribes*, *Plumbago zeylanica* and bark of *Holarrhena antidysenterica* cures hemorrhoids (*Ci.* 8: 33).
4. Buttermilk fortified with powders of *Sphaeranthus indicus*, *Ferula foetida* and *Plumbago zeylanica* cures hemorrhoids (*Ci.* 8: 36).
5. Buttermilk with low acidity is mixed with powders of *Coriandrum sativum*, *Elettaria cardamomum*, *Sphaeranthus indicus*, *Piper longum*, *Piper chaba*, *Foeniculum vulgare*, *Piper brachystachyum*, *Kaempferia galanga*, *Trachyspermum ammi*, *Plumbago zeylanica* and *Trachyspermum roxburghianum*. The fortified buttermilk is then poured into an earthen pot, the inside of which is smeared with clarified butter. The fermented buttermilk can be consumed when acidic and pungent tastes become prominent. This beverage improves digestion and cures hemorrhoids with edema, itching and pain in anus (*Ci.* 8: 45–48).
6. Consumption of buttermilk mixed with powders of rock salt, *Plumbago zeylanica* root, *Holarrhena antidysenterica* seeds, *Holoptelia integrifolia* bark and *Azadirachta indica* bark cures hemorrhoids in seven days (*Ci.* 8: 161).
7. Powders of *Cyclea peltata*, *Salmalia malabarica*, *Cyperus rotundus*, *Woodfordia fruticosa*, *Aegle marmelos* and *Zingiber officinale* are to be suspended in buttermilk sweetened with jaggery. Consumption of this medicated buttermilk cures even very obstinate dysentery with predominance of *kapha* (*Ci.* 9: 109–110).
8. Consumption of sweet-tasting, low fat buttermilk is recommended in all types of enlargement of abdomen. In *vāta*-dominant condition, it is advised to be consumed with *Piper longum* and rock salt (*Ci.* 15: 126–127).
9. Buttermilk containing powder of *Piper nigrum* and sugar is recommended in enlargement of abdomen with *pitta* dominance (*Ci.* 15: 127).
10. Fortification of buttermilk with *Trachyspermum ammi*, rock salt, *Cuminum cyminum*, *Trikaṭu* and honey is recommended in *kapha*-dominant enlargement of abdomen (*Ci.* 15: 127).
11. Enlargement of abdomen with predominance of *tridōṣa* can be pacified by buttermilk mixed with *Trikaṭu*, *yavakṣāra* and rock-salt (*Ci.* 15: 128).
12. Patient suffering from splenomegaly should consume buttermilk mixed with honey, sesame oil, *Acorus calamus*, *Zingiber officinale*, *Anethum graveolens*, *Saussurea lappa* and rock salt (*Ci.* 15: 128).

13. Buttermilk fortified with *Sphaeranthus indicus*, *Trachyspermum ammi*, rock salt and Cuminum *cyminum* is to be consumed by patients suffering from enlargement of abdomen caused by intestinal obstruction (*Ci.* 15: 129).
14. Buttermilk containing *Piper longum* and honey is ideal for consumption by patients suffering from enlargement of abdomen caused by perforation of intestine (*Ci.* 15: 129).
15. Ascites can be pacified by buttermilk containing *Trikaṭu* powder (*Ci.* 15: 129).
16. Patient suffering from edema, experiencing dull appetite and voiding of *āma*-containing stool with obstruction should consume regularly buttermilk mixed with sonchal salt, *Trikaṭu* powder and honey (*Ci.* 17: 4–5).
17. One part of *Holarrhena antidysenterica* seeds, two parts of *Piper longum* fruits, three parts of *Plumbago zeylanica* roots and four parts of *Amorphophallus commutatus* tuber are to be powdered and suspended in buttermilk. Consumption of this medicated buttermilk pacifies hemorrhoids with edema and pain (*Ci.* 8: 34).

7.3.5 Fortified Butter

1. AH describes an imaginative method to fortify butter with herbal bioactives. For example, milk boiled with bark powders of *Ficus benghalensis*, *Ficus religiosa*, *Ficus glomerata*, *Ficus retusa*, *Shorea robusta*, *Callicarpa macrophylla*, apical leaves of *Borassus flabellifer*, *Ziziphus mauritiana*, *Buchanania lanzan*, *Prunus cerasoides* and *Terminalia paniculata* is curdled and churned. The butter isolated is to be administered to patients suffering from injury to thorax and sexual debility (*Ci.* 3: 89–91).
2. Milk boiled with *Daśamūla* herbs is to be curdled and churned. The butter isolated is consumed with powder of *Piper longum* mixed with honey. It is good for curing headache, pain in flanks, pain in arm, cough, respiratory distress and fever and for improving voice quality in patients of kingly consumption (*Ci.* 5: 19–20).
3. Milk boiled with *Mahat pañcamūla*, *Hrasva pañcamūla*, *Madhyama pañcamūla*, *Jīvana pañcamūla* and *Tṛṇapañcamūla* is to be curdled and churned. The butter obtained is good for improving voice quality of patients suffering from kingly consumption (*Ci.* 5: 19–20).
4. Latex of *Euphorbia neriifolia* is added to milk and boiled. On cooling, the milk is churned and the cream isolated. Consumption of this cream with the latex cures enlargement of abdomen (*Ci.* 15: 31–32).
5. Butter is to be isolated form milk boiled with *Randia dumetorum* and curdled. Consumption of this butter is useful in inducing emesis in emaciated persons and those with dull appetite due to enragement of *kapha* (*Ka.* 1: 14).
6. Milk boiled with roots of *Withania somnifera* is curdled and the butter separated by churning. Consumption of this butter with sugar cures emaciation (*Ci.* 5: 25).
7. Dehulled seeds of *Psoralea corylifolia* are powdered and boiled with milk, which is thereafter curdled and the butter isolated. Consumption of this butter mixed with unequal quantity of honey cures leprosy and other skin diseases. Disfigured fingers and nose regenerate, just as tender leaves appear on a tree (*Ut.* 39: 109–110).

7.3.6 Fortified Sweets

Sweets form an integral part of traditional Indian food, *Pāyasam* (sweet pudding) and *Mōdaka* being representative examples. *Pāyasam* is prepared by cooking grains or legumes in milk. Sugar, jaggery or honey is added to enhance its acceptability. *Mōdaka* is cooked grains mixed with jaggery, grated coconut and rolled into balls. Such sweet preparations also form food matrices for delivering herbal constituents.

1. Consumption of rice cooked with milk, clarified butter and powder of *Glycyrrhiza glabra* pacifies lassitude of voice caused by aggravation of *pitta* (*Ci.* 5: 40).
2. *Mōdaka* prepared with *Operculina turpethum*, *Emblica officinalis*, *Terminalia belerica*, *Terminalia chebula*, *Piper longum*, sugar and honey cures hemorrhages of obscure origin caused by aggravation of *tridōṣa*, edema and fever (*Ci.* 2: 10–11). *Mōdaka* prepared with sugar, *Piper longum* and *Operculina turpethum* may also be consumed for pacifying these conditions (*Ci.* 2: 11).
3. Weaned babies are fed with *Mōdaka* made up of kernels of *Buchanania lanzan*, *Glycyrrhiza glabra*, honey, parched rice and sugar (*Ut.* 1: 38).
4. Similar *Mōdaka* is prepared with unripe *Aegle marmelos* fruit, *Elettaria cardamomum*, sugar and parched rice. It improves digestion (*Ut.* 1: 39).
5. *Mōdaka* incorporating *Woodfordia fruticosa*, sugar and parched rice is an absorbent (*samgrāhi*) that hardens stool of weaned babies (*Ut.* 1: 39).
6. *Sarpirguḍam* is a sweet ball made use of in the treatment of cough. It is good for lean and emaciated patients. Starch of *Maranta arundinacea*, powders of *Piper longum* fruits and parched rice are turned into a thick dough by adding sufficient quantity of honey. The dough is rolled into balls. Consumption of this *Sarpirguḍam* with milk as post-prandial drink improves stamina and sexual vigor (*Ci.* 3: 112–113).
7. Pudding prepared with rice, decoction of *Triphala* and milk is allowed to cool. Thereafter, it is to be consumed daily mixed with honey and sugar. This treatment cures cataract (*Ut.* 13: 18).

7.3.7 Fortified Cakes

Cakes are invariably prepared with grains. Soaked rice is ground to thick paste, rolled into lemon-shaped balls, filled with or without coconut-jaggery mixture and dropped into boiling water. In a short time, the ball gets cooked and is ready to be served. Such a cake is called *Utkārika*. Cakes like *Apūpa* are prepared by placing a ball of rice paste on a hotplate, smeared with sesame oil, and gently pressing it with fingers to assume a thick and flat shape. It is served with clarified butter poured over it. Such cakes are also made use of in AH to serve as fortified food.

1. Consumption of *Utkārika* incorporating starch of *Maranta arundinacea*, *Piper longum*, *Chonemorpha fragrans*, *Zingiber officinale* and clarified butter cures respiratory distress and hiccup having association with *pitta* (*Ci.* 4: 34).
2. In respiratory distress and hiccup associated with aggravation of *vāta*, the *Utkārika* should be prepared with flesh and blood of porcupine (*Hystrix indica*), meat of rabbit, *Piper longum* and clarified butter (*Ci.* 4: 35).
3. Pericarp of *Terminalia chebula* and *Piper longum* fruits should be soaked and dried several times in latex of *Euphorbia neriifolia*. Consumption of *Utkārika* incorporating powder of the treated herbs cures enlargement of abdomen (*Ci.* 15: 44).
4. Consumption of rice cakes prepared with paste of tender leaves of *Merremia emarginata* cures worm infestation (*Ci.* 20: 29).
5. Similar cakes prepared with tender leaves of *Neolamarckia cadamba*, *Eclipta alba*, *Vitex negundo* and powder of *Embelia ribes* seeds may be consumed for curing worm infestation (*Ci.* 20: 30–31).
6. *Utkārika* prepared with leaf paste of *Symplocos spicata* sautéed in clarified butter and mixed with sugar, cures vomiting, excessive thirst, cough and *āmātisāra* (*Ci.* 3: 175).
7. Cakes prepared with seeds of *Bambusa bambos* can be consumed by patients of *pramēha* (Ci. 12: 10–11).
8. Consumption of cakes incorporating *Triphala* powder cures cataract (*Ut.* 13: 18).

7.3.8 Functional Chutneys

1. Chutney is a spicy preparation that forms an accompaniment to a main dish. AH recommends a few medicinal chutneys that can be included in the menu of patients. Specific quantities of rock salt, sonchal salt, *Zingiber officinale*, fruits of *Garcinia gummi-gutta*, *Punica granatum*; *Ocimum kilimandsharicum*, *Piper nigrum*, *Cuminum cyminum*, *Coriandrum sativum* and jaggery are powdered separately and then blended. This medicated chutney can be consumed along with cooked rice, pancakes and gruels. It is tasty, carminative and improves stamina. It also cures cough, respiratory distress and pain in flanks (*Ci*. 3: 141–144).
2. Specific quantities of *Coriandrum sativum*, *Cuminum cyminum*, *Trachyspermum ammi*, *Punica granatum*, *Garcinia gummi-gutta*, *Zingiber officinale*, unripe fruits of *Feronia elephantum* and rock salt are powdered separately. The powders are blended with sugar. This chutney can be consumed along with various food (*Ci*. 3: 144–146).
3. Specific quantities of sugar, sonchal salt, *Cuminum cyminum*, *Tamarindus indica*, *Solena amplexicaulis*, *Cinnamomum zeylanicum*, *Elettaria cardamomum* and *Piper nigrum* are powdered separately and blended. Known as *Aṣṭāṅgalavaṇaṃ*, this powder improves digestive efficiency by clearing obstructions in body channels and is useful in pacifying alcoholism aggravated by *kapha*. It also serves as a chutney (*Ci*. 7: 40–41).
4. Bark of *Holoptelia integrifolia*, roots of *Plumbago zeylanica* and *Solanum xanthocarpum* are boiled in water to yield a decoction. Jaggery and honey are dissolved in it. This is followed by addition of powders of *Elettaria cardamomum*, *Cinnamomum zeylanicum*, *Cinnamomum tamala*, *Trikaṭu*, root of *Piper chaba*, rind of *Punica granatum*, *Rotula aquatica* root, *Commiphora mukul* gum resin, *Inula racemosa*, *Coriandrum sativum*, *Piper longum* root, *Sphaeranthus indicus* and *Solena amplexicaulis* to the mixture. Thereafter, fresh fruits of *Vitis vinifera*, *Citrus medica*, pieces of fresh *Zingiber officinale* rhizomes and *Saccharum officinarum* stems are added to the mixture, stirred well and stored in a pot for a month. Consumption of this fermented preparation improves digestion and cures hemorrhoids, morbid pallor, enlargement of abdomen, *gulma*, splenomegaly, bloating of abdomen, urinary stone and dysuria (*Ci*. 8: 145–148).
5. *Tamarindus indica*, *Solena amplexicaulis*, *Ziziphus mauritiana*, *Punica granatum* and *Trikaṭu* are powdered finely and blended with sugar. This dry chutney can be consumed along with cooked rice, lentil soups and dry curries. This formulation cures *grahaṇi*, cough, indigestion, respiratory distress, heart disease and pain in flanks (*Ci*. 10: 6–7).
6. Specific quantities of *Trachyspermum ammi*, *Tamarindus indica*, *Solena amplexicaulis*, *Zingiber officinale*, *Punica granatum*, *Ziziphus mauritiana*, *Coriandrum sativum*, sonchal salt, *Cuminum cyminum*, *Cinnamomum zeylanicum*, *Piper longum* and *Piper nigrum* are powdered. The powder is blended with sugar and stored. It is pleasant in taste and constipating. It can be added to the food of patients suffering from kingly consumption. Consumption of this dry chutney also cures cough, heart disease, pain in flanks and hemorrhoids (*Ci*. 5: 55–58).

7.3.9 Medicinal Curries

1. Curries are dishes prepared with vegetables, sauces and seasoned with spices. A recipe of an interesting curry is described in the chapter on treatment of fever. *Momordica charantia* fruit, *Momordica dioica* fruit, *Raphanus sativus*, *Oldenlandia corymbosa*, *Solanum indicum* fruit, *Azadirachta indica* flowers, *Trichosanthes cucumerina* (fruit, tender leaf) and meat of rabbit are cooked in water and the soup saved. Fruits of *Solanum xanthocarpum*, *Phoenix pusilla*, *Premna serratifolia*, *Vitis vinifera*, *Emblica officinalis* and *Punica granatum* are added to the soup. Finally, powders of *Zingiber officinale*, *Piper longum*,

Coriandrum sativum, *Cuminum cyminum* and rock salt are added to the curry and stirred well. Sugar and honey may be added to improve taste. This curry is recommended to be included in the food of patients suffering from fever (*Ci.* 1: 75–77).

2. Ingredients of this medicinal curry are tender leaves of *Chenopodium murale*, *Plumbago zeylanica*, *Operculina turpethum*, *Baliospermum montanum*, *Cyclea peltata*, *Oxalis corniculata*, *Ferula foetida*, *yamaka snēha* (combination of any two lipids such as sesame oil and clarified butter or coconut oil and mustard oil), green *Coriandrum sativum*, *Pañcakōla* herbs, whey, juice of *Punica granatum*, pieces of fresh *Zingiber officinale*, powders of *Cuminum cyminum*, *Piper nigrum*, *vida* salt and sonchal salt. Patients of hemorrhoids with constipation and poor digestion can have food along with thus curry (*Ci.* 8: 80–85).
3. Equal quantities of pastes of unripe *Aegle marmelos* fruit and *Sesamum indicum* seeds are mixed with sour whey and clarified butter. The mixture is brought to boil and then used. Consumption of this curry cures *pravāhika* (*Ci.* 9: 25).
4. *Piper nigrum*, *Coriandrum sativum*, *Cuminum cyminum*, *Tamarindus indica* (fruit pulp), *Kaempferia galanga*, vida salt, *yavakṣāra*, *Punica granatum*, *Woodfordia fruticosa*, *Cyclea peltata*, *Terminalia belerica*, *Terminalia chebula*, *Emblica officinalis*, *Pañcakōla* herbs, *Trachyspermum ammi* and unripe fruits of *Feronia elephantum*, *Mangifera indica* and *Syzygium cumini* are ground to fine paste. One part of this paste is mixed with six parts of paste of unripe *Aegle marmelos* fruits. Curd, green gram soup, jaggery, sesame oil, clarified butter, coconut oil and mustard oil are added and mixed well. Known as *Aparājita* (invincible), this curry improves digestive efficiency and cures *pravāhika* (*Ci.* 9: 26–28).
5. Pastes of *Ziziphus mauritiana*, unripe fruits of *Aegle marmelos*, red rice, barley, green gram and sesame seeds are cooked in water to form a soup. Curd, pomegranate juice, sesame oil, coconut oil, clarified butter and mustard oil are added to it and stirred well. Cooked rice and this curry are to be served to patient suffering from dysentery and dryness of throat due to excessive voiding of bowels (*Ci.* 9: 29–30).
6. Soup of goat meat is mixed with pomegranate juice, *Coriandrum sativum*, *Zingiber officinale* and clarified butter. This mixture is cooked to form a curry. Consumption of cooked red rice with this curry pacifies all discomforts arising from excessive voiding of stool (*Ci.* 9: 33–35).
7. Curry prepared with curd, pomegranate juice, bark of *Oroxylum indicum*, *Glycyrrhiza glabra*, *Callicarpa macrophylla* flowers and tender leaves of *Punica granatum* cures dysentery (*Ci.* 9: 65–66).
8. Paste of *Embelia* ribes, *Piper nigrum*, *Feronia elephantum* and *Zingiber officinale* is mixed with buttermilk. Juices of *Oxalis corniculata* and *Ziziphus mauritiana* are added to enhance the acidic taste. Consumption of this curry along with food cures dysentery with predominance of *kapha* (*Ci.* 9: 116).
9. Meat is mixed with copious quantities of dry ginger, powder of black pepper and green ginger. It is then cooked in juice of *Citrus medica* till most of the water evaporates. Bamboo shoots and *Carissa carandas* fruits are added along with some vegetables. Finally, it is blended with *Aṣṭāṅgalavaṇaṃ* and served with *Mādhava* wine (prepared from honey) and served to patients suffering from *kapha*-dominant alcoholism, whose appetite is improved (*Ci.* 7: 38–41).
10. Meat soups cooked with leafy vegetables are recommended to be included in the food of patients suffering from hemorrhoids. Meat of cow, monitor lizard, goat, camel and carnivorous animals are to be the ingredients of such curries (*Ci.* 8: 85).
11. Food of patients suffering from dysentery is to be fortified in several ways as described in the chapter on treatment of dysentery. Many combinations of herbs are suggested:

    i. *Aegle marmelos* unripe fruit, *Kaempferia galanga*, *Coriandrum sativum*, *Ferula foetida*, *Garcinia gummi-gutta* and *Punica granatum*
    ii. *Butea monosperma*, *Sphaeranthus indicus*, *Cuminum cyminum*, *Trachyspermum ammi*, vida salt and rock salt

iii. Roots of *Hrasva pancamula*
iv. *Pañcakōla* herbs
v. *Cyclea peltata* (root) (*Ci.* 9: 11–13)
vi. *Kaempferia galanga* (tender leaves)
vii. *Raphanus sativus* (tender root)
viii. *Cyclea peltata* (tender leaves)
ix. *Ziziphus mauritiana* (tender leaves)
x. *Marsilea quadrifolia* (whole herb)
xi. *Trachyspermum ammi* (tender leaves)
xii. *Benincasa hispida* (tender leaves)
xiii. *Hemidesmus indicus* (tender leaves)
xiv. *Cucumis sativus* (tender leaves)
xv. *Basella alba* (tender leaves)
xvi. *Holostemma annulare* (tender leaves)
xvii. *Psoralea corylifolia* (tender leaves)
xviii. *Chenopodium murale* (tender leaves)
xix. *Linum usitatissimum* (tender leaves)
xx. *Pistia stratiotes* (whole herb)
xxi. *Portulacca oleracea* (whole herb) (*Ci.* 9: 20–22)

7.3.10 Medicated Broth

Medicated broths (*yūṣa*) are prepared by cooking one part of legumes, grains or vegetables in 18 parts of water and filtering the broth. *Yūṣa* of pea (*Pisum sativum*), horse gram (*Dolichos biflorus*), chickpea (*Cicer arietinum*), green gram (*Vigna radiata*) (*Ci.* 1: 71–74), black gram (*Phaseolus mungo*), velvet bean (*Mucuna pruriens*) (*Ci.* 3:19), pastes of barley (*Hordeum vulgare*), *Śāli* (red) rice, *Ziziphus mauritiana* fruit, unripe *Aegle marmelos* fruit, sesame seeds (*Sesamum indicum*) (*Ci.* 9: 19), bitter gourd (*Momordica charantia*) (*Ci.* 1: 96) radish (*Raphanus sativus*), *Momordica dioica* fruit, *Oldenlandia corymbosa* whole herb, *Solanum melongena* fruit, *Azadirachta indica* flowers and *Trichosanthes cucumerina* fruit and tender leaves (*Ci.* 1: 75) are recommended in AH for the preparation of *yūṣa*.

1. *Vigna radiata* seeds are cooked in juice of *Solanum xanthocarpum* fruits. The resultant broth is made slightly sour by adding juice of *Emblica officinalis*. Consumption of this *yūṣa* with clarified butter pacifies all types of cough (*Ci.* 3: 176).
2. *Yūṣa* prepared with legumes and *vāta*-pacifying herbs is to be administered to patients suffering from cough due to emaciation (*kṣayakāsa*) (*Ci.* 3: 177).
3. Cough, respiratory distress and hiccup are cured by consumption of *yūṣa* prepared with *Dolichos biflorus*, *Daśamūla* herbs and rabbit meat (*Ci.* 4: 20).
4. *Yūṣa* prepared with leaves of *Moringa oleifera*, *Solanum indicum*, *Senna occidentalis*, *Adhatoda vasica* and *Raphanus sativus* root can be administered for curing cough, respiratory distress and hiccup (*Ci.* 4: 20).
5. Tender leaves of *Azadirachta indica*, *Trichosanthes cucumerina*, *Solanum indicum* and *Citrus medica* can also be utilized for preparing *yūṣa* that cures cough, respiratory distress and hiccup (*Ci.* 4: 21).
6. *Yūṣa* prepared with leaves of *Solanum xanthocarpum*, *Tragia involucrata*, *Pistacia integerrima*, *Tribulus terrestris* and unripe *Aegle marmelos* fruits cures cough, respiratory distress and hiccup (*Ci.* 4: 21).
7. When a patient of alcoholism with aggravation of *kapha* experiences good appetite, he should consume *yūṣa* of dried *Raphanus sativus* roots, processed with *Solena amplexicaulis*, *Garcinia gummi-gutta*, *Trichosanthes cucumerina*, *Trikaṭu* and *Punica granatum* (*Ci.* 7: 37).

8. *Yūṣa* prepared with fruits of *Gmelina arborea*, mixed with a small quantity of juice of *Citrus medica*, is to be administered with sugar. It cures *raktātisāra* (*Ci.* 9: 85).
9. Patient suffering from traumatic wound should consume food along with fat-free *yūṣa* of *Hordeum vulgare*, *Ziziphus mauritiana* and *Dolichos biflorus* (*Ut.* 26: 39).
10. Consumption of *yūṣa* prepared with sugar, legumes, *Ziziphus mauritiana*, *Vitis vinifera*, *Hemidesmus indicus*, *Cyperus rotundus* and *Santalum album* cures fever associated with excessive thirst and vomiting (*Ci.* 1: 33–34).
11. *Saussurea lappa* and *Abies webbiana* are to be ground and suspended in *yūṣa* of *Dolichos biflorus*. Consumption of this *yūṣa* induces expulsion of placenta following childbirth (*Śā.* 1: 87).
13. *Yūṣa* prepared with green gram and herbs of *Jīvanīya pañcamūla*, mixed with small amounts of lemon juice, rock salt and seasoned with mustard seeds in hot clarified butter is recommended in the treatment of excessive thirst (*Ci.* 6: 64–65).
14. *Yūṣa* prepared with dried radish or seeds of *Dolichos biflorus* is to be served to *kāmala* patients experiencing poor appetite, hiccup, pain in flanks, distaste for food, pain and fever (*Ci.* 16: 49).

7.3.11 Medicated Wine

AH describes the medicinal value of various kinds of wine such as *Sura*, *Vāruṇi*, *Vaibhītaki*, *Ariṣṭa*, *Mārdvīka*, *Khārjūraṃ*, *Śārkara*, *Sīdhu* and *Madhvāsava* (*Sū.* 5: 62–80).

1. Consumption of *Sīdhu* or *Gauḍa* wines mixed with powders of *Plumbago zeylanica* and *Zingiber officinale* cures hemorrhoids (*Ci.* 8: 61).
2. Consumption of *Sura* wine mixed with powders of *Sphaeranthus indicus*, *Cyclea peltata* and sonchal salt cures hemorrhoids (*Ci.* 8: 62).
3. *Madira* wine is to be consumed after mixing it with kernel of *Ziziphus mauritiana* for curing cough with *vāta* predominance (*Ci.* 3: 17).
4. Consumption of *Madira* wine mixed with paste of *Piper longum* and rock salt sautéed in clarified butter cures *vāta*-dominant cough (*Ci.* 3: 17).
5. *Piṣṭa madya* brewed with rice powder, *Citrus medica*, *Garcinia gummi-gutta*, *Ziziphus mauritiana*, *Punica granatum*, *Trachyspermum roxburghianum*, *Trachyspermum ammi*, *Sphaeranthus indicus*, *Cuminum cyminum*, *Trikaṭu*, *Zingiber officinale*, rock-salt, sonchal salt and vida salt pacifies *vāta*-dominant alcoholism (*Ci.* 7: 12–13).
6. *Vāruṇi* wine fortified with *Coriandrum sativum* leaves, green *Zingiber officinale* and vinegar is beneficial in the treatment of *vāta*-dominant alcoholism (*Ci.* 7: 15–16).
7. Patient suffering from *pitta*-dominant alcoholism should be served with *Śārkara* wine containing juices of *Punica granatum*, *Phoenix dactylifera*, *Averrhoa carambola*, *Vitis vinifera* and *Phoenix pusilla* (*Ci.* 7: 19–20).
8. When a patient suffering from *kapha*-dominant alcoholism is freed of indigestion and has good appetite, he should consume *Śārkara* wine, *Ariṣṭa* wine or *Sīdhu* wine mixed with *Trachyspermum ammi* and *Zingiber officinale* (*Ci.* 7: 34–35).
9. Paste of *Vacādi gaṇa* herbs and rock salt are to be mixed in wine and administered for improving digestion in patients suffering from *grahaṇi* (*Ci.* 10: 9).
10. Powders of *Piper longum*, *Zingiber officinale*, *Cyclea peltata*, *Hemidesmus indicus*, *Solanum indicum*, *Solanum xanthocarpum*, *Plumbago zeylanica*, *Holarrhena antidysenterica*, *yavakṣāra*, rock salt, sonchal salt, vida salt, sea salt and salt prepared from vegetable alkali are to be suspended in *Sura* wine and consumed. Digestive efficiency will be improved in *grahaṇi* patients (*Ci.* 10: 12–14).
11. Powders of *Trichosanthes cucumerina*, *Azadirachta indica*, *Bacopa monnieri*, *Andrographis paniculata*, *Oldenlandia corymbosa*, *Holarrhena antidysenterica*, *Chonemorpha fragrans*,

Moringa concanensis, *Acorus calamus*, *Berberis aristata*, *Prunus cerasoides*, *Vetiveria zizanioides*, *Trachyspermum ammi*, *Cyperus rotundus*, *Santalum album*, *Aconitum heterophyllum*, *Trikaṭu*, *Cinnamomum zeylanicum*, *Cinnamomum tamala* and *Cedrus deodara* are to be suspended in wine and consumed. It cures heart disease, morbid pallor, *grahaṇi*, *gulma*, distaste for food and fever (*Ci.* 10: 34–37).

12. *Ferula foetida*, juices of *Citrus medica* and *Punica granatum*, vida salt and rock salt are to be mixed in *Sura* wine. Consumption of this medicated wine cures *vāta-gulma* (*Ci.* 14: 41).
13. Consumption of *Sura* wine mixed with powder of *Piper longum* (fruit, root), *Plumbago zeylanica*, *Cuminum cyminum* and rock salt cures *gulma* very quickly (*Ci.* 14: 113).
14. Paste of *Trachyspermum roxburghianum*, *Plumbago zeylanica*, *Acorus calamus*, *Physalis angulata*, *Lepidium sativum*, *Piper longum*, *Adhatoda vasica*, *yavakṣāra*, sugar and rock salt is to be suspended in *Prasanna* wine, along with a small quantity of clarified butter. Diseases of vagina, pain in the flanks, heart disease, *gulma* and hemorrhoids will be cured by the consumption of this medicated wine (*Ut.* 34: 30–31).
15. *Adhatoda vasica* root, *Citrus medica* root and *Woodfordia fruticosa* flowers are to be ground to paste and mixed with any wine. Consumption of the medicated wine cures pricking pain in vagina (*Ut.* 34: 32).
16. Similarly, wine fortified with *Piper longum* and *Physalis angulata* also cures pricking pain in vagina (*Ut.* 34: 32).

7.3.12 Fortified Beverages

1. AH recommends the consumption of beverages like *Rasāḷa*, *Khāṇḍava* and *Pañcasāram*. *Pañcasāram* is prepared by mixing equal volumes of juices of grapes (*Vitis vinifera*), *Madhūka* flowers (*Madhuca longifolia*), dates (*Phoenix dactylifera*), *Kāśmarya* fruits (*Gmelina arborea*) and *Parūṣaka* fruits (*Grewia asiatica*). This is a beverage recommended to patients suffering from hemorrhages of obscure origin (*Ci.* 2: 15).
2. *Pānakam* is another medicinal beverage. It is prepared with *Cinnamomum zelanicum*, *Mesua ferrea*, *Piper longum*, *Piper nigrum*, *Cuminum cyminum*, *Coriandrum sativum*, fruits of *Phoenix pusilla*, *Madhuca longifolia*; *Elettaria cardamomum*, *Cedrus deodara*, juice of *Feronia elephantum* fruits and sugar. It improves taste, appetite and digestive efficiency. *Pānakam* can be administered to patients of all kinds of alcoholism (*Ci.* 7: 44–46).
3. Consumption of decoction of *Baliospermum montanum* roots mixed with juices of *Gmelina arborea* and *Vitis vinifera* fruits cures morbid pallor (*Ci.* 16: 6).
4. *Rasāḷa* is a beverage that is tasty, diuretic, roborant and virilific. It pacifies tiredness, hunger and thirst (*Sū.* 6: 34). Water is removed from curd by straining through muslin cloth. Clarified butter, honey, sugar, powders of *Piper longum* fruits, dry ginger, *Mesua ferrea*, *Elettaria cardamomum*, *Cinnamomum zeylanicum* and *Cinnamomum tamala* are added and mixed well. The *Rasāḷa* is stored in a pot perfumed with camphor, and the mouth of the pot is closed tightly with cloth. It can be consumed after three hours (Vaidya, 1936). It is recommended as a refreshing drink in summer season (*Sū.* 3: 30). *Rasāḷa* is extolled as an excellent remedy for common cold (*pratiśyāya*) (*Ut.* 40: 51).

7.4 CONCLUSION

AH gives great importance to food and beverages. Moderate amount of compatible food is to be consumed at the appropriate time, by one who wants to lead a life free from diseases. Adjustments also need to be made in dietary patterns during various seasons. Modern research shows that dietary habits influence disease risk. While some foods trigger chronic health conditions, others offer strong protective qualities (Kim et al., 2018; Sanchez et al., 2019). Therefore, it is not surprising that AH makes use of food and beverages as vehicles for transporting medication. Many foods or food ingredients such as gruels, churned gruels, milk, curd, buttermilk, butter, sweets, cakes, chutneys, curries, broth, wine and beverages are considered as matrices for delivering the therapeutic agents. Administering medicinal food is a natural way, as the patient does not feel that he is taking a medicine. Giving greater importance to medicinal foods and beverages in clinical practice will enhance the role of Ayurveda in contemporary healthcare.

REFERENCES

Arvanitoyannis, I.S. and M. van Houwelingen-Koukaliaroglou. 2005. Functional foods—A survey of health claims, pros and cons and current legislation. *Critical Reviews in Food Science and Nutrition* 45:385–404.

Kim, H., L.E. Caulfield and C.M. Rebholz. 2018. Healthy plant-based diets are associated with lower risk of all-cause mortality in US Adults. *The Journal of Nutrition* 148:624–631.

Kumar, D.S. 2016. Food and beverages fortified with phytonutrients. In *Herbal Bioactives and Food Fortification: Extraction and Formulation*, 173–238. Boca Raton: CRC Press.

Madhulika. 2013. *Pāka Darpaṇam*, 1–111. Varanasi: Chowkhamba Orientalia.

Sanchez, A., A. Mejia, J. Sanchez, E. Runte, S. Brown-Fraser and R.L. Bivens. 2019. Diets with customary levels of fat from plant origin may reverse coronary artery disease. *Medical Hypotheses* 122:103–105.

Upadhyaya, Y. 1975. *Pramēha cikitsa* (Treatment of polyuria). In *Aṣṭāṅgahṛdaya*, 372–374. Varanasi: Chowkhamba Sanskrit Sansthan.

Vaidya, K.M. 1936. *Ashtanga Hridaya Kosha with the Hridaya Prakasha (A Critical and Explanatory Commentary)*, 473–474. Trichur: The Mangalodayam Press.

Vaidyan, P.M.G. 2017a. *Annasvarūpa vijñānīyam* (Chapter on food ingredients). In *Aṣṭāṅgahṛdaya, Sūtrasthānam*, 134–192. Kodungallur: Devi Book Stall.

Vaidyan, P.M.G. 2017b. *Ṛtucarya* (Chapter on seasonal regimen). In *Aṣṭāṅgahṛdaya, Sūtrasthānam*, 63–87. Kodungallur: Devi Book Stall.

8 Roadmap for the Future

8.1 INTRODUCTION

Aṣṭāṅgahṛdaya is an authoritative treatise of Ayurveda, continuing to be popular among students, physicians and scholars. It is the first treatise that synthesizes the teachings of Caraka and Suśruta. Beauty of its verses, masterly style of condensation, logical arrangement of subjects, clarity of description and several other merits endear it to votaries of Ayurveda. *Aṣṭāṅgahṛdaya* has been translated into all major Indian languages, Arabic, Persian, Tibetan and German. In 7,462 quatrains spread over 120 chapters distributed in six sections such as *Sūtrasthānam*, *Śārīrasthānam*, *Nidānasthānam*, *Cikitsāsthānam*, *Kalpasthānam* and *Uttarasthānam*, Vāgbhaṭa presents the sacred knowledge to stay away from diseases and to treat the various diseases that appear due to endogenous or exogenous causes. Well-founded on the *tridōṣa* doctrine, *Aṣṭāṅgahṛdaya* provides the key to diagnose and treat effectively, even diseases that may appear in the future (*Sū.* 12: 64).

8.2 HEALTHY FOOD HABITS

Nowadays, the incidence of lifestyle diseases is on the rise. These are non-communicable diseases caused by several factors including unhealthy eating. They include heart disease, stroke, obesity, Alzheimer's disease, arthritis and so on (Mathur and Mascarenhas, 2019). AH provides valuable information on food habits favorable to prevention of these diseases. According to AH, certain food substances are to be used with caution. For example, curd is not to be used every day. It is also to be avoided at night. Curd should be consumed neither in warm condition nor in summer season. It should also not to be used without adding green gram, honey, clarified butter, sugar or gooseberries. Transgression of this instruction causes fever, hemorrhages of obscure origin, cellulitis, skin diseases and morbid pallor (*Sū.* 5: 31–33). Honey turns to be toxic if ingested warmed and blended with ingredients having hot potency or in warm season. It is also forbidden to persons with warm body, as in the case of individuals of *pitta* constitution (*Sū.* 5: 53). Milk is not to be consumed with acidic substances and all acidic fruits (*Sū.* 7: 31). Chicken meat or deer meat and curd should not be consumed together (*Sū.* 7: 33). If honey and clarified butter are to be used, they should be mixed in unequal quantities (*Sū.* 7: 39). Chapter 8 entitled *Mātrāśitīyam* (quantitative dietetics) provides a long list of substances that are not to be consumed regularly. They include dairy products such as *Kīlāṭa*, curd and *Kūcika*, alkalis, vinegar, green radish, meat of small animals, dried meat, meat of pig, goat, buffalo, cow; fish, black gram, *Dolichos lablab* (*Niṣpāva*), *Alocasia macrorhizos* (*Śālūka*), leaf stalk of lotus, delicacies prepared with rice powder, buds of herbs, dried pot-herbs, the rice variety *Yavaka* and thickened or congealed sugarcane juice (*Sū.* 8: 40–41). Nevertheless, AH advises that certain food substances can be consumed every day. They include *Śāli* rice, *Ṣaṣṭika* rice, wheat, barley, meat of *jāṅgala* animals, tender radish, *Vāstukam* (*Chenopodium murale*), *Suniṣaṇṇakam* (*Marsilea quadrifolia*), grapes, green gram, jaggery and pomegranate fruits (*Sū.* 8: 42–43).

It is commonly believed that ayurvedic herbs and medicines can be used every day by healthy individuals, as they are of natural origin. However, this is a false belief. They are intended only for treatment of diseases. Nevertheless, herbs such as *Terminalia chebula*, *Holostemma annulare* and *Emblica officinalis* can be consumed every day to conserve health. Similarly, daily consumption of milk, clarified butter, honey and rock salt is also advised (*Sū.* 8: 42–43). *Harītaki* (*Terminalia*

DOI: 10.1201/9781003148296-8

chebula) is considered to be an exceptionally valuable herb. A quatrain composed by an anonymous physician extolls its virtues:

> Daśa vaidya samā patni
> Daśa patni samō ravi
> Daśa sūrya samā mātā
> Daśa mātṛ harītaki

(Ten physicians are equal to a wife. Ten wives are equal to sun. Ten suns are equal to a mother and ten mothers are equal to *Harītaki*!) (Menon, 1931).

As far as medicinal formulations are concerned, *Triphala* powder mixed with honey and clarified butter is to be consumed every night for conserving eye health (*Sū.* 8: 44). Similarly, *Triphala* powder subjected to *bhāvana* with decoctions of *Acacia catechu* and *Pterocarpus marsupium* is to be consumed daily, mixed with honey and clarified butter. He who does so enjoys long life devoid of diseases (*Ut.* 39: 152). *Vasiṣṭha rasāyana* is a reputed medicine recommended in the treatment of cough. It can be used without any dietary or behavioral restrictions. Healthy persons can consume it in all seasons (*Ci.* 3: 133–141). Same is the case with *Sukumāra ghṛta* that cures around 13 diseases. It can be consumed without any behavioral restrictions regarding exposure to wind and hot sum, travelling by foot or in carriages. Daily consumption of this *rasāyana* improves complexion and stamina (*Ci.* 3: 41–47). One who performs *nasya* every day is said to develop well-formed high neck, shoulders and chest. He will be free from greying and will have handsome face and complexion. All sense organs will stay healthy. *Aṇu taila* is a medicated oil that can be instilled into nasal passages every day (*Sū.* 20: 37–39).

8.3 NOVEL CONCEPTS IN AH

8.3.1 Mind and Causation of Diseases

AH teaches that mind is the seat of all diseases. Lust (*kāma*), anger (*krōdha*), greed for wealth (*lōbha*), delusion (*mōha*), arrogance (*mada*) and envy (*mātsarya*) are the six pollutants of mind. Residing in the mind and all parts of body, they generate anxiety (*autsukya*), mental confusion (*mōha*) and disinclination (*arati*), which are the causes of all diseases. Anxiety, mental confusion and disinclination cause behavioral changes that finally induce the appearance of diseases. To drive home the importance of mind in health and disease, Vāgbhaṭa begins AH after paying homage to the great physician, the Almighty, who has conquered the six pollutants of mind or *rāgādidōṣān* (*Sū.* 1: 2). Dysentery, insanity, epilepsy and headache also stem from fear, grief, worry and anxiety (*Ni.* 8: 1, *Ut.* 6: 1, 15–16, *Ut.* 7: 1, *Ut.* 23: 1).

The influence of mind over life processes is lately being recognized in modern research. This has given rise to a new discipline called psychoneuroimmunology, which studies the interaction between psychological, nervous and immune processes of the human body.

Research has shown that a variety of transient and chronic stress-inducing situations are associated with increased release of adrenocorticotropic hormone along with β-endorphin and with elevated cortisol levels in the circulation. These increases may amplify the effect of the pituitary–adrenal axis on other tissues. Stress activates the sympathetic nervous system and promotes the release of norepinephrine and epinephrine from nerve terminals and the adrenal medulla, respectively. Acetyl choline release from the parasympathetic nervous system is decreased, because parasympathetic activity remains suppressed for the most part by the stress. As classical neurotransmitters such as norepinephrine and epinephrine are co-localized with various peptides such as the enkephalins, there is a possibility that such peptides modulate tissue responses to stress as well. Studies have documented that stress-inducing conditions cause widespread activation of structures within the central nervous system that play a role in emotion, responses of autonomic nervous system and endocrine regulation. Stress also produces distinct patterns of immune responsiveness such as enhancement of innate defenses, general activation of both innate and adaptive responsiveness, suppression of adaptive cell-mediated responses and a general suppression of both innate and adaptive immunity (Daruna, 2012).

8.3.2 Chronobiology of *Tridōṣa*

AH advances several concepts which are way ahead of its time. For example, AH states that *vāta*, *pitta* and *kapha* follow diurnal and circannual rhythms. If life span, day, night and the period between consumption of food and its digestion are divided into three equal parts, the first part will be dominated by *kapha*, the middle by *pitta* and the end by *vāta* (*Sū.* 1: 8). In addition to diurnal rhythms, the *tridōṣa* exhibit circannual rhythms as well. *Vāta* increases in its own site in summer, spreads in monsoon season and decreases in autumn. *Pitta* increases in monsoon, spreads to other parts of body in autumn and returns to basal level in pre-winter season. *Kapha* increases in winter, spreads in spring and returns to normal level in summer (*Sū.* 3: 1–58, *Sū.* 12: 24–25). The cyclical pattern of *tridōṣa* described in AH is nothing but chronobiology, a modern discipline in biology.

Chronobiology is a field that studies cyclical phenomena in living organisms, such as their adaptation to solar- and lunar-related rhythms known as biological rhythms (Webb, 1994). The French scientist Jean-Jacques d'Ortous de Mairan observed for first time a circadian cycle in the movement of leaves (Johansson and Köster, 2019). Swedish botanist and pioneer of binomial nomenclature, Carl von Linné (1707–1778) observed over several years that the flowers of many plants opened and closed periodically and that these times were species-specific. For example, the hawk's beard plant opened its flowers at 6.30 A.M. and the hawkbit did not open its flowers until 7 A.M. Arranged in sequence of flowering over the day, they constituted a kind of flower clock or *Horologium Florae* as Linnaeus called it in his *Philosophia Botanica* (Gardiner, 1987). The symposium on biological clocks held from the 5th through the 14th of June, 1960, at Cold Spring Harbor Laboratory, New York, laid the foundation of the field of chronobiology (Bünning, 1960). Franz Halberg of the University of Minnesota coined the word *circadian* and is generally considered the "father of chronobiology" (Halberg et al., 2003).

The circadian and circannual rhythms of *tridōṣa* are considered in diagnosis and therapeutics also, just as chronobiology guides western medicine. *Kapha* increased in winter season is to be eliminated from body by emesis and nasal instillation using piercing medicines; light, oil-free food; exercise; and powder massage. Similarly, *vāta*, *pitta* and *kapha* vitiated in monsoon season need to be eliminated using medicinal enema with herbal decoctions, emesis and purgation. Purgation and bloodletting are the measures to be carried out in autumn season to eliminate the *pitta* increased in body. Consumption of food and beverages having specific tastes is also recommended to keep the *tridōṣa* in steady states during the various seasons (*Sū.* 3: 1–58).

Chronobiological aspects of *tridōṣa* are also taken into consideration while administering medicines. Curative medicines are administered in ten different modes such as without food, in the beginning of eating, in the middle of eating, soon after completion of eating, in between each morsel of food, concealed in each morsel of food, mixed frequently with food, before and after food and at night. Administering medicines before retiring to bed is recommended in the treatment of diseases above the collar bone (*Sū.* 13: 37–41).

8.3.3 Diseases Caused by Subtle Beings

AH devotes one chapter to the treatment of infants afflicted with spiritual possession. Twelve *graha*, spirits or subtle beings are the causative agents. Out of them, five *graha* have masculine appearance and are called *Skanda*, *Viśākha*, *Mēṣāsya*, *Śvagraha* and *Pitṛgraha*. Another group consists of seven *graha* such as *Śakuni*, *Pūtana*, *Śītapūtana*, *Dṛṣṭipūtana*, *Mukhamaṇḍatika*, *Rēvati* and *Śuṣkarēvati*, which have feminine appearance. Possessions by them can be diagnosed on the basis of distinct symptoms. Treatment consists of fumigation of the afflicted child's body with various herbs and hides of animals, administration of medicated clarified butters and bathing with decoctions of herbs (*Ut.* 3: 1–60).

According to AH, insanity is caused by disturbance of *tridōṣa* and exogenous agents such as subtle beings or *graha*. Eighteen *graha* such as *Dēvagraha*, *Asuragraha*, *Gandharvagraha*, *Sarpagraha*, *Yakṣagraha*, *Brahmarākṣasagraha*, *Rākṣasagraha*, *Piśācagraha*, *Prētagraha*, *Kūśmāṇḍakagraha*, *Niṣādagraha*, *Aukiraṇagraha*, *Vētāḷagraha*, *Pitṛgraha*, *Gurugraha*, *Vṛddhagraha*, *Ṛṣigraha* and *Siddhagraha* cause characteristic behavioral disturbances. Afflictions caused by these *graha* are cured by fumigation of body using herbs, hides, gall bladder, teeth and claw of animals, nasal instillation of medicines, application of collyria in eyes, application of paste over body, pouring of water boiled with herbs, administration of *ghṛta* and propitiatory rites (*Ut.* 4: 1–44, *Ut.* 5: 1–53).

Afflictions caused by *graha* and the unusual treatments suggested are considered by some as figments of imagination of the ancient author. However, results of recent research offer a rational explanation. *Toxoplasma gondii* is an obligate parasitic protozoan that infects virtually all warm-blooded animals and causes toxoplasmosis. Domestic cats are the only known definitive hosts in which the parasite undergoes sexual reproduction (Di Genova et al., 2019). *T. gondii* alters the behavior of infected rodents in ways that increase the rodents' chances of being preyed upon by cats (Berdoy et al., 2000; Webster, 2007; Webster et al., 2013). Support for this "manipulation hypothesis" comes from studies indicating that *T. gondii*-infected rats have a decreased aversion to cat urine (Berdoy et al., 2000). Rats that frequent cats' habitations will more likely become their prey. *Toxoplasma gondii* infection in mice lowers general anxiety, increases explorative behavior and increases loss of aversion to predators. A positive correlation between the severity of the behavioral alterations and the cyst load is observed, reflecting indirectly the level of inflammation during brain colonization. These results indicate unspecific and immune-related changes in the infected mice brains associated with altered behavior (Boillat et al., 2020).

There is some evidence that *T. gondii*, while infecting humans, alters the behavior of the host in ways similar to what happens in rodents, linking toxoplasmosis to schizophrenia. Toxoplasmosis causes specific differences in the personality of infected subjects when compared to non-infected subjects. Lindová et al. (2012) studied the personality of infected subjects mostly using the Cattell's questionnaire. They searched for the association between toxoplasmosis and the personality by screening a population of students with the NEO–PI–R questionnaire. It was found that *Toxoplasma*-infected male and female students had significantly higher extraversion and lower conscientiousness. The conscientiousness negatively correlated with the length of infection in men, suggesting that the toxoplasmosis-associated differences were more probably the result of slow cumulative changes induced by latent toxoplasmosis. This correlation also supports the hypothesis that *Toxoplasma gondii* influences the personality of the infected human subjects.

A fascinating example of parasite-induced host manipulation is that of the hairworm *Paragordius tricuspidatus*. It induces a spectacular "suicide" water-seeking behavior in its terrestrial insect host *Nemobius sylvestris*. Crickets infected by horsehair worms exhibit light-seeking behavior and increased walking speed, leading them to open spaces and ponds. The crickets will eventually find and enter a water body, where the worm will wriggle out of the cricket's abdomen and swim away. Ponton et al. (2011) examined the effect of hairworm infection on different behavioral responses of the host when stimulated by light to record responses from uninfected, infected and ex-infected crickets. They observed that hairworm infection fundamentally modifies cricket behavior by inducing directed responses to light, a condition from which they mostly recover once the parasite is released.

Grahabādha-like conditions have been documented in psychology research. Wikström (1982) described the experiences of a group of four individuals who, sincerely convinced that they were possessed by demons or evil spirits, sought psychiatric or clinical psychological aid. All of them were abnormally active in religious programs and were diagnosed to suffer from manic-depressive psychosis, schizophrenia, hysteria and extreme anxiety. This type of personality transformation is defined in modern psychiatric terminology as cacodemonomania (Schendel and Kourany, 1980).

In recent years, mental disorders that involve experiencing a disconnection or lack of continuity between thoughts, memory, environment, actions and identity are termed "dissociative disorders". Spirit possession is considered to be a dissociative phenomenon. This phenomenon is reported in Oman, a country situated on the south-eastern coast of the Arabian Peninsula, spanning the mouth of the Persian Gulf. According to Omani belief, there are two kinds of spirit possession: intermittent dissociative phenomenon and transitory dissociative phenomenon. The possessed individual appears to have been taken over by a benevolent or malevolent spirit entity or other supernatural forces continuously in the intermittent dissociative phenomenon. In the transitory dissociative phenomenon, the individual is possessed for a short duration, often associated with stressful events. Al-Adawi et al. (2019) remark that there is a neuropsychological basis for these two types of spirit possession.

In the light of this knowledge, a newer interpretation can be given to the concepts of *grahabādha* and *unmāda* detailed in AH. *Grahabādha* in infants may be infections by microbes. *Unmāda* is insanity caused by endogenous factors (*tridōṣa*), whereas *grahabādha* in adults can be considered to be insanity brought about by exogenous causative agents such as protozoans, viruses, bacteria or hitherto unidentified "subtle beings". Viruses have already been implicated in the causation of schizophrenia (Kneeland and Fatemi, 2013). That *grahabādha* in adults may be schizophrenia or other psychic disorders caused by microbes is strengthened by the observation that fumigation of body, topical application of herbal pastes and pouring of medicated water are the major measures to cure insanity caused by *graha*. Therefore, the Ayurveda world should initiate studies to unravel the link between microbes and insanity. Ayurveda experts and psychiatry researchers should join hands to investigate *grahabādha* vis-à-vis dissociative disorders. In the era of artificial intelligence, transdisciplinary studies incorporating inputs from medicine and information technology can unravel Vāgbhaṭa's thoughts in the context of *grahabādha*.

8.3.4 Benefits of Fasting

Fasting (*laṅghana*) is a major therapeutic measure advocated in AH for the treatment of many diseases such as dysentery, vomiting, *gulma*, fever, alcoholism, hemorrhages of obscure origin, enlargement of abdomen and so on. Scientific studies reveal the health benefits of fasting. The first scientific study on diet restriction and its ability to extend life-span was published by McCay et al. (1935). They showed that feeding rats with a diet containing indigestible cellulose dramatically extended both mean and maximum life-span of the animal (McCay et al., 1935). This result was confirmed and extended to other species of animals (Martin et al., 2006). More recently, intermittent fasting or every other day feeding has also been shown to extend life span, with beneficial health effects (Sohal and Weindruch, 1996; Goodrick et al., 1982; Ingram and Reynolds, 1987). Both caloric restriction and intermittent fasting diminish the severity of risk factors for diseases such as diabetes and cardiovascular disease in rodents (Anson et al., 2003; Wan et al., 2003). It is observed that maintenance of rats on alternate day by caloric restriction feeding for two to four months results in resistance of hippocampal neurons to chemically induced degeneration. This reduced damage to hippocampus is also correlated with a striking preservation of learning and memory in water maze spatial learning task (Bruce-Keller et al., 1999). Therefore, it is possible that these dietary regimes may have a significant benefit for debilitating neurodegenerative disorders such as Alzheimer's disease, Huntington's disease and Parkinson's disease (Martin et al., 2006).

Fasting or caloric restriction initiates in body a molecular process called autophagy, which is the natural removal of unnecessary, surplus or dysfunctional cellular components through a lysosome-dependent regulated mechanism. It degrades and recycles cellular components in an orderly way (Mizushima and Komatsu, 2011; Kobayashi, 2015). Autophagy functions as cell's defense against intracellular pathogens. It is involved in recognition of a pathogen, its destruction and the development of a specific adaptive immune response to it. Additionally, autophagy controls cell homeostasis and modulates the activation of many immune cells, including macrophages, dendritic cells and lymphocytes. It performs specific functions such as pathogen killing or antigen processing and presentation (Valdor and Macian, 2012). Autophagy plays an antiviral role against viruses such as mammalian vesicular stomatitis virus (Shelly et al., 2009). Many types of viruses including herpes simplex virus 1 and Sindbis virus have been observed inside autophagic compartments for degradation (Orvedahl et al., 2007). These beneficial effects of autophagy justify the pivotal position that fasting occupies in Ayurveda therapy.

8.3.5 Colon-Targeted Drug Delivery

Targeted delivery of drugs into colon is now being adopted for local treatment of many bowel diseases such as Crohn's disease, ulcerative colitis and colonic cancer (Philip et al., 2009; Oluwatoyin and John, 2005). Colon-specific drug delivery system protects the drug on its way to the colon. Drug release and absorption do not occur in the stomach as well as the small intestine. The bioactive molecule does not get degraded in either of the dissolution sites. It is released and absorbed only when the drug system reaches the colon (Akala et al., 2003). The colon is a suitable absorption site for peptide and protein drugs because of less diversity and intensity of digestive enzymes, and comparatively lower proteolytic activity of colon mucosa than that is observed in the small intestine. Therefore, colon-targeted delivery system protects peptide drugs from hydrolysis and enzymatic degradation that takes place in duodenum and jejunum. Eventually the system releases the drug into colon, leading to greater systemic bioavailability (Chourasia and Jain, 2003). Both synthetic and naturally occurring polymers are used as drug carriers for drug delivery to the colon (Ashord et al., 1993).

AH makes use of an equally versatile colon-targeted delivery system known as *vasti* (medicinal enema). Decoctions of herbs, milk, clarified butter, sesame oil, visceral fat, fat from bone marrow, honey, sugarcane juice, rock-salt, cow urine and meat soups are administered as *vasti* after conditioning the body of the patient with oleation and sudation. *Vasti* is also performed in many alternating combinations of decoctions and oleaginous substances (see Chapter 3). The superiority of *vasti* in curing diseases can be attributed partly to the high water-absorption capacity of the colon. While colon-targeted drug delivery system of Western medicine is adopted in the treatment of a few bowel diseases, *vasti* is capable of curing an array of diseases. In fact, AH concludes the chapter on medicinal enema with the advice that *vasti* can be considered to be the sole panacea for all diseases (*Sū.* 19: 87).

8.3.6 Versatile Therapeutic Measures

Various therapeutic measures described in AH are multifaceted and serve many purposes. *Abhyaṅga* is one such measure. It is the application of medicated oil (*taila*) all over the body and head. *Abhyaṅga* is recommended in the treatment of diseases such as *vāta* diseases, *vāta*-induced blood disorders, enlargement of abdomen, skin diseases, *gulma* and so on (see Table 5.8). In a pilot study, Basler (2011) assessed the single-dose effect of *abhyaṅga*, in terms of its impact on subjective stress experience. Ten healthy women and ten healthy men underwent a one-hour *abhyaṅga* massage treatment. The study employed a repeated-measures design for the collection of stress data as well as values of heart rate and blood pressure, pre- and post-intervention. Subjects showed high statistically and clinically significant reductions in subjective stress experience. There was reduction in heart rate also. Pre-hypertensive subgroup showed reduction in blood pressure. *Abhyaṅga* was beneficial in lowering heart rate in all, and blood pressure in pre-hypertensive subjects.

Śirōdhāra is another treatment measure recommended in AH for treating headache, pityriasis, burning sensation, suppuration and sores. It is the process of pouring in a steady stream, medicated oil, milk or buttermilk on the forehead in a specific manner, with the patient lying in supine position (*Sū.* 22: 24–26). To address the problem of not being able to use the same reproducible stimulation in ayurvedic oil treatments, Uebaba et al. (2005) developed a healing robot which conducts *śirōdhāra* in a computerized, reproducible manner. They studied the physio-psychological changes and estimated psychological experiences during *śirōdhāra* by psychometric studies of anxiety and altered states of consciousness. Sixteen healthy adult females participated in the study. Blood pressure and intermittent blood pressure, electrocardiogram R-R intervals, expired gas analysis, impedance cardiography and electroencephalogram (EEG) were monitored during *śirōdhāra*. In active studies, *śirōdhāra* was performed by a machine with a pumping and heating system. The subject's feelings during *śirōdhāra* showed deep restfulness with less anxiety, as if the subjects were between sleep and wakeful states. *Śirōdhāra* induced bradycardia and lowered sympathetic tone. Expired gas analysis showed a decreased tidal volume. EEG showed the slowing of the /spl alpha/wave, an increase in / spl alpha/ and /spl theta/ activity and an increase in right-left coherence. These findings support the reported experiences of relaxed and low metabolic states during *śirōdhāra*. Physiological changes happening during *śirōdhāra* were similar to those of meditation. These findings indicated a change in the function of the frontal lobe, limbic system, brain stem and autonomic nervous system. On the other hand, the EEGs of the technicians of the manual *śirōdhāra* showed an increase in their stressful condition, justifying the utility of the healing robot as an assistant for the technicians.

The anxiolytic and altered states of consciousness-inducing effects of *śirōdhāra* were further studied by Uebaba et al. (2008) using a robotic system. In the first experiment for determining the most appropriate conditions of *śirōdhāra*, 16 healthy females underwent a thirty-minute treatment. In the second study, another 16 healthy females were assigned to either the *śirōdhāra* treatment or controls in supine position for thirty minutes. Their physiological, biochemical, immunological and psychometric parameters including anxiety and altered states of consciousness were monitored.

Subjects receiving *śirōdhāra* treatment showed lowered levels of anxiety and higher levels of altered states of consciousness than those in the control position. Plasma noradrenaline and urinary serotonin excretion showed significant decrease more after *śirōdhāra* treatment than in the control. Plasma levels of thyrotropin-releasing hormone, dopamine and natural killer cell activity were different between control and *śirōdhāra* treatment. The correlation between anxiolysis and the depth of altered states of consciousness was significant in the *śirōdhāra* treatment group whereas no correlation was observed in the controls. The increase in foot skin temperature after *śirōdhāra* showed a significant correlation with anxiolysis and depth of trance of altered states of consciousness. Natural killer cell activity after *śirōdhāra* treatment showed a significant correlation with anxiolysis and the depth of trance of altered states of consciousness. Results of the study indicate that *śirōdhāra* has anxiolytic and altered states of consciousness-inducing effects. It promotes a decrease of noradrenaline and exhibits a sympatholytic effect, resulting in the activation of peripheral foot skin circulation and immunopotentiation (Uebaba et al., 2008).

The mind-calming effect of *śirōdhāra* was further reported by Dhuri et al. (2013). They evaluated the psychological and physiological effects of *śirōdhāra* in healthy volunteers by monitoring the rate of mood, levels of stress, electrocardiogram, electroencephalogram (EEG) and selected biochemical markers of stress. There was a significant improvement in mood scores and the level of stress, accompanied by significant decrease in rate of breathing and reduction in diastolic pressure along with reduction in heart rate. Concomitant with relaxed alert state, alpha rhythm in EEG increased after *śirōdhāra*. The study concluded that a standardized *śirōdhāra* leads to a state of alert calmness similar to the relaxation response observed in meditation. The clinical benefits observed with *śirōdhāra* in anxiety neurosis, hypertension and stress aggravation due to chronic degenerative diseases can be explained on the basis of these adaptive physiological effects.

Massage is another treatment measure applied in many conditions. Kumar et al. (2017) studied the effectiveness of ayurvedic massage in chronic low-back pain, using *Sahacarādi taila*. Sixty-four patients with chronic low back pain were randomly assigned to a two-week massage group with six hours of ayurvedic massage or to a two-week local thermal therapy group. The study lasted for four weeks, consisting of a two-week intervention phase followed by a two-week follow-up phase. Mean back pain after treatment was significantly reduced in the massage group and the standard thermal therapy group. Beneficial effects on pain-related annoyance and psychological well-being were also apparent. The study showed that ayurvedic massage treatment is effective for pain relief in chronic low back pain.

Support for the effectiveness of ayurvedic massage comes from several published reports (Moyer et al., 2004). Massage may stimulate vagal activity, leading to a reduction of stress hormones, physiological arousal and a subsequent parasympathetic response of the autonomic nervous system (Field, 1998; Ferrell-Torry and Glick, 1993). By stimulating a parasympathetic response, massage may promote reduction in anxiety, depression and pain leading to a state of calmness. The same mechanism may produce several condition-specific benefits resulting from massage, such as increased immune system response in HIV-positive individuals (Diego et al., 2001). Massage may increase levels of serotonin in blood (Field et al., 1996; Ironson et al., 1996), which may inhibit the transmission of noxious nerve signals to the brain (Field, 1998). Manipulations such as rubbing or applying pressure may stimulate the release of endorphins into the bloodstream, providing pain relief or feeling of well-being by altering body chemistry (Andersson and Lundeberg, 1995; Oumeish, 1998). The manipulations and pressure applied on body may break down subcutaneous adhesions and prevent fibrosis (Donnelly and Wilton, 2002). They may also promote the circulation of blood and lymph (Weerapong et al., 2005), which may lead to reduction in pain associated with injury or strenuous exercise.

Changes in body chemistry may take place in individuals deprived of deep sleep, and this may lead to increase in pain. In the absence of deep sleep, levels of substance P increase and levels of somatostatin decrease. Both of these changes have influence on the experience of pain. Sunshine et al. (1996) reached the conclusion that massage may have promoted deeper, less disturbed sleep in fibromyalgia sufferers who experienced a reduction in pain during the course of massage.

8.3.7 Medicinal Products from Cow

AH describes the therapeutic application of medicinal products from cow. These include milk, clarified butter, curd, cow urine, cow dung, hide, gall bladder, teeth, hoof and hair (see Table 6.38). Clarified butter processed with cow urine, juice of cow dung, milk, curd (*Pañcagavya ghṛta*) and many herbs (*Mahatpañcagavya ghṛta*) are indicated in insanity, epilepsy, fever, jaundice, enlargement of abdomen, anal fistula, hemorrhoids and morbid pallor (*Ut.* 7: 18–24). Cow urine is an important ingredient of fluids intended for medicinal enema (*vasti*) (*Ka.* 4: 1–73).

Cow milk occupies a unique position among many foods, as it contains a high concentration of nutrients essential for health and formation of bone mass. It is frequently included as an important element in a healthy and balanced diet. Epidemiologic studies confirm the nutritional importance of milk in human diet and reinforce the possible role of its consumption in preventing several chronic conditions like cardiovascular diseases, some forms of cancer, obesity and diabetes (Pereira, 2014).

Clarified butter prepared by heating butter is also considered a good source of lipophilic ingredients such as conjugated linoleic acid and vitamins (A and E) that exhibit several nutraceutical actions, both *in vitro* and *in vivo* (Rani and Kansal, 2011; Pena-Serna and Restrepo-Betancur, 2020). Curd is an important food ingredient. Fat present in clarified butter and curd is responsible for their nutritive value (Shori and Baba, 2014; Antony et al., 2018).

Cow urine contains many essential micronutrients like calcium, phosphorous, vitamins B1, B2, C) and enzymes such as amylase, acid phosphatase, lactate and lactate dehydrogenase (Ganguly and Prasad, 2010; Mohanvel et al., 2017; Ketan et al., 2020). Cow urine is also rich in urea, creatinine, carbolic acid, phenols, calcium and manganese, which explain its antimicrobial and germicidal properties. Presence of amino acids and peptides may enhance the bactericidal effect of cow urine by increasing bacterial cell surface hydrophobicity (Randhawa and Sharma, 2015). Kumar et al. (2021) identified 15 peptides displaying in-silico anti-cancer, anti-hypertensive, anti-microbial and anti-inflammatory activities.

Numerous reports confirm the beneficial biological effects of products from cow (Prasad and Kothari, 2022). Some representative reports are presented here. Cow milk increases immunoglobulin concentration (IgG and IgM), chitotriosidase activity and complement system activity (Hernández-Castellano et al., 2015). Fermented milk containing conjugated linoleic acid decreases levels of fasting blood glucose, serum insulin, and leptin and increased oral glucose tolerance and insulin tolerance (Song et al., 2016). Curd has the ability to inhibit α-amylase and α-glucosidase (Shori and Baba, 2014). Clarified butter alters membrane lipid composition and downregulates the activities of enzymes responsible for carcinogen activation in the liver. It also upregulates carcinogen detoxification activities in liver and mammary tissues (Rani and Kansal, 2011). Fresh cow urine causes reduction in tumor incidence, tumor yield, tumor burden and cumulative number of papillomas (Raja and Agrawal, 2010). Stimulation of peripheral use of blood glucose by fresh urine is also reported (Jarald et al., 2008). Fresh urine reduces body mass index, abdominal circumference, obesity index, atherogenic index, total cholesterol, triglycerides, LDL-cholesterol and VLDL-cholesterol while increasing HDL-cholesterol level (Sharma et al., 2017). Fresh urine lowers levels of urine oxalate, serum creatinine, blood urea and calcium oxalate depositions and restores kidney weight (Shukla et al., 2013). These reports justify the use of products from cow as medicinal substances endorsed by AH.

8.3.8 Therapeutic Heating of Body

Warmth is connected to tranquility and relaxation. Therefore, it is logical that AH employs sudation (*svēdana*) in general treatment regimens and as a prerequisite before performing *pañcakarma*. Several modes of therapeutic heating body are recommended (see Chapter 3). Various research reports unravel the physiological mechanisms underlying these measures.

Topical application of heat increases local blood flow directly by stimulating vasodilation and indirectly through increasing the metabolic rates in underlying tissues (Delisa, 1983; Hale et al., 1985; Tepperman and Devlin, 1983). In the healing phase of the clinical condition, these effects on circulation and metabolism enhance the removal and repair aspects of the healing process (Delisa, 1983; Tepperman and Devlin, 1983; Nanneman, 1991). Locally applied heating modalities promote relaxation, provide pain relief, increase blood flow, facilitate tissue healing, decrease muscle spasm and decrease tissue tightness and joint stiffness (Rennie and Michlovitz, 2016).

8.3.9 Fecal Microbiota Transplantation

The human gastrointestinal tract is colonized by numerous bacteria, which provide aid in digestion, assist in nutrition, facilitate maturation of colonic epithelium and offer protection from pathogens. The gut microbiota is implicated in many bowel diseases such as infectious gastroenteritis, inflammatory bowel disease, obesity and functional gastrointestinal disorders. A wide variety of therapeutic strategies are adopted to correct gut dysbiosis. But a great majority of them do not cause satisfactory clinical results, except fecal microbiota transplantation, which is the transfer of stool from a healthy donor into another patient's gastrointestinal tract to normalize the recipient's gut microbiota (Wang et al., 2019).

Centuries ago, AH had put forth the concept of fecal microbiota transfer, not using human stool, but with dung and droppings of creatures such as bull calf, camel, cow, donkey, elephant, goat, horse, pig, sheep and pigeon. Sugar, clarified butter and honey are the adjuvants recommended in such fecal microbiota transplantations. Interestingly, fecal microbiota transplants are indicated not in bowel diseases, but in diseases such as cough, respiratory distress, discomforts arising from spider sting and hemorrhages of obscure origin (see Table 6.38). Dung of cow, camel, donkey, elephant, goat, horse, pig, sheep and droppings of pigeon are rich in microbiota (Girija et al., 2013; He et al., 2019; Xing et al., 2020; Ilmberger et al., 2014; Li et al., 2019; Kauter et al., 2019; Wylensek et al., 2020; Chang et al., 2020; Grond et al., 2019). No reports are available on the transplantation of microbiota from animals into humans. However, from the knowledge point of view, it will be interesting to study this aspect.

8.3.10 Cream as Ayurvedic Dosage Form

Cream is a semisolid dosage form employed in orthodox medicine, for topical application. It contains more than 20% water, additives and actives. Creams are defined as viscous, semi-solid emulsions of either the oil-in-water or water-in-oil type dosage forms. They are prepared by mixing the ingredients and stirring them continuously (Chauhan and Gupta, 2020). Vāgbhaṭa was aware of the preparation of creams, as is evident from a formula described in chapter on treatment of *vāta*-induced blood disorders. *Taila* prepared from 4 l of fermented rice gruel, 1 l sesame oil and 200 g of gum-resin from *Shorea robusta* is to be churned with "a large volume of water". Topical application of the butter-like product pacifies fever, burning sensation and pain associated with *vātarakta* (*Ci.* 22: 21).

8.3.11 Fortification of Food and Beverages

Functional foods have emerged as an effective way of prevention of diseases in recent times. Thus, fortified bread, sausages, biscuits, tea, coffee, soups and chocolates are becoming increasingly popular among consumers (Kumar, 2016). The concept of food fortification was already put into practice by Vāgbhaṭa. Consumption of medicated gruels, milk, curd, buttermilk, butter, sweets, cakes, chutneys, curries, broth, wine and beverages is recommended for curing many diseases (see Chapter 7).

8.4 AH AS A CULTURAL WINDOW

8.4.1 Male-Dominant Society

When viewed from the perspective of sociology, AH opens a window to the culture prevalent in north and north-western India of 7th century AD. Over the centuries, Ayurveda was broadly classified into eight branches such as *kāyacikitsa* (general medicine), *śalyatantra* (surgery), *śālākya tantra* (knowledge of diseases of supra-clavicular region), *kaumārabhṛtya* (gynecology, obstetrics, pediatrics), *agada tantra* (treatment of homicidal poisoning and poisonous bites), *bhūtavidya* (knowledge of subtle beings or spirits), *rasāyana* (rejuvenation) and *vājīkaraṇa* (virilification) (*Sū.* 1: 35–48). Among them, *vājīkaraṇa* is exclusively for the treatment of men. Chapter 40 of *Uttarasthānam* advocates several therapeutic measures and medicinal formulations for improving stamina, vigor and libido of men. Interestingly, no such medicines or measures intended for women are described, even though women are equally important in the furtherance of human race. This suggests that AH was composed in a male-dominant society.

Gender bias is evident in some of the teachings of AH. The human body is said to be composed of seven tissue elements such as *rasa*, *rakta*, *māmsa*, *mēdas*, *asthi*, *majja* and *śukḷa*. This implies that all of these seven *dhātu*s exist in women as well. Nevertheless, while discussing the "low" state of *śukḷa dhātu*, AH mentions delayed ejaculation, blood-stained semen, pricking pain in scrota and burning sensation in phallus as the corresponding signs (*Sū.* 11: 20). Increased interest in women and *śukḷāśmari* (crystals of semen?) are the only signs mentioned in the context of "high" state of *śukḷa dhātu* (*Sū.* 11: 12). These signs are discernible only in men. Signs related to women are totally lacking in these descriptions.

Some pieces of evidence suggesting gender bias are available in *Nidānasthānam* also. Both men and women are afflicted with dysuria, urinary stone and polyuric diseases. However, pricking pain in phallus and feeling of heaviness in phallus are some of the signs of dysuria (*Ni.* 9: 4–5). Pressing of phallus due to excruciating pain is a symptom of urinary stone with *vāta* dominance (*Ni.* 9: 11). Heaviness of phallus is a symptom of *mūtrōtsaṅga* also (*Ni.* 9: 30), and pain in perineal area is one of the symptoms of urinary stone (*Ni.* 9: 9). *Pittaja pramēha* is characterized by several symptoms including pricking pain in the phallus and pain in scrota (*Ni.* 10: 23). These anomalies need correction.

Importance is always given to a son, rather than a daughter, in classical texts of Ayurveda. Many medicinal formulations are said to bless the user with a son, as in the case of *Mahāmāyūra ghṛta* described in *Caraka Samhita* (Sharma, 1983). AH also endorses this view. While describing the benefits of consuming *Amṛtaprāśa ghṛta*, it is said that this *ghṛta* cures several diseases and also helps in begetting a son (*Ci.* 3: 94–101). Similarly, *vasti* performed with *Jīvantyādi yamaka* is reputed to cause the birth of a son (*Ka.* 4: 59–62). Similar effect is attributed to *Brahmī ghṛta* (*Ut.* 6: 25) and *Mahāmāyūra ghṛta* (*Ut.* 24: 56). Caressing a son and watching him playing with staggering walk and unclear words are considered to be some of the benefits of having an ideal offspring (*Ut.* 40: 10–11).

8.4.2 Sexual Liberalism

Cohabitation with beautiful and voluptuous women was popular as a therapeutic measure in ancient days. Embrace of lascivious, inebriated, warm and sweet-smelling women is recommended as a measure to ward off cold in pre-winter season (*Sū.* 3: 15). A similar piece of advice is given in the treatment of *śītajvara* (fever with shivering fits). Lovely, young, inebriated, buxom women should embrace the patient. Because of the ensuing excitement, the patient will feel very warm. On confirming this effect, the women should be separated from him (*Ci.* 1: 146–147). Alcoholism with predominance of *vāta* is treated with a similar technique. Along with other measures, young, attractive women with heavy breasts, thighs and buttocks are to embrace the patient and caress his body (*Ci.* 7: 17–19). With their soft touch, cool-bodied buxom women ward off burning sensation, pain and tiredness associated with *vātaśōṇita* (*Ci.* 22: 27).

Sexual intercourse itself is recommended as a treatment mode in several diseases. For example, excessive sexual intercourse is advised as a general measure to pacify the effects caused by aggravation of *kapha* (*Sū.* 13: 11). *Ūrustambha* (stiffness of thighs) is a disease caused by increased *kapha*, *mēdas* and *vāta*, rendering the thighs cold, heavy and immobile (*Ni.* 15: 47–51). To reduce *kapha*, the patient should get used to doing some light exercises. He should also indulge in sexual intercourse (*Ci.* 21: 53–54). Keeping awake at night by indulging in sexual intercourse in the proper way is also known to cure the effects of *madātyaya* caused by *kapha* (*Ci.* 7: 42–43).

Śukḷāśmari is a disease manifesting in elderly men. It is caused by restraining of sexual urge (*Ni.* 9: 16–19). This disorder is cured by preliminary measures such as *uttaravasti* (urethral enema) and consumption of aphrodisiac meat, including chicken meat. Thereafter, the patient should indulge in sexual intercourse *ad libitum* with young and lascivious women (*Sū.* 4: 19–21, *Ci.* 11: 41–43). While describing the benefits of *Sukumāra ghṛta*, AH states that it is suitable for soft-bodied persons, princes and "men who keep many wives" (*Ci.* 13: 41–47). Drinking of wine in small measure is advised while entertaining women at night (*Ci.* 7: 86). These statements suggest that polygamy was not taboo in many parts of the country at that time.

8.4.3 Consumption of Alcoholic Beverages

A critical study of AH reveals that drinking of *madya* (alcoholic beverages) was widespread in north and north-western India, at least in the region where the text was composed. *Madātyaya* (alcoholism) is considered as a major disease, and 115 quatrains are devoted to the description of its treatment. Vāgbhaṭa extolls the virtues of *madya* in the chapter on treatment of alcoholism. *Madya* (wine) is the supreme benefactor which gave glory to the Aśvinīkumāra, power to Sarasvatī Dēvi, prowess to Indra, pre-eminence to Viṣṇu and arrows to Makarakētu, the god of love. It is said to be a substance praised by deities (*dēva*), demoniac beings (*asura*), celestial musicians (*gandharva*), benevolent spirits (*yakṣa*), cannibalistic beings (*rākṣasa*) and humans. It exists in various forms such as *Madhu* (*Sōma*), *Mādhava* (from honey), *Mairēya* (spicy wine made out of fruits and flowers), *Sīdhu* (from sugarcane juice), *Gauḍa* (from molasses) and *Āsava* (derived from juices of fruits or herbs). It makes even noble women elated and when consumed in plenty impels soldiers to lay down their lives in combat. It saves people from grief, anxiety, apathy, fear and even fear of death. Therefore, wine is to be enjoyed according to a virtuous code (*Ci.* 7: 54–67).

AH offers a code for drinking of wine in moderation. After having a bath and paying obeisance to deities, scholars and masters, one should enter the ante-room adjacent to the dining hall, sprinkled with scented water. He should then recline on a soft couch surrounded by servants, friends, beautiful women and panegyrists who praise him. Songs, dances, music, colorful attire and women with heavy breasts and buttocks heighten the pleasantness of the situation. Wine should be cooled by fans and then enriched with fragrant herbs and flowers. The wine should be served in vessels made of glass or shell. Before beginning to drink, the individual should consume medicinal powders such as *Tālīsādi* or *Ēlādi*, offer wine to others in the company and keep a small quantity in a clean place, to please the invisible beings that arrive to partake of the wine. Thereafter, he should drink wine in moderate volume, accompanied with pleasing dishes. After drinking two glasses of wine, he should enter the dining hall and take food in front of a good physician. Food should include meats, various cakes, clarified butter and vegetables. While taking food and thereafter, he may drink two or three times. After concluding the session, he should retire to bed (*Ci.* 7: 75–88). One who observes such a code will be free from diseases that arise from aggravation of *mēdas* (adipose tissue), *vāta* and *kapha*, which turn incurable in those who transgress this code (*Ci.* 7: 68). Anyone who follows this code will acquire wealth and prosperity and the one who does not enjoy this pleasure leads a worthless life. Such a rich man who leads a poor life is comparable to the legendary spirit guarding a treasure (*Ci.* 7: 89–91). This code is intended for affluent persons. Those who strive to get rich should enjoy wine according to the resources at their disposal (*Ci.* 7: 93).

Before drinking wine, a *vātaja*-type person should condition his body with *abhyaṅga*, *udvartana*, fumigation with fragrant herbs and smearing of fragrant pastes on body. Individuals of *pittaja* constitution should enjoy wine after consuming food that is sweet, unctuous and of cold potency. *Kaphaja*-type person should consume warm food prepared with barley, wheat and *jāṅgala* meat seasoned with pepper powder (*Ci.* 7: 95–97).

Wines brewed from rice powder and jaggery are suitable for persons of *vātaja* constitution. *Pittaja*-type individuals should drink wine mixed with water. Wines brewed from grapes, juices of herbs or other fruits and honey are ideal for *kaphaja*-type persons. Drinking wine before food is recommended for *kaphaja*-type individuals, whereas *pittaja* type should drink after food. *Vātaja*-type person should drink when he is half way through the meal. One of *tridōṣaja* constitution need not follow any such rule; he can drink at any time (*Ci.* 7: 98–99).

It was a common practice in those days to administer copious volumes of wine to patients, from whose body deeply embedded sharp objects were to be extricated by surgery. Before subjecting a patient to surgery, he was fed with appropriate food. Strong wine was administered to those who were habituated to drinking wine and who were unable to bear pain (*Sū.* 29: 14–15). Wine was also administered to stupefy patients before subjecting them to chemical and thermal cautery (*Ci.* 7: 72).

Wine was also used as a medicinal substance. In the event of abortion of fetus, very strong wine was administered to the mother to cleanse the uterus and to allay pain (*Śā.* 2: 9–10). Consumption of wine, except the one brewed from sugarcane juice, is favorable for production of breast milk (*Ut.* 1: 18). People suffering from insomnia are advised to drink wine regularly (*Sū.* 7: 66). Regular use of wine and meat is recommended to patients convalescing from kingly consumption (*Ci.* 5: 83). For improving digestive efficiency, consumption of *Mārdvīkā* wine is advised in treatment of morbid pallor (*Ci.* 16: 56).

The popularity of wine drinking prevalent in ancient days is attested by the statement in *Nidānasthānam*, that those who are habituated to its use and those born in the families of wine drinkers do not get inebriated easily (*Ni.* 6: 11–12). Drinking wines such as *Āsava*, *Ariṣṭa*, *Sīdhu*, *Mārdvīkā* and *Mādhava* in the company of friends is a part of regimen to be adopted in spring season (*Sū.* 3: 22). While describing the treatment of mishaps met with in pregnancy, it is stated that "if the woman is not habituated to drinking wine (*amadyapā*)", a decoction of herbs is to be administered to her (*Śā.* 2: 10–11). This suggests that wine drinking was popular among women as well. In spite of all the words of appreciation, AH instructs that one should not brew, drink, offer or sell wine (*Sū.* 2: 40). This is rather puzzling.

8.4.4 Consumption of Cow Meat

Ahiṃsa (nonviolence) is a cardinal principle of contemporary Hinduism, which encourages vegetarianism. Present-day Hindu society views cow as a sacred being, and cow slaughter is legally banned in many states of India. However, AH recommends cow meat as a remedy for several diseases. Chapter 6 (*Annasvarūpavijnanīyam*) describes the qualities of cow meat. It is indicated in dry cough, exhaustion resulting from physical exertion and overindulgence in sexual activity, excessive appetite, irregular fever, rhinitis, leaning of body and *vāta* diseases without association of *pitta* or *kapha* (*Sū.* 6: 65) (Table 8.1). Chapter 2 of *Śārīrasthānam* dealing with mishaps in pregnancy recommends the administration of soup of cow meat, mixed with copious amount of clarified butter, as a remedy for *līna*, a mishap in pregnancy (*Śā.* 2: 18–19). These are interesting points which suggest that consumption of cow meat was not at all a taboo in Vāgbhaṭa's time and vegetarianism was a later introduction to Hindu society, most probably through the teachings of Śrī Śaṅkara (Tenzin, 2006).

8.4.5 Formulae of Poisons

The chapter on incompatible food (*viruddhāhāra*) describes some food preparations that cause death. For example, meat of *Hārīta* bird (green pigeon) barbecued over a fire set with *Hāridra* wood and using skewer sticks made of *Hāridra* wood is said to cause immediate death (*Sū.* 7: 44). Similar lethal effect is attributed to meat of birds such as *Tittiri* (grey partridge), peacock, *Lāva* (bustard quail) and *Kapiñjala* (black partridge) cooked in castor oil over a fire set with wood of castor plant (*Sū.* 7: 43). Poisons were used in ancient India to eliminate rivals and to punish those suspected of betrayal and treachery. One example is the attempt on the life of the Mauryan king Chandragupta, a contemporary of Alexander the Great, in the 4th century BC, by using a "poison maiden" (*viṣakanyaka*). This appellation was applied to women who administered to their victims,

TABLE 8.1
Cow Meat as Food and Medicinal Substance

Sl. No.	Particulars	Reference
1	Consumption of cow meat pacifies dry cough, tiredness, excessive appetite, irregular fever and catarrh	*Sū.* 6: 65
2	Cow meat is not to be consumed regularly	*Sū.* 8: 40
3	To improve stamina, cooked cow meat is to be served to patient suffering from kingly consumption, saying that it is his most favorite meat	*Ci.* 5: 6–7
4	Curry cooked with soup of cow meat and vegetables is to be served to patients suffering from hemorrhoids	*Ci.* 8: 85
5	Powder of detoxified *Danti* or *Dravanti* is to be suspended in the soup of cow meat and administered to patients suffering from morbid pallor, worm infestation or anal fistula, to cause purgation	*Ka.* 2: 53–55
6	Powder of cow meat, soaked in animal fat, is to be applied over *ajaka* (eye disease) subjected to surgery	*Ut.* 11: 56
7	Soup of cow meat is recommended in *līna*, a mishap encountered in pregnancy	*Śā.* 2: 18–19
8	Fumigation with cow meat is recommended in the treatment of insanity caused by aggravation of *vāta* and *kapha*	*Ut.* 6: 44

poison concealed in food or drink (Arnold, 2016). Vāgbhaṭa surreptitiously described these poisonous foods, perhaps as an information which can have some non-medical applications.

8.5 OUTLOOK FOR FUTURE

8.5.1 Redaction of AH in the Light of Modern Knowledge

There are eight Sanskrit commentaries on AH. Among them, Aruṇadatta's *Sarvāṅgasundarā* is the only commentary published in full. Parts of Hēmādri's commentary are also published (Paradakara, 2018). The others still remain in manuscript form. Ayurveda enthusiasts should strive to publish the remaining commentaries also. Finally, all the eight commentaries should be reviewed and another one composed so as to make AH truly modernized. A few pieces of outdated information are available in AH. For example, it is stated that riding on horses continuously can cause *vātaśōṇita* (*Ni.* 16: 5). This is irrelevant in modern times, as that mode of travel is now obsolete. Riding motor cycles or operating bulldozers and tractors regularly can be considered as relevant causative factors, rather than horse riding.

8.5.2 Objective Way of Teaching AH

AH is taught at present with the help of texts with Sanskrit verses and their translations in English, Hindi or other regional languages. These texts do not have information from the Sanskrit quatrains presented wherever possible, in the form of figures, tables, flowcharts or diagrams, which make learning easier. There is a need for well-organized translations of AH, free from the defects mentioned above and incorporating, in addition, photographs related to many of the diseases such as *arśas, vidradhi, vṛddhi, gulma, udara, śvayathu, visarpa, kuṣṭha* and *śvitra*. Teaching of therapeutic measures can be made more effective and student-friendly by including photographs, which explain how *vamana, virēcana, vasti, nasya, dhūmapāna, gaṇḍūṣa, āścyōtana, tarpaṇa, sirāvēdha* and others are performed.

Instead of encouraging students to just memorize Sanskrit quatrains, they should be taught AH objectively. Concepts like *prakṛti* and *dōṣakōpa* should be taught with the aid of practical classes. At present in examinations, the knowledge of students is judged on the basis of their ability to recite Sanskrit quatrains correctly. Instead of that, the examiner should evaluate their skill in identifying *prakṛti* of individuals and assessing degree of *dōṣakōpa*. It goes without saying that greater emphasis needs to be given to the identification of symptoms and signs. Students will be able to diagnose diseases on the basis of ayurvedic principles better, once they are taught AH in such a progressive way.

8.5.3 Thrust Areas

In contemporary Ayurveda clinical practice, greater attention is paid only to certain diseases of *kāyacikitsa* (general medicine). The other seven areas are not given much importance. To usher in progress and to demonstrate the value of Ayurveda, clinical practice needs to be initiated in the other areas and very specially, mental disorders, eye diseases especially cataract (*timira*), epilepsy, wound healing, poisoning, rejuvenation and virilification. Cataract is a part of aging process, and the underlying cause is the accumulation of protein in the lens, reducing the transmission of light to the retina. Western medicine considers cataract as an incurable condition, calling for surgical measures (Asbell et al., 2005). Nevertheless, Ayurveda treats it as a curable one. AH recommends a ten-pronged approach to the treatment of *timira*, in relation to *dōṣakōpa*. It includes (i) oleation; (ii) bloodletting; (iii) purgation; (iv) nasal instillation; (v) application of collyria; (vi) smearing of cold potency pastes in eyes, over face and head; (vii) application of medicinal oil on scalp (*śirōvasti*); (viii) medicinal enema; (ix) refreshing of eyes (*tarpaṇa*); and (x) pouring of medicinal fluids (*sēka*)

TABLE 8.2
Some Medicines Recommended in the Treatment of Cataract

Sl. No.	Name of Medicine	Reference
I	**Internal Medicine**	
1	Cakes and powder of parched rice (*lāja*) mixed with *Triphalā cūrṇa*	*Ut.* 13: 18
2	*Drākṣādi ghṛta*	*Ut.* 13: 4–6
3	*Ghṛta* prepared with *Daśamūla* decoction, milk and *Triphala* powder	*Ut.* 13: 49
4	*Jīvantyādi ghṛta*	*Ut.* 13: 2–4
5	*Mahātraiphala ghṛta*	*Ut.* 13: 12–14
6	*Paṭōlādi ghṛta*	*Ut.* 13: 6–10
7	Powder of *Terminalia chebula* with grapes, sugar or honey, in early morning or before food	*Ut.* 13: 19
8	*Traiphala ghṛta*	*Ut.* 13: 10–11
9	*Triphala* decoction with clarified butter	*Ut.* 13: 17
10	*Triphalā cūrṇa* mixed with honey and clarified butter	*Ut.* 13: 16–17
II	**Collyria**	
1	*Akṣabījādi guḷika*	*Ut.* 13: 43
2	*Bhāskara cūrṇa*	*Ut.* 13: 28–31
3	*Māmsyādi cūrṇāñjanam*	*Ut.* 13: 23–24
4	*Maricādi cūrṇa*	*Ut.* 13: 25
5	*Srōtōñjana*	*Ut.* 13: 26–27
6	*Tutthāñjanam*	*Ut.* 13: 33
III	**Nasal Instillation**	
1	*Jīvantyādi taila*	*Ut.* 13: 51–54
2	*Sitairaṇḍādi taila*	*Ut.* 13: 54–55
3	*Taila* prepared with sesame oil, oil of *Terminalia belerica* seeds, *Eclipta alba* and decoction of *Pterocarpus marsupium* heartwood	*Ut.* 13: 46
IV	***Ghṛta* for *tarpaṇa***	
	Śatāhvādi ghṛta	*Ut.* 13: 58–59
V	**Medicine for Purgation**	
1	*Śarkarādi virēcana*	*Ut.* 13: 64
2	*Pūgādi virēcana*	*Ut.* 13: 68–69

(*Ut.* 13: 47). The patient also needs to be fed with unctuous and cold potency food (*Ut.* 13: 97). Some of the important medicines are presented in Table 8.2.

Surgery is one of the branches of the eight-fold Ayurveda. Surgical intervention in Ayurveda is recommended as a last resort, and it is to be performed after obtaining permission from the king (*Ci.* 11: 43–44). However, there is no point in reviving ayurvedic surgery, as developing an effective ayurvedic anesthetic is next to impossible, in the current state of affairs. As Western medicine is already equipped with an array of anesthetics and is permitted by the state to perform surgery, it is better to seek surgical assistance from that discipline. Instead, Ayurveda community should expeditiously revive the emergency medicine advised in the texts. Emergency medicine was applied in ancient days in the treatment of fever, cough, vomiting, dysentery and many other diseases. With proper emergency measures, even a condition like appendicitis can be managed without surgery.

The increase or decrease of *vāta*, *pitta* and *kapha* in disease conditions present themselves in 63 combinations (*Sū.* 12: 74–78). The six taste modalities also exist in 63 combinations. AH states that the combination of tastes of the medicinal substances should match with the combination of *tridōṣa* responsible for the disease condition, so as to bring them back to normalcy (*Sū.* 10: 44). There is an urgent need to discover the key to achieve this taste-*dōṣa* equilibrium in therapy. Similarly, the soundness of the concept of taste influencing drug action, as defined in AH also remains to be demonstrated with the help of taste sensitivity studies. In addition to establishing the relationship between tastes of matter (medicinal substances) and *tridōṣa*, such studies will help to identify medicinal herbs of AH on the basis of their taste(s) and thereby clear the confusion existing around the identity of some of them (Kumar, 2021).

8.5.4 Diagnosis in Future

Vāgbhaṭa's insights into the diagnosis of a disease primarily focus on the role of the five-fold methods of disease assessment – *rōgaparīkṣa* (*nidānapañcaka*) such as *nidāna*, *pūrvarūpa*, *lakṣaṇa*, *upaśaya* and *samprāpti*. Adding strength to this dictum, clinical examination of a patient for the final diagnosis is solely structured on the methods of *darśana*, *sparśana* and *praśna*. Classically, this algorithm is in place over centuries, and ayurvedic diagnosis could be contemplated reasonably and substantially on these pointers to plan the treatment and management of diseases. In the 21st century of computers, artificial intelligence, finer understanding of human anatomy, upfront laboratory and imaging modalities, can the principles and practices of diagnosis in Ayurveda be updated? Is there enough space left in AH to be expanded, so that intelligent inclusion of Western medical investigative modalities will strengthen diagnosis in Ayurveda? In other words, is there scope for Vāgbhaṭa's methodology of disease diagnosis further "confirmed" by modalities of modern diagnostics?

In the fast-changing medical landscape, Ayurveda is often regarded (by the Western medical fraternity) as weak in precise clinical diagnosis. Most Ayurveda physicians resort to exclusive ayurvedic diagnosis, completely ignoring modern methods and tools. Many *vaidyas* are confused how to incorporate modern diagnostic methods with ayurvedic aphorisms. A close review of the diagnosis of certain diseases mentioned in AH and reinforcement possible by the wise application of modern diagnostics reveal that Vāgbhaṭa's approach to diagnosis is open-ended, and there is enough space for "open diagnosis" that would eventually offer better treatment options in Ayurveda. The synthesis of ayurvedic and Western approaches to diagnosis is discussed with three examples.

8.5.4.1 Examination and Assessment of Diseases

a. *Nidāna* (etiology) represents the causative factors, both intrinsic, extrinsic or both. It includes indulgence in unwholesome diet, and physical or mental activities, abuse of sense organs, abnormal sleep patterns, non-adaptation to seasonal changes, intense climatic factors and so on.
b. *Pūrvarūpa* – Premonitory symptoms specific to diseases.
c. *Rūpa* – Symptoms and clinical features classified and described in the texts.
d. *Upaśaya* – Clinical response to prescribed medication, diet or activity.
e. *Samprāpti* (pathogenesis) – Monitoring the dynamics of *dōṣa* and evolution of the disease.

8.5.4.2 Examination of the Patient

a. *Darśana* – Examination with the tool of vision
b. *Sparśana* – Examination with tools of touch and palpation
c. *Praśna* – Interrogation of the patient

Arguably, both *rōgaparīkṣa* (assessment of the disease) and *rōgīparīkṣa* (examination of the patient), put together on exclusive parameters of Ayurveda, are comprehensive in their own manner. But on a critical review, it can be said that contemporary Western medical methods of patient examination and investigative modalities can strengthen ayurvedic methodology to have a better understanding of the very ayurvedic tenets. The following examples support this argument.

8.5.4.3 Aśmari

The term *aśmari* (urinary calculi) refers to urinary concretions, usually manifested as painful urinary symptoms, regional abdominal symptoms and some systemic symptoms. This clinical condition, referred to as a *mahārōga,* meaning "difficult to cure", is discussed in the chapter on *mūtrāghāta* (*Ni.* 9: 1–40). Vāgbhaṭa's diagnosis of *aśmari* depends partly on symptoms (*nidānapañcaka*) and partly on the morphology of the calculus that passes out by chance. For example, *aśmari* with predominance of *vāta* is black in color and with spines. *Pittāśmari* has shape of seed of *Semecarpus anacardium* and is red, yellow or black in color. *Kapha*-dominant *aśmari,* on the other hand, will be large, smooth, honey-colored or white (*Ni.* 9: 11–14). However, it is not possible always to verify these characteristics in a patient. Modern diagnostics can help in expanding the scope of *darśana.* With a modern sophisticated microscope, one can view the crystals and red blood corpuscles in the urine under a high-power field. X-rays reveal radio-opaque stones. Excretory urography enables a physician to detect filling defects in the ureters. Presence and size of the calculus can be ascertained with ultrasonography, ruling out other pathologies with similar clinical features (Alshoabi et al., 2020; Jinzaki et al., 2011). Judicious adoption of modern diagnostic tools can empower contemporary Ayurveda physician to arrive at the most confirmatory diagnosis in cases of *aśmari,* even in terms of size and location in the urinary tract.

8.5.4.4 Arśas

The disease group *arśas* (often referred to as hemorrhoids or piles) is primarily a cluster of anorectal conditions manifested as bleeding *per anum,* prolapse of soft pile masses during defecation, difficulty in regular bowels and many other related colorectal symptoms. This complex clinical condition is again referred to as a *mahārōga* and discussed in seventh chapter of *Nidānasthānam.* Classified as being *sahaja* (hereditary) and *uttarōtthāna* (acquired), *śuṣka* (dry) and *srāvi* (discharging), detailed anatomical details are portrayed in this context to facilitate the diagnosis in terms of *bāhya* (external) and *ābhyantara* (internal) types. Vāgbhaṭa's diagnosis of *arśas* is based on *pūrvarūpa* (prodromal symptoms), manifested symptoms and other clinical features present (*Ni.* 7: 1–59). Added to this, the scope of *darśana* is expanded by Vāgbhaṭa himself recommending a specially designed instrument called *arśōyantra* (anoscope or rectoscope) to confirm the diagnosis and to aid in the surgical or para-surgical treatments of *arśas* (*Sū.* 25: 16–18).

Modern diagnostics include expanding the scope of *darśana* by using a perfect design illuminated proctoscope, flexible sigmoidoscope, fiberoptic video-assisted colonoscope to view the whole of colorectum, ultrasonography and computerized tomography to detect various pathologies mentioned under the spectrum of *arśas* as described by Vāgbhaṭa (Mori et al., 2018; Abdulazeez et al., 2020; Klang et al., 2020; Liang et al., 2020). Prudent use of modern diagnostic tools can help the Ayurveda physician to examine the various possibilities of the underlying pathology and to arrive at the correct diagnosis of *arśas* with precision. It will also help in the exploration of hidden pathologies that are obscure and incomprehensible.

8.5.4.5 Raktapitta

The disease *raktapitta* refers to bleeding disorders, manifested as bleeding through upper orifices (nose, eyes, ears, mouth), lower orifices (anus, urethra, vagina) or extensively mixed type, involving both upper and lower orifices, and also through the skin at times. Apart from bleeding, *raktapitta* is associated with many regional and generalized systemic symptoms. This clinical condition is elaborated in the chapter *Raktapittakāsanidānādhyāyaḥ* (third chapter of *Nidānasthānam*), including prognosis with reasoning (*Ni.* 3: 1–17). Vāgbhaṭa's diagnosis of *raktapitta* depends partly on symptoms (*nidānapañcaka*) and partly on the bleeding points. Modern diagnostics include clinical laboratory values of bleeding time, clotting time and Vitamin K; tests for hemophilia, radiography, ultrasonography and computerized tomography to investigate diseases which precipitate bleeding disorders; gastro-duodenoscopy which is the gold standard to assess as well as to treat upper gastrointestinal tract bleeding; otoscope; laryngoscope; bronchoscope and so on, which are of value in confirming the final diagnosis (Honeybrook et al., 2020; Meng et al., 2020; Teh et al., 2020; Wang et al., 2020). A wise choice of modern laboratory, imaging and other diagnostic tools can empower contemporary Ayurveda physician to obtain the most confirmatory diagnosis in varied manifestations of *raktapitta*. AH is precise (though not explicit) in the futuristic application for understanding of diseases, their diagnosis and treatment options. Methodology of diagnosis of the diseases in AH can be seen as primarily based on the fundamental aspects of Ayurveda but has left vast scope for the physician to apply the knowledge of contemporary biomarkers and modern tools of diagnosis. It is a task to be undertaken by the fraternity of Ayurveda for sustainable practice in future.

8.6 CONCLUSION

Ayurveda appears to be an expensive healing system, when one views its treatment modes, therapeutic measures and medicines critically. Seasonal regimen calls for various kinds of massages and topical application of pastes of expensive, fragrant substances such as musk, sandalwood and saffron. Sophisticated blankets like *Prāvāra*, *Ājina*, *Kauśēya*, *Pravēṇi* and *Kaucava* are recommended for use in pre-winter and winter seasons to ward off cold. Regimen for spring and summer stipulates that the subject spends time in picturesque gardens, woods and cool homes. Such vacations are affordable only to the affluent class. Elimination measures are required in certain seasons for effective removal of vitiated *dōṣa*s (*Sū*. 3: 1–58). Elimination measures are to be preceded invariably by oleation and sudation. This demands the expenditure of considerable sums of money. Moreover, these are time-consuming measures as well. Some medicated oils are prepared by adding every day milk and paste of a specific herb to the sesame oil added initially. *Balā taila* prepared in this way by repeating the process 100 or 1,000 times is recommended in the treatment of *vātavyādhi* and *vātaśōṇita*, which are common diseases that affect the elderly (*Ci*. 22: 45–46). *Yaṣṭhyāhvā śatapāka taila* prepared by adding 100 times milk and paste of *Glycyrrhiza glabra* roots is recommended for patients suffering from heart disease (*Ci*. 6: 38–39). *Ghṛta* of *Āragvadha* (*Cassia fistula*) prepared by repeating 100 times the step of addition of root paste and milk to the initially added clarified butter is prescribed in the treatment of leprosy and other skin diseases (*Ci*. 19: 13–14).

Gold, silver and precious stones, such as topaz (*puṣyarāgam*), cat's eye (*vaiḍūryam*), *gōmēdakam* (hessonite); animal horns; and teeth are also the ingredients of many collyria used in the treatment of eye diseases (*Ut*. 11: 36, *Ut*. 13: 22, 45). *Avagāha* and *sēka* are two therapeutic measures applied in the treatment of certain diseases. *Taila*, *ghṛta*, decoction of herbs, milk, vinegar or fermented wheat-water mixture is poured into a large cauldron, and the patient sits immersed in it for a specific period of time (see Table 5.10). The same fluid in warm or cold condition is poured over the whole body or affected body part in a continuous stream. This measure is called *sēka*, *pariṣēka* or *dhāra* (see Table 5.9). Large volumes of these medicinal fluids, including *taila* or *ghṛta*, are required in both *avagāha* and *sēka* (*Ci*. 1: 133–134, *Ci*. 5: 77, *Ci*. 8: 12–13, *Ci*. 21: 28–29). It is needless to state that these are very expensive therapeutic measures. Substantial amount of money is, therefore, to be spent on treatments involving such medicines. One should at this juncture recollect the advice of Vāgbhaṭa that the patient accepted for treatment should be rich (*Sū*. 1: 29).

One way of making Ayurveda more affordable to the common man is by limiting *śōdhana* (elimination) therapy and widening the scope of *śamana* (curative) therapy. Measures such as *svēdana*, *vamana*, *virēcana*, *vasti*, *nasya* and *raktamōkṣa* require appreciable amount of money and time. On the other hand, the classical seven-fold *śamana* therapy involves *dīpana*, *pācana* medicines, fasting (*upavāsa*), restricting water intake (*tṛṣṇānigraha*), exercise (*vyāyāma*), exposure to hot sun (*ātapa sēvana*) and exposure to breeze (*māruta sēvana*) (*Sū*. 14: 6–7). Coupled with this approach, medicated food and beverages such as gruels, churned gruels, broth, curd, milk, butter, buttermilk, wines, chutneys, curries and cakes can also be included in treatment.

AH describes numerous single-herb remedies. All major diseases such as fever, hemorrhages of obscure origin, cough, respiratory distress, kingly consumption, vomiting, excessive thirst, hemorrhoids, dysentery, *grahaṇi*, urinary stone, polyuric diseases, abscess, *gulma*, enlargement of abdomen, morbid pallor, edema, cellulitis, skin diseases, *vātavyādhi*, *vāta*-induced blood disorders, epilepsy, eye diseases, heart disease and sexual debility can be successfully treated with these preparations (Table 6.35). Adoption of these single-herb remedies reduces the cost of treatment drastically. Thus, by imbibing the true spirit of AH and making minor variations in treatment protocols, Ayurveda can be made more affordable to all sections of society. Rediscovering the soul of AH should, therefore, be the mission of all students and practitioners of Ayurveda.

REFERENCES

Abdulazeez, Z., N. Kukreja, N. Qureshi and S. Lascelles. 2020. Colonoscopy and flexible sigmoidoscopy for follow-up of patients with left-sided diverticulitis. *Annals of the Royal College of Surgeons of England* 102:744–747.

Akala, E.O., O. Elekwachi, V. Chase, H. Johnson, M. Lazarre and K. Scott. 2003. Organic redox-initiated polymerization process for the fabrication of hydrogels for colon-specific drug delivery. *Drug Development and Industrial Pharmacy* 29:375–386.

Al-Adawi, S., Y. Al-Kalbani, S.M. Panchatcharam et al. 2019. Differential executive functioning in the topology of spirit possession or dissociative disorders: An explorative cultural study. *BMC Psychiatry* 19:379. https://doi.org/10.1186/s12888-019-2358-2.

Alshoabi, S.A., D.S. Alhamodi, M.B. Gameraddin, M.S. Babiker, A.M. Omer and S.A. Al-Dubai. 2020. Gender and side distribution of urinary calculi using ultrasound imaging. *Journal of Family Medicine and Primary Care* 9:1614–1616.

Andersson, S. and T. Lundeberg. 1995. Acupuncture—From empiricism to science: Functional background to acupuncture effects in pain and disease. *Medical Hypotheses* 45:271–281.

Anson, R.M., Z. Guo, R. de Cabo et al. 2003. Intermittent fasting dissociates beneficial effects of dietary restriction on glucose metabolism and neuronal resistance to injury from calorie intake. *The Proceedings of the National Academy of Sciences USA* 100:6216–6220.

Antony, B., S. Sharma, B.M. Mehta, K. Ratnam and K.D. Aparnathi. 2018. Study of Fourier transform near infrared (FT-NIR) spectra of ghee (anhydrous milk fat). *International Journal of Dairy Technology* 71:484–490.

Arnold, D. 2016. The social life of poisons. In *Toxic Histories*, 17–40. Cambridge: Cambridge University Press.

Asbell, P.A., I. Dualan, J. Mindel, D. Brocks, M. Ahmad and S. Epstein. 2005. Age-related cataract. *Lancet* 365:599–609.

Ashord, M., J.T. Fell, D. Attwood, H. Sharma and P. Woodhead. 1993. An evaluation of pectin as a carrier for drug targeting to the colon. *Journal of Controlled Release* 26:213–220.

Basler, A.J. 2011. Pilot study investigating the effects of ayurvedic Abhyanga massage on subjective stress experience. *Journal of Alternative and Complementary Medicine* 17:435–440.

Berdoy, M., J.P. Webster and D.W. Macdonald. 2000. Fatal attraction in rats infected with *Toxoplasma gondii. Proceedings of the Royal Society of London B: Biological Sciences* 267:1591–1594.

Boillat, M., P.M. Hammoudi, S.K. Dogga et al. 2020. Neuroinflammation-associated aspecific manipulation of mouse predator fear by *Toxoplasma gondii. Cell Reports* 30:320–334.

Bruce-Keller, A.J., G. Umberger, R. McFall and M.P. Mattson. 1999. Food restriction reduces brain damage and improves behavioral outcome following excitotoxic and metabolic insults. *Annals of Neurology* 45:8–15.

Bünning, E. 1960. Opening address: Biological clocks. In *Biological Clocks*, ed. A. Chovnick, 1–9. New York: Cold Spring Harbor Laboratory.

Chang, J., X. Yao, C. Zuo, Y. Qi, D. Chen and W. Ma. 2020. The gut bacterial diversity of sheep associated with different breeds in Qinghai province. *BMC Veterinary Research* 16:254. https://doi.org/10.1186/s12917-020-02477-2.

Chauhan, L. and S. Gupta. 2020. Creams: A review on classification, preparation methods, evaluation and its applications. *Journal of Drug Delivery and Therapeutics* 10:281–289.

Chourasia, M.K. and S.K. Jain. 2003. Pharmaceutical approaches to colon targeted drug delivery systems. *Journal of Pharmaceutical Sciences* 6:33–66.

Daruna, J.H. 2012. Psychosocial stress: Neuroendocrine and immune effects. In *Introduction to Psychoneuroimmunology*. London: Academic Press, 131–151.

Delisa, J.A. 1983. Practical use of therapeutic physical modalities. *American Family Physician* 27:129–138.

Dhuri, K.D., P.V. Bodhe and A.B. Vaidya. 2013. Shirodhara: A psycho-physiological profile in healthy volunteers. *Journal of Ayurveda and Integrative Medicine* 4:40–44.

Diego, M.A., T. Field, M. Hernandez-Reif, K. Shaw, L. Friedman and G. Ironson. 2001. HIV adolescents show improved immune function following massage therapy. *International Journal of Neuroscience* 106:35–45.

Di Genova, B.M., S.K. Wilson, J.P. Dubey, L.J. Knoll. 2019. Intestinal delta-6-desaturase activity determines host range for *Toxoplasma* sexual reproduction. *PLoS Biology* 17(8):e3000364. doi: 10.1371/journal.pbio.3000364.

Donnelly, C.J. and J. Wilton. 2002. The effect of massage to scars on active range of motion and skin mobility. *British Journal of Hand Therapy* 7:5–11.

Ferrell-Torry, A.T. and O.J. Glick. 1993. The use of therapeutic massage as a nursing intervention to modify anxiety and the perception of cancer pain. *Cancer Nursing* 16:93–101.

Field, T.M. 1998. Massage therapy effects. *American Psychologist* 53:1270–1281.

Field, T., N. Grizzle, F. Scafidi and S. Schanberg. 1996. Massage and relaxation therapies' effects on depressed adolescent mothers. *Adolescence* 31:903–911.

Ganguly, S. and A. Prasad. 2010. Role of plant extracts and cow urine distillate as immunomodulator in comparison to levamisole – a review. *Journal of Immunology and Immunopathology* 12:91–94.

Gardiner, B.G. 1987. Linnaeus' floral clock. *The Linnean* 3:26–29.

Girija, D., K. Deepa, F. Xavier, I. Antony and P.R. Shidhi. 2013. Analysis of cow dung microbiota – A metagenomic approach. *Indian Journal of Biotechnology* 12:372–378.

Goodrick, C.L., D.K. Ingram, M.A. Reynolds, J.R. Freeman and N.L. Cider. 1982. Effects of intermittent feeding upon growth and life span in rats. *Gerontology* 28:233–241.

Grond, K., J.M. Perreau, W.T. Loo, A.J. Spring, C.M. Cavanaugh and S.M. Hird. 2019. Correction: Longitudinal microbiome profiling reveals impermanence of probiotic bacteria in domestic pigeons. *PLoS One* 14(7):e0220347. https://doi.org/10.1371/journal.pone.0220347.

Halberg, F., G. Cornélissen, G. Katinas et al. 2003. Transdisciplinary unifying implications of circadian findings in the 1950s. *Journal of Circadian Rhythms* 1:2. http://www.JCircadianRhythms.com/content/1/1/2.

Hale, J.R., C. Jessen, A.A. Fawcett and R.B. King. 1985. Arteriovenous anastomosis and capillary dilation and constriction induced by local heating. *Pflugers Archives* 404:203–207.

He, J., L. Hai, K. Orgoldol et al. 2019. High-throughput sequencing reveals the gut microbiome of the Bactrian camel in different Ages. *Current Microbiology* 76:810–817.

Hernández-Castellano, L.E., I. Moreno-Indias, A. Morales-delaNuez et al. 2015. The effect of milk source on body weight and immune status of lambs. *Livestock Science* 175:70–76.

Honeybrook, A., W. Lee and S. Cohen. 2020. A novel laryngoscope with an adjustable distal tip. *Laryngoscope* 130:2859–2862.

Ilmberger, N., S. Güllert, J. Dannenberg et al. 2014. A comparative metagenome survey of the fecal microbiota of a breast- and a plant-fed Asian elephant reveals an unexpectedly high diversity of glycoside hydrolase family enzymes. *PLoS One* 9(9):e106707.

Ingram, D.K. and M.A. Reynolds. 1987. The relationship of bodyweight to longevity within laboratory rodent species. *Basic Life Sciences* 42:247–282.

Ironson, G., T. Field, F. Scafidi et al. 1996. Massage therapy is associated with enhancement of the immune system's cytotoxic capacity. *International Journal of Neuroscience* 84:205–218.

Jarald, E.E., S. Edwin, V. Tiwari, R. Garg and E. Toppo. 2008. Antidiabetic activity of cow urine and a herbal preparation prepared using cow urine. *Pharmaceutical Biology* 46:789–792.

Jinzaki, M., K. Matsumoto, E. Kikuchi et al. 2011. Comparison of CT Urography and excretory urography in the detection and localization of urothelial carcinoma of the upper urinary tract. *American Journal of Roentgenology* 196:1102–1109.

Johansson, M. and T. Köster. 2019. On the move through time – A historical review of plant clock research. *Plant Biology* 21:13–20.

Kauter, A., L. Epping, T. Semmler et al. 2019. The gut microbiome of horses: current research on equine enteral microbiota and future perspectives. *Animal Microbiome* 1:14. https://doi.org/10.1186/s42523-019-0013-3.

Ketan, H., S. Sanjay, D. Ashwini, P. Kiran, K. Sravani and G. Rupesh. 2020. Prevention of apparent adiposity by fractions of distilled cow urine: A non-invasive approach. *Indian Drugs* 57: 73–78.

Klang, E., T. Sobeh, M.M. Amitai, S. Apter, Y. Barash and N. Tau. 2020. Post hemorrhoidectomy complications: CT imaging findings. *Clinical Imaging* 60:216–221.

Kneeland, R.E. and S. H. Fatemi. 2013. Viral infection, inflammation and schizophrenia. *Progress in Neuro-psychopharmacology and Biological Psychiatry* 42:35–48.

Kobayashi, S. 2015. Choose delicately and reuse adequately: The newly revealed process of autophagy. *Biological and Pharmaceutical Bulletin* 38:1098–1103.

Kumar, D.S. 2016. Food and beverages fortified with phytonutrients. In *Herbal Bioactives and Food Fortification: Extraction and Formulation*, 173–238. Boca Raton: CRC Press.

Kumar, D.S. 2021. Ayurveda renaissance- *Quo vadis*. In *Ayurveda in the New Millennium: Emerging roles and future challenges*, ed. D.S. Kumar, 241–272. Boca Raton: CRC Press.

Kumar, R., S.A. Ali, S.K. Singh et al. 2021. Peptide profiling in cow urine reveals molecular signature of physiology-driven pathways and ins-ilico predicted bioactive properties. *Scientific Reports* 11:12427. https://doi.org/10.1038/s41598-021-91684-4.

Kumar, S., T. Rampp, C. Kessler et al. 2017. Effectiveness of ayurvedic massage (*Sahacharadi Taila*) in patients with chronic low back pain: A randomized controlled trial. *The Journal of Alternative and Complementary Medicine* 23(2). https://doi.org/10.1089/acm.2015.0272.

Li, B., K. Zhang, C. Li, X. Wang, Y. Chen and Y. Yang. 2019. Characterization and comparison of microbiota in the gastrointestinal tracts of the Goat (*Capra hircus*) during preweaning development. *Frontiers in Microbiology*. https://doi.org/10.3389/fmicb.2019.02125.

Liang, W.J., D.Q. Wu, Z.J. Lyu et al. 2020. Application of indocyanine green fluorescence proctoscope in rectal cancer surgery. *Zhonghua Wei Chang Wai Ke Za Zhi* 23:1104–1105.

Lindová, J., L. Příplatová and F. Jaroslav. 2012. Higher extraversion and lower conscientiousness in humans infected with *Toxoplasma*. *European Journal of Personality* 26:285–291.

Martin, B., M.P. Mattson and S. Maudsley. 2006. Caloric restriction and intermittent fasting: Two potential diets for successful brain aging. *Ageing Research Reviews* 5:332–353.

Mathur, P. and L. Mascarenhas. 2019. Lifestyle diseases: Keeping fit for a better tomorrow. *Indian Journal of Medical Research* 149(Supplement):129–135.

McCay, C.M., M.F. Crowell and L.A. Maynard. 1935. The effect of retarded growth upon the length of life-span and upon the ultimate body size. *Journal of Nutrition* 10:63–79.

Meng, X., Z. Dai, C. Hang and Y. Wang. 2020. Smartphone-enabled wireless otoscope-assisted online tele-medicine during the COVID-19 outbreak. *American Journal of Otolaryngology* 41(3):102476. doi: 10.1016/j.amjoto.2020.102476.

Menon, R.G. 1931. *Terminalia chebula*. In *Bhāratīya Ouṣadhacceṭikaḷ (Medicinal Plants of India)*, 161–162. Trichur: Rāmānuja Mantrālaya.

Mizushima, N. and M. Komatsu. 2011. Autophagy: Renovation of cells and tissues. *Cell* 147:728–741.

Mohanvel, S.K., S.K. Rajasekharan, T. Kandhari, B.P.K.G. Doss and Y. Thambidurai. 2017. Cow urine distillate as a bioenhancer for antimicrobial and antiproliferative activity and redistilled cow urine distillate as an anticlastogenic agent. *Asian Journal of Pharmacy and Clinical Research* 10:273–277.

Mori, Y., S.E. Kudo and K. Mori. 2018. Potential of artificial intelligence-assisted colonoscopy using an endocytoscope (with video). *Digestive Endoscopy* 30 Suppl. 1:52–53.

Moyer, C.A., J. Rounds and J.W. Hannum. 2004. A meta-analysis of massage therapy research. *Psychological Bulletin* 130:3–18.

Nanneman, D. 1991. Thermal modalities- heat and cold: A review of physiologic effects with clinical applications. *AAOHN Journal* 39:70–75.

Oluwatoyin, A.O. and T. F. John. 2005. *In vitro* evaluation of khaya and albizia gums as compression coating for drug targeting to the colon. *Journal of Pharmacy and Pharmacology* 57:63–168.

Orvedahl, A., D. Alexander, Z. Talloczy et al. 2007. HSV-1 1CP34.5 confers neurovirulence by targeting the Beclin 1autophagy protein. *Cell Host & Microbe* 1:23–35.

Oumeish, O.Y. 1998. The philosophical, cultural, and historical aspects of complementary, alternative, unconventional, and integrative medicine in the Old World. *Archives of Dermatology* 134:1373–1386.

Paradakara, H.S.S. 2018. *Aṣṭāṅgahṛdaya, with commentaries Sarvāṅgasundarā of Aruṇadatta and Āyurvedarasāyana of Hēmādri*. Varanasi: Chaukhamba Sanskrit Sansthan.

Pena-Serna, C. and L.F. Restrepo-Betancur. 2020. Chemical, physicochemical, microbiological and sensory characterization of cow and buffalo ghee. *Food Science Technology* 40:444–450.

Pereira, P.C. 2014. Milk nutritional composition and its role in human health. *Nutrition* 30:619–627.

Philip, A.K., S. Dabas and K. Pathak. 2009. Optimized prodrug approach: A means for achieving enhanced anti-inflammatory potential in experimentally induced colitis. *Journal of Drug Targeting* 17:235–241.

Ponton, F., F. Ota´lora-Luna, T. Lefe`vre et al. 2011. Water-seeking behavior in worm-infected crickets and reversibility of parasitic manipulation. *Behavioral Ecology* 22:392–400.

Prasad, A. and N. Kothari. 2022. Cow products: Boon to human health and food security. *Tropical Animal Health and Production* 54:12. https://doi.org/10.1007/s11250-021-03014-5.

Raja, W. and R.C. Agrawal. 2010. Chemopreventive potential of cow urine against 7, 12 dimethyl benz(a) anthracene-induced skin papilloma genesis in mice. *Academic Journal Cancer Research* 3:7–10.

Randhawa, G.K. and R. Sharma. 2015. Chemotherapeutic potential of cow urine: A review. *Journal of Intercultural Ethnopharmacology* 4:180–186.

Rani, R. and V.K. Kansal. 2011. Study on cow ghee versus soybean oil on 7, 12-dimethylbenz (a)- anthracene induced mammary carcinogenesis and expression of cyclooxygenase-2 and peroxisome proliferators-activated receptor-γ in rats. *Indian Journal of Medical Research* 761:133–497.

Rennie, S. and S.L. Michlovitz. 2016. Therapeutic heat. In *Modalities for Therapeutic Intervention*, 6th Edition, eds. J.W. Bellew, S.L. Michlovitz and T.P. Nolan Jr., 61–88. Philadelphia: F.A. Davis Company.

Schendel, E. and R.F. Kourany. 1980. Cacodemonomania and exorcism in children. *Journal of Clinical Psychology* 41:119–123.

Sharma, P.V. 1983. *Tṛmarmīya cikitsitam* (Treatment of diseases of three vital organs). In *Caraka Samhita*, Volume 2, 419–455. Varanasi: Chowkhamba Orientalia.

Sharma, S., K. Hatware, A. Deshpande, P. Dande and S. Karri. 2017. Antiobesity potential of fresh cow urine and its distillate-a biomedicine for tomorrow. *Indian Journal of Pharmaceutical Education and Research* 51:712–721.

Shelly, S., N. Lukinova, S. Bambina, A. Berman and S. Cherry. 2009. Autophagy is an essential component of *Drosophila* immunity against vesicular stomatitis virus. *Immunity* 30:588–598.

Shori, A.B. and A.S. Baba. 2014. Comparative antioxidant activity, proteolysis and *in vitro* α-amylase and α-glucosidase inhibition of *Allium sativum*-yogurts made from cow and camel milk. *Journal of Saudi Chemical Society* 18:456–463.

Shukla, A.B., D.R. Mandavia, M.J. Barvaliya, S.N. Baxi and C.B. Tripathi. 2013. Anti-urolithiatic effect of cow urine Ark on ethylene glycol-induced renal calculi. *International Brazilian Journal of Urology* 39:565–571.

Sohal, R.S. and R. Weindruch. 1996. Oxidative stress, caloric restriction, and aging. *Science* 273:59–63.

Song, K., I.B. Song and H.J. Gu et al. 2016. Anti-diabetic effect of fermented milk containing conjugated linoleic acid on type II diabetes mellitus. *Korean Journal for Food Science of Animal Resources* 36:170–177.

Sunshine, W., T.M. Field and O. Quintino et al. 1996. Fibromyalgia benefits from massage therapy and transcutaneous electrical stimulation. *Journal of Clinical Rheumatology* 2:18–22.

Teh, J.L., A. Shabbir, S. Yuen and J.B.Y. So. 2020. Recent advances in diagnostic upper endoscopy. *World Journal of Gastroenterology* 26:433–447.

Tenzin, K. 2006. Shankara: A Hindu revivalist or a crypto-Buddhist? Thesis, Georgia State University. doi: 10.57709/1062066.

Tepperman, P.S. and M. Devlin. 1983. Therapeutic heat and cold. *Postgraduate Medicine* 73:69–76.

Uebaba, K., F.H. Xu, M. Tagawa et al. 2005. Using a healing robot for the scientific study of shirodhara. *IEEE Engineering in Medicine and Biology Magazine* 24:69–78.

Uebaba, K., F.H. Xu, H. Ogawa et al. 2008. Psychoneuroimmunologic effects of ayurvedic oil-dripping treatment. *The Journal of Alternative and Complementary Medicine* 14:1189–1198.

Valdor, R. and F. Macian. 2012. Autophagy and the regulation of the immune response. *Pharmacological Research* 66:475–483.

Wan, R., S. Camandola and M.P. Mattson. 2003. Intermittent food deprivation improves cardiovascular and neuroendocrine responses to stress in rats. *Journal of Nutrition* 133:1921–1929.

Wang, J.W., C.H. Kuo, F.C. Kuo et al. 2019. Fecal microbiota transplantation: Review and update. *Journal of the Formosan Medical Association* 118:S23–S31.

Wang, C., M. Oda, Y. Hayashi et al. 2020. A visual SLAM-based bronchoscope tracking scheme for bronchoscopic navigation. *International Journal of Computer Assisted Radiology and Surgery* 15:1619–1630.

Webb, W.B. 1994. Sleep as a biological rhythm: A historical review. *Sleep* 17:188–194.

Webster, J.P. 2007.The effect of *Toxoplasma gondii* on animal behavior: Playing cat and mouse. *Schizophrenia Bulletin* 33:752–756.

Webster, J.P., M. Kaushik, G.C. Bristow and G.A. McConkey. 2013. *Toxoplasma gondii* infection, from predation to schizophrenia: Can animal behaviour help us understand human behaviour? *The Journal of Experimental Biology* 216:99–112.

Weerapong, P., P.A. Hume and G.S. Kolt. 2005. The mechanisms of massage and effects on performance, muscle recovery and injury prevention. *Sports Medicine* 35:235–256.

Wikström, O. 1982. Possession as a clinical phenomenon: A critique of the medical model. *Scripta Instituti Donneriani Aboensis* 11:87–102.

Wylensek, D., T.C.A. Hitch, T. Riedel et al. 2020. A collection of bacterial isolates from the pig intestine reveals functional and taxonomic diversity. *Nature Communications* 11:6389. https://doi.org/10.1038/s41467-020-19929–w.

Xing, J., G. Liu and X. Zhang et al. 2020. The composition and predictive function of the fecal microbiota differ between young and adult donkeys. *Frontiers in Microbiology* 11. doi: 10.3389/fmicb.2020.596394.

Glossary of Sanskrit Terms

Abhighātaja śvayathu: Edema due to fracture and physical trauma
Abhyaṅga: Application of oil all over the body followed by massage
Adhidantakam: Eruption of too many teeth
Adhijihva: Edematous swelling appearing below tongue, causing heaviness, burning sensation, itching, pain and obstruction of deglutition and speech
Adhimāṃsam: Edema below molar teeth, causing pain in jaw and ear, making deglutition difficult
Adhimantha: Glaucoma
Āḍhyavāta: Rheumatic palsy on the loins
Āgantu: Of exogenous origin
Āgantuja śvayathu: *Śvayathu* (edema) arising from exogenous causes
Agnimāndya: Lowering of abdominal fire or *jaṭharāgni*
Agnivisarpa: Black, blue or red, spreading blisters resembling scalded skin
Ajīrṇa: Indigestion
Akṣipāka: Inflammation of eyes
Alaji: Eye disease affecting the sclero-corneal junction equated with keratitis marginalis, keratitis marginalis profunda, diciform keratitis and phlyctenular keratoconjunctivitis; also means red eruptions with intolerable pain appearing on penis
Āmadōṣa: Discomforts caused by improper digestion of food
Āmāśaya: The receptacle (*āśaya*) that holds undigested food (*āma*) in stomach
Āmātisāra: Dysentery caused by vitiated mucus in abdomen and characterized by hard and fetid stool
Āmajvara: Fever arising from impaired digestion
Amḷavidagdha: Hazy vision
Aṅgula: One *aṅgula* is equal to a finger's breadth
Añjana: Eye disease characterized by a small boil of the size of a seed of green gram appearing on the eyelid, associated with itching, warmth and copper color; also means antimony sulphide
Antarāyāma: Muscle spasm resulting in bending of body inward (decorticate posture)
Antardāha: Internal burning sensation
Ānūpa dēśam: Land blessed with plenty of water, trees and hills
Anuvāsanavasti: *Vasti* performed with oleaginous substances
Anyatōvāta: Neuralgia of the fifth cranial nerve
Apabāhuka: Paralysis of arms
Apatānaka: Convulsion
Apaci: Cervical glandular swelling
Apcaravarga: Birds that live around or on the surface of water
Apīnasa: Catarrh with pain, exudation of slimy, yellow and foul-smelling phlegm
Ardhāvabhēdaka: Hemicrania
Ariṣṭa: Medicines prepared by fermenting decoctions of herbs; wine brewed from decoctions of herbs mixed with jaggery
Arma: Pterygium
Arūṃṣika: Pityriasis
Āsava: Medicines prepared by fermenting juice of herbs or fruits
Aṣṭāṅgahṛdaya: Heart of the eight limbs (summary of the eight branches of Ayurveda)
Aṣṭhīla: Hard and round swelling below the navel
Aṣṭhīlika: Hard, round boil with uneven surface, appearing at base of penis

Atharvavēda: *Atharvavēda* is the collection of procedures for everyday life. It is said to represent a popular religion, incorporating not only formulas for magic, but also the daily rituals for initiation into learning (*upanayana*), marriage and funerals. Royal rituals and the duties of court priests are also included in *Atharvavēda*.

Atyagni: Excessive appetite

Avagāha: Sitting immersed in warm medicinal fluid

Avapīḍaka ghṛta: Any medicated clarified butter consumed after dinner

Avapīḍaka nasya: Nasal instillation of penetrating substances such as paste of herbs, with the intention of eliminating pathogenic substances lodged inside throat and sinuses

Ayurveda: Sacred knowledge of longevity

Baddhōdara: Enlargement of abdomen caused by intestinal obstruction

Bahaḷa vartma: An inflammatory condition of the eyelids which can be correlated with meibomitis

Bāhyāyāma: Muscle spasm resulting in bending of body backward

Bhāvana: A specific procedure in which a powdered drug of herbal or mineral origin is homogeneously mixed with expressed juice, decoction of herbs or other specific liquids and dried, so as to increase its potency

Bisavartma: Xanthelasma or a type of xanthoma found in upper or lower eyelid as an irregular-shaped yellowish patch due to the deposition of cholesterol in the skin

Chidrōdara: Enlargement of abdomen caused by perforation of intestine

Cikitsāsthānam: Section of a Sanskrit textbook on Ayurveda that describes the treatment of diseases

Dadru kuṣṭha: Tinea corporis

Dantabhēdam: Toothache

Dantacāla: Looseness of teeth

Dantaharṣam: Dentine hypersensitivity

Dantanāḷi: Dental abscess

Dantapupputa: Edema of the size of a seed of *Badara* (*Ziziphus mauritiana*) seed appearing below two or three teeth, causing suppuration and excruciating pain

Dantaśarkara: Tartar or dental calculus

Dantasuṣiraṃ: Periodontitis

Dantavidarbha: Traumatic injury to gums

Dantavidradhi: Dental abscess

Daśamūla: A famous grouping of ten roots, consisting of *Hrasva pañcamūla* and *Mahat pañcamūla* (Table 6.36)

Dhamani: Artery

Dhanvantari: Hindu deity of medicine and an incarnation of Viṣṇu. Dhanvantari is also identified with King Divōdāsa of Kāśi

Dhānyāmḷa: Rice gruel fermented with grains and herbs

Dharma: Proper conduct conforming to one's duty and nature

Dhūmara: Smoky vision

Dōṣāndha: Night blindness

Duṣṭapīnasa: Chronic catarrh

Duṣṭavraṇa: Obstinate ulcers

Gaḷagaṇḍa: Swelling in neck

Gaḷaśuṇṭhika: Pharyngitis

Gaḷavidradhi: Swelling in neck with pain and exudation of pus

Gaṇḍālaji: A fixed swelling in cheek marked by fever and burning sensation

Gara/Viṣa: Poisons administered surreptitiously through food and beverages for influencing individuals subliminally or for homicide

Gauḍa: Wine brewed from jaggery

Gharṣanaṃ: Act of rubbing or grinding
Gilāyu: Globus pharyngeus
Grahaṇi: Bowel disease characterized by voiding of hard and loose stool alternatively, before the ingested food is digested
Granthi: Glandular swelling
Granthi visarpa: A type of skin disease with red and painful blisters, arranged in the form of beads
Gulma: Phantom tumor; abdominal mass characterized by tumor-like hard, round mass, unstable in size and consistency, moving or immobile, situated in the bowel
Halīmaka: A subtype of *pāṇḍu* characterized by green, dark-brown or yellow complexion, giddiness, thirst, fever, loss of stamina and lowering of abdominal fire
Hikka: Hiccup
Jālakagardabha: Lymphangitis
Jalārbudam: Bubble-like edema on lips
Jalōdara: Ascites
Jāṅgala dēśam: Land endowed with less water, vegetation and hills
Jaṭharāgni: Abdominal fire
Jihvākaṇṭakam: Glossitis
Jvarātisāra: Fever with diarrhea
Kalpasthānam: Section of a Sanskrit textbook on Ayurveda that describes herbs, other medicinal substances and mode of preparation of medicines
Kāmala: Jaundice
Kanṭharōhiṇi: Fast-growing polyp appearing in the throat, below the tongue
Kapālika: Hardening of teeth followed by breaking
Kapha: Principle of conservation operating in a biological system
Kaphōtkḷiṣṭa: Disease of eyelid characterized by stickiness of lid margin and heaviness of lid
Kardama vartma: Black or mud-colored eyelids
Kardama visarpa: Dirty, swollen, suppurative and foul-smelling blisters appearing on skin
Karṇārbuda: Tumor in ear
Karṇārśas: Polyp in ear
Karṇavidradhi: Edematous swelling in ear with pain
Kēvalavāta: *Vāta*-dominant disease, devoid of involvement with other *dōṣas*
Khārjūraṃ: Wine brewed from dates
Kīlāṭa: Mixture of more buttermilk and less milk
Kōṭha: Skin rash
Kṛchrōnmīla: Eye disease characterized by symptoms such as stiff and painful eyelids, sensation of eye filled with sand, difficulty in opening eyes and moving the eyelid upward
Kṛmi: Worm infestation
Kṛmidantam: Dental caries
Kṛmigranthi: Small boil appearing in the inner or outer canthus, associated with pain, warmth, exudation of pus and loss of eyelashes
Kṛmikarṇa: Worms or lice in the ear
Kṛmiśūla: Pricking pain in teeth
Kṛṣṇasarpa: Black lance-hooded cobra
Kṣata kāsa: Cough due to trauma to thorax
Kṣataśukḷa: Small growth on cornea, turning the sclera red, with pricking pain and lachrymation
Kṣīravasti: Medicinal enema performed with milk
Kṣīrī vṛkṣa: Four trees with milky latex – *Ficus benghalensis, Ficus religiosa, Ficus glomerata* and *Ficus retusa*
Kūcika: Mixture of buttermilk and warm milk
Kukūṇaka: Neonatal conjunctivitis
Kumbhakāmala: Advanced stage of jaundice, characterized by an enlarged abdomen (*kumbha* = pot)

Kumbhīkā vartma: Meibomian cyst
Kuṣṭha: Leprosy and other skin diseases
Lagaṇa: Chalazion or non-infectious granulomatous inflammatory disease due to obstruction of meibomian gland secretions present on upper or lower lid on both lids in single or both eyes.
Lakṣaṇa: Symptoms and signs of a disease
Lēpa: Paste of herbs to be applied topically
Līna: Mishap in pregnancy characterized by lack of pulsation in fetus
Madya: Intoxicating beverage, wine
Madātyaya: Alcoholism
Madhvāsava: Wine brewed from honey
Mahāmṛga: Large animals
Mahāsnēha: Mixture of four oleaginous substances, an example being combination of sesame oil, clarified butter, coconut oil and bone marrow
Mantha: A liquid obtained by churning of any grain or millet, after adding 14 times of water
Manthādi krama: Beginning with *mantha* and gradually ending in solid food
Mārdvīka: Wine brewed from grape juice
Matsyavarga: Animals that live under water
Mēdōbhava gaḷagaṇḍam: Swelling in neck caused by increase in adipose tissue and causing lassitude of voice
Mēdōvarti: Fold of peritoneum connecting the stomach with other abdominal organs
Mōraṇa: Beverage prepared by mixing equal parts of milk and buttermilk
Mṛduvirēcana: Mild laxation
Mṛga: Grazing animals
Mukhapākam: Inflammation of mouth
Mūtrāghāta: Syndrome of obstructive urinary pathology due to deranged function of *vāta*
Mūtragranthi: A round and acute tumor in urinary bladder, causing excruciating pain
Mūtrajaṭharaṃ: Accumulation of urine with pain, bloating of abdomen and constipation following restrainment of micturition reflex
Mūtraja vṛddhi: Disease manifested in those who habitually restrain the urge to void urine. When walking the testicles will shake as if filled with water. Pain and dysuria are also evident.
Mūtrakṣayaṃ: Voiding of very scanty urine, with pain and burning sensation
Mūtrasādaṃ: Difficult voiding of white or blood-colored urine
Mūtraśukḷaṃ: Voiding of cloudy urine caused by indulging in sexual intercourse while the bladder is full
Mūtrātītaṃ: Obstruction of urine or discontinuous urination with pain, preceded by restrainment of micturition reflex for long time
Mūtrōtsaṅgaṃ: Oozing of urine with or without pain and heaviness of penis
Nāḷīvraṇa: Ulcers turned into sinus
Nāsārbuda: Nasal tumor
Nāsārśas: Nasal polyp
Nasya: Nasal instillation of *taila*, *ghṛta*: juice, paste and decoction of herbs; sap of shrubs and trees, honey, rock-salt, *āsava*, soup of *jāṅgala* meat, blood, milk, water and *cūrṇa* to eliminate *dōṣa*s lodged in parts above the collar bone
Nidāna: Intrinsic and extrinsic causative factors of diseases
Nidānapañcaka: The five aspects of diagnosis of a disease such as *nidāna*, *pūrvarūpa*, *lakṣaṇa*, *upaśaya* and *samprāpti*
Nidānasthānam: Section of a Sanskrit textbook on Ayurveda that explains causative factors, symptoms, signs and classification of diseases
Nirūhavasti: Medicinal enema performed with decoction of herbs, same as *kaṣāya vasti*

Ōṣṭhakōpam: Inflammation of lips
Pakṣāghāta: Hemiplegia
Pakṣmasadanaṃ: Madarosis
Pakvātisāra: Chronic dysentery
Pañcakarma: The five elimination measures such as emesis, purgation, medicinal enema, nasal instillation of medicinal liquids and removal of blood
Pāṇḍu: Morbid pallor
Parvaṇī: Copper-colored eruption of the size of a seed of green gram, appearing on sclera, with warmth and pain
Pēyādi krama: Beginning with *pēya* or gruel with more fluid, less grains and gradually ending in solid food
Phāṇitam: Thickened sugarcane juice
Picchāvasi: Mucilaginous enema
Pillarōga: A group of 18 eye diseases which persist for a long period
Pīnasa: Common cold
Pitta: Principle of transformation operating in a biological system
Pittaduṣṭayōni: Vaginal disease characterized by heat, suppuration, fever and voiding of foul-smelling, blue, black or red menstrual blood
Pittajihvākaṇṭakam: Glossitis due to vitiation of *pitta*
Pittavidagdha: Seeing all objects in yellow color
Pittōtkḷiṣṭa: Eye disease characterized by warmth, pain and pricking pain inside and outside eyelids which turn red
Pḷīhōdara: Enlargement of spleen
Pōṭaḷa svēda: Sudation using chopped herbs, balls of rice paste cooked in boiling water, powders, grains, powder of cow dung, sand and bran packed in boluses and belonging to the variety of *ūṣmasvēda*
Pōthakī: White-colored eruptions on eyelids of the size of mustard seeds, associated with edema, pain, itching, sliminess and lachrymation
Pramēha: Diabetes mellitus and other polyureic diseases
Prasahavarga: Creatures that grab and tear off their food
Pratināha: Obstruction of nose
Pratūdavarga: Birds that peck and gobble their food
Pravāhika: A disease characterized by voiding of semi-solid stool with mucus and tenesmus
Purāṇaghṛta: Aged clarified butter that is more than five years old
Purīṣōttha kṛmi: Worms that evolve from intestine
Pūrvarūpa: Premonitory symptoms (prodromes) specific to diseases
Pūtikarṇa: Chronic suppurative otitis media
Pūtināsa: Foul-smelling nasal discharge arising from roof of palate, caused by aggravation of *vāta* and *kapha*
Pūyālasa: Dacryocystitis, which is an infection of the tear sac, typically due to an obstruction of the nasolacrimal duct
Rajas: Innate tendency or quality that drives motion, energy and activity
Rājayakṣma: Kingly consumption
Raktamōkṣa: An important measure included in *pañcakarma*. It is carefully controlled removal of small volumes of impure blood, so as to detoxify the body.
Raktātisāra: Dysentery with bleeding
Raktagulma: Abdominal tumors appearing in women
Rasāñjana: Concentrated aqueous extract of *Berberis aristata*
Rasāyana: Medicinal substance that cures diseases and rejuvenates body
Rasāyana vidhi: Protocol for administering *rasāyana*

Rōgaparīkṣa: Diagnosis of disease
Rōpaṇa: That which heals wounds
Rūkṣa: Dryness, one of the qualities of matter
Rūkṣaṇa: The process of inducing dryness
Saktu: Powder of parched rice (*lāja*), rice powder
Śalāka: Spatula
Śālūka: Adenoids
Sāmkhya: A dualistic theistic school of Indian philosophy that considers reality and human experience as being constituted by two independent ultimate principles of *puruṣa* (consciousness or spirit) and *prakṛti* (cognition, mind and emotions, and nature or matter)
Samprāpti: Etiopathogenesis
Sannyāsa: Coma
Śārīrasthānam: Section of an Ayurveda textbook that explains fundamental aspects of human fertility, treatment of infertility, conception and anatomy of human body
Sarjikṣāra: Caustic alkali prepared from *Fagonia cretica* or *Tragia involucrata*
Śārkara: Wine brewed from sugar
Sattva guṇa: Quality of goodness, positivity, truth, serenity, balance, peacefulness and virtuousness that is drawn toward duty and knowledge
Sēka/Pariṣēka: Continuous pouring in a thin stream of oil, warm water, decoctions, *taila* and *ghṛta* over body parts
Sidhma kuṣṭha: A variety of skin diseases with raised thick scaly skin lesions all over the body, especially on the trunk and limbs, with chronic itching, pain, burning sensation and occasional oozing. The disease is caused by irregular food pattern, bowel habits and sleep.
Sīdhu: Wine brewed from sugarcane juice
Sira: Vein
Sirāgranthi: Varicose vein
Sirāharṣa: Redness and inflammation of eye, leading to impaired vision
Śiraḥkampa: Shivering of head
Śiraḥstāpa: Headache
Śirōpiṭaka: Eruptions appearing on scalp
Śītāda: Gingivitis
Śītadantam: Cracked teeth
Śītajvara: Fever with shivering fits
Śḷēṣmagranthi: Glandular swelling with predominance of *kapha*
Śḷīpada: Elephantiasis
Snēha: Oleaginous substances like various vegetable oils, butter, clarified butter, bone marrow and visceral fat
Snēhavarti: Cotton wick impregnated with medicated oil or clarified butter
Snēhavasti: Medicinal enema using oleaginous substances
Snēhavirēcana: Purgation with unctuous ingredients
Stambhanam: Stiffness or cessation of movement
Śuddhaśukra: Eye disease characterized by white cornea and slight pain
Śukḷarōga: Diseases of sclera
Śukḷāśmari: Disease caused in men by restrainment of the urge to ejaculate seminal plasma
Sūkṣma: Subtle
Śuṣkākṣipāka: Eye disease characterized by pricking pain, difficulty to open and close eyes, swollen lids, dryness and suppuration
Sura: Wine brewed from rice powder
Sūryāvarta: Migraine
Sūtrasthānam: Introductory section of a Sanskrit textbook that describes the basic principles of Ayurveda

Śvāsa: Respiratory distress
Śyāva vartma: Reddish-black eyelids with pain, exudation of fluid and edema
Tālupākam: Inflammation of tongue
Tamas: Quality of inertia, inactivity, dullness or lethargy
Tīkṣṇa: Penetrating
Tridōṣa: Collective name of *vāta*, *pitta* and *kapha*
Tridōṣaja: Arising from destabilization of *tridōṣa*
Trikaṭu: Combination of *Piper longum*, *Piper nigrum* and *Zingiber officinale*
Tṛvṛtsnēha: Mixture of three oleaginous substances
Tuṇḍikēri: Tonsilitis
Udāvarta: Abdominal disease due to retention of feces
Udvartana: Massaging with powder of herbs
Unmantha: Edema with itching in ear lobule
Upadamśa: Venereal ulcer
Upajihva: Edema appearing on the surface of tongue, causing pain, itching, heaviness and burning sensation
Upakuśa: Chronic gingivitis
Upanāha: Application of thick herbal paste on any part of body, covering it up with leaves, cloth or hide and keeping for a specific period of time; also means an eye disease characterized by the appearance of a bubble in the inner canthus, with itching but having no suppuration or pain
Upaśaya: Confirming diagnosis with temporary adoption of medicine, food or measures
Upaśīrṣaka: Painless swelling of the same color of scalp caused by *vāta* vitiation while the fetus remains *in utero*
Ūrustambha: Stiffness of thighs
Uṣṇa: Hot in nature
Uṣṇavātaṃ: Voiding of warm, yellow, red urine with difficulty, accompanied by pain and burning sensation in genital area
Uṣṇavidagdha: Hazy vision during day time and blindness at night
Utkārika: Balls of rice paste cooked in boiling water
Utkḷiṣṭa vartma: Red-colored eruptions with lines over them and excruciating pain appearing on eyelids
Utsaṅga: Eye disease characterized by red eruptions inside eyelids
Uttarasthānam: The sixth and final section of *Aṣṭāṅgahṛdaya* that describes the diagnosis and treatment of pediatric diseases, mental disorders, epilepsy, diseases of eye, ear, nose and so on
Uttaravasti: Urethral and vaginal enema
Vaibhītaki: Wine brewed from decoction of bark of *Terminalia belerica*
Vaiśēṣika: One of the six Hindu philosophies that discusses qualities of matter
Vamana: Emesis, one of the five actions (*pañcakarma*)
Vāruṇi: Wine brewed from diluted honey
Vasti: Medicinal enema
Vāta: Principle of movement operating in a biological system
Vātaja gaḷagaṇḍam: Swelling in neck due to enragement of *vāta*
Vātakuṇḍalika: Condition characterized by pain and heaviness in urinary bladder with voiding of urine and stool in drops
Vātaparyaya: Ophthalmic neuralgia
Vātāṣṭhīla: Cylindrical, hard and immobile tumor in urinary bladder, causing bloating of abdomen and obstruction to urine and stool
Vātaśōṇita: *Vāta*-induced blood disorders
Vātavasti: Disease characterized by voiding of urine in drops and accompanied with pain, burning sensation, pulsations and feeling of getting pressed

Vātavyādhi: Diseases caused by the aggravation of *vāta*
Vātōṣṭhakōpa: Inflammation of lips caused by *vāta* enragement
Vāṭya: Gruel made of roasted and powdered barley
Vēśavāram: Food as well as medicinal ingredient prepared by mixing paste of cooked boneless meat with powder of *Trikaṭu*, jaggery and clarified butter
Vicarcika: Dry and weeping eczema
Vidāha: Food turning acidic in stomach
Viḍvighātaṃ: Voiding of foul-smelling urine
Vilēśayavarga: Animals that live in burrows
Vimḷāpana: Mutilation of swelling by massaging
Vipāka: Post-digestive transformation of tastes
Virecana: Purgation
Viṣama: Irregular
Viṣkiravarga: Birds that scatter their food and peck
Viṣṭambha: Feeling of fullness of abdomen due to constipation
Viṣūcika: Cholera
Vraṇa: Wounds, sores, ulcers
Vṛnda: Edema on either side of neck associated with fever
Vyaṅga: Black discoloration on face
Vyavāyi: Spreading
Yakṣa: Nature spirit, generally benevolent, but sometimes mischievous or capricious, associated with woods, trees, treasures and wilderness
Yamaka snēha: Mixture of two oleaginous substances
Yāpana vasti: Medicinal enema performed using paste of *Cyperus rotundus* suspended in a mixture of honey, sesame oil, meat soup and clarified butter
Yavakṣāra: Alkali from barley straw
Yūṣa: Broth

Index

Note: **Bold** page numbers refer to tables; *italic* page numbers refer to figures.

Printed in the United States
by Baker & Taylor Publisher Services

Printed in the United States
by Baker & Taylor Publisher Services